Repair & Remodeling Estimating Methods *Fourth Edition*

*Edward B. Wetherill and
R.S. Means Engineering Staff*

WILEY
John Wiley & Sons, Inc.

RSMeans

RSMeans

Repair &
Remodeling
Estimating
Methods *Fourth Edition*

Edward B. Wetherill and
RSMeans Engineering Staff

Managing Editor: Mary Greene. Editors: Robert Mewis and Danielle Georges. Production Manager: Michael Kokernak. Production Coordinator: Marion Schofield. Composition Supervisor: Paula Reale-Camelio. Composition: Jonathan Forgit. Proofreader: Robin Richardson. Book and cover design: Norman R. Forgit.

For general information on our other products and services, or technical support, please contact our Customer Care Department within the United States at 800-762-2974, outside the United States at 317-572-3993 or fax 317-572-4002.

Wiley also publishes its books in a variety of electronic formats. Some content that appears in print may not be available in electronic books. For more information about Wiley products, visit our web site at www.wiley.com.

Library of Congress Cataloging-in-Publication Data:

ISBN: 978-0-876-29661-5

10 9 8 7 6 5 4

Table of Contents

Foreword

This fourth edition has been updated throughout, with current costs in the estimating examples and cost pages from *Means Repair & Remodeling Cost Data*. The book also addresses significant changes that have occurred in the MasterFormat and UNIFORMAT systems for organizing estimate data. A separate section addresses the special job planning and cost estimating considerations of a unique type of repair work—reconstruction following fire, flood, hurricane or other disaster. This section was written by one of the nation's leading disaster reconstruction contractors. Merl Vandervort, CR, CGR, CMB is the president of Kansas City-based K.C. Renovators, Inc., a team of specialists in restoration, offering complete services with state of the art equipment and emergency response coverage.

In new construction, the developer, architect, and estimator have the advantage of being able to plan and analyze a project completely on paper before estimating and construction. There are relatively few unknown variables. Renovation, remodeling, and repair pose an entirely different set of challenges, and the estimator must approach this type of work with a unique set of cautions. Without proper analysis and estimating, such projects can easily wind up with cost overruns and time delays.

The object of this book is to establish proper methods to analyze and estimate repair and remodeling projects. Part I of the book, "The Estimating Process," focuses on types of estimates and on the analysis of a project. Part II, "Estimating by CSI Division," provides a step-by-step explanation of the estimate from takeoff to pricing. The final chapter in Part II is a detailed explanation of how to use the annual cost reference, *Means Repair & Remodeling Cost Data*.

Part III contains two complete sample estimates. The first estimate, based on unit prices, is arranged according to the 16 major divisions of the Construction Specifications Institute's (CSI) MasterFormat. The second example is a Systems, or Assemblies, estimate; it uses a format that groups all the functional elements of a building into 7 sequential construction divisions, in accordance with the UNIFORMAT II. This fourth edition includes changes to reflect the new organization system in UNIFORMAT II.

This book is written for contractors, architects and interior designers, engineers, owners and facility managers, and developers. If all parties to a project gain a better understanding of the repair and remodeling process, a basis is established for a more efficient and cost effective project—a goal that serves everyone's interests.

All prices and construction costs used in this book are found in *Means Repair & Remodeling Cost Data 2002*. Many pages of this annual publication have been reproduced to show the origin and development of the cost data. *Means Repair & Remodeling Cost Data* is revised annually with new information on materials, equipment and installation methods, and with updated costs for these components and for total installation including overhead and profit.

Part I

Estimating Process

Part I

Estimating Process

In new construction, the estimator is provided with a set of plans and specifications from which a complete estimate can be prepared. Repair and remodeling, however, requires that the estimator have greater knowledge and experience. Even the best set of plans and specifications for a renovation project cannot anticipate or include all restrictive, existing conditions or possible pitfalls that contractors and design professionals encounter when renovating an existing structure. As a result, a repair and remodeling project poses the greatest challenge to the estimator in the exercise of skill and professional judgment.

In this book, *estimator* is used as an all-inclusive title that may include the contractor, architect, owner, or any person who might require costs for a remodeling project. The estimator's first priority must be analysis of the structure. Existing conditions must be identified in terms of their effect on the work to be performed, and therefore on the estimate. The estimator must also decide what type of estimate is appropriate for the proposed renovation project. This decision should be based on:

- The amount of information supplied to the estimator
- The amount of time allowed to perform the estimate
- The purpose of the estimate

How many times have clients given estimators a small sketch on scrap paper or an oral description of a project, and said, "I need to know how much this will cost, right away?" Although this situation may be extreme, it does happen to everyone at some point. On the other hand, the client may give the estimator 40 sheets of plans, a specification book two inches thick, and a month to complete the estimate. Most remodeling projects fall somewhere between these two extremes. Depending on the above factors, the estimator must choose the type of estimate best suited for the project. This choice will affect how much time will be spent on the estimate as well as how accurate the result will be.

Chapter 1

Types of Estimates

Every cost estimate involves three basic steps. Without all three, the estimate cannot be completed. These steps are:

1. Designation of a "unit" of measure
2. Determination of quantity of units
3. Establishment of a reasonable cost per unit

First, the estimator must designate an appropriate **unit** of measure. The units chosen will depend on the level of detail required (or known) at this stage of the project. In construction, such units could be as detailed as a square foot of drywall or as all-encompassing as a square foot of floor area. In auto repair, a unit could be an individual spark plug or a complete engine. Depending upon the estimator's intended use, the designation of the unit may imply only an isolated entity, or may describe the unit as *in place*. In building construction, these units are described as *material only* or *installed*, respectively. The installed unit includes both material and labor.

Second, the estimator must determine the **quantity of units**. This is an actual counting process: how many square feet of drywall, how many spark plugs, and so forth.

In construction, the process of determining the quantity of units is called the "quantity takeoff." In order to perform this function successfully, the estimator should have a working knowledge of construction materials and methods. This knowledge helps to ensure that each quantity is correctly tabulated, and that items are not forgotten or omitted. The estimator with a sound construction knowledge is more likely to account for all required work in the estimate. Experience is, therefore, invaluable.

Finally, and perhaps the most difficult step to carry out, is the determination of a reasonable **cost for each unit.** If costs for units were to remain constant and there were no variables in construction, then there would be no need for estimates or estimators. Prices do fluctuate, however, and labor productivity varies; no two projects are exactly alike. It is this third step, determination of unit costs, that is most responsible for variations in estimating. For example, even though material costs for framing lumber may be the same for competing contractors, the labor costs for installing that material may vary because of a difference in productivity. One factor may be the use of specialized equipment, which can decrease installation time and, therefore, cost. Labor rates may also vary according to pay scales in different areas.

Generally, material prices fluctuate within the market. Cost differences occur from city to city and even from supplier to cross-town supplier. It is the experienced and well-prepared estimator who can keep track of these variations and fluctuations and use them to best advantage when preparing accurate estimates.

What is the correct, or accurate, cost of a given construction project? Is it the total price that the owner pays to the contractor? Might not another reputable contractor perform the same work for a different cost, whether higher or lower? There is no one correct estimated cost for a given project because there are too many variables in construction. At best, the estimator can determine a very close approximation of what the final costs to the owner will be. The resulting accuracy of this approximation is directly affected by the amount of detail provided and by the amount of time spent on the estimate.

Estimating for building construction, especially remodeling and renovation, is certainly not as simple as proceeding through three simple steps and arriving at a "magic figure." Detailed estimates for large projects require many weeks of hard work. The purpose of this text is to make the estimating process easier and more organized for the experienced estimator, and to provide those who are less experienced with a basis for sound estimating practice.

We begin with a discussion of the different types of estimates. All estimates break a construction project down into various stages of detail. By determining the quantities involved and the cost of each item, the estimator can complete an estimate. The units used to determine the quantities can be large in scale, like the number of apartments in a housing renovation, or very detailed, like the number of square feet of drywall for a repair job. The type of estimate used is based on the amount and detail of information supplied to the estimator, the amount of time available to complete the estimate, and the purpose of the estimate, such as for a preliminary feasibility study or for a final construction bid. Depending on these criteria, there are four basic choices of estimate types:

1. **Unit Price** The Unit Price estimate requires working drawings and specifications. They are the most detailed and take the greatest amount of time to complete. The relative accuracy of a unit price estimate for repair and remodeling can be plus (+) or minus (−) 10%.

2. **Assemblies,** or **Systems** A Systems estimate is used when certain parameters of a renovation project are known, such as building size, general type of construction, type of heating system, etc. Relative accuracy can be + or − 15%.

3. **Square Foot and Cubic Foot** This type of estimate is used when only the size and proposed use for a renovation are known. These estimates are much faster to complete than unit price or systems estimates and can provide relative accuracy of + or − 20%.

4. **Order of Magnitude** The Order of Magnitude estimate is used for planning future renovation projects. Relative accuracy can be + or −25%.

Variables in both the estimating process and the projects themselves are responsible for the differences in estimate accuracy. Figure 1.1 shows the relationship between the time spent preparing an estimate and the relative accuracy for each type of estimate. The accuracy described is based on an "average of reasonable bids," assuming that there are a number of competitive bids for a given project.

Unit Price Estimates

Unit Price estimates are the most accurate and detailed of the four estimate types, and take the most time to complete. Detailed working drawings and specifications should be available to the unit price estimator. Therefore, all decisions regarding materials and methods for the remodeling project should have been made. This effectively reduces the number of variables and "educated guesses" that can decrease the accuracy of an estimate.

The working drawings and specifications are needed to determine the quantities of material, equipment, and labor. Current and accurate costs for these items, in the form of unit prices, are also necessary to complete the estimate. These costs can come from different sources. Wherever possible, the estimator should use prices based on experience or costs from recent, similar projects. No two renovation projects are alike, however. Prices may, in some cases, be determined using an up-to-date industry source book, such as *Means Repair and Remodeling Cost Data*.

Because the preparation of unit price estimates requires a great deal of time and expense, this type of estimating is best suited for construction bids. It can also be effective for determining certain detailed costs in conceptual budgets or during design development.

Most construction specification manuals and cost reference books, including Means' annual cost data books, allocate all construction components into the 16 MasterFormat divisions developed by the Construction Specifications Institute, Inc. They are listed on the following pages.

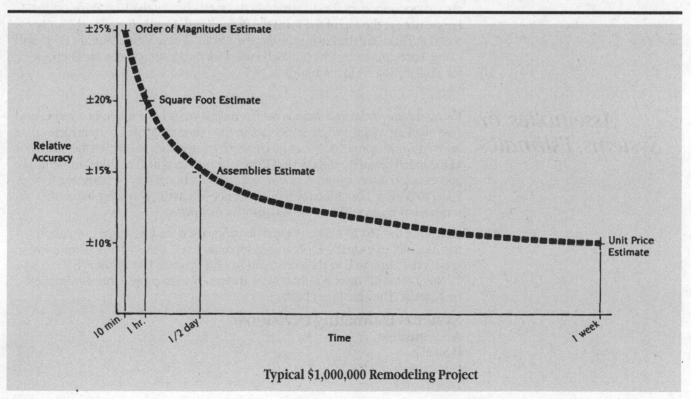

Typical $1,000,000 Remodeling Project

Figure 1.1

CSI MasterFormat Divisions

Division 1–General Requirements
Division 2–Site Construction
Division 3–Concrete
Division 4–Masonry
Division 5–Metals
Division 6–Wood & Plastics
Division 7–Thermal & Moisture Protection
Division 8–Doors & Windows
Division 9–Finishes
Division 10–Specialties
Division 11–Equipment
Division 12–Furnishings
Division 13–Special Construction
Division 14–Conveying Systems
Division 15–Mechanical
Division 16–Electrical

This method of categorizing construction components provides a standard of uniformity that is widely used in the construction industry. Since a great number of architects use this system for specifications, it makes sense to base estimates and cost data on the same format. Figure 1.2 shows a page listing types of drywall from the Unit Price section of *Means Repair and Remodeling Cost Data*. Each unit price page contains a wealth of information that can be used for Unit Price estimates. The type of work to be performed is described in detail; it outlines what kind of crew is needed, how long it takes to perform the unit of work, and separate costs for material, labor, and equipment. Total costs are extended to include the installer's overhead and profit.

Figure 1.3 is a Means Condensed Estimate Summary form (from *Means Forms for Building Construction Professionals*). This form can be used to summarize the information that has been separately estimated by division. This form provides a checklist to ensure that estimators have included all divisions. It is also a concise means for determining the total costs of the Unit Price estimate. Please note at the top of the form the items *Total Area/Volume*, and *Cost per S.F./C.F.* These are the basic units for the Order of Magnitude, Square Foot, and Cubic Foot estimates, which may be used for budgeting or as a cross-check for similar projects in the future.

Assemblies, or Systems, Estimates

Rising design and construction costs in recent years have made budgeting and cost efficiency increasingly important in the planning stages of remodeling and renovation projects. Never has it been so important to involve the estimator in the initial planning stages. Unit Price estimating, which requires more time and detailed information, is not an appropriate budgetary or planning tool. A faster and more cost effective method is needed at the planning stage of a renovation project. This is the *Assemblies Estimate*.

Instead of the 16 CSI MasterFormat divisions used for Unit Price estimating, the Assemblies estimate reorganizes separate trade items to reflect a logical, sequential approach to the construction of a project. The result is 8 "UNIFORMAT II" divisions that break the renovation project into assemblies, or systems. They are listed below:

Systems Estimating Divisions:

A Substructure
B Shell
C Interior
D Services

09250	Gypsum Board	CREW	DAILY OUTPUT	LABOR-HOURS	UNIT	2002 BARE COSTS				TOTAL INCL O&P	
						MAT.	LABOR	EQUIP.	TOTAL		
200 0080	3' x 5' x 1/2" sheets	2 Carp	525	.030	S.F.	1.04	.91		1.95	2.66	**200**
0090	3' x 6' x 1/2" sheets		525	.030		.82	.91		1.73	2.42	
0100	3' x 4'x 5/8" sheets		525	.030		1.08	.91		1.99	2.70	
0110	3' x 5' x 5/8" sheets		525	.030		1.06	.91		1.97	2.68	
0120	3' x 6' x 5/8" sheets		525	.030		1.07	.91		1.98	2.68	
0150	On wall, 3' x 4'x 1/2" sheets		350	.046		1.04	1.37		2.41	3.42	
0160	3' x 5' x 1/2" sheets		350	.046		1.04	1.37		2.41	3.42	
0170	3' x 6' x 1/2" sheets		350	.046		.82	1.37		2.19	3.18	
0180	3' x 4'x 5/8" sheets		350	.046		1.08	1.37		2.45	3.46	
0190	3' x 5' x 5/8" sheets		350	.046		1.06	1.37		2.43	3.44	
0200	3' x 6' x 5/8" sheets		350	.046		1.07	1.37		2.44	3.44	
0250	On counter, 3' x 4'x 1/2" sheets		180	.089		1.04	2.67		3.71	5.55	
0260	3' x 5' x 1/2" sheets		180	.089		1.04	2.67		3.71	5.55	
0270	3' x 6' x 1/2" sheets		180	.089		.82	2.67		3.49	5.35	
0300	3' x 4'x 5/8" sheets		180	.089		1.08	2.67		3.75	5.60	
0310	3' x 5' x 5/8" sheets		180	.089		1.06	2.67		3.73	5.60	
0320	3' x 6' x 5/8" sheets	▼	180	.089	▼	1.07	2.67		3.74	5.60	
300 0010	**BLUEBOARD** For use with thin coat										**300**
0100	plaster application (see division 09210-900)										
1000	3/8" thick, on walls or ceilings, standard, no finish included	2 Carp	1,900	.008	S.F.	.22	.25		.47	.66	
1100	With thin coat plaster finish		875	.018		.29	.55		.84	1.23	
1400	On beams, columns, or soffits, standard, no finish included		675	.024		.25	.71		.96	1.46	
1450	With thin coat plaster finish		475	.034		.33	1.01		1.34	2.04	
3000	1/2" thick, on walls or ceilings, standard, no finish included		1,900	.008		.22	.25		.47	.66	
3100	With thin coat plaster finish		875	.018		.29	.55		.84	1.23	
3300	Fire resistant, no finish included		1,900	.008		.22	.25		.47	.66	
3400	With thin coat plaster finish		875	.018		.29	.55		.84	1.23	
3450	On beams, columns, or soffits, standard, no finish included		675	.024		.25	.71		.96	1.46	
3500	With thin coat plaster finish		475	.034		.33	1.01		1.34	2.04	
3700	Fire resistant, no finish included		675	.024		.25	.71		.96	1.46	
3800	With thin coat plaster finish		475	.034		.33	1.01		1.34	2.04	
5000	5/8" thick, on walls or ceilings, fire resistant, no finish included		1,900	.008		.26	.25		.51	.71	
5100	With thin coat plaster finish		875	.018		.33	.55		.88	1.27	
5500	On beams, columns, or soffits, no finish included		675	.024		.30	.71		1.01	1.51	
5600	With thin coat plaster finish		475	.034		.38	1.01		1.39	2.09	
6000	For high ceilings, over 8' high, add		3,060	.005			.16		.16	.26	
6500	For over 3 stories high, add per story	▼	6,100	.003	▼		.08		.08	.13	
9000	Minimum labor/equipment charge	1 Carp	2	4	Job		120		120	199	
500 0010	**CEILINGS** Gypsum drywall, fire rated, finished										**500**
0100	Screwed to grid, channel or joists, 1/2" thick	2 Carp	765	.021	S.F.	.22	.63		.85	1.28	
0200	5/8" thick		765	.021		.22	.63		.85	1.28	
0300	Over 8' high, 1/2" thick		615	.026		.22	.78		1	1.53	
0400	5/8" thick	▼	615	.026	▼	.22	.78		1	1.53	
0600	Grid suspension system, direct hung										
0700	1-1/2" C.R.C., with 7/8" hi hat furring channel, 16" O.C.	2 Carp	600	.027	S.F.	.84	.80		1.64	2.25	
0800	24" O.C.		900	.018		.77	.53		1.30	1.73	
0900	3-5/8" C.R.C., with 7/8" hi hat furring channel, 16" O.C.		600	.027		.89	.80		1.69	2.31	
1000	24" O.C.	▼	900	.018		.78	.53		1.31	1.74	
700 0010	**DRYWALL** Gypsum plasterboard, nailed or screwed	R09250-100									**700**
0100	to studs unless otherwise noted										
0150	3/8" thick, on walls, standard, no finish included	2 Carp	2,000	.008	S.F.	.19	.24		.43	.61	
0200	On ceilings, standard, no finish included		1,800	.009		.19	.27		.46	.65	
0250	On beams, columns, or soffits, no finish included		675	.024		.22	.71		.93	1.42	
0300	1/2" thick, on walls, standard, no finish included	▼ ▼	2,000	.008	▼	.21	.24		.45	.63	

Figure 1.2

CONDENSED ESTIMATE SUMMARY

PROJECT							SHEET NO.	
							ESTIMATE NO.	
LOCATION			TOTAL AREA/VOLUME				DATE	
ARCHITECT			COST PER S.F./C.F.				NO. OF STORIES	
PRICES BY:			EXTENSIONS BY:				CHECKED BY:	

DIV.	DESCRIPTION	MATERIAL	LABOR	EQUIPMENT	SUBCONTRACT	TOTAL
1.0	General Requirements					
2.0	Site Construction					
3.0	Concrete					
4.0	Masonry					
5.0	Metals					
6.0	Wood & Plastics					
7.0	Thermal & Moisture Protection					
8.0	Doors & Windows					
9.0	Finishes					
10.0	Specialties					
11.0	Equipment					
12.0	Furnishings					
13.0	Special Construction					
14.0	Conveying Systems					
15.0	Mechanical					
16.0	Electrical					
	Subtotals					
	Sales Tax %					
	Overhead %					
	Subtotal					
	Profit %					
	Contingency %					
	Adjustments					
	TOTAL BID					

Figure 1.3

E Equipment & Furnishings
F Special Construction & Demolition
G Building Sitework

Each of these UniFormat divisions may incorporate items from different unit pricing divisions. For example, when estimating an interior partition using the Assemblies approach, the estimator uses Division C – Interiors. (See Figure 1.4, a page from *Means Repair and Remodeling Cost Data*.) When estimating the same interior partition using the Unit Price approach, the estimator would refer to CSI MasterFormat Division 6 (for wood studs and baseboard), Division 7 (for insulation), and Division 9 (for drywall, taping, and paint).

Conversely, a particular unit price item like cast-in-place concrete (Division 3) may be included in different Assemblies divisions: Division A – Substructure, Division B – Shell. The Assemblies method better reflects the way the contractor views the construction of a renovation project. Although it does not allow for the detail provided by the Unit Price approach, it is a faster way to develop costs.

By using the Assemblies estimate, the estimator/designer has the advantage of being able to vary components of a renovation project in order to quickly determine the cost differential. The owner can then anticipate accurate budgetary requirements before final design, details, and dimensions are established.

In the Unit Price estimate, final details regarding the renovation project are available to the estimator. When using the Assemblies estimate, particularly for renovation projects, estimators must draw on their experience and knowledge of building code requirements, design options, and the ways in which existing conditions limit and restrict the proposed building renovations.

The Assemblies estimate should not be used as a substitute for the Unit Price estimate. While the Assemblies approach can be invaluable in the planning and budgeting stages of a renovation, the Unit Price method should be used when greater accuracy is required.

Square Foot and Cubic Foot Estimates

Square Foot and Cubic Foot estimates are appropriate when a building owner wants to know the cost of a renovation before the plans or even sketches are available. Often these costs are needed to determine whether it is economically feasible to proceed with a project, or to determine the best use (apartments, offices, etc.) for an existing structure.

Square foot costs for new construction can be found in the annually updated *Means Square Foot Costs*. However, in remodeling and renovation, each existing building and each project is unique. Therefore, square foot estimating is not always effective for this kind of work. One building might have a two-year-old heating system requiring little work, while another may need all new equipment. One building might need a new roof, while another requires only patching. These variations challenge the estimator and make a site visit to determine and evaluate existing conditions critical to the square foot and cubic foot estimating process. The site visit and evaluation are covered in Chapter 2.

The best data available to the estimator is from past projects. (Please refer back to Figure 1.3, the Condensed Estimate Summary form, which provides a convenient reference of such historical data, listing costs per S.F./C.F.) This data, together with the estimator's experience with the variables between projects, enables the estimator to use Square Foot and Cubic Foot estimates effectively for repair and remodeling work.

C1010 Partitions

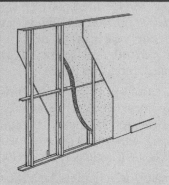

This page illustrates and describes a load bearing metal stud wall system including metal studs, sheetrock–taped and finished, insulation, baseboard and painting. Lines within System Components give the unit price and total price per square foot for this system. Prices for alternate load bearing metal stud wall systems are on Line Items C1010 138 1500 thru 2500. Both material quantities and labor costs have been adjusted for the system listed.

Factors: To adjust for job conditions other than normal working situations use Lines C1010 138 2900 thru 4000.

Example: You are to install the system using temporary shoring and bracing. Go to Line C1010 138 3900 and apply these percentages to the appropriate MAT. and INST. costs.

System Components	QUANTITY	UNIT	COST PER S.F.		
			MAT.	INST.	TOTAL
Load bearing, 18 ga., 3-5/8", galvanized metal studs, 24" O.C.,including Top and bottom runners, 1/2" drywall, taped, finished and painted 2 Faces, 3" insulation, and painted baseboard, wall 10' high.					
Metal studs, 24" O.C., 18 ga., 3-5/8" wide, galvanized	1.000	S.F.	.43	.76	1.19
Gypsum drywall 1/2" thick	2.000	S.F.	.46	.80	1.26
Taping and finishing	2.000	S.F.	.08	.80	.88
Insulation, 3-1/2" fiberglass batts	1.000	S.F.	.37	.25	.62
Baseboard	.200	L.F.	.25	.33	.58
Paint baseboard, primer + 2 coats	.200	L.F.	.03	.21	.24
Painting, roller 2 coats	2.000	S.F.	.30	.92	1.22
TOTAL		S.F.	1.92	4.07	5.99

C1010 138	Partitions, Metal Stud, LB	COST PER S.F.		
		MAT.	INST.	TOTAL
1400	For alternate metal stud systems:			
1500	Load bearing, 18 ga., 24" O.C., 2-1/2" wide	1.85	4.06	5.91
1600	6" wide	2.07	4.10	6.17
1700	16" O.C. 2-1/2" wide	1.94	4.24	6.18
1800	3-5/8" wide	2.03	4.25	6.28
1900	6" wide	2.21	4.29	6.50
2000	16 ga., 24" O.C., 2-1/2" wide	1.91	4.16	6.07
2100	3-5/8" wide	1.99	4.18	6.17
2200	6" wide	2.16	4.22	6.38
2300	16" O.C., 2-1/2" wide	2.01	4.37	6.38
2400	3-5/8" wide	2.12	4.39	6.51
2500	6" wide	2.32	4.44	6.76
2600				
2700				
2900	Cut & patch to match existing construction, add, minimum	2%	3%	
3000	Maximum	5%	9%	
3100	Dust protection, add, minumum	1%	2%	
3200	Maximum	4%	11%	
3300	Material handling & storage limitation, add, minimum	1%	1%	
3400	Maximum	6%	7%	
3500	Protection of existing work, add, minimum	2%	2%	
3600	Maximum	5%	7%	
3700	Shift work requirements, add, minimum		5%	
3800	Maximum		30%•	
3900	Temporary shoring and bracing, add, minimum	2%	5%	
4000	Maximum	5%	12%	

Figure 1.4

Order of Magnitude Estimates

Order of Magnitude estimates require the least amount of time to complete and provide the lowest level of estimate accuracy. The information required is the **proposed use of the building** and the **number of units** (apartments, hospital beds, etc.). Figure 1.5, which shows an example of this level of information, is a page from *Means Building Construction Cost Data*. Note that under *Hospitals* and *Housing,* a cost per unit is assigned. Order of Magnitude estimates are used primarily for planning purposes and primarily for new construction. The complexities of remodeling and renovation make the Order of Magnitude estimate ineffective unless costs from similar projects in similar existing buildings arc available.

17100 | S.F. & C.F. Costs			UNIT	UNIT COSTS			% OF TOTAL			
				1/4	MEDIAN	3/4	1/4	MEDIAN	3/4	
400	0500	Masonry	R17100-100 / S.F.	5.65	11.10	17.20	6%	10%	15.50%	**400**
	2720	Plumbing		2.71	5.20	9.80	3.60%	6.70%	8%	
	2730	Heating & ventilating		4.64	6.70	12.95	6.20%	7.40%	13.50%	
	2900	Electrical		4.46	6.60	10.10	6.30%	8%	11.70%	
	3100	Total: Mechanical & Electrical		11.30	20.50	33.50	15.20%	25.50%	32.70%	
410	0010	**GARAGES, PARKING**	R17100-100 / S.F.	23	33	59				**410**
	0020	Total project costs	C.F.	2.21	3	4.37				
	2720	Plumbing	S.F.	.61	1.03	1.59	2.60%	3.40%	3.90%	
	2900	Electrical		.96	1.46	2.14	4.30%	5.20%	6.30%	
	3100	Total: Mechanical & Electrical		1.28	3.77	4.83	6.50%	9.40%	12.80%	
	3200									
	9000	Per car, total cost	Car	9,900	12,500	16,100				
430	0010	**GYMNASIUMS**	R17100-100 / S.F.	65.50	83.50	106				**430**
	0020	Total project costs	C.F.	3.26	4.17	5.40				
	1800	Equipment	S.F.	1.38	2.81	5.60	2.10%	3.40%	6.70%	
	2720	Plumbing		4.14	5.10	6.10	5.40%	7.30%	7.90%	
	2770	Heating, ventilating, air conditioning		4.45	6.80	13.60	9%	11.10%	22.60%	
	2900	Electrical		4.95	6.15	8.15	6.60%	8.30%	10.70%	
	3100	Total: Mechanical & Electrical		15.35	21	28.50	20.60%	26.20%	29.40%	
	3500	See also division 11480								
460	0010	**HOSPITALS**	R17100-100 / S.F.	135	157	233				**460**
	0020	Total project costs	C.F.	9.50	11.75	16.95				
	1800	Equipment	S.F.	3.19	6.15	10.55	2.50%	3.80%	6%	
	2720	Plumbing		11.25	15.10	19.45	7.80%	9.40%	11.80%	
	2770	Heating, ventilating, air conditioning		15.85	21	28.50	8.40%	14.60%	17%	
	2900	Electrical		13.50	17.55	27.50	9.90%	12%	14.50%	
	3100	Total: Mechanical & Electrical		38.50	51	83.50	26.60%	33.10%	39%	
	9000	Per bed or person, total cost	Bed	48,400	103,000	162,000				
	9900	See also division 11700 & 11780								
480	0010	**HOUSING** For the Elderly	R17100-100 / S.F.	62	78	95.50				**480**
	0020	Total project costs	C.F.	4.38	6.10	7.80				
	0100	Site work	S.F.	4.30	6.80	9.80	5.10%	8.20%	12.10%	
	0500	Masonry		1.88	7.05	10.25	2.10%	7.10%	12.20%	
	1800	Equipment		1.48	2.02	3.27	1.90%	3.20%	4.40%	
	2510	Conveying systems		1.50	2.02	2.74	1.80%	2.30%	2.90%	
	2720	Plumbing		4.58	5.90	7.85	8.10%	9.70%	10.90%	
	2730	Heating, ventilating, air conditioning		2.35	3.33	4.98	3.30%	5.60%	7.20%	
	2900	Electrical		4.60	6.20	7.95	7.30%	8.50%	10.50%	
	3100	Total: Mechanical & Electrical		15.85	18.60	24.50	18.10%	22%	29.10%	
	9000	Per rental unit, total cost	Unit	57,500	67,000	74,500				
	9500	Total: Mechanical & Electrical	"	12,300	14,700	17,200				
500	0010	**HOUSING** Public (Low Rise)	R17100-100 / S.F.	52	72	94				**500**
	0020	Total project costs	C.F.	4.12	5.75	7.15				
	0100	Site work	S.F.	6.60	9.50	15.40	8.40%	11.70%	16.50%	
	1800	Equipment		1.41	2.30	3.67	2.30%	3%	5.10%	
	2720	Plumbing		3.53	4.95	6.25	6.80%	9%	11.60%	
	2730	Heating, ventilating, air conditioning		1.88	3.65	4	4.20%	6%	6.40%	
	2900	Electrical		3.14	4.68	6.50	5.10%	6.60%	8.30%	
	3100	Total: Mechanical & Electrical		14.90	19.20	21.50	14.50%	17.60%	26.50%	
	9000	Per apartment, total cost	Apt.	57,000	65,000	81,500				
	9500	Total: Mechanical & Electrical	"	12,200	15,000	16,600				
510	0010	**ICE SKATING RINKS**	R17100-100 / S.F.	46	80	114				**510**
	0020	Total project costs	C.F.	3.26	3.34	3.85				
	2720	Plumbing	S.F.	1.66	3.11	3.18	3.10%	5.60%	6.70%	
	2900	Electrical		4.75	7.30	7.70	6.70%	15%	15.80%	

Figure 1.5

Chapter 2

The Site Visit and Evaluation

In repair and remodeling, it is of the utmost importance that the estimator conduct a site visit and evaluation before performing the estimate. To create a reliable estimate, the estimator must become familiar with each project. This requires an inspection of the existing conditions.

The extent and detail of the plans and specifications supplied to the estimator have an important influence on the site visit and on the proper completion of the estimate. As stated previously, this information may range from minimal to extensive. Before conducting the site visit, the estimator should examine all information provided. As important as it is to perform the site evaluation before making the estimate, it is just as important to know what the project entails before visiting the site. For example, the specifications may state:

> *"All existing mortar joints shall be scraped to a depth of at least 1/2", using hand tools only, before tuck pointing individual joints. Care shall be taken not to allow new mortar to stain existing porous bricks."*

or:

> *"Clean and patch existing masonry joints."*

Clearly, the two statements could have entirely different implications.

Generally, the architect involved in the planning and supervision of the project will answer most questions about methods and materials. When an architect is not involved in the supervision of the project, then the estimator is responsible for the interpretation of the plans and specifications. However, estimators should not make assumptions or interpretations in areas where they lack experience or knowledge of specific conditions.

When there is no architect involved in the project, it is often the estimator's responsibility to make sure that all items are included in the estimate and that the roles of the subcontractors do not overlap. This is especially important in renovation, where omissions or overlaps are more likely to occur. For example, wood backing for plumbing fixtures could be included in both carpentry and plumbing subcontracts. At the same time, providing holes through existing walls for sprinkler piping could be omitted, because each trade assumes that this item is the responsibility of another. This type of problem is less likely to occur in new construction, where all parties are familiar with established precedents. It cannot be stressed enough, however, that each remodeling project is different and requires special attention and diligence.

If possible, subcontractors and engineers should accompany the estimator on the site visit. Undoubtedly, questions will arise specific to certain trades. The estimator should know how existing conditions will affect the work to be performed.

Knowing the applicable Building, Fire, and Energy Codes also helps the estimator to anticipate any requirements that may not be included in the plans and specifications. It should be noted that Building, Fire, and Energy Codes for existing structures are often different from those that apply to new buildings.

The estimator for remodeling projects should also be familiar with older types of construction, such as mill-type wood construction built before extensive use of steel and concrete, or the terra cotta-encased structural steel method introduced with the advent of fire-resistant construction. Each of these older types of construction requires a different approach to remodeling.

This chapter concerns the site visit and evaluation. Although one text cannot possibly describe all the variables in commercial renovation, the chapter will present an effective approach for evaluating an existing building, describing what to look for, how to look, and what to anticipate.

The best way to evaluate and analyze an existing structure is from the bottom up. By beginning the inspection of existing conditions at the bottom, the estimator can observe the structure as it was built. Structural elements can be identified and followed up through the building. The mechanical and electrical systems can also be traced from below to better understand the distribution above.

Throughout the site visit and evaluation, the estimator should keep in mind the renovation plans and the building's ultimate use, noting any conditions or potential problems that may not have been incorporated into the plans and specifications. The architect or owner should be consulted on any questionable conditions.

The following evaluation is based on the 7 UNIFORMAT II divisions used in Assemblies estimating and explained and listed in Chapter 1. The UNIFORMAT II divisions represent the chronological order of construction and, thereby, provide a good format for the site visit and evaluation, no matter what estimating method is used.

Throughout the discussion of the site visit and evaluation, the person performing the inspection is referred to as the *estimator.* This reference is used primarily for consistency. The inspector could, in fact, be the building owner, the architect, engineer, contractor, or any other professional involved in the renovation project. The project is enhanced if all involved parties perform a site visit to gain a better understanding of the renovation process.

Substructure

Materials

The foundation of an existing building may consist of stone rubble, concrete, wood piles, or other material. The foundation material and, if possible, its thickness, should be noted. This information is important in planning remodeling work. For example, if new utilities are to be provided, the estimator should know if the subcontractor has to drill through 12″ of concrete block or hand chisel through 54″ of stone and mortar.

Moisture Problems

In cases of water or moisture problems, the plans and specifications may call for foundation waterproofing. Depending on the amount of information provided for the work, the estimator may have to determine whether the chosen method is the best solution to the problem. The solution can be as simple as repairing an undiscovered, buried, water pipe leak or as extensive as total exterior excavation, and sealing and coating of the entire foundation. If possible, the source of the problem should be determined, and the solution estimated before work begins.

The inside surface of the foundation should be inspected. The condition of the masonry, mortar, or concrete may reveal subsurface conditions. For example, spalling, or deteriorated mortar, in only a small area suggests a localized problem.

The exposed portion of the foundation's exterior should also be inspected. Downspout locations should be noted and examined for adequate runoff, away from the foundation. Potential site drainage problems, low spots, standing water, and clogged catch basins and sewer drains should be noted. If the source of the water problem cannot be found readily, the estimator should consult the architect or owner to discuss the problem and possible solutions before starting the work.

Settling

The estimator should note cracks or any signs of settling at the foundation, interior bearing walls, or columns. If conditions suggest unusual amounts of settling, or if there are any questions about the foundation's structural integrity, an engineer should be consulted.

Excavation and Access Problems

If specified work is to be performed on the foundation, the estimator must determine how to do the work and what equipment to use. If workers must excavate or place concrete, is there adequate access for machinery, or will the work be performed by hand? Workers must often create access for equipment and required materials. When exterior foundation work is to be done, an adjoining property may be involved, particularly if the building is close to lot lines. In such cases, space in lots adjacent to renovation projects may be rented, for actual work access or for material storage. This arrangement can, however, become an expensive item, one which may or may not have been included in the original estimate.

Another item to consider for excavation is sheet piling, which may be necessary to protect adjacent areas and/or to prevent the collapse of trenches. See Figure 2.1 for an illustration.

For interior foundation work, where much of the work and delivery of materials must be performed manually, workers sometimes gain access by cutting holes in floors or walls. Expenses include not only the cost of opening the holes but also the cost of labor and materials to patch the opening.

The estimator should try to envision the work to be done on the project, including all intermediate steps and those items that may not be directly specified or obvious. Such an approach will help to prevent omissions, reducing the chance of unforeseen costs and change orders.

Different Construction Methods

When inspecting the foundation, the estimator is well advised to take nothing for granted. Older methods of construction may be very different from those used today. Above grade, the differences can be seen; below, they may be a mystery. Figure 2.2 is an example of a situation in which looks may be deceiving. Note that the basement slab, obviously added after the building was constructed, is actually below the base of the foundation (with no footing). This condition, depending on the proposed use of the building, is a potential problem and a costly "extra" to rectify. Again, the estimator should consult an engineer if there is any question about structural integrity and how it might affect the work or the use of the proposed building.

Also included in the substructure is the basement floor or the slab on grade. Older buildings may lack concrete slabs, having only soil floors. In many older buildings, basement floors were constructed of brick or wood placed directly on the soil. When concrete slabs do exist in older buildings, they may have been installed at some point after the building was constructed (as in Figure 2.2). In such cases, there may be no reinforcing.

Modern compaction methods and some of the materials used in subgrading today were not as widely used in older buildings. For this reason, settling and cracking is more evident in older buildings. Cracks should be inspected for excessive movement, and the general condition of the slab should be checked for level and for signs of deterioration. Any existing penetrations through the slab should be examined as an indication of the thickness. Remember, especially in older buildings, slab thickness may not be constant.

As is the case with foundations, water seepage may be a problem at floor slabs. Placing a new, reinforced slab with a vapor barrier is an effective, but expensive and not always feasible, solution. If the instructions provided are not specific, the estimator may have to determine the source of the problem and determine how best to repair the existing slab for less cost.

The estimator should note the time of year during which the site visit is conducted and the amount of precipitation over recent weeks. If there has been a dry spell, potential water problems may not be evident. Any water marks should be noted. Cracks are another indication of a moisture problem.

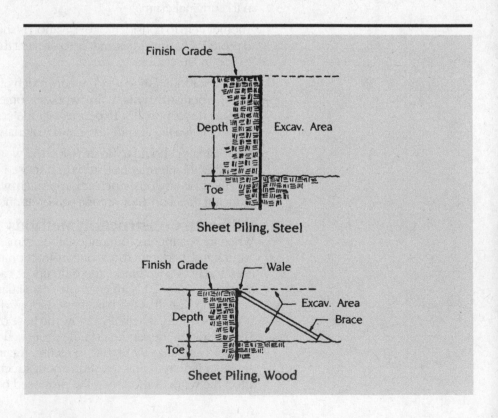

Figure 2.1

Drainage

Any existing interior drainage should be checked for clogs or broken pipes. If a new drainage system is to be installed, the estimator should check the proposed path and ultimate outlet to ensure that existing conditions will not inhibit the installation. Particular drainage requirements may be specified by Building Codes for boilers and hot water heaters.

Access

If the plans and specifications call for a new basement floor slab or concrete pads for equipment, builders face the same problems of access for equipment and materials as they do when performing foundation work. Hand work is often the only way to excavate and compact the subgrade.

Utilities and Mechanical Work

The estimator should be especially aware of the work involved in new utility and mechanical installations. Clearances for the installation of new equipment

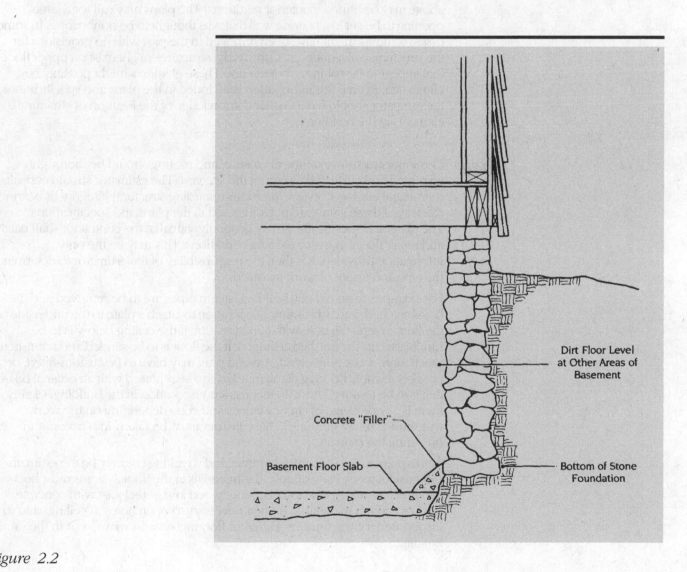

Figure 2.2

should be verified. Presumably, a structural engineer will have checked the bearing capacities of existing slabs. If not, and there is any question regarding heavy equipment, an analysis should be done.

A specified, new electrical service may pose many problems. In urban areas, space for exterior transformers may not be available. Consequently, transformer vaults are required in basements. The contractor must provide the utility with access for the transformers. Utility companies may not permit installation of utility equipment and service switch-overs during normal working hours, in which case this work must be performed at night or on weekends. This situation is especially likely when the building remains partially occupied during the renovation. For these reasons, overtime work must be anticipated.

Other Considerations

Before beginning the estimate, the estimator must determine from this evaluation of the substructure how the specified work is to be performed. Particular attention should be given to access for materials and equipment.

If an architect or engineer has not thoroughly analyzed the building, the estimator should examine and check the location of existing structural elements, as seen in the basement columns, beams, and interior bearing walls. These are important factors in the analysis of the structure above. On upper floors, the structural members may not be exposed or readily evident. Columns above may be buried in interior partitions. The plans may call for a large opening to be cut in a bearing wall that was thought to be nonbearing. In some cases, columns might have been removed in the past with no provisions for the structural consequences. Conversely, structures might exist on upper floors that appear to be columns, but are not. These elements might possibly be eliminated. Even if this information is included in the plans and specifications, the estimator should have firsthand knowledge of the location of structural elements in the building.

Shell

The superstructure, exterior, enclosure, and roofing should be thoroughly inspected and evaluated as part of the site visit. The estimator should consult a structural engineer on any questions regarding structural integrity or bearing capacity, if these issues are not addressed in the plans and specifications. The plans and specifications may generally state that the contractor shall patch and repair floors to match existing conditions. This may be the only information provided. It is then the responsibility of the estimator to determine the type and extent of work involved.

For example, in an old mill building, steam pipes are to be removed and the holes patched. It may be a simple operation to attach a plate to the underside of the floor and fill the hole with concrete. But, if the ceiling below is to be sandblasted as the finished ceiling, or if the floor is to be sanded and refinished, much more work is involved. A wood plug may have to be custom-fitted, or possibly a whole flooring plank may have to be replaced with an original board that is to be removed from an inconspicuous location in the building. Clearly, there is a wide range of circumstances and costs depending on the work that must actually be done. These factors must be taken into account in preparing the estimate.

At this point in the site visit, a hammer and wrecking bar may be the estimator's most useful tools. For example, the materials in the floor systems must be identified as mill-type wood planking, wood joists, steel joist with concrete slab, or another material. It is often necessary to open holes in ceilings and to inspect under floor finishes. Different flooring systems may occur in the same building.

Floor and Ceiling Systems

Floors in older buildings are often out of level, sometimes extremely so. Settling is common in buildings built before modern engineering standards and Building Codes were established. The plans and specifications may call for the floors to be leveled with a lightweight "self-leveling" concrete. Not only is the placement of the concrete often difficult, but the determination of quantities involved can also be a problem. It often takes more concrete than originally thought, and if a large amount is required, a structural engineer may have to be consulted to determine whether or not the existing structure can support it.

When new floor systems are to be added in existing buildings, the estimator must be able to visualize the complete sequence of construction. Depending on the floor system specified in the plans, placement of new structural members can be difficult at best. Figure 2.3 shows three typical new floor systems that may be used in remodeling and renovation projects. Instead of installing larger pieces (such as steel beams) directly from the truck, with large equipment, workers may have to handle materials manually, moving them two or three times, in order to position the pieces before erecting the system. If the job calls for very long structural members, some fabrication may be needed on-site in order to accommodate restricted access. Depending on the bearing capacity of existing structural members, the existing structure may require structural reinforcement or a new structural support system, beginning with new footings.

Adding new floors in commercial renovation may be very costly, and may represent a significant percentage of the total cost of a renovation project. Consequently, the owner or architect's final decision to add a new floor is often not final until the estimator has determined the costs. Thus, it is the estimator's responsibility to be thorough and to include all costs that may result from such work.

Other Considerations

The estimator should look for patched or repaired areas in the flooring and ceilings, as these may be signs of significant damage. Fire or water damage may have been covered up. Where possible, existing steel should be inspected for corrosion. Throughout the site evaluation, the estimator should recall the proposed renovations and note areas where the existing conditions will affect the work.

Much commercial renovation that was performed before the relatively recent general acceptance of uniform Building Codes would not be acceptable under today's standards. Such poor quality work must be uncovered and dealt with properly in buildings that are to undergo remodeling. Openings in the floors of these older buildings have often been filled with materials of weaker structural capacity than the original surrounding floor system. These factors all have an effect on the renovation to be performed, and the estimator should be aware of the consequences. The owner or architect should be notified immediately of any such conditions.

Exterior Closure

Having determined the work to be performed, based on the plans and specifications, the estimator must thoroughly inspect the entire exterior of a building. If the building is tall, the exterior should be checked from upper-story windows. Much can be overlooked from a ground level inspection. For example, when the project calls for repointing of any deteriorated or loose masonry joints, the estimator must determine the area of wall surface where repointing is required. Without close inspection, areas of deteriorated mortar may go undetected until the work has begun. All sides of the building should be

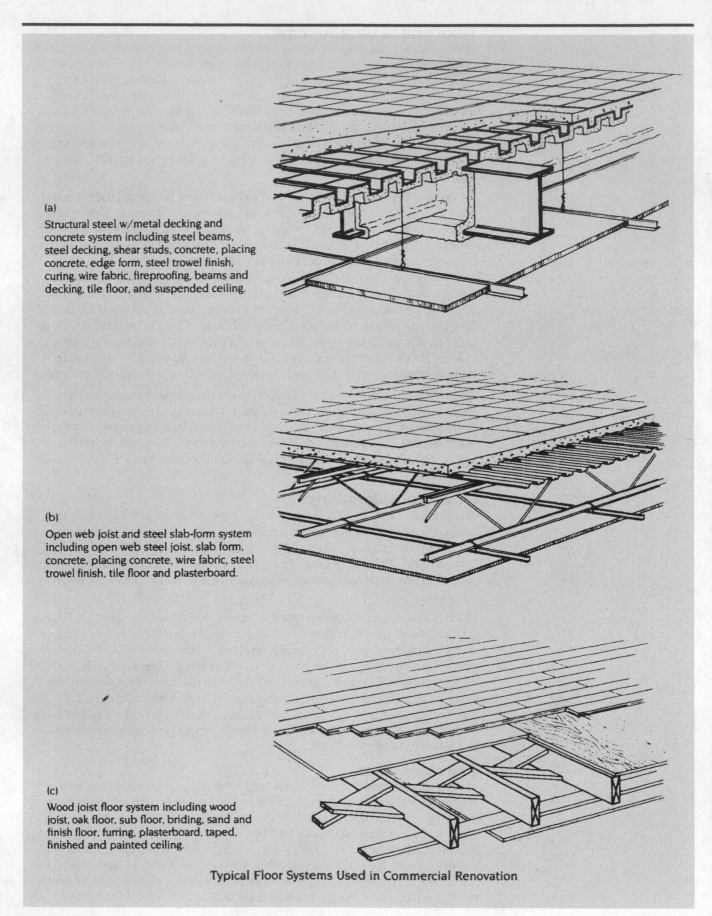

(a)

Structural steel w/metal decking and concrete system including steel beams, steel decking, shear studs, concrete, placing concrete, edge form, steel trowel finish, curing, wire fabric, fireproofing, beams and decking, tile floor, and suspended ceiling.

(b)

Open web joist and steel slab-form system including open web steel joist, slab form, concrete, placing concrete, wire fabric, steel trowel finish, tile floor and plasterboard.

(c)

Wood joist floor system including wood joist, oak floor, sub floor, briding, sand and finish floor, furring, plasterboard, taped, finished and painted ceiling.

Typical Floor Systems Used in Commercial Renovation

Figure 2.3

checked. Different weather exposures will exhibit varied amounts of deterioration. Conditions will also vary over a particular wall surface.

Materials

A great number of older commercial buildings have some form of masonry exterior, whether stone or brick. In addition, cast iron, copper, and bronze were often used for storefronts, cornices, and facades. The properties of these materials differ greatly from the concrete, manufactured brick, and aluminum that are so common in new construction. Therefore, a knowledge of these different materials, along with installation methods, is crucial to both estimator and architect when renovating older buildings.

When inspecting the masonry, the estimator should note the type of brick or stone, and if possible, the type of mortar used. Also, note the joint size and conditions, paying particular attention to the condition of joints between dissimilar materials (e.g., brick-stone, brick-wood). Due to differences in thermal expansion and contraction of different materials, this type of joint is most susceptible to deterioration. The information gathered from such an inspection will be used to determine the best way to perform the specified renovation work, and when calculating costs.

Cleaning

When the plans call for the exterior of a brick building to be cleaned (often before repair), the estimator must carefully read the specifications before deciding the appropriate cleaning method. Older brick is usually very porous and softer than modern brick. Sandblasting and harsh chemical washing may cause extensive damage to such exteriors, and architects and preservation officials often reject these methods. High pressure water with a mild chemical cleaner may be effective, but this approach is much more time consuming. Hand scrubbing, the most labor-intensive choice, may be necessary. Therefore, the estimator and architect alike must be aware of the materials involved and of the appropriate working methods.

When exterior repairs or replacement are specified, materials to match the existing surrounding conditions are often required. These materials may have to be taken from the existing exterior. If existing materials must be reused elsewhere, the estimator should include extra labor expense for special handling to prevent damage to these materials.

Windows and Doors

When the project calls for new windows, the estimator should determine whether the windows can be installed easily from the interior, or whether scaffolding is necessary. Some types of replacement windows require only the removal of the existing sash. This type is installed around or within the existing frame. If the existing frames must be removed, more labor is involved in both demolition and installation.

In most renovation projects, doors and windows must be custom-made to fit existing openings, or the openings must be altered to accept standard sizes. Both options generally require a good deal of time, work, and money. Figures 2.4 and 2.5 show two buildings with very different types of windows. The building in Figure 2.4 has a different size and shape of window at each floor. The windows in the building in Figure 2.5 are all rectangular and are of only two sizes. The estimator must remember that like these examples, every job is different and must be evaluated carefully and independently.

Historic Preservation Guidelines

Especially on older buildings, the exterior closure can be a significant factor in commercial renovation. Today, a heightened appreciation for fine old architecture and a rising interest in historic preservation present new considerations in dealing with older buildings. The exterior should be thoroughly inspected, and if an architect is not involved, the estimator should determine whether local preservation officials have any jurisdiction over renovations to the exterior of the building. Renovation projects have been stopped legally because historic guidelines were not being met.

The building in Figure 2.4 was renovated according to required historic preservation guidelines. Under these guidelines the existing windows could have been replaced only by custom-made wood windows exactly replicating the existing windows. The millwork subcontractor's estimate for this work was so great that the owner chose to have the existing windows repaired. The estimator bid the repair work at 30% of the replacement cost. However, when the repair work was completed, the eventual cost was over 50% of the cost of replacement. A more careful and thorough evaluation would have prevented this expensive error.

Other Considerations

When working on the exterior of an existing building, estimators must take into account factors other than actual labor and materials. For example, scaffolding often must be erected along the entire exterior perimeter. This, in turn, involves pedestrian protection, sidewalk permits, and depending on the weather, tarpaulins to enclose the scaffolding (see Figure 2.6). If the exterior work is lengthy, the rental expense for these items can be significant. When possible, the estimator should determine whether swing staging might be an adequate, less costly alternative.

The estimator should carefully read and understand the drawings and specifications for every remodeling project. Required methods of work and materials will vary from project to project and will be greatly influenced by the age of the building and by the role of historic preservation in the project.

Roofing

In commercial renovation, the estimator will encounter many different types and combinations of roofing structures and materials. For example, older houses may have two layers of asphalt shingles over cedar shingles. A flat, built-up roof on an old building may have patches over patches. If the project includes roof work, the estimator must know what the type and condition of the existing roofing is in order to make a proper estimate of the work. If access is possible, the estimator should closely examine the existing roof structure, materials, flashing, and penetrations.

Insulation and Venting

When examining the roof structure, the estimator should note any existing insulation. The specifications may call for an "R" value of 19, leaving the methods and materials up to the estimator. As seen in Figure 2.7, 2″ Perlite roofdeck insulation and 1″ extruded polystyrene have approximately the same "R" value, but differ greatly in cost.

A wet sprinkler system in an attic space makes it necessary to install insulation above the piping to prevent freezing. If the architect has made no provisions for the ventilation that should be included in this newly insulated space, then the estimator must address this issue. Without proper ventilation, the addition of insulation can cause condensation problems.

Assessing Damage

The roof structural system is often hidden from view by a ceiling. In such cases, inspection holes should be made or access gained in order to examine the structure. The structural members, and especially the roof decking, should be checked for water marks, damage, or deterioration. When older roofing materials are removed for replacement, portions of existing decking or

Figure 2.4

Figure 2.5

sheathing (sometimes over the whole roof) must also be replaced. If possible, such conditions should be uncovered before work starts.

By examining the ceiling under the roof structure and under any eaves, the estimator can also detect signs of water damage or leaks. Knowledge of the locations of problems on the underside of the roof will be helpful when inspecting above.

Inspecting the Roof Surface

The estimator should try to gain access to the roof surface for close examination of the roofing materials. Built-up roofing should be inspected for visible cracks and bubbles. If the specifications call only for patching of the existing material, the estimator must determine the limits of this work. The amount of required patching may be so extensive that the estimator concludes that a new roof will cost little more than the repair. In this case, the client may benefit from the estimator's experienced opinion and suggested alternatives to the proposed work. Not only might the client get a better job, but the estimator may get more work for providing better service.

Flashing, Gutters, and Downspouts

Flashing, gutters, and downspouts should be inspected in order to determine materials and to evaluate their condition. Replacement of roofing, but not

Figure 2.6

flashing materials, will require extra care and labor expense to protect the flashing. If the materials are very old, disturbing the system may cause problems. When copper or lead gutters and flashing must be repaired by soldering, the estimator must account for the additional cost of adequate fire protection.

This work requires caution, even to the point of hiring personnel to remain at the end of the day to ensure that there are no smoldering materials.

Penetrations

Even when no general roofing work is specified, the estimator should examine the plans to see whether any existing plumbing vents or other penetrations are to be removed or relocated, thereby requiring incidental patching. He should also note whether new penetrations, vents, equipment curbs, smoke hatches, and skylights will require flashing and roofing work that is not specifically shown on the drawings. In this case, the estimator must know what conditions exist and, therefore, what work will be required.

The Importance of an Inspection

Roofing repair and patching must be planned and performed with extra care. Just one or two call-backs to repair leaks can severely erode the contractor's profit margin. Knowing what is involved and what to expect before beginning the work will help to eliminate future problems.

Interiors ## Restoration

Every existing building is unique in some way and should be given thorough and independent examination. The restoration of existing interior features, such as brick walls, plaster cornices, and wood floors, in older buildings is becoming more popular as a design concept in commercial renovation. Restoration of these features requires expertise, in estimating as well as in construction, whether the work involves repair, patching, or even duplication of existing features. If existing doors or moldings are to be matched, they may have to be custom milled. Plaster cornice work, almost unheard of in new construction, may have to be repaired or recreated. Clearly, this is demanding work, for estimator and contractor alike.

Complications

Interior construction in commercial renovation, from the initial delivery of materials to final installation, can be more difficult than new construction. For example, existing windows may have to be removed to allow for delivery of new materials by lift truck. When the building is more than three stories high, workers must place material, either by crane or by hand, carrying it up stairs or in an elevator. When an existing elevator is used, materials such as sheets of drywall may have to be cut to fit into the elevator. These restrictions entail the extra costs for material handling, as well as for the extra labor involved, such as when it is necessary to tape more joints upon installation. Only in very large renovation projects is it cost-effective to install a temporary exterior construction elevator.

During the site visit, the estimator should anticipate material handling problems and check clearances for large items which cannot be broken down, like large sheets of glass. The timing of deliveries is critical as well. Large items must be on-site before construction restricts access.

Variations in Construction Standards

Every building to be renovated is different, and the new interior construction must be built around existing conditions. Ceiling heights in older buildings are often higher than those found in new construction and can vary from floor

B30 Roofing

B3010 Roof Coverings

B3010 160	Selective Price Sheet	COST PER S.F.		
		MAT.	INST.	TOTAL
0100	Roofing, built-up, asphalt roll roof, 3 ply organic/mineral surface	.39	1.05	1.44
0200	3 plies glass fiber felt type iv, 1 ply mineral surface	.59	1.14	1.73
0300	Cold applied, 3 ply		.39	.39
0400	Coal tar pitch, 4 ply tarred felt	1.12	1.36	2.48
0500	Mopped, 3 ply glass fiber	.93	1.50	2.43
0600	4 ply organic felt	1.12	1.36	2.48
0700	Elastomeric, hypalon, neoprene unreinforced	2.26	2.51	4.77
0800	Polyester reinforced	2.28	2.97	5.25
0900	Neoprene, 5 coats 60 mils	5	8.80	13.80
1000	Over 10,000 S.F.	4.68	4.55	9.23
1100	PVC, traffic deck sprayed	1.46	4.55	6.01
1200	With neoprene	1.55	1.84	3.39
1300	Shingles, fiber cement, strip, 14" x 30", 325#/sq.	2.68	1.74	4.42
1500	Shake, 9.35" x 16" 500#/sq.	2.43	1.74	4.17
1600				
1700	Asphalt, strip, 210-235#/sq.	.31	.70	1.01
1800	235-240#/sq.	.41	.77	1.18
1900	Class A laminated	.44	.85	1.29
2000	Class C laminated	.54	.96	1.50
2100	Slate, buckingham, 3/16" thick	5.95	2.20	8.15
2200	Black, 1/4" thick	7.90	2.20	10.10
2300	Wood, shingles, 16" no. 1, 5" exp.	1.66	1.59	3.25
2400	Red cedar, 18" perfections	1.76	1.45	3.21
2500	Shakes, 24", 10" exposure	1.52	1.59	3.11
2600	18", 8-1/2" exposure	1.07	1.99	3.06
2700	Insulation, ceiling batts, fiberglass, 3-1/2" thick, R11	.25	.25	.50
2800	6" thick, R19	.36	.29	.65
2900	9" thick, R30	.66	.35	1.01
3000	12" thick, R38	.84	.40	1.24
3100	Mineral, 3-1/2" thick, R13	.29	.25	.54
3200	Fiber, 6" thick, R19	.43	.25	.68
3300	Roof deck, fiberboard, 1" thick, R2.78	.37	.48	.85
3400	Mineral, 2" thick, R5.26	.75	.48	1.23
3500	Perlite boards, 3/4" thick, R2.08	.36	.48	.84
3600	2" thick, R5.26	.63	.55	1.18
3700	Polystyrene extruded, R5.26, 1" thick,	.25	.26	.51
3800	2" thick R10	.40	.31	.71
3900	40 PSI compressive strength, 1" thick R5	.39	.26	.65
4000	Tapered for drainage	.55	.27	.82
4100	Foamglass, 1 1/2" thick R4.55	1.61	.48	2.09
4200	3" thick R9.00	3.43	.55	3.98
4300	Ceiling, plaster, gypsum, 2 coats	.38	2.17	2.55
4400	3 coats	.54	2.55	3.09
4500	Perlite or vermiculite, 2 coats	.40	2.55	2.95
4600	3 coats	.66	3.19	3.85
4700	Gypsum lath, plain 3/8" thick	.43	.48	.91
4800	1/2" thick	.43	.51	.94
4900	Firestop, 3/8" thick	.43	.58	1.01
5000	1/2" thick	.48	.62	1.10
5100	Metal lath, rib, 2.75 lb.	.28	.54	.82
5200	3.40 lb.	.40	.58	.98
5300	Diamond, 2.50 lb.	.18	.54	.72
5400	3.40 lb.	.28	.68	.96
5500	Drywall, taped and finished standard, 1/2" thick	.28	1.04	1.32
5600	5/8" thick	.29	1.04	1.33
5700	Fire resistant, 1/2" thick	.29	1.04	1.33
5800	5/8" thick	.29	1.04	1.33
5900	Water resist., 1/2" thick	.29	1.04	1.33

Figure 2.7

to floor. The two foot module of modern materials was not a standard in the past. A 10'-3" high ceiling necessitates buying 12' studs and drywall for partitions. The estimator should check all existing ceiling heights if they are not shown clearly on the drawings.

Labor Considerations

The labor involved in constructing a relatively simple stud wall can almost double when working around existing conditions. For example, the upper plate may have to be secured with mastic and toggle bolts to an existing, hollow, terra-cotta ceiling. Or, the new wall may have to be fitted around ceiling beams. If the floor is extremely out of level, each stud may have to be cut to a different length. Extra inside and outside corners are necessary when workers have to box existing piping.

Without evaluating the existing conditions, the estimator cannot assume that a wall is just a simple straight run. Similarly, installation of ceilings, whether acoustical grid-type, drywall, or plaster, may require soffits to accommodate existing conditions or to enclose new mechanical and electrical systems.

Analysis Based on Experience

Typical instructions in plans and specifications for commercial renovation are as follows: "Prepare existing surfaces to receive new finishes" or "Refinish existing surfaces." When the existing surface is a plaster wall, the work may be as simple as light sanding or as extensive as complete replacement of the wall by hand-scraping and replastering. In this example, the estimator must determine the soundness of the existing conditions in order to estimate the amount of work involved. When scraping loose plaster, what was thought to be a minor patching job may easily spread to include large areas that were not anticipated. If the condition of the existing surfaces is questionable or very deteriorated, the estimator may suggest alternatives to the architect or owner, such as covering an unsound plaster wall with drywall. It is in the estimator's best interest to call upon his experience and ingenuity to suggest comparable, but less costly, alternatives, especially if the project is being competitively bid.

Protection of Adjacent Work and Surroundings

When performing remodeling and renovation work, the estimator must take into account many indirect factors. For example, extensive precautions must be taken to control the vast amounts of dust created when interior surfaces are sandblasted. Large, high-volume window fans are required to exhaust the dust. However, pedestrians must also be protected from the exhaust. Consequently, sandblasting may have to occur at night or on weekends, and it may require special permits.

If the building under renovation is partially occupied, floors and ceilings must also be covered or sealed to prevent dust dispersion. Adjacent surfaces must be masked or otherwise protected. Often, the costs involved in preparation exceed the actual cost of the sandblasting.

Preparation costs can also become a major factor when wood floors are being refinished. Repair of damaged areas or replacement of random individual boards may involve a great deal of handwork for the removal of existing boards and custom fitting of new pieces. If the original floor was installed with exposed nails, each nail must be hand set before sanding.

Stairs

Particularly under today's Building Codes, stairways are an extremely important focus of commercial renovation. Often in older buildings, only one *unenclosed* stairway exists where two *enclosed* stairways will be required. If no architect is involved in the project, the estimator should consult local building inspectors as early as possible to make sure that the renovation plans comply with access and egress requirements.

When new stairways are specified as part of remodeling work, installation becomes a major factor. For example, it may not be possible to preassemble large sections to be lowered into place. In such cases, workers must carry smaller pieces to the site and assemble them by hand. Railings often must be built in place. These factors require more labor, and greater expense than does new construction.

Existing stairways, as well as new ones, will most likely have to be enclosed in firewalls. Achieving the required continuity of firewalls may be very difficult in existing buildings. When the existing ceilings are suspended, or made of plaster or another material, they must be cut. The firewall must then be constructed on the underside of the deck above, and the existing structure patched to meet the new work. The firewalls also must be extended through attic spaces to the underside of the roof deck. Again, the estimator must anticipate the extra labor involved.

Firewalls

Installation of required firewalls at stairways and at tenant separations may involve additional expense due to the layout of the existing building. Most Building Codes require vertical continuity of firewalls and fireproofing of structural members that support the firewalls. Vertical continuity is not always possible and may involve fireproofing ceilings, beams, or columns outside of the stairway enclosure. While these elements are almost always addressed in the plans and specifications, the estimator should pay particular attention to them during the site visit, and calculate the labor involved in constructing firewalls.

The elements covered in this section for interior construction demonstrate the importance of thoroughly understanding the existing conditions and the work to be performed. The estimator must use every applicable resource, including experience and knowledge, to prepare a complete and proper estimate.

Services

For the installation of an elevator in commercial renovation, the best estimates are clearly competitive bids from elevator subcontractors. In other than budget pricing, such bids are necessary. The estimator should closely review the elevator subcontractors' bids, which may involve more exclusions than inclusions. It is the estimator's responsibility to make sure that such items as construction and preparation of the shaft, structural supports for rails and equipment, access for drilling machinery (if required), and construction of the machine room are included in the estimate. In other words, the estimator may have to include—and price—all work other than the actual, direct installation of the elevator equipment.

Often in older buildings there is no shaft, or the existing shaft for an old freight or passenger elevator does not conform to requirements for the proposed new elevator. Pre-engineered, standard-size elevators often must be altered or custom-built to conform to existing conditions. In either case, the installing subcontractor should be consulted, and the costs included in the estimate. Where new floor openings are to be cut or existing openings altered, structural details in the plans and specifications may be general. The estimator should determine whether the structural configuration is similar at all levels. As stated previously, floor systems and materials may vary in existing buildings. If there are questions regarding the structure, the estimator should consult an engineer.

Hydraulic Elevators

Unless there is adequate overhead structure, and space for pulleys and machinery, at the top of the shaft, hydraulic elevators are usually specified in commercial renovation. (See Figure 2.8.) Machinery for hydraulic elevators is usually placed adjacent to and at the base of the shaft. While standard hydraulic elevators eliminate the requirements at the top of the shaft, they do require a piston shaft to be drilled, often to a depth equal to the height of elevator travel. In new construction, drilling is a relatively easy operation, because it is usually performed before erection of the structure. The tailings, waste, and mud from drilling can often be disposed of at the site. In renovation, however, the drilling can become a difficult and expensive proposition. The drilling rig often must be dismantled and reassembled by hand within the existing structure, and the debris created by this operation may be a problem. While this work is usually performed as part of the elevator subcontract, the estimator must be aware of the associated costs and include them in the estimate.

The following example illustrates the importance of a thorough site visit and analysis, followed by proper planning and provisions. In the building shown in Figure 2.5, an interesting problem occurred during the installation of a hydraulic elevator. An existing freight elevator shaft was to be used for the new passenger elevator. A metal pan was at the base of the shaft. The estimator had included the cost of cutting a hole in the pan for drilling, as required by the elevator subcontract. After two hours of attempting to cut the metal with an acetylene torch, workers determined that the metal was cast iron and would not melt at the temperature of the torch. Attempts to break the brittle metal with a sledgehammer also failed. After many telephone calls, project managers located a welding subcontractor with a high temperature heli-arc welder capable of cutting what was found to be 2-1/2″ thick cast iron. The cost for the work of the welding subcontractor alone was $1,600. This amount does not include the cost of labor for the five hours of previous attempts to cut the hole. Determination of responsibility for these extra costs—the owner or the contractor—led to costly litigation.

This example is obviously isolated and unique, but it appropriately demonstrates how important the site evaluation can be. While every conceivable problem may not be discovered in a site evaluation, the estimator's thorough inspection and evaluation may reduce the probability that such hidden problems will arise.

Disposal of Drilling Waste

The costs of collecting, draining, and removing drilling waste (usually 90% water, 10% solids) must be included in the estimate. Especially in urban settings, the contractor is not allowed to dispose of this material in public sewers. The waste often must be somewhat filtered and hauled to an appropriate dump site. Provisions must be made for the costs and logistics of these operations.

Other Considerations

There are many other factors involved in the preparation of an existing building for the installation of elevators. Shaft enclosures are usually of fire-rated drywall or "shaftwall" masonry, or a combination. For example, elevator pits and roof-top enclosures with smoke hatches are often required. When contractors dig pits, water may be encountered, calling for the installation of pumps. The estimator should carefully examine the plans and specifications for such requirements, (including the elevator subcontractor's own requirements) and evaluate existing conditions in order to prepare a proper estimate.

Other conveying systems often encountered in commercial renovation are wheelchair lifts and pneumatic tube systems. While these are specialty items,

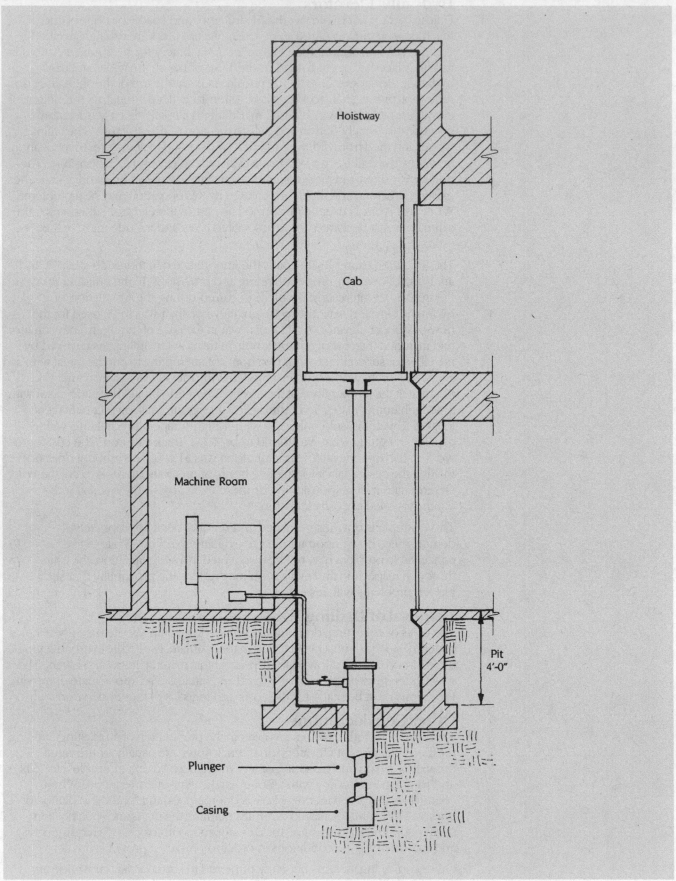

Hoistway

Cab

Machine Room

Pit
4'-0"

Plunger

Casing

Figure 2.8

usually installed by the supplier, the estimator still must anticipate the work involved in preparing the existing structure for such installations.

Until recently, building design and construction did not address accessibility for the handicapped. Almost all Building Codes today require modifications for the handicapped when a commercial building is renovated. Electric wheelchair lifts are often specified when there is no room for exterior or sidewalk ramps, especially in urban locations. The estimator should carefully evaluate the installation, which usually requires extensive cutting and patching.

The existing conditions at locations for pneumatic tube installation should also be closely inspected. Invariably, structural members will occur in places, preventing straight-run installations of the tubes and requiring offsets. Unless these problems are accounted for in the estimate, the contractor may have to bear the extra costs for irregular installations.

When conveying systems have been installed, the finished work must be protected during the remainder of the construction period. In commercial renovation, where construction elevators are usually not provided, the brand new passenger elevator can be damaged and abused if workers use it to carry materials, equipment, and tools. The estimator must account for this needed protection in estimating the total project costs.

Safety

During construction of an elevator shaft, and throughout the entire renovation project, the estimator *must* plan for the requirements and costs of job safety. Temporary railings and toe boards must be securely erected and maintained at floor openings. When welding equipment or cutting torches are being used, an adequate number of fire extinguishers must be supplied and maintained. Time should be allotted to monitor fire safety during hazardous operations. These are just some of the arrangements that must be made and costs that must be included for this essential element of the project.

The best estimates for conveying systems are competitive bids by the installing subcontractors. Nevertheless, the estimator must be aware of and must anticipate the work and costs involved.

Mechanical

When estimating mechanical systems, the estimator should read and study the architectural, structural, and mechanical plans and specifications before visiting the site. In this way, the estimator is sure to include all items and can examine the site to determine how the proposed work will be affected by existing conditions. The estimator should make as few assumptions as possible, directing any questions to the architect or engineer. Mechanical plans are often prepared by consulting engineers hired by the architect. Plans prepared by different parties should be compared for possible variation of scale or inconsistencies regarding proposed locations for installation. Symbols used on the plans should be checked to ensure conformity with the estimator's interpretations.

In new construction, the mechanical systems are incorporated and designed into the planning of the structure. Architectural consideration is given to make mechanical installations as practical and efficient as possible. In renovation, however, mechanical systems must be designed to suit existing systems.

Utilities

The basement, if there is one, is the best place to begin the inspection of the mechanical systems. The existing utility connections should be noted and located on a plan. The estimator should verify sizes of sewer connections and of water and gas services. If the building has sprinklers, the estimator should check for two water services, domestic and fire protection.

The estimator should also inspect the exterior of the building for shut-offs, manholes, and any other indications of utility connections. Where new utilities are specified, closely examine the proposed locations and surrounding conditions. For example, a building with an individual sewage disposal system (septic tank and leaching field) is to be tied into a public sewer. The estimator should be aware of the location of the proposed connection, requirements of the local public works department, and the contractor responsible for trenching and backfilling. Many municipalities require that the existing septic or holding tanks be removed and backfilled. If no engineer is involved, the estimator should investigate all of these possibilities.

Piping

The estimator should inspect the general condition of existing piping that is to remain in place. Corrosion or pitting of the exposed piping may indicate similar conditions at those areas where piping is concealed in walls or ceilings. Where old piping is to remain in place, the estimator should, if possible, try to inspect the inside of the pipes for scaling. Often old pipes, especially galvanized steel, may be encrusted so that flow is severely restricted. When such conditions are found, the owner or architect should be notified to effect a solution before work begins.

Plumbing Installations

Where existing plumbing fixtures are to remain, gaskets and seals should be checked for general condition. If existing fixtures are to be replaced, rough-in dimensions should be measured to ensure compliance with the new fixtures.

The proposed plumbing installation should be followed visually in the building inspection. In renovation, the typical efficient "stacked bathrooms" of new construction are not always feasible. Consequently, piping may require many jogs and extra fittings, and there may be substantially more labor involved than in new construction. If a tenant is occupying the space below a proposed plumbing installation, the estimator must take into account the work that will be required below the floor. If the ceiling below is concealed spline or plaster, large sections may have to be replaced. This work may have to be done at off-hours, on an overtime basis, and existing finished space may require protection.

Sprinkler Systems

Proposed sprinkler installations require the same thorough preparation and evaluation as the other plumbing systems. Existing sprinkler valves should be checked. If no regular maintenance has been performed, the valves may not be operable. If the existing sprinkler system is connected to a fire alarm system, the estimator may have to provide for extra labor to keep the system operational during renovations. Each morning, the fire alarm monitoring agency must be notified and the system drained. Similarly, at the end of each day, workers must notify the monitoring agency again to refill and activate the system. These necessary operations may require up to two hours or 25% of each working day.

If there is no engineer involved in the project, it may become the estimator's responsibility to bring an old sprinkler system into compliance with current codes and regulations. Along with local Fire and Building Code officials, the owner's insurance company should be consulted on sprinkler system requirements. The following items should be noted:

- The age of the sprinkler heads (usually engraved on the fused link)
- The location of the heads
- Sizes and condition of piping

- Existing water pressure

A possible extreme example: If the existing water pressure is not sufficient to meet regulation, a supplementary electric fire pump may be required. The installation of a fire pump has many ramifications. For example, the existing electrical service must be adequate to handle what is usually a large horsepower motor. Codes may also call for an emergency power generator, which in turn includes many indirect costs. Clearly, when a job becomes this involved, a mechanical engineer should design the system. Building owners, however, often wish to avoid such costs and request that the contractor does the planning.

When an engineer designs and specifies a new sprinkler system, the estimator must compare the plans with the existing conditions. New plans are often prepared from old as-built plans, which may not be accurate or current. What are shown as proposed straight pipe runs may, in fact, require many offsets, fittings, and more labor to conform to existing conditions. The estimator should walk through the space to be renovated, keeping the proposed system in mind, and anticipating any installation problems.

Heating and Cooling Systems

Similar design and installation problems may occur with heating and cooling systems. The engineer may provide information on the type of system and its capacity, as well as diffuser or register locations. However, only line diagrams may be provided for piping or ductwork. In these cases, the contractor is usually required to furnish shop drawings that show the actual configuration of the installation. Especially with ductwork, the estimator should examine accompanying architectural and structural drawings to ensure that there is adequate space for the proposed work. Proposed equipment locations should also be inspected. Any questions about bearing capacities for equipment, such as roof-top compressors, should be directed to the structural engineer.

Other Considerations

The estimator should visualize all of the proposed mechanical installations as a whole. Even if one engineering firm has provided complete mechanical drawings and specifications, it is likely that different individuals designed each system, and conflicts may occur. For example, sprinkler pipes have been inadvertently run through planned ductwork. Such conflicts should be discovered and rectified as early as possible. When following the paths of the proposed work, the estimator must be aware of any obstacles that may restrict or complicate installation. Penetrations through masonry walls require much hand work. Penetrations of ductwork through firewalls is sometimes not allowed by local codes. If they are allowed, fire dampers will be required.

Even if the specifications do not require shop drawings, the estimator may find that sketches or drawings of the proposed work, including all mechanical systems, will be helpful in preventing problems and in preparing a complete estimate.

Mechanical work in commercial renovation can be challenging and even fun, requiring ingenuity, innovation, and the resources of the experienced estimator.

Electrical

As with mechanical systems, the electrical plans and specifications should be thoroughly reviewed so that the estimator understands the full extent of the work involved before the site visit.

The electrical estimator should have adequate knowledge of both national and local Electric Codes. Even when an engineer prepares the plans and specifications, the following clause is often included.

"Perform all electrical work in compliance with applicable codes and ordinances, even when in conflict with the drawings and specifications."

If inspectors reject work for noncompliance with codes, it may become the contractor's responsibility to rectify the condition. When analyzing the plans and specifications, the estimator should make notes on questionable installations and advise the architect or engineer as soon as possible.

The electrical drawings will show wiring layouts and circuiting, but often do not show wire types. Different localities have varying requirements regarding the uses of nonmetallic, sheathed cable ("Romex"), BX cable (flexible metallic), EMT (conduit), and galvanized steel conduit.

Figure 2.9 shows the differences in the cost of wiring similar devices using different materials. Note the cost of a duplex receptacle for each type of wiring. The lowest to highest costs vary by over 100%. This variation shows why it is so important to know what type of wiring is required in particular localities. Similarly, aluminum wiring was once widely used as a less expensive alternative to copper, but it is now often disallowed as wiring material. Since copper prices have been fluctuating greatly in recent years, the estimator must keep abreast of suppliers' current prices for this type of wiring.

Service Connections

The site evaluation should begin at the existing, incoming electric service. The estimator should attempt to determine the size and condition of wiring, switches, meter trim, and other components. If the service is to be revamped or increased, the estimator may be required to determine what materials to reuse and what to replace. In older buildings, all electrical wiring and equipment often must be replaced. If the existing wiring is not very old, some equipment may be rebuilt or reused, thereby saving costs. These determinations require experience and ingenuity. Local electrical inspectors should also be consulted.

For a new electric service, the estimator should contact the electric utility company as soon as possible. Scheduling for new connections should be established. Often, adjacent buildings may be affected by an electric service switch over, and the work must therefore be performed at night or on weekends. If the existing service must be removed before connecting the new, the estimator may have to include the costs of a temporary generator in the estimate.

The Distribution System

The entire electrical installation should be visualized and kept in mind by the estimator conducting the site visit and analysis. If there is no engineer involved in the project, the estimator must determine the best distribution system. If the building is to have many separately metered areas, there may be several options available. Some local electric utilities require that all meters be installed in one location. This approach requires expensive, and possibly long, individual feeders to each tenant. On the other hand, if remote meter locations are allowed, only one main feeder is required to each remote multiple meter location. Distribution feeders may be much shorter in this case, and the system may be less expensive. (See Figure 2.10). The estimator should be able to analyze the options to determine the best system for each renovation project.

Wiring

At some point in new construction, all spaces are established, and yet they are still open and accessible for the electrical installation. This point can be reached before concrete slabs are poured or before walls and ceilings are closed. In the renovation of existing buildings, however, gaining access to spaces for wiring may be difficult, costly, and time-consuming.

D50 Electrical

D5020 Lighting and Branch Wiring

D5020 180	Selective Price Sheet	COST EACH		
		MAT.	INST.	TOTAL
0100	Using non-metallic sheathed, cable, air conditioning receptacle	13.05	43.50	56.55
0200	Disposal wiring	10.45	48.50	58.95
0300	Dryer circuit	32	79.50	111.50
0400	Duplex receptacle	13.05	33.50	46.55
0500	Fire alarm or smoke detector	66.50	43.50	110
0600	Furnace circuit & switch	19.60	73	92.60
0700	Ground fault receptacle	42	54.50	96.50
0800	Heater circuit	12.30	54.50	66.80
0900	Lighting wiring	12.95	27.50	40.45
1000	Range circuit	65.50	109	174.50
1100	Switches single pole	12.70	27.50	40.20
1200	3-way	17.05	36.50	53.55
1300	Water heater circuit	19.10	87.50	106.60
1400	Weatherproof receptacle	104	73	177
1500	Using BX cable, air conditioning receptacle	23.50	52.50	76
1600	Disposal wiring	20.50	58.50	79
1700	Dryer circuit	42.50	95	137.50
1800	Duplex receptacle	23.50	40.50	64
1900	Fire alarm or smoke detector	66.50	43.50	110
2000	Furnace circuit & switch	31	87.50	118.50
2100	Ground fault receptacle	52.50	66	118.50
2200	Heater circuit	20	66	86
2300	Lighting wiring	21	33	54
2400	Range circuit	91	132	223
2500	Switches, single pole	23.50	33	56.50
2600	3-way	26.50	43.50	70
2700	Water heater circuit	35	104	139
2800	Weatherproof receptacle	112	87.50	199.50
2900	Using EMT conduit, air conditioning receptacle	26	65.50	91.50
3000	Disposal wiring	24	73	97
3100	Dryer circuit	38.50	118	156.50
3200	Duplex receptacle	26	50.50	76.50
3300	Fire alarm or smoke detector	75.50	65.50	141
3400	Furnace circuit & switch	32.50	109	141.50
3500	Ground fault receptacle	66	81	147
3600	Heater circuit	22.50	81	103.50
3700	Lighting wiring	22	41	63
3800	Range circuit	66	162	228
3900	Switches, single pole	26	41	67
4000	3-way	24.50	54.50	79
4100	Water heater circuit	29	129	158
4200	Weatherproof receptacle	114	109	223
4300	Using aluminum conduit, air conditioning receptacle	29.50	87.50	117
4400	Disposal wiring	27.50	97	124.50
4500	Dryer circuit	49	156	205
4600	Duplex receptacle	28.50	67.50	96
4700	Fire alarm or smoke detector	83	87.50	170.50
4800	Furnace circuit & switch	38.50	146	184.50
4900	Ground fault receptacle	63.50	109	172.50
5000	Heater circuit	28	109	137
5100	Lighting wiring	30.50	54.50	85
5200	Range circuit	80	219	299
5300	Switches, single pole	36.50	54.50	91
5400	3-way	35	73	108
5500	Water heater circuit	39	175	214
5600	Weatherproof receptacle	129	146	275
5700	Using galvanized steel conduit	31	93	124
5800	Disposal wiring	26.50	104	130.50

Figure 2.9

If existing walls and ceilings are to remain, access holes must be made in order to snake wiring, and then the holes must be patched. Not only can cutting and patching become a major expense in renovation electrical work, but wiring often must be installed in circuitous routes to avoid obstacles. This situation involves more labor and more material. In renovation wiring, the shortest route from point A to point B is not always a straight line. An example is when duplex receptacles are being installed in existing plaster walls. To wire horizontally from one outlet box to the next would involve removing large

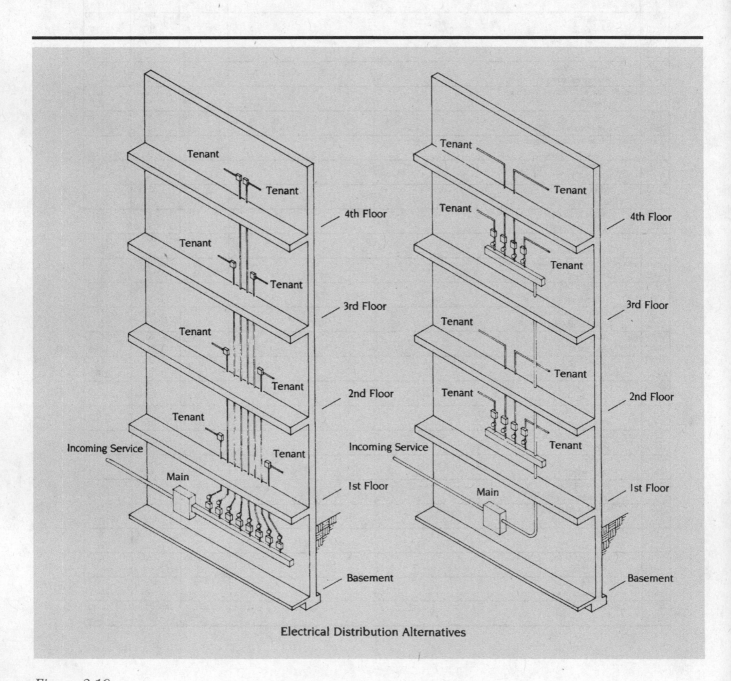

Electrical Distribution Alternatives

Figure 2.10

amounts of plaster, drilling each stud, and then patching. Instead, each box must be individually fed from a junction box, perhaps at the ceiling. When rigid conduit is needed for fire alarms or other purposes, many more offsets and bends may be necessary in order to adapt the wiring to existing conditions. Similarly, when wiring is to be exposed, rigid conduit or surface-mounted raceways are often specified and must be run at right angles to the existing structure.

Temporary Lighting and Power

The estimate should also include temporary lighting and power. The estimator must thoroughly evaluate the plans and specifications, and examine the site to ensure that adequate provisions are made for this item. For example, if scaffolding is to be erected at the exterior, weatherproof temporary lighting is necessary for pedestrian protection. Certain construction equipment, like mortar mixers and welders, may have special temporary electrical needs.

Special Construction & Demolition

Special items that are usually included in renovations are kitchen cabinets, toilet accessories, toilet partitions, wood or coal stoves, and fireplaces. These items are often clearly specified, and materials prices may be easily obtained. During the site visit, the estimator should inspect the proposed locations for these installations. If the installation is to be performed on existing surfaces, the estimator must be sure that there is adequate backing or space for specified, recessed equipment. Labor costs can increase quickly when workers have to remove existing wall finishes, install backing, and patch the area.

When installing wood-burning stoves, or metal or masonry fireplaces, contractors should consult a structural engineer to determine whether there is adequate bearing capacity at the proposed location for these items, or whether the structure is to be modified. The estimator should also examine the existing framing that is to be penetrated by the flue or chimney to determine the extent and type of alterations needed. An inspection of these elements may show that the proposed location is at a bearing structural member that cannot be removed.

The estimator should be aware of local Fire Safety Codes and ordinances regulating the installation of stoves and fireplaces. These codes are constantly being updated because of the increasing popularity of wood stove use. If there are any questions, an engineer or local fire official should be consulted. Again, the estimator must know exactly what is specified and how the existing conditions will affect the work.

Demolition

Demolition may be the most challenging aspect of estimating for commercial renovation. When the job calls for extensive demolition, it may be best to obtain prices from local subcontractors who are familiar with local regulations, hauling requirements, and dump-site locations. Local demolition subcontractors are also the most familiar with the values of and procedures for disposing of salvageable materials, a factor that may reduce costs of demolition. When the specifications call for the removal of asbestos or other hazardous materials, the estimator should always consult with local authorities and a licensed subcontractor. Not only does asbestos present a health hazard, but EPA and OSHA regulations impose stiff fines and penalties when hazardous materials are improperly handled and discarded.

Demolition in commercial renovation may be divided into three phases:

1. The actual dismantling of the existing structures, including labor and equipment.
2. Handling the debris. This includes the transport of material to an

on-site container or truck, and may include the installation and rental of a trash chute and/or dumpsters.

3. Hauling the rubbish to an approved dump site.

Identifying Materials

It is important to be accurate when identifying the materials and determining the limits of demolition. Figure 2.11 shows a page from *Means Repair and Remodeling Cost Data*. Note Section 02225-840, "Wall Partition Demolition." Removal of gypsum plaster on metal lath ($1.04 per square foot) is three times more costly than the removal of nailed drywall ($.31 per square foot). If the estimator does not take the time to identify the material, expensive mistakes may occur.

Establishing the Limits

Determining the limits of demolition is often the estimator's responsibility. Even if a demolition drawing is included in the plans and specifications (and this is often not the case), it may have been based on obsolete plans. Or, the specifications may state, "Contractor is to remove and dispose of all materials not to remain as part of the work." This is a vague statement that leaves all responsibility to the estimator.

If there is a demolition plan, the estimator should walk through the renovation site to verify the location and dimensions and to identify the materials to be removed. If there is no plan of the existing layout, the estimator should make a sketch, using the *proposed* floor plan as a reference. To help determine quantities, the estimator should measure ceiling heights. Measurements should be written on the sketch at the site. The estimator should leave as little as possible to memory when preparing the estimate.

Protecting Adjacent Materials

When certain existing materials are to remain, the estimator must choose the method of work that offers effective protection for these components. Sometimes skilled labor is necessary for the removal of certain items while retaining adjacent materials. The estimator may also have to determine when it is more economical to remove existing material completely and to build new, or when it is cheaper to cut, patch, and alter existing work to conform to new specifications.

Removal of Debris

Once the materials marked for demolition have been dismantled, the estimator must decide on the best method for moving the debris to a dumpster or truck for removal from the site. If the building is not tall, a covered slide may be constructed with relative ease. If the building is many stories above grade, however, alternatives must be examined. An existing freight elevator might be available, but if it can transport only small loads, the process is very time-consuming. A trash chute is another possibility. Figure 2.12 (from *Means Repair and Remodeling Cost Data*) shows the costs involved in erecting this device. In addition to these costs, the estimator should include the costs of support scaffolding and the possibility that workers will need a crane to erect it.

In determining the number of dumpsters or truckloads required, the estimator must also draw on experience with previous projects. Most disposal contractors include landfill costs in the dumpster rental fees, but the estimator should verify what this service includes. If the estimator decides that hauling by truck is the best alternative, he should contact local landfills to determine the cost of dumping and to make sure that they will accept construction materials.

The demolition and removal of materials is rarely well defined in plans and specifications. It is a tricky facet of commercial renovation, and one where the

		02225	Selective Demolition	CREW	DAILY OUTPUT	LABOR-HOURS	UNIT	2002 BARE COSTS				TOTAL INCL O&P	
								MAT.	LABOR	EQUIP.	TOTAL		
760	5020		On roof	1 Carp	250	.032	L.F.		.96		.96	1.59	760
	9000		Minimum labor/equipment charge	A-1	2	4	Job		94	30.50	124.50	189	
	9950		See also div. 02210-320 core drilling										
790	0010		**TORCH CUTTING** Steel, 1" thick plate	1 Clab	32	.250	L.F.	.18	5.85		6.03	9.90	790
	0040		1" diameter bar	"	210	.038	Ea.		.89		.89	1.48	
	1000		Oxygen lance cutting, reinforced concrete walls										
	1040		12" to 16" thick walls	1 Clab	10	.800	L.F.		18.75		18.75	31	
	1080		24" thick walls	"	6	1.333	"		31.50		31.50	52	
	1090		Minimum labor/equipment charge	A-1	1	8	Job		188	60.50	248.50	375	
	1100		See also division 05090-920										
840	0010		**WALLS AND PARTITIONS DEMOLITION** R02220 -510										840
	0100		Brick, 4" to 12" thick	B-9C	220	.182	C.F.		4.34	.81	5.15	8.10	
	0200		Concrete block, 4" thick		1,000	.040	S.F.		.95	.18	1.13	1.78	
	0280		8" thick	▼	810	.049			1.18	.22	1.40	2.19	
	0300		Exterior stucco 1" thick over netting	B-9	3,200	.013			.30	.06	.36	.55	
	1000		Drywall, nailed	1 Clab	1,000	.008			.19		.19	.31	
	1020		Glued and nailed		900	.009			.21		.21	.35	
	1500		Fiberboard, nailed		900	.009			.21		.21	.35	
	1520		Glued and nailed		800	.010			.23		.23	.39	
	2000		Movable walls, metal, 5' high		300	.027			.63		.63	1.04	
	2020		8' high	▼	400	.020			.47		.47	.78	
	2200		Metal or wood studs, finish 2 sides, fiberboard	B-1	520	.046			1.11		1.11	1.84	
	2250		Lath and plaster		260	.092			2.23		2.23	3.69	
	2300		Plasterboard (drywall)		520	.046			1.11		1.11	1.84	
	2350		Plywood	▼	450	.053			1.29		1.29	2.13	
	3000		Plaster, lime and horsehair, on wood lath	1 Clab	400	.020			.47		.47	.78	
	3020		On metal lath		335	.024			.56		.56	.93	
	3400		Gypsum or perlite, on gypsum lath		410	.020			.46		.46	.76	
	3420		On metal lath		300	.027	▼		.63		.63	1.04	
	3800		Toilet partitions, slate or marble		5	1.600	Ea.		37.50		37.50	62	
	3820		Hollow metal	▼	8	1	"		23.50		23.50	39	
	5000		Wallcovering, vinyl	1 Pape	700	.011	S.F.		.31		.31	.50	
	5040		Designer	"	480	.017	"		.45		.45	.73	
	9000		Minimum labor/equipment charge	1 Clab	4	2	Job		47		47	77.50	
850	0010		**WINDOW DEMOLITION** R02220 -510										850
	0200		Aluminum, including trim, to 12 S.F.	1 Clab	16	.500	Ea.		11.75		11.75	19.45	
	0240		To 25 S.F.		11	.727			17.05		17.05	28.50	
	0280		To 50 S.F.		5	1.600			37.50		37.50	62	
	0320		Storm windows, to 12 S.F.		27	.296			6.95		6.95	11.50	
	0360		To 25 S.F.		21	.381			8.95		8.95	14.80	
	0400		To 50 S.F.		16	.500	▼		11.75		11.75	19.45	
	0600		Glass, minimum		200	.040	S.F.		.94		.94	1.55	
	0620		Maximum		150	.053	"		1.25		1.25	2.07	
	1000		Steel, including trim, to 12 S.F.		13	.615	Ea.		14.45		14.45	24	
	1020		To 25 S.F.		9	.889			21		21	34.50	
	1040		To 50 S.F.		4	2			47		47	77.50	
	2000		Wood, including trim, to 12 S.F.		22	.364			8.55		8.55	14.15	
	2020		To 25 S.F.		18	.444			10.40		10.40	17.25	
	2060		To 50 S.F.	▼	13	.615			14.45		14.45	24	
	5020		Remove and reset window, minimum	1 Carp	6	1.333			40		40	66.50	
	5040		Average		4	2			60		60	99.50	
	5080		Maximum	▼	2	4	▼		120		120	199	
	9000		Minimum labor/equipment charge	1 Clab	4	2	Job		47		47	77.50	

Figure 2.11

		02225	Selective Demolition	CREW	DAILY OUTPUT	LABOR-HOURS	UNIT	MAT.	2002 BARE COSTS LABOR	EQUIP.	TOTAL	TOTAL INCL O&P	
690	5000		Siding, metal, horizontal	1 Clab	444	.018	S.F.		.42		.42	.70	690
	5020		Vertical		400	.020			.47		.47	.78	
	5200		Wood, boards, vertical		400	.020			.47		.47	.78	
	5220		Clapboards, horizontal		380	.021			.49		.49	.82	
	5240		Shingles		350	.023			.54		.54	.89	
	5260		Textured plywood		725	.011			.26		.26	.43	
	9000		Minimum labor/equipment charge		2	4	Job		94		94	155	
720	0010	**DISPOSAL ONLY** Urban buildings with salvage value allowed											720
	0020		Including loading and 5 mile haul to dump										
	0200		Steel frame	B-3	430	.112	C.Y.		2.86	3.80	6.66	8.85	
	0300		Concrete frame		365	.132			3.37	4.47	7.84	10.40	
	0400		Masonry construction		445	.108			2.76	3.67	6.43	8.55	
	0500		Wood frame		247	.194			4.97	6.60	11.57	15.35	
730	0010	**RUBBISH HANDLING** The following are to be added to the											730
	0020		demolition prices										
	0400		Chute, circular, prefabricated steel, 18" diameter	B-1	40	.600	L.F.	26	14.45		40.45	52.50	
	0440		30" diameter	"	30	.800	"	35	19.30		54.30	70.50	
	0725		Dumpster, weekly rental, 1 dump/week, 20C.Y. capacity (8Tons)				Week					425	
	0800		30 C.Y. capacity (10 Tons)									640	
	0840		40 C.Y. capacity (13 Tons)									775	
	0900		Alternate pricing for dumpsters										
	0910		Delivery, average for all sizes				Ea.					50	
	0920		Haul, average for all sizes									125	
	0930		Rent per day, average for all sizes									3	
	0940		Rent per month, average for all sizes									30	
	0950		Disposal fee per ton, average for all sizes				Ton					45	
	1000		Dust partition, 6 mil polyethylene, 4' x 8' panels, 1" x 3" frame	2 Carp	2,000	.008	S.F.	.17	.24		.41	.58	
	1080		2" x 4" frame	"	2,000	.008	"	.30	.24		.54	.73	
	2000		Load, haul to chute & dumping into chute, 50' haul	2 Clab	24	.667	C.Y.		15.65		15.65	26	
	2040		100' haul		16.50	.970			22.50		22.50	37.50	
	2080		Over 100' haul, add per 100 L.F.		35.50	.451			10.55		10.55	17.50	
	2120		In elevators, per 10 floors, add		140	.114			2.68		2.68	4.44	
	3000		Loading & trucking, including 2 mile haul, chute loaded	B-16	45	.711			17.30	10.45	27.75	40	
	3040		Hand loading truck, 50' haul	"	48	.667			16.25	9.80	26.05	37.50	
	3080		Machine loading truck	B-17	120	.267			6.80	4.43	11.23	15.90	
	3120		Wheeled 50' and ramp dump loaded	2 Clab	24	.667			15.65		15.65	26	
	5000		Haul, per mile, up to 8 C.Y. truck	B-34B	1,165	.007			.17	.40	.57	.72	
	5100		Over 8 C.Y. truck	"	1,550	.005			.13	.30	.43	.54	
740	0010	**DUMP CHARGES** Typical urban city, tipping fees only											740
	0100		Building construction materials				Ton					55	
	0200		Trees, brush, lumber									45	
	0300		Rubbish only									50	
	0500		Reclamation station, usual charge									80	
760	0010	**SAW CUTTING**, Asphalt, up to 3" deep	B-89	1,050	.015	L.F.	.25	.41	.29	.95	1.25	760
	0020		Each additional inch of depth		1,800	.009		.06	.24	.17	.47	.63	
	0400		Concrete slabs, mesh reinforcing, up to 3" deep		980	.016		.34	.44	.31	1.09	1.42	
	0420		Each additional inch of depth		1,600	.010		.11	.27	.19	.57	.76	
	0800		Concrete walls, hydraulic saw, plain, per inch of depth	B-89B	250	.064		.31	1.73	1.98	4.02	5.30	
	0820		Rod reinforcing, per inch of depth		150	.107		.43	2.89	3.30	6.62	8.70	
	1200		Masonry walls, hydraulic saw, brick, per inch of depth		300	.053		.31	1.44	1.65	3.40	4.46	
	1220		Block walls, solid, per inch of depth		250	.064		.32	1.73	1.98	4.03	5.30	
	2000		Brick or masonry w/hand held saw, per inch of depth	A-1	125	.064		.26	1.50	.49	2.25	3.30	
	5000		Wood sheathing to 1" thick, on walls	1 Carp	200	.040			1.20		1.20	1.99	

R02220 -510

Figure 2.12

estimator may easily underestimate the costs involved. The whole process of the work must be well thought out and planned.

Building Sitework

Excavation

Usually, site work in commercial renovation is relatively limited. The economy of scale that can be applied to site work for new construction is often not applicable for renovation work. Trenching and excavation around an existing building usually involves cutting and patching concrete or asphalt. If the particular job is small or if there is inadequate space for the use of large equipment, the work may have to be performed by hand. To prevent damage, workers must take extra care when digging near existing buried utilities or piping.

The estimator must try to gather as much information as possible when inspecting for proposed site work. Any holes or test pits should be examined for soil type and stability. Shutoffs and manholes for utilities should be located or verified on a site plan. If site work is to be performed on public sidewalks or in the street, the estimator should consult local authorities to determine what requirements must be met. For example, a renovation project calls for a new sewer connection in the street. After a visual examination, the estimator concludes that the job involves only cutting and patching of the concrete sidewalk and asphalt pavement, as well as trenching, and backfilling. Upon further investigation, however, the estimator learns that the existing sidewalk contains metal reinforcing, that the street has a six-inch, reinforced concrete base that must be cut and patched to match the existing work, and that under local regulations, a policeman must be hired for traffic control. Furthermore, low overhead wires severely restrict the use of large equipment. This example demonstrates that the project may become much more involved than originally planned if it is not properly investigated and evaluated at the start.

General Conditions

The estimator must include all the "hard costs," or materials and labor, to produce the actual finished project, as well as the "soft costs," those indispensable items that are required for the performance of the work. The following is a partial list of items that are included in the general conditions, or general requirements:

- Overhead and main office expenses
- Profit
- Sales and employer's taxes
- Bonds
- Insurance
- Architectural and engineering fees
- Testing & borings
- General working conditions
- Temporary requirements & construction
- Contractor equipment

Added together, these items may represent a large percentage of the cost of the proposed renovation project. Incorporating these items into the estimating process will be discussed in Chapter 9. The following section deals with how the estimator anticipates the cost of general conditions while evaluating the site and analyzing the existing conditions.

The first seven items mentioned above are determined and included while preparing the actual estimate. Items 8 through 10, however, must be determined in part while inspecting the building to be renovated.

Figure 2.13 shows a page from *Means Repair and Remodeling Cost Data*. Please note section 01250-400, "FACTORS." The factors listed there are those that most commonly increase the costs of a renovation project. The estimator must visualize the proposed work and keep in mind how existing conditions will affect the work.

Cutting and Patching

Almost every renovation or remodeling job involves a good deal of cutting and patching to match existing conditions. Workers must open access holes in existing walls to install wire and pipe. Often, original components, moldings, and trim must be carefully removed and then replaced. Damaged pieces must be recreated or repaired, at some expense.

For example, a job calls for a new plumbing vent to be installed through an old slate shingle roof. A copper sleeve and flashing must be custom-made to match the existing work. Cutting the hole for the pipe may be difficult. Chances are that shingles at the hole and elsewhere near the work will break and need replacement. The scaffolding or roof jacks must be installed with extreme care to prevent such breakage. Costs of performing such work may be many times greater than that for similar work on an asphalt shingle roof. The estimator must be familiar with the materials in order to determine how the work is to be performed and what it will cost.

Dust Protection

When a renovation project is to be performed in a partially occupied building, the estimator is often required to protect from dust, noise, and general disturbance the adjacent areas that are not involved in the project. Workers must employ certain methods, while performing the work, to prevent the production and dispersal of dust. These precautions may slow production.

When the main lobby and only elevator in a building must be used for materials and equipment while it is in constant use by tenants, the estimator must provide for constant cleaning of these areas. Door and window openings at the work site must be sealed to prevent dust dispersion. If a dust problem occurs in adjacent areas, the contractor may be required to hire an outside cleaning contractor.

Especially in urban areas, dumpsters have to be covered, and laborers must use trash chutes instead of throwing materials out of a window. These requirements are not only for pedestrian protection but also for the reduction of dust problems on the site. Covering a dumpster has the added benefit of saving the contractor from having to haul the entire neighborhood's garbage, which has a tendency to mysteriously appear in dumpsters each morning.

Equipment Usage

In new construction, work may be planned and scheduled to take full advantage of the most efficient and productive construction equipment. For example, hydraulic elevator piston shafts are drilled before erection of the superstructure to allow the mobile drilling rig free access. In commercial renovation, however, such access is often severely curtailed. Work must be performed with small, less-effective equipment, or even by hand. This approach may add substantially to the labor costs of the project.

The use of equipment may also be restricted because of noise. In a partially occupied building, the specifications may state that the contractor must cease a noisy operation when a tenant disturbance results in a complaint. The available options then become (1) substitute use of the equipment for more time-consuming, quieter methods, or (2) perform the work during overtime

01100 | Summary

01103 | Models & Renderings

			CREW	DAILY OUTPUT	LABOR-HOURS	UNIT	2002 BARE COSTS				TOTAL INCL O&P	
							MAT.	LABOR	EQUIP.	TOTAL		
500	0010	**RENDERINGS** Color, matted, 20" x 30", eye level,										500
	0020	1 building, minimum				Ea.	1,700			1,700	1,875	
	0050	Average					2,850			2,850	3,125	
	0100	Maximum					4,000			4,000	4,400	
	1000	5 buildings, minimum					3,500			3,500	3,850	
	1100	Maximum					6,800			6,800	7,475	

01107 | Professional Consultant

			CREW	DAILY OUTPUT	LABOR-HOURS	UNIT	MAT.	LABOR	EQUIP.	TOTAL	TOTAL INCL O&P	
100	0011	**ARCHITECTURAL FEES** R01107 -010										100
	0020	For new construction										
	0060	Minimum				Project					4.90%	
	0090	Maximum									16%	
	0100	For alteration work, to $500,000, add to fee									50%	
	0150	Over $500,000, add to fee									25%	
200	0011	**CONSTRUCTION MANAGEMENT FEES**										200
	0060	For work to $10,000				Project					10%	
	0070	To $25,000									9%	
	0090	To $100,000									6%	
	0100	To $500,000									5%	
	0110	To $1,000,000									4%	
300	0010	**ENGINEERING FEES** R01107 -030										300
	0020	Educational planning consultant, minimum				Project					.50%	
	0100	Maximum				"					2.50%	
	0400	Elevator & conveying systems, minimum				Contrct					2.50%	
	0500	Maximum									5%	
	1000	Mechanical (plumbing & HVAC), minimum									4.10%	
	1100	Maximum									10.10%	
	1200	Structural, minimum				Project					1%	
	1300	Maximum				"					2.50%	
700	0010	**SURVEYING** Conventional, topographical, minimum	A-7	3.30	7.273	Acre	16	236		252	405	700
	0100	Maximum	A-8	.60	53.333			48	1,700		1,748	2,800	
	0300	Lot location and lines, minimum, for large quantities	A-7	2	12		25	390		415	660	
	0320	Average	"	1.25	19.200		45	620		665	1,050	
	0400	Maximum, for small quantities	A-8	1	32		72	1,025		1,097	1,725	
	1100	Crew for building layout, 2 person crew	A-6	1	16	Day		540		540	875	
	1200	3 person crew	A-7	1	24			780		780	1,275	
	1300	4 person crew	A-8	1	32			1,025		1,025	1,650	

01200 | Price & Payment Procedures

01250 | Contract Modification Procedures

			CREW	DAILY OUTPUT	LABOR-HOURS	UNIT	2002 BARE COSTS				TOTAL INCL O&P	
							MAT.	LABOR	EQUIP.	TOTAL		
200	0010	**CONTINGENCIES** for estimate at conceptual stage				Project					25%	200
	0050	Schematic stage									20%	
	0100	Preliminary working drawing stage (Design Dev.)									15%	
	0150	Final working drawing stage									8%	
400	0010	**FACTORS** Cost adjustments R01250 -010										400
	0100	Add to construction costs for particular job requirements										
	0500	Cut & patch to match existing construction, add, minimum				Costs	2%	3%				
	0550	Maximum					5%	9%				

Figure 2.13

hours when the building is not occupied. Both result in higher construction costs. The estimator must anticipate restrictions such as these and include the appropriate costs in the estimate.

Material Handling and Storage

Limitations regarding the delivery, storage, and handling of construction materials may also add to the cost of renovation projects. The specifications may impose a restriction that allows delivery and placement of materials only at certain off-peak times of the day. Local authorities may further restrict the use of the street, or the building may limit the use of loading areas or elevators. All parties with an influence on such restrictions should be consulted before the estimate is prepared.

If delivery and handling are restricted to certain time periods, it may be necessary to have all personnel stop work in order to assist with unloading and placement of materials during the permissible times. This stopping and starting of work during the course of the day, possibly adding to the number of required deliveries, may significantly reduce productivity.

Inside the building, there may be insufficient bearing capacities or lack of area for material storage. Materials stored outside must be adequately protected from the elements. Architects usually specify that materials should be stored in the environment where they are to be used for a stated period of time before placement. This is a requirement for proper installation. If interior storage areas are limited, only small amounts of material can be placed at a time. Often, materials must be carried up stairs or cut to fit inside elevators.

Protection of Adjacent Areas

When materials are delivered or work is performed in a building where certain areas are finished or occupied, contractors must protect the finished work. For example, the contractor may be responsible for providing temporary coverings, and moving pads or plywood for elevators and floor runners for carpeted areas. The estimator must determine appropriate protection measures and must be sure to include the costs in the estimate.

Shift Work

Shift work or overtime may be required in order to provide some kinds of protection for adjacent work. In such cases, the estimator should discuss the conditions surrounding the shift work or overtime with the architect and owner. He must be sure that all requirements are met and that appropriate costs are included in the estimate. For smaller projects, even a minor amount of unanticipated overtime work may severely erode the profit margin.

Temporary Shoring and Bracing

When contractors perform structural work in an existing building, temporary shoring or bracing may be necessary. If bracing is not included in the plans and specifications, the estimator should consult the engineer on the structural requirements of such work.

Temporary Structures

Temporary requirements for construction may add substantially to the costs of renovation. For example, a field office, either inside the building or in a trailer, may be necessary. The field office may require a telephone, heat, and lighting for the duration of the project.

Barricades, scaffolding, temporary railings, and fences are other temporary items that may not be directly specified. In new construction, these items may be included in the estimate as established costs or percentages determined from previous similar jobs. Each commercial renovation project is unique, however, and has particular requirements.

Temporary Nonconstruction Personnel

Hiring nonconstruction personnel on a temporary basis may also be necessary. For example, local authorities in cities may require a police officer to direct traffic or to protect pedestrians when work or material deliveries are performed at the street level. Watchmen, sometimes with guard dogs, may be necessary to prevent theft and vandalism. Estimators will find that it is helpful to be familiar with the particular nature of the neighborhood in which the renovation project will take place. With this knowledge, they can anticipate and include appropriate indirect costs in the estimate.

Coordination of Subcontractors

Coordination of subcontractors is another intangible item that affects the cost of a commercial renovation. The division of labor and responsibility is often unclear in this kind of construction work, because existing conditions may make it impossible to use the standard new construction approach. The estimator must anticipate potential misunderstandings by visualizing the complete renovation process, including appropriate supervision of each part of the project. With information gathered from the site visit, the estimator must determine how many workers will be needed and what equipment may or may not be used.

Other appropriate items are shown in Figure 2.14 from *Means Repair and Remodeling Cost Data*. Section 01250-500, "Job Conditions," lists more general items that affect the estimate. Please note that the items discussed in Figures 2.13 and 2.14 all involve percentages of project costs to be added or deducted. The estimator must use his own judgment, based on experience, to incorporate these figures into the estimate without losing accuracy.

Summary

The most important point to be made regarding the site evaluation is that the estimator must have a thorough understanding of the proposed work and must perform a detailed inspection of the site to determine how the work will be affected by the existing conditions. The size or age of an existing building should not determine how thorough the site visit should be. In fact, on a smaller project, a particular error or oversight may have a larger percentage effect on the accuracy of the estimate than would be the case for a larger job.

The estimator should always verify dimensions and feasibility, and should be familiar with the types of construction and materials in the existing building. Application of previous experience, knowledge, and all available resources is key to the successful preparation of complete and accurate estimates for commercial renovation.

01200 | Price & Payment Procedures

01250 | Contract Modification Procedures

			CREW	DAILY OUTPUT	LABOR-HOURS	UNIT	2002 BARE COSTS				TOTAL INCL O&P	
							MAT.	LABOR	EQUIP.	TOTAL		
400	0800	Dust protection, add, minimum R01250-010				Costs	1%	2%				**400**
	0850	Maximum					4%	11%				
	1100	Equipment usage curtailment, add, minimum					1%	1%				
	1150	Maximum					3%	10%				
	1400	Material handling & storage limitation, add, minimum					1%	1%				
	1450	Maximum					6%	7%				
	1700	Protection of existing work, add, minimum					2%	2%				
	1750	Maximum					5%	7%				
	2000	Shift work requirements, add, minimum						5%				
	2050	Maximum						30%				
	2300	Temporary shoring and bracing, add, minimum					2%	5%				
	2350	Maximum					5%	12%				
	2400	Work inside prisons and high security areas, add, minimum						30%				
	2450	Maximum						50%				
500	0010	**JOB CONDITIONS** Modifications to total										**500**
	0020	project cost summaries										
	0100	Economic conditions, favorable, deduct				Project					2%	
	0200	Unfavorable, add									5%	
	0300	Hoisting conditions, favorable, deduct									2%	
	0400	Unfavorable, add									5%	
	0500	General Contractor management, experienced, deduct									2%	
	0600	Inexperienced, add									10%	
	0700	Labor availability, surplus, deduct									1%	
	0800	Shortage, add									10%	
	0900	Material storage area, available, deduct									1%	
	1000	Not available, add									2%	
	1100	Subcontractor availability, surplus, deduct									5%	
	1200	Shortage, add									12%	
	1300	Work space, available, deduct									2%	
	1400	Not available, add									5%	

01290 | Payment Procedures

			CREW	DAILY OUTPUT	LABOR-HOURS	UNIT	MAT.	LABOR	EQUIP.	TOTAL	INCL O&P	
800	0010	**TAXES** Sales tax, State, average R01100-090				%	4.70%					**800**
	0050	Maximum					7%					
	0200	Social Security, on first $80,400 of wages R01100-100						7.65%				
	0300	Unemployment, MA, combined Federal and State, minimum						2.10%				
	0350	Average						7%				
	0400	Maximum						8%				

01300 | Administrative Requirements

01310 | Project Management/Coordination

			CREW	DAILY OUTPUT	LABOR-HOURS	UNIT	2002 BARE COSTS				TOTAL INCL O&P	
							MAT.	LABOR	EQUIP.	TOTAL		
150	0010	**PERMITS** Rule of thumb, most cities, minimum				Job					.50%	**150**
	0100	Maximum				"					2%	
350	0010	**INSURANCE** Builders risk, standard, minimum R01100-040				Job					.22%	**350**
	0050	Maximum									.59%	
	0200	All-risk type, minimum R01100-060									.25%	
	0250	Maximum									.62%	
	0400	Contractor's equipment floater, minimum				Value					.50%	
	0450	Maximum				"					1.50%	

Figure 2.14

Chapter 3

The Quantity Takeoff

The quantity takeoff is the process in which detailed lists are compiled, based on drawings and specifications, of all the material and equipment necessary to construct a project. Quantity takeoff is the basis of estimating. This phase of the estimate is regarded as so important in England that those who perform the takeoff are registered and licensed as *Quantity Surveyors*. Not only is the takeoff essential for the application of prices, but it can also be used effectively in other aspects of a construction project.

The quantity takeoff should be organized so that the information gathered can be used to future advantage. For example, scheduling can be made easier if items are taken off and listed by construction phase, or by floor. Material purchasing will similarly benefit from consistent organization. Units for each item should be used consistently throughout the whole project—from takeoff to cost control. In this way, the original estimate can be equitably compared to any progress reports and final cost reports. The result is better record keeping and monitoring.

When the estimator works with the plans during the quantity takeoff, consistency is very important. If each job is approached in the same manner, a pattern will develop, such as moving from the lower floors to the top, clockwise or counterclockwise. The particular method is not important, as long as it is used consistently. By maintaining a regular pattern, duplications as well as omissions and errors are minimized.

Using Preprinted Forms

Preprinted forms provide an excellent means for developing consistent patterns. Figures 3.1 to 3.3 are examples of such forms. The *Quantity Sheet* (Figure 3.1) is designed purely for quantity accumulation. Note that one list of materials and dimensions can be used for up to four different areas or segments of work. Figure 3.2 shows a *Cost Analysis* form, which can be used in conjunction with a Quantity Sheet. Totals of quantities are transferred to this form for pricing and extensions. Figure 3.3, a *Consolidated Estimate* form, is designed to incorporate onto one form both quantity takeoff and pricing. Refer to Part III of this book (the sample estimates) for information on how these forms may be used effectively.

Custom-designed forms are another option. If all employees use the same types of forms, communication and coordination of the estimating process will proceed more smoothly. One estimator will be able to understand more easily the work of another. R.S. Means has published two collections of full-sized

QUANTITY SHEET

					SHEET NO.
PROJECT					ESTIMATE NO.
LOCATION		ARCHITECT			DATE
TAKE OFF BY		EXTENSIONS BY:			CHECKED BY:

DESCRIPTION	NO.	DIMENSIONS				UNIT		UNIT		UNIT		UNIT

Figure 3.1

COST ANALYSIS

PROJECT	SHEET NO.
ARCHITECT	ESTIMATE NO.
	DATE

TAKE OFF BY: QUANTITIES BY: PRICES BY: EXTENSIONS BY: CHECKED BY:

DESCRIPTION	SOURCE/DIMENSIONS			QUANTITY	UNIT	MATERIAL		LABOR		EQ./TOTAL	
						UNIT COST	TOTAL	UNIT COST	TOTAL	UNIT COST	TOTAL

Figure 3.2

CONSOLIDATED ESTIMATE

										MATERIAL			LABOR		EQUIPMENT		TOTAL	

SHEET NO.

CLASSIFICATION

ESTIMATE NO.

PROJECT

LOCATION

ARCHITECT

DATE

TAKE OFF BY

QUANTITIES BY

PRICES BY

EXTENSIONS BY

CHECKED BY

DIMENSIONS

QUANTITIES

NO.

DESCRIPTION

UNIT

UNIT COST

TOTAL

UNIT COST

TOTAL

UNIT COST

TOTAL

UNIT COST

TOTAL

UNIT

MATERIAL

LABOR

EQUIPMENT

Figure 3.3

52

and ready-to-use forms, *Means Forms for Building Construction Professionals* and *Means Forms for Contractors*. Each book includes examples and instructions for use. Estimating software currently linked to *Means Repair & Remodeling Cost Data* allows the user to print report forms of varying detail.

Another important type of form is the schedule. The two most important forms the interior estimator can use, *Door and Frame Schedule* and *Room Finish Schedule,* are shown in Figures 3.4 and 3.5, respectively. These schedules help to efficiently organize project data in the following ways:

- The designer can be sure that all work is included in the drawings and specifications.
- The estimator can be sure that all items are included in the estimate.
- The contractor can be sure of proper installation instructions.

Appropriate and easy-to-use forms are the first and most important "tools of the trade" for estimators. Other tools useful to the estimator may include scales, digitizers, highlighters, rotometers, mechanical counters, and colored pencils.

Time-saving Measures

A number of shortcuts may be used for the quantity takeoff. If approached logically and systematically, these techniques help to save time without sacrificing accuracy. Consistent use of accepted abbreviations saves the time of writing things out. An abbreviations list might be posted in a conspicuous place to provide a consistent pattern of definitions for use within an office.

All dimensions—whether printed, measured, or calculated—that can be used for determining quantities of more than one item should be listed on a separate sheet and posted for easy reference. Posted, overall dimensions may also be useful as a quick check for order of magnitude errors.

Rounding off, or decreasing the number of significant digits, should be done only when it will not statistically affect the resulting product. The estimator must use good judgment to determine instances where rounding may be appropriate. An overall two- or three-percent variation in a competitive market may often be the difference between winning a contract and losing a job, or between profit and no profit. The estimator should establish rules for rounding to achieve a consistent level of precision. In general, it is best not to round numbers until the final summary of quantities.

The final summary is also the place to convert units of measure into standards for practical use, for example, square feet of wallcovering to number of rolls. This is done to keep the quantities in the same units as they are purchased, handled, and recorded in cost reports.

Be sure to quantify (count) and to include "labor only" items that are not shown on the plans. Such items may or may not be indicated in the specifications and might include clean-up, special labor for handling materials, and furniture setup.

Summary of Takeoff Guidelines

The following list is a summary of the previous discussion, along with some additional guidelines for the quantity takeoff.

- Use preprinted forms.
- Transfer carefully when copying numbers from one sheet to the next.
- List dimensions (width, length, height) in consistent order.
- Verify the scale of drawings before using them as a basis for measurement.
- Mark drawings neatly and consistently as quantities are counted.
- Be alert to changes in scale, or notes such as "N.T.S." (not to scale). Sometimes drawings have been photographically reduced.
- Include required items which may not appear on the drawings or in the specifications.

The four most important points:

- Print legibly.
- Be organized.
- Use common sense.
- Be consistent.

DOOR AND FRAME SCHEDULE

PROJECT _____ ARCHITECT _____ PAGE ___ OF ___

LOCATION _____ OWNER _____ DATE _____ BY _____

DOOR NO.	SIZE			DOOR						FRAME					FIRE RATING		HARDWARE		REMARKS
	W	H	T	MAT.	TYPE	GLASS	LOUVER	MAT.	TYPE	DETAILS			LAB	CON	SET NO.	KEYSIDE ROOM NO.			
										JAMB	HEAD	SILL							

Figure 3.4

55

01833 | Landscape Maintenance

			Crew	Daily Output	Labor-Hours	Unit	Mat.	Labor	Equip.	Total	Total In-House	Total Incl O&P	
510	0010	**EDGING**											**510**
	0020	Hand edging, at walks	1 Clam	16	.500	C.L.F.		9.25		9.25	11.40	14.25	
	0030	At planting, mulch or stone beds		7	1.143			21		21	26	32.50	
	0040	Power edging, at walks		88	.091			1.68		1.68	2.07	2.59	
	0050	At planting, mulch or stone beds	▼	24	.333	▼		6.15		6.15	7.60	9.50	
520	0010	**FLOWER, SHRUB & TREE CARE**											**520**
	0020	Flower or shrub beds, bark mulch, 3" deep hand spreader	1 Clam	100	.080	S.Y.	1.63	1.48		3.11	3.62	4.32	
	0030	Peat moss, 1" deep hand spreader		900	.009		1.67	.16		1.83	2.04	2.34	
	0040	Wood chips, 2" deep hand spreader		220	.036	▼	1.70	.67		2.37	2.70	3.16	
	0050	Cleaning		1	8	M.S.F.		148		148	182	228	
	0060	Fertilizing, dry granular, 3 #/C.S.F.		85	.094		9.55	1.74		11.29	12.65	14.60	
	0070	Weeding, mulched bed		20	.400			7.40		7.40	9.10	11.40	
	0080	Unmulched bed		8	1	▼		18.50		18.50	23	28.50	
	0090	Trees, pruning from ground, 1-1/2" caliper		84	.095	Ea.		1.76		1.76	2.17	2.71	
	0100	2" caliper		70	.114			2.11		2.11	2.60	3.26	
	0110	2-1/2" caliper		50	.160			2.96		2.96	3.65	4.56	
	0120	3" caliper	▼	30	.267			4.93		4.93	6.10	7.60	
	0130	4" caliper	2 Clam	21	.762			14.10		14.10	17.35	21.50	
	0140	6" caliper		12	1.333			24.50		24.50	30.50	38	
	0150	9" caliper		7.50	2.133			39.50		39.50	48.50	61	
	0160	12" caliper	▼	6.50	2.462			45.50		45.50	56	70	
	0170	Fertilize, slow release tablets	1 Clam	100	.080		.32	1.48		1.80	2.18	2.68	
	0180	Pest control, spray		24	.333		14.85	6.15		21	24	28	
	0190	Systemic		48	.167		13.80	3.08		16.88	19	22	
	0200	Watering, under 1-1/2" caliper		34	.235			4.35		4.35	5.35	6.70	
	0210	1-1/2" to 4" caliper		14.50	.552			10.20		10.20	12.55	15.70	
	0220	4" caliper and over		10	.800	▼		14.80		14.80	18.25	23	
	0230	Shrubs, prune, entire bed		7	1.143	M.S.F.		21		21	26	32.50	
	0240	Per shrub, 3' height		190	.042	Ea.		.78		.78	.96	1.20	
	0250	4' height		90	.089			1.64		1.64	2.03	2.53	
	0260	6' height		50	.160			2.96		2.96	3.65	4.56	
	0270	Fertilize, dry granular, 3#/M.S.F.		85	.094	M.S.F.	.91	1.74		2.65	3.15	3.82	
	0280	Watering, entire bed		7	1.143	"		21		21	26	32.50	
	0290	Per shrub	▼	32	.250	Ea.		4.63		4.63	5.70	7.10	
530	0010	**LAWN CARE**											**530**
	0020	Mowing lawns, power mower, 18"-22"	1 Clam	80	.100	M.S.F.		1.85		1.85	2.28	2.85	
	0030	22"-30"		120	.067			1.23		1.23	1.52	1.90	
	0040	30"-32"	▼	140	.057			1.06		1.06	1.30	1.63	
	0050	Self propelled or riding mower, 36"-44"	A-16	300	.027			.49	.24	.73	.84	1.02	
	0060	48"-58"		480	.017			.31	.15	.46	.53	.64	
	0070	Tractor, 3 gang reel, 7' cut		930	.009			.16	.08	.24	.27	.33	
	0080	5 gang reel, 12' cut	▼	1,200	.007	▼		.12	.06	.18	.21	.25	
	0090	Edge trimming with weed whacker	1 Clam	5,760	.001	L.F.		.03		.03	.03	.04	
540	0010	**LAWN RENOVATION**											**540**
	0020	Lawn renovations, aerating, 18" walk behind cultivator	1 Clam	95	.084	M.S.F.		1.56		1.56	1.92	2.40	
	0030	48" tractor drawn cultivator	A-18	750	.011			.20	.29	.49	.53	.62	
	0040	72" tractor drawn cultivator	"	1,100	.007			.13	.20	.33	.36	.42	
	0050	Fertilizing, dry granular, 4#/M.S.F., drop spreader	1 Clam	24	.333		1.89	6.15		8.04	9.70	11.85	
	0060	Rotary spreader	"	140	.057		1.89	1.06		2.95	3.38	3.99	
	0070	Tractor drawn 8' spreader	A-18	500	.016		1.89	.30	.43	2.62	2.88	3.29	
	0080	Tractor drawn 12' spreader	"	800	.010		1.89	.19	.27	2.35	2.58	2.94	
	0090	Overseeding, utility mix, 7#/M.S.F., drop spreader	1 Clam	10	.800		11	14.80		25.80	30.50	36.50	
	0100	Tractor drawn spreader	A-18	52	.154	▼	11	2.85	4.16	18.01	19.75	22.50	

2003 Bare Costs (Mat., Labor, Equip., Total columns)

III-5

Figure 3.5

Chapter 4

Pricing the Estimate

When the quantities have been counted, values in the form of unit costs must be applied and markups (e.g., overhead and profit) added in order to determine the total "selling" price (the quote). Depending upon the chosen estimating method and the level of detail required, these unit costs may be direct, or "bare," costs. Or, they may include overhead, profit, or contingencies. In unit price estimating, the unit costs most commonly used are bare, or "unburdened." Items such as overhead and profit are usually added to the total bare costs at the estimate summary. (Refer to Chapter 1 for a discussion of the different types of estimates.)

Sources of Cost Information

Determining accurate and reliable bare cost data is one of the most difficult aspects of the estimator's job. Sources for such data are varied, but they can be categorized in terms of their relative reliability.

In-house Historical Cost Data

The most reliable of any cost information is contained in the accurate, up-to-date, well-kept records of completed work by the estimator's own company. There is no better cost reference for a particular construction item than the *actual* cost of that item to the contractor during another recent job, modified (if necessary) to meet the requirements of the project being estimated.

Subcontractor Bids

Bids from responsible subcontractors are the next most reliable source of cost data. Any estimating inaccuracies are essentially absorbed by the subcontractor. A subcontract bid is a known, fixed cost prior to the project. Whether the price is "right" or "wrong" does not matter (as long as it is a responsible competitive bid with no apparent errors). The bid price is what the appropriate portion of work will cost. No estimating is required, except for possible verification of the quote and comparison with other subcontractors' quotes.

Vendor Quotes

Quotations by vendors for material costs are, for the same reasons, as reliable as subcontract bids. In this case, however, the estimator must apply estimated labor costs. Thus the "installed" price for a particular item may be more variable.

Published Cost Data

When the estimator has no cost records for a particular item and is unable to obtain a quotation, he might call upon another reliable source of price information, current construction cost books such as *Means Repair and Remodeling Cost Data*. Means presents all such data in the form of national averages. These figures can be adjusted to local conditions, a procedure that will be explained later in this book. In addition to being a source of primary costs, current annual construction cost data books may be useful as reference tools (for productivity data, estimating methods, and design criteria), or as cross-checks for verifying costs obtained from unfamiliar sources.

Lacking cost information from any of the above-mentioned sources, the estimator may have to rely on experience and personal knowledge of the field to develop costs.

No matter what source of cost information is used, the system and sequence of pricing should be the same as that used for the quantity takeoff. This consistent approach should continue through both accounting and cost control during work on the project.

Obtaining and Recording Information

Whenever possible, all price quotations from vendors or subcontractors should be confirmed in writing. Qualifications and exclusions should be clearly stated. The requirements and times quoted should be checked to be sure that they are complete and as specified. One way to check these items is to prepare a standard form to be used by all subcontractors and vendors submitting quotations. This form, generally called a "Request for Quote," should include all of the appropriate questions, so that the estimator can obtain all needed information. The form also provides a standard format for organizing the information.

The above procedures are ideal. In the realistic haste of estimating and bidding, however, quotations are often received verbally, either in person or by telephone. Under such circumstances, the need to gather all pertinent information becomes even more crucial, because omissions are more likely. Using a preprinted form, such as the one shown in Figure 4.1, is essential to ensure that all required information and qualifications are obtained and understood. How often has the subcontractor or vendor said, "I didn't know that I was supposed to include that?" Use of forms such as these helps to ensure that the appropriate questions are asked and that bids are better understood and more complete.

Types of Costs

All costs included in a unit price estimate can be divided into two types, *direct* and *indirect*. Direct costs are those dedicated solely to the physical construction of a specific project. Material, labor, equipment, and subcontract costs, as well as project overhead costs, are all direct.

Indirect costs are usually added to the estimate at the summary stage, and they are most often calculated as percentages of the direct costs. Indirect costs include such items as taxes, insurance, overhead, profit, and contingencies. It is the indirect costs that account for the greatest variation in estimates among different bidders.

Types of Costs in a Construction Estimate

Direct Costs	Indirect Costs
Material	Taxes
Labor	Insurance
Equipment	Office Overhead
Subcontractors	Profit
Project Overhead	Contingencies
Sales Tax	
Bonds	

TELEPHONE QUOTATION

DATE _____

PROJECT _____ TIME _____

FIRM QUOTING _____ PHONE (   )

ADDRESS _____ BY _____

ITEM QUOTED _____ RECEIVED BY _____

WORK INCLUDED	AMOUNT OF QUOTATION

DELIVERY TIME	**TOTAL BID**	
DOES QUOTATION INCLUDE THE FOLLOWING	If ☐ NO is checked, determine the following	
STATE & LOCAL SALES TAXES ☐ YES ☐ NO	MATERIAL VALUE	
DELIVERY TO THE JOB SITE ☐ YES ☐ NO	WEIGHT	
COMPLETE INSTALLATION ☐ YES ☐ NO	QUANTITY	
COMPLETE SECTION AS PER PLANS & SPECIFICATIONS ☐ YES ☐ NO	DESCRIBE BELOW	

EXCLUSIONS AND QUALIFICATIONS

ADDENDA ACKNOWLEDGEMENT	**TOTAL ADJUSTMENTS**	
	ADJUSTED TOTAL BID	

ALTERNATES

ALTERNATE NO.	
ALTERNATE NO.	
ALTERNATE NO.	
ALTERNATE NO.	
ALTERNATE NO.	
ALTERNATE NO.	
ALTERNATE NO.	

Figure 4.1

A clear understanding of direct and indirect cost factors is a fundamental part of pricing the estimate. The following chapters address the components of direct and indirect costs in detail.

Chapter 5

Direct Costs

Direct costs may be defined as those necessary for the completion of the project, in other words, the "hard," or unburdened costs. Material, labor, and equipment are among the more obvious items in this category. While subcontract costs include the subcontractor's overhead and profit (indirect costs), these costs are direct costs to the primary contractor. Also included in this category are certain project overhead costs for items that may be necessary for construction, for example, rolling scaffolding, tools, and temporary power and lighting. Sales tax and bonds are additional direct costs, because they are essential for the completion of the project.

Material

When careful attention is given to quantity takeoff and pricing, estimates of material cost are usually quite accurate. For a high level of accuracy, the material unit prices (especially for expensive finish materials) must be reliable and current. The most reliable source of material costs is a quotation from a vendor for the particular job in question. Ideally, the vendor will have access to the plans and specifications for verification of quantities and specified products. Material quotes and specified submittals for approval are services that most suppliers can readily provide.

Analyzing Material Quotations

While material pricing seems straightforward, there are certain considerations that the estimator must address when analyzing material quotations. For example, the reputation of the vendor is a significant factor. An important question to ask is the following: "Can the vendor 'deliver,' both figuratively and literally?" Estimators may choose not to rely on a "competitive" lower price from an unknown vendor. Instead, they may use a slightly higher price from a known, reliable vendor. Experience is the best judge for decisions such as these.

There are many other questions that the estimator should ask when analyzing material quotations. How long is the price guaranteed? When does the price guarantee begin? Is there an escalation clause? Does the price include delivery charges or sales tax, if required? Where is the point of FOB? (This can be an extremely important factor.) Do product guarantees and warranties comply with the specification requirements? Will there be adequate and appropriate storage space available? If not, can staggered shipments be made? Note that some of these questions are included on the form in Chapter 4, Figure 4.1.

More information should be obtained, however, to ensure that a quoted price is accurate and competitive. A written quotation should always follow a verbal one.

The estimator must be sure that the quotation or obtained price is for the materials according to the plans and specifications. Architects, engineers, and designers may write the following information into the specifications:

- A particular type or brand of product must be used, with no substitution allowed. (This is known as a *proprietary specification*.)
- A particular type or brand of product is specified, but alternate brands of equal quality and performance may be accepted *upon approval*.
- No particular type or brand is specified. (This is a *generic specification*.)

Depending on the specifications, the estimator may be able to find an acceptable, less expensive alternative. In some cases, these substitutions may substantially lower the cost of a project. Note also that many specification packages require that "catalogue cuts," or published product data, are submitted for approval for certain materials as part of the bid proposal. In this case, there is pressure on the estimator to obtain the lowest possible price on materials that will meet the specified criteria.

There are still other considerations that should have a bearing on the final choice of a vendor. For example, lead time—the amount of time between order and delivery—must be determined and considered. It does not matter how competitive or low a quote is if the material cannot be delivered to the job site in time to support the schedule. Similarly, the estimator should be sure to determine that there is a guarantee or a penalty clause for late delivery.

The estimator must also consider the possibility of unusual payment requirements. For example, a company's cash flow may be severely affected if a large material purchase, thought to be payable within 30 days, is delivered C.O.D. When this happens, truck drivers may not allow unloading until payment is received. The requirements for financing in unexpected situations must be determined at the estimating phase so that the cost of borrowing money, if necessary, may be included.

If the estimator is unable to obtain a quotation from a vendor from whom the material would be purchased, he has other sources for obtaining material prices. These include, in order of reliability:

1. *Current* price lists from manufacturers' catalogs. Be sure to check that the list is for "contractor prices."
2. Cost records from previous jobs. Historical costs must be updated to reflect present market conditions.
3. Reputable and current annual unit price cost books, such as *Means Repair & Remodeling Cost Data*. Such books usually represent national averages and must be factored to local markets.

No matter which price source is used, the estimator must be sure to include an allowance for any burdens, such as delivery costs, taxes, or finance charges, over the actual cost of the material.

Labor

In order to determine the installation cost for each item of construction, the estimator must know two pieces of information: (1) the *labor rate* (hourly wage or salary) of the worker, and (2) how much time a worker will need to complete a given unit of the installation (in other words, the *productivity* or output). Wage rates are usually known going into a project, but productivity may be very difficult to predict before work gets under way.

Labor Rates

To estimators working for contractors, the construction labor rates that the contractor pays will be known, well documented, and constantly updated. Projected labor rate increases for construction jobs of long duration should also be anticipated. Estimators for owners, architects, or engineers must determine labor rates from outside sources. Annual unit price data books, such as *Means Repair & Remodeling Cost Data*, provide national average labor wage rates. The unit costs for labor are based on these averages. Figure 5.1 shows national average *union* rates for the construction industry based on January 1, 2002. Figure 5.2 lists national average *nonunion* rates, again based on January 1, 2002 (from *Means Open Shop Building Construction Cost Data*).

If more localized union labor rates are required, the following sources are available to the estimator. Union locals can provide rates (as well as negotiated increases) for a particular location. Employer bargaining groups can usually provide labor cost data as well. R. S. Means Co., Inc. publishes *Means Labor Rates for the Construction Industry* on an annual basis. This book lists the union labor rates by trade for over 300 U.S. and Canadian cities.

Determination of nonunion, or "open shop" rates is much more difficult. In larger cities, there are often employer organizations that represent nonunion contractors. These organizations may have records of local pay scales, but ultimately, wage rates are determined by individual contractors.

Productivity

Productivity is the least predictable of all factors for repair, remodeling, and renovation. Nevertheless, it is important to determine as accurately as possible the prevailing productivity. The best source of labor productivity information or labor units (and therefore labor costs) is the estimator's well-kept records from previous projects. If there are no company records for productivity, cost data books such as *Means Repair & Remodeling Cost Data*, and productivity reference books, such as *Means Productivity Standards for Construction*, can be valuable.

The designation of a suggested crew makeup is included with the Means listing for each individual construction item. A *crew* is the minimum grouping of workers who may be expected to accomplish a task efficiently. Figure 5.3, a typical page from *Means Repair & Remodeling Cost Data*, includes this crew information and indicates the labor-hours as well as the average daily output of the designated crew for each construction item.

The estimator who has access to neither company records nor the alternative sources described in this section must put together the appropriate crews and determine the expected output or productivity. This type of estimating should only be attempted based on strong experience and considerable exposure to construction methods and practices. There are rare occasions when this approach is necessary to estimate a particular item or a new technique. Even then, the new labor units are often extrapolated from existing figures for similar work, rather than being created from scratch.

Equipment

Construction equipment used for interior projects may be a very expensive item. For example, on a sixth-floor job, the elevator may be too small to transport the drywall. Consequently, a crane may have to be used, a window removed, and other work essentially halted during delivery. These procedures

Installing Contractor's Overhead & Profit

Below are the **average** installing contractor's percentage mark-ups applied to base labor rates to arrive at typical billing rates.

Column A: Labor rates are based on union wages averaged for 30 major U.S. cities. Base rates including fringe benefits are listed hourly and daily. These figures are the sum of the wage rate and employer-paid fringe benefits such as vacation pay, employer-paid health and welfare costs, pension costs, plus appropriate training and industry advancement funds costs.

Column B: Workers' Compensation rates are the national average of state rates established for each trade.

Column C: Column C lists average fixed overhead figures for all trades. Included are Federal and State Unemployment costs set at 7.0%; Social Security Taxes (FICA) set at 7.65%; Builder's Risk Insurance costs set at 0.34%; and Public Liability costs set at 1.55%. All the percentages except those for Social Security Taxes vary from state to state as well as from company to company.

Columns D and E: Percentages in Columns D and E are based on the presumption that the installing contractor has annual billing of $1,500,000 and up. Overhead percentages may increase with smaller annual billing. The overhead percentages for any given contractor may vary greatly and depend on a number of factors, such as the contractor's annual volume, engineering and logistical support costs, and staff requirements. The figures for overhead and profit will also vary depending on the type of job, the job location, and the prevailing economic conditions. All factors should be examined very carefully for each job.

Column F: Column F lists the total of Columns B, C, D, and E.

Column G: Column G is Column A (hourly base labor rate) multiplied by the percentage in Column F (O&P percentage).

Column H: Column H is the total of Column A (hourly base labor rate) plus Column G (Total O&P).

Column I: Column I is Column H multiplied by eight hours.

		A		B	C	D	E	F		H	I
		Base Rate Incl. Fringes		Workers' Comp. Ins.	Average Fixed Overhead	Overhead	Profit	Total Overhead & Profit		Rate with O & P	
Abbr.	Trade	Hourly	Daily					%	Amount	Hourly	Daily
Skwk	Skilled Workers Average (35 trades)	$30.95	$247.60	16.8%	16.5%	16.0%	15%	64.3%	$19.90	$50.85	$406.80
	Helpers Average (5 trades)	22.75	182.00	18.5				66.0	15.00	37.75	302.00
	Foreman Average, Inside ($.50 over trade)	31.45	251.60	16.8				64.3	20.20	51.65	413.20
	Foreman Average, Outside ($2.00 over trade)	32.95	263.60	16.8				64.3	21.20	54.15	433.20
Clab	Common Building Laborers	23.45	187.60	18.1				65.6	15.40	38.85	310.80
Asbe	Asbestos/Insulation Workers/Pipe Coverers	33.45	267.60	16.2				63.7	21.30	54.75	438.00
Boil	Boilermakers	36.25	290.00	14.7				62.2	22.55	58.80	470.40
Bric	Bricklayers	30.50	244.00	16.0				63.5	19.35	49.85	398.80
Brhe	Bricklayer Helpers	23.50	188.00	16.0				63.5	14.90	38.40	307.20
Carp	Carpenters	30.00	240.00	18.1				65.6	19.70	49.70	397.60
Cefi	Cement Finishers	28.70	229.60	10.6				58.1	16.65	45.35	362.80
Elec	Electricians	35.45	283.60	6.7				54.2	19.20	54.65	437.20
Elev	Elevator Constructors	37.10	296.80	7.7				55.2	20.50	57.60	460.80
Eqhv	Equipment Operators, Crane or Shovel	32.35	258.80	10.6				58.1	18.80	51.15	409.20
Eqmd	Equipment Operators, Medium Equipment	31.20	249.60	10.6				58.1	18.15	49.35	394.80
Eqlt	Equipment Operators, Light Equipment	29.80	238.40	10.6				58.1	17.30	47.10	376.80
Eqol	Equipment Operators, Oilers	26.65	213.20	10.6				58.1	15.50	42.15	337.20
Eqmm	Equipment Operators, Master Mechanics	32.80	262.40	10.6				58.1	19.05	51.85	414.80
Glaz	Glaziers	30.00	240.00	13.8				61.3	18.40	48.40	387.20
Lath	Lathers	28.75	230.00	11.1				58.6	16.85	45.60	364.80
Marb	Marble Setters	30.10	240.80	16.0				63.5	19.10	49.20	393.60
Mill	Millwrights	31.75	254.00	10.6				58.1	18.45	50.20	401.60
Mstz	Mosaic & Terrazzo Workers	29.25	234.00	9.8				57.3	16.75	46.00	368.00
Pord	Painters, Ordinary	27.15	217.20	13.8				61.3	16.65	43.80	350.40
Psst	Painters, Structural Steel	27.90	223.20	48.4				95.9	26.75	54.65	437.20
Pape	Paper Hangers	27.10	216.80	13.8				61.3	16.60	43.70	349.60
Pile	Pile Drivers	29.80	238.40	24.9				72.4	21.60	51.40	411.20
Plas	Plasterers	28.10	224.80	15.8				63.3	17.80	45.90	367.20
Plah	Plasterer Helpers	23.70	189.60	15.8				63.3	15.00	38.70	309.60
Plum	Plumbers	35.95	287.60	8.3				55.8	20.05	56.00	448.00
Rodm	Rodmen (Reinforcing)	34.25	274.00	28.3				75.8	25.95	60.20	481.60
Rofc	Roofers, Composition	26.60	212.80	32.6				80.1	21.30	47.90	383.20
Rots	Roofers, Tile & Slate	26.75	214.00	32.6				80.1	21.45	48.20	385.60
Rohe	Roofers, Helpers (Composition)	19.80	158.40	32.6				80.1	15.85	35.65	285.20
Shee	Sheet Metal Workers	35.10	280.80	11.7				59.2	20.80	55.90	447.20
Spri	Sprinkler Installers	36.20	289.60	8.7				56.2	20.35	56.55	452.40
Stpi	Steamfitters or Pipefitters	36.20	289.60	8.3				55.8	20.20	56.40	451.20
Ston	Stone Masons	30.65	245.20	16.0				63.5	19.45	50.10	400.80
Sswk	Structural Steel Workers	34.25	274.00	39.8				87.3	29.90	64.15	513.20
Tilf	Tile Layers	29.15	233.20	9.8				57.3	16.70	45.85	366.80
Tilh	Tile Layers Helpers	23.35	186.80	9.8				57.3	13.40	36.75	294.00
Trlt	Truck Drivers, Light	24.30	194.40	14.9				62.4	15.15	39.45	315.60
Trhv	Truck Drivers, Heavy	25.00	200.00	14.9				62.4	15.60	40.60	324.80
Sswl	Welders, Structural Steel	34.25	274.00	39.8				87.3	29.90	64.15	513.20
Wrck	*Wrecking	23.45	187.60	41.2				88.7	20.80	44.25	354.00

*Not included in averages

Figure 5.1

Installing Contractor's Overhead & Profit

Below are the **average** installing contractor's percentage mark-ups applied to base labor rates to arrive at typical billing rates.

Column A: Labor rates are based on average open shop wages for 7 major U.S. regions. Base rates including fringe benefits are listed hourly and daily. These figures are the sum of the wage rate and employer-paid fringe benefits such as vacation pay, and employer-paid health costs.

Column B: Workers' Compensation rates are the national average of state rates established for each trade.

Column C: Column C lists average fixed overhead figures for all trades. Included are Federal and State Unemployment costs set at 7.0%; Social Security Taxes (FICA) set at 7.65%; Builder's Risk Insurance costs set at 0.34%; and Public Liability costs set at 1.55%. All the percentages except those for Social Security Taxes vary from state to state as well as from company to company.

Columns D and E: Percentages in Columns D and E are based on the presumption that the installing contractor has annual billing of $1,000,000 and up. Overhead percentages may increase with smaller annual billing. The overhead percentages for any given contractor may vary greatly and depend on a number of factors, such as the contractor's annual volume, engineering and logistical support costs, and staff requirements. The figures for overhead and profit will also vary depending on the type of job, the job location, and the prevailing economic conditions. All factors should be examined very carefully for each job.

Column F: Column F lists the total of Columns B, C, D, and E.

Column G: Column G is Column A (hourly base labor rate) multiplied by the percentage in Column F (O&P percentage).

Column H: Column H is the total of Column A (hourly base labor rate) plus Column G (Total O&P).

Column I: Column I is Column H multiplied by eight hours.

		A		B	C	D	E	F		G	H	I
		Base Rate Incl. Fringes		Work-ers' Comp. Ins.	Average Fixed Over-head	Over-head	Profit	Total Overhead & Profit			Rate with O & P	
Abbr.	Trade	Hourly	Daily					%	Amount		Hourly	Daily
Skwk	Skilled Workers Average (35 trades)	$21.25	$170.00	16.8%	16.5%	27.0%	10%	70.3%	$14.95		$36.20	$289.60
	Helpers Average (5 trades)	15.65	125.20	18.5		25.0		70.0	10.95		26.60	212.80
	Foreman Average, Inside ($.50 over trade)	21.75	174.00	16.8		27.0		70.3	15.30		37.05	296.40
	Foreman Average, Outside ($2.00 over trade)	23.25	186.00	16.8		27.0		70.3	16.35		39.60	316.80
Clab	Common Building Laborers	15.25	122.00	18.1		25.0		69.6	10.60		25.85	206.80
Asbe	Asbestos/Insulation Workers/Pipe Coverers	22.10	176.80	16.2		30.0		72.7	16.05		38.15	305.20
Boil	Boilermakers	23.95	191.60	14.7		30.0		71.2	17.05		41.00	328.00
Bric	Bricklayers	21.35	170.80	16.0		25.0		67.5	14.40		35.75	286.00
Brhe	Bricklayer Helpers	16.45	131.60	16.0		25.0		67.5	11.10		27.55	220.40
Carp	Carpenters	21.00	168.00	18.1		25.0		69.6	14.60		35.60	284.80
Ccfi	Cement Finishers	20.10	160.80	10.6		25.0		62.1	12.50		32.60	260.80
Elec	Electricians	23.75	190.00	6.7		30.0		63.2	15.00		38.75	310.00
Elev	Elevator Constructors	24.85	198.80	7.7		30.0		64.2	15.95		40.80	326.40
Eqhv	Equipment Operators, Crane or Shovel	22.30	178.40	10.6		28.0		65.1	14.50		36.80	294.40
Eqmd	Equipment Operators, Medium Equipment	21.55	172.40	10.6		28.0		65.1	14.05		35.60	284.80
Eqlt	Equipment Operators, Light Equipment	20.55	164.40	10.6		28.0		65.1	13.40		33.95	271.60
Eqol	Equipment Operators, Oilers	18.40	147.20	10.6		28.0		65.1	12.00		30.40	243.20
Eqmm	Equipment Operators, Master Mechanics	22.65	181.20	10.6		28.0		65.1	14.75		37.40	299.20
Glaz	Glaziers	21.30	170.40	13.8		25.0		65.3	13.90		35.20	281.60
Lath	Lathers	20.15	161.20	11.1		25.0		62.6	12.60		32.75	262.00
Marb	Marble Setters	21.05	168.40	16.0		25.0		67.5	14.20		35.25	282.00
Mill	Millwrights	22.25	178.00	10.6		25.0		62.1	13.80		36.05	288.40
Mstz	Mosaic & Terrazzo Workers	20.50	164.00	9.8		25.0		61.3	12.55		33.05	264.40
Pord	Painters, Ordinary	19.30	154.40	13.8		25.0		65.3	12.60		31.90	255.20
Psst	Painters, Structural Steel	19.80	158.40	48.4		25.0		99.9	19.80		39.60	316.80
Pape	Paper Hangers	19.25	154.00	13.8		25.0		65.3	12.55		31.80	254.40
Pile	Pile Drivers	20.85	166.80	24.9		30.0		81.4	16.95		37.80	302.40
Plas	Plasterers	19.65	157.20	15.8		25.0		67.3	13.20		32.85	262.80
Plah	Plasterer Helpers	16.60	132.80	15.8		25.0		67.3	11.15		27.75	222.00
Plum	Plumbers	23.75	190.00	8.3		30.0		64.8	15.40		39.15	313.20
Rodm	Rodmen (Reinforcing)	22.95	183.60	28.3		28.0		82.8	19.00		41.95	335.60
Rofc	Roofers, Composition	18.35	146.80	32.6		25.0		84.1	15.45		33.80	270.40
Rots	Roofers, Tile & Slate	18.45	147.60	32.6		25.0		84.1	15.50		33.95	271.60
Rohe	Roofers, Helpers (Composition)	13.65	109.20	32.6		25.0		84.1	11.50		25.15	201.20
Shee	Sheet Metal Workers	23.15	185.20	11.7		30.0		68.2	15.80		38.95	311.60
Spri	Sprinkler Installers	23.90	191.20	8.7		30.0		65.2	15.60		39.50	316.00
Stpi	Steamfitters or Pipefitters	23.90	191.20	8.3		30.0		64.8	15.50		39.40	315.20
Ston	Stone Masons	20.85	166.80	16.0		25.0		67.5	14.05		34.90	279.20
Sswk	Structural Steel Workers	22.95	183.60	39.8		28.0		94.3	21.65		44.60	356.80
Tilf	Tile Layers	20.40	163.20	9.8		25.0		61.3	12.50		32.90	263.20
Tilh	Tile Layers Helpers	16.35	130.80	9.8		25.0		61.3	10.00		26.35	210.80
Trlt	Truck Drivers, Light	17.25	138.00	14.9		25.0		66.4	11.45		28.70	229.60
Trhv	Truck Drivers, Heavy	17.75	142.00	14.9		25.0		66.4	11.80		29.55	236.40
Sswl	Welders, Structural Steel	22.95	183.60	39.8		28.0		94.3	21.65		44.60	356.80
Wrck	*Wrecking	15.70	125.60	41.2		25.0		92.7	14.55		30.25	242.00

*Not included in averages

Figure 5.2

			CREW	DAILY OUTPUT	LABOR-HOURS	UNIT	2002 BARE COSTS				TOTAL INCL O&P	
	06220	**Millwork**					MAT.	LABOR	EQUIP.	TOTAL		
800	3170	4-1/2" wide	1 Carp	5.30	1.509	Opng.	24	45.50		69.50	101	**800**
	3200	Glass beads, stock pine, 1/4" x 11/16"		285	.028	L.F.	.30	.84		1.14	1.73	
	3250	3/8" x 1/2"		275	.029		.36	.87		1.23	1.85	
	3270	3/8" x 7/8"		270	.030		.40	.89		1.29	1.91	
	4850	Parting bead, stock pine, 3/8" x 3/4"		275	.029		.29	.87		1.16	1.77	
	4870	1/2" x 3/4"		255	.031		.36	.94		1.30	1.96	
	5000	Stool caps, stock pine, 11/16" x 3-1/2"		200	.040		1.31	1.20		2.51	3.43	
	5100	1-1/16" x 3-1/4"		150	.053		1.97	1.60		3.57	4.82	
	5300	Threshold, oak, 3' long, inside, 5/8" x 3-5/8"		32	.250	Ea.	6.20	7.50		13.70	19.30	
	5400	Outside, 1-1/2" x 7-5/8"		16	.500	"	25	15		40	52.50	
	5900	Window trim sets, including casings, header, stops,										
	5910	stool and apron, 2-1/2" wide, minimum	1 Carp	13	.615	Opng.	14.85	18.45		33.30	47	
	5950	Average		10	.800		25.50	24		49.50	68.50	
	6000	Maximum		6	1.333		38	40		78	108	
	9000	Minimum labor/equipment charge		4	2	Job		60		60	99.50	
900	0010	**SOFFITS** Wood fiber, no vapor barrier, 15/32" thick R06110-030	2 Carp	525	.030	S.F.	.75	.91		1.66	2.34	**900**
	0100	5/8" thick		525	.030	"	.81	.91		1.72	2.40	
	0300	As above, 5/8" thick, with factory finish		525	.030	S.F.	.83	.91		1.74	2.42	
	0500	Hardboard, 3/8" thick, slotted		525	.030		1	.91		1.91	2.61	
	1000	Exterior AC plywood, 1/4" thick		420	.038			1.14		1.14	1.89	
	1100	1/2" thick		420	.038			1.14		1.14	1.89	
	1150	For aluminum soffit, see division 07460-750										
	9000	Minimum labor/equipment charge	2 Carp	5	3.200	Job		96		96	159	

		06250	**Prefinished Paneling**									
200	0010	**PANELING, HARDBOARD**										**200**
	0050	Not incl. furring or trim, hardboard, tempered, 1/8" thick	2 Carp	500	.032	S.F.	.31	.96		1.27	1.93	
	0100	1/4" thick		500	.032		.39	.96		1.35	2.02	
	0300	Tempered pegboard, 1/8" thick		500	.032		.39	.96		1.35	2.02	
	0400	1/4" thick		500	.032		.42	.96		1.38	2.05	
	0600	Untempered hardboard, natural finish, 1/8" thick		500	.032		.33	.96		1.29	1.95	
	0700	1/4" thick		500	.032		.32	.96		1.28	1.94	
	0900	Untempered pegboard, 1/8" thick		500	.032		.33	.96		1.29	1.95	
	1000	1/4" thick		500	.032		.37	.96		1.33	2	
	1200	Plastic faced hardboard, 1/8" thick		500	.032		.54	.96		1.50	2.18	
	1300	1/4" thick		500	.032		.72	.96		1.68	2.38	
	1500	Plastic faced pegboard, 1/8" thick		500	.032		.51	.96		1.47	2.15	
	1600	1/4" thick		500	.032		.63	.96		1.59	2.28	
	1800	Wood grained, plain or grooved, 1/4" thick, minimum		500	.032		.48	.96		1.44	2.12	
	1900	Maximum		425	.038		1.01	1.13		2.14	2.98	
	2100	Moldings for hardboard, wood or aluminum, minimum		500	.032	L.F.	.33	.96		1.29	1.95	
	2200	Maximum		425	.038	"	.91	1.13		2.04	2.87	
	9000	Minimum labor/equipment charge	1 Carp	2	4	Job		120		120	199	
500	0010	**PANELING, PLYWOOD** R06160-020										**500**
	2400	Plywood, prefinished, 1/4" thick, 4' x 8' sheets										
	2410	with vertical grooves. Birch faced, minimum	2 Carp	500	.032	S.F.	.74	.96		1.70	2.40	
	2420	Average		420	.038		1.12	1.14		2.26	3.12	
	2430	Maximum		350	.046		1.64	1.37		3.01	4.07	
	2600	Mahogany, African		400	.040		2.10	1.20		3.30	4.30	
	2700	Philippine (Lauan)		500	.032		.90	.96		1.86	2.58	
	2900	Oak or Cherry, minimum		500	.032		1.76	.96		2.72	3.53	
	3000	Maximum		400	.040		2.70	1.20		3.90	4.96	
	3200	Rosewood		320	.050		3.83	1.50		5.33	6.70	
	3400	Teak		400	.040		2.70	1.20		3.90	4.96	
	3600	Chestnut		375	.043		3.99	1.28		5.27	6.50	

Figure 5.3

may cost thousands of dollars. Estimators must carefully consider the issue of equipment and related expenses, when required. Equipment costs may be divided into two categories:

- **Rental, lease, or ownership costs.** These costs may be determined based on hourly, daily, weekly, monthly, or annual increments. These fees or payments only buy the "right" to use the equipment and do not include operating costs.
- **Operating Costs.** Once the "right" of use is obtained, costs are incurred for actual use or operation. These costs may include fuel, lubrication, maintenance, and parts.

Equipment costs, as described above, do not include the labor expense of operators. However, some cost books and suppliers may include the operator in the quoted price for equipment as an "operated" rental cost. In other words, the equipment is priced as if it were a subcontract cost.

Equipment ownership costs apply to both leased and owned equipment. The operating costs of equipment, whether rented, leased or owned, are available from the following sources (listed in order of reliability):

1. The company's own records
2. Annual cost books containing equipment operating costs, such as *Means Repair & Remodeling Cost Data*
3. Manufacturers' estimates
4. Reference books dealing with equipment operating costs

Operating costs include fuel, lubrication, expendable parts replacement, and minor maintenance. For estimating purposes, the equipment ownership and operating costs should be listed separately. In this way, the decision to rent, subcontract, or purchase can be decided on a project-by-project basis.

Allocating Equipment Costs

There are two commonly used methods for including equipment costs in a construction estimate. First, the equipment may be thought of as part of the construction task for which it will be used. In this case, costs are included in each line item as a separate unit price. The advantage of this method is that costs are allocated to the division or task that actually incurs the expense. As a result, more accurate records are kept for each installed component. One disadvantage of this method occurs in the pricing of equipment that may be used for many different tasks. In this instance, duplication of costs may occur. Another disadvantage of this method occurs in budgeting. For the following reason, the budget may be left short: The estimate may reflect only two hours for a crane truck when the minimum cost of such a crane is usually a daily (8-hour) rental charge.

The second method for including equipment costs in the estimate is the following: Keep all such costs separate and include them in Division 1 as a component of Project Overhead. The advantage of this method is that all equipment costs are grouped together and machines that may be used for several tasks are included (without duplication). One disadvantage of the method is that, for future estimating purposes, equipment costs will be known only on a job basis and not per installed unit.

No matter which method for including equipment costs is used, the estimator must be consistent and must be sure that all equipment costs are included and not duplicated. The estimating method should be the same as that chosen for cost monitoring and accounting. In this way, the data will be available for monitoring the project costs and for bidding future projects.

Subcontractors

Subcontractors may account for a large percentage of a remodeling or renovation project. When subcontractors are used, quotations should be solicited and analyzed in the same way as material quotes. Of primary concern are the following: (1) that the bid covers the work according to the plans and specifications, and (2) that all appropriate work alternates and allowances, if any, are included. Any inclusions should be clearly stated and explained.

If the bid is received orally, a form such as Figure 4.1 in Chapter 4 will help to ensure that it is documented accurately. Any unique scheduling or payment requirements must be noted and evaluated. Such requirements might affect (restrict or enhance) the normal progress of the project, and should, therefore, be known in advance.

The estimator should know how long the subcontract bid will be honored. This time period usually varies from 30 days to 90 days, and it is often included as a condition in complete bids.

The estimator should also know or verify the bonding capability and capacity of unfamiliar subcontractors, when required. This precaution may be necessary when bidding in a new location. Other than by word of mouth, these inquiries may be the only way to confirm subcontractor reliability.

Project Overhead

Project overhead represents those construction costs that are usually included in Division 1–General Requirements. Typical project overhead items are supervisory personnel, cleanup labor, and temporary heat and power. While these items may not be directly part of the physical structure, they are a part of the project. Project overhead, like all other direct costs, can be separated into material, labor, and equipment components.

Figures 5.4a and 5.4b are examples of a form that may help ensure that all appropriate costs are included. This form is for general construction, but it may be adapted easily to the specific requirements of repair and remodeling projects.

Some estimators may not agree that certain items (such as equipment or scaffolding) should be included as Project Overhead, preferring to list such items in another division. Ultimately, it is not important *where* each item is incorporated into the estimate. However, *every item must be included somewhere*.

Project Overhead often includes time-dependent items, for example, equipment rental, supervisory labor, and temporary utilities. The cost for these items depends on the duration of the project. A preliminary schedule should, therefore, be developed *prior* to completion of the estimate, so that time-sensitive items are properly counted. This topic is covered in greater detail in Chapter 8, "Pre-Bid Scheduling."

Bonds

Although bonds are really a type of "direct cost," they are priced and based on the total "bid" or "selling price." For that reason, they are generally figured after indirect costs have been added. Bonding requirements for a project will be specified in Division 1–General Requirements and included in the construction contract.

Sales Tax

Sales tax varies from state to state, and often from city to city within a state. Larger cities may have a sales tax in addition to the state sales tax. Some localities also impose separate sales taxes on labor and equipment.

When bidding takes place in an unfamiliar location, the estimator should check with local agencies regarding the amount and the method of payment

PROJECT
OVERHEAD SUMMARY

| PROJECT | | | SHEET NO. | |
| LOCATION | | ARCHITECT | | DATE |

| QUANTITIES BY: | PRICES BY: | EXTENSIONS BY: | CHECKED BY: |

DESCRIPTION	QUANTITY	UNIT	MATERIAL/EQUIPMENT		LABOR		TOTAL COST	
			UNIT	TOTAL	UNIT	TOTAL	UNIT	TOTAL
Job Organization: Superintendent								
Project Manager								
Timekeeper & Material Clerk								
Clerical								
Safety, Watchman & First Aid								
Travel Expense: Superintendent								
Project Manager								
Engineering: Layout								
Inspection/Quantities								
Drawings								
CPM Schedule								
Testing: Soil								
Materials								
Structural								
Equipment: Cranes								
Concrete Pump, Conveyor, Etc.								
Elevators, Hoists								
Freight & Hauling								
Loading, Unloading, Erecting, Etc.								
Maintenance								
Pumping								
Scaffolding								
Small Power Equipment/Tools								
Field Offices: Job Office								
Architect/Owner's Office								
Temporary Telephones								
Utilities								
Temporary Toilets								
Storage Areas & Sheds								
Temporary Utilities: Heat								
Light & Power								
Water								
PAGE TOTALS								

Page 1 of 2

Figure 5.4a

DESCRIPTION	QUANTITY	UNIT	MATERIAL/EQUIPMENT		LABOR		TOTAL COST	
			UNIT	TOTAL	UNIT	TOTAL	UNIT	TOTAL
Totals Brought Forward								
Winter Protection: Temp. Heat/Protection								
Snow Plowing								
Thawing Materials								
Temporary Roads								
Signs & Barricades: Site Sign								
Temporary Fences								
Temporary Stairs, Ladders & Floors								
Photographs								
Clean Up								
Dumpster								
Final Clean Up								
Punch List								
Permits: Building								
Misc.								
Insurance: Builders Risk								
Owner's Protective Liability								
Umbrella								
Unemployment Ins. & Social Security								
Taxes								
City Sales Tax								
State Sales Tax								
Bonds								
Performance								
Material & Equipment								
Main Office Expense								
Special Items								
TOTALS:								

Figure 5.4b

of sales tax. Local authorities may require owners to withhold payments to out-of-state contractors until payment of all required sales tax has been verified. Sales tax, often taken for granted or even omitted, may be as much as 7.5% of material costs. This percentage may represent a significant portion of a project's total cost. Conversely, some clients, and/or their projects, may be tax exempt. If this fact is unknown to the estimator, a large dollar amount for sales tax might be needlessly included in a bid.

Chapter 6

Indirect Costs

Indirect costs are the "costs of doing business." These expenses are sometimes referred to as "burden" to the project. Indirect costs may include certain fixed, or known, expenses and percentages, as well as costs that are variable and may be subjectively determined. For example, government authorities require payment of certain taxes and insurance that are usually based on labor costs and are determined by trade. This expense is a type of *fixed* indirect cost. Office overhead, if well understood and established, is also thought of as a relatively fixed percentage. Profit and contingencies, however, are more variable and subjective. Often, these figures are determined on the basis of judgment and on the discretion of the person responsible for the company's growth and success.

If the direct costs for the same project have been carefully determined, they should not vary significantly from one estimator to another. The indirect costs often carry the responsibility for variations among bids.

The direct costs of a project must be itemized, tabulated, and totaled before the indirect costs can be applied to the estimate. The most prevalent indirect costs include:

- Taxes and Insurance
- Office or Operating Overhead
- Profit
- Contingencies

Taxes and Insurance

The taxes and insurance included as indirect costs are most often related to the costs of labor and/or to the type of work. This category may include Workers' Compensation, Builder's Risk, and Public Liability Insurance, as well as employer-paid social security tax and federal and state unemployment insurance. By law, the employer must pay these expenses. Rates are based on the type and salary of the employees, as well as on the location and for the type of business.

Office or Operating Overhead

Office overhead, or the cost of doing business, is perhaps one of the main reasons why so many contractors are unable to realize a profit, or even to stay in business. The problem shows up in two ways. Either a company does not know its true overhead cost and therefore fails to mark up its costs enough to recover them, or management does not restrain or control overhead costs effectively and fails to remain competitive.

If a contractor does not know the costs of operating the business, it is more than likely that these costs will not be recovered. Many companies survive, and even turn a profit, by simply adding an arbitrary percentage for overhead to each job, without knowing how the percentage is derived or what it includes. When annual volume changes significantly, by increase or decrease, or when office staff and/or expenses increase, the previously used percentage for overhead may no longer be valid. When such a change occurs, the owner often finds that the company is not doing as well as before and cannot determine the reasons. Chances are, overhead costs are not being fully recovered. The following items may be included when determining office overhead costs.

Owner: This item should include only a reasonable base salary and does not include profits. An owner's salary is **not** a company's profit.

Engineer/Estimator/Project Manager: Since the owner may be primarily on the road getting business, the project manager runs the daily operation of the company and is responsible for estimating. In some operations, the estimator who successfully wins a bid becomes the "project manager" and is responsible to the owner for that project's profitability.

Secretary/Receptionist/Bookkeeper: This person manages office operations and handles paperwork. A talented individual in this position can be a tremendous asset.

Office Staff Insurance & Taxes: These costs are for main office personnel and should include, but are not limited to, the following:
- Workers' Compensation insurance
- FICA
- Unemployment insurance
- Medical & other insurance
- Profit sharing, pension, etc.

Physical Plant Expenses: Whether the office, warehouse and/or yard are rented or owned, roughly the same costs are incurred. Telephone and utility costs will vary, depending on the size of the building and on the type of business. Office equipment includes items such as a copy machine, typewriters, computer, and furniture.

Professional Services: Accountant fees may be primarily for quarterly audits and annual tax return preparation. Legal fees go towards collecting receivables and for contract disputes. In addition, a prudent contractor will have every contract read by his lawyer, prior to signing.

Miscellaneous: There are many expenses that may be placed in this category. Association dues, seminars, travel, and entertainment may be included. Advertising includes the *Yellow Pages*, promotional materials, etc.

Uncollected Receivables: This amount may vary greatly, and it is often affected by the overall economic climate. Depending on the timing of "uncollectibles," cash flow may be severely restricted and may cause serious financial problems, even for large companies. Sound cash planning and anticipation of such possibilities are essential.

In order to stay in business *without losses* (profit is not yet a factor), a company must pay all direct construction costs and must recover, during the year, all additional overhead costs necessary to operate the office. The most common method for recovering these costs is to apply this percentage to each job over the course of the year. The percentage may be calculated and applied in two ways:

- Office overhead applied as a percentage of *labor costs only*. This method requires that labor and materials costs are estimated separately.
- Office overhead applied as a percentage of *total project costs*. This method is appropriate when material and labor costs are not estimated separately.

The estimator must also remember that, if volume changes significantly, the percentage for office overhead should be recalculated to reflect current conditions. Since salaries are the major portion of office overhead costs, any changes in office staff should also be figured into an accurate percentage for overhead.

An additional percentage is commonly applied to material costs, for handling, regardless of the method of recovering overhead costs. This percentage is more easily calculated when material costs are estimated and listed separately.

Profit

Determining a fair and reasonable percentage to be included for profit is not an easy task. This responsibility is usually left to the owner or chief estimator. Experience is crucial in anticipating what profit the market will bear. The economic climate, competition, knowledge of the project, and familiarity with the architect, designer, or owner all affect the way in which profit is determined.

Contingencies

Like profit, contingencies may also be difficult to quantify. The addition of a contingency, especially appropriate in preliminary budgets, is intended to protect the contractor as well as to give the owner a realistic estimate of potential project costs.

A contingency percentage should be based on the number of "unknowns" in a project, or on the level of risk involved. This percentage should be inversely proportional to the amount of planning that has been done for the project and to the detail of available information. If complete plans and specifications are supplied and the estimate is thorough and precise, there is little need for a contingency. Figure 6.1 from *Means Repair & Remodeling Cost Data* lists suggested contingency percentages that may be added to an estimate, based on the stage of planning and development.

If an estimate is priced and each individual item is rounded up, or "padded," the estimator is in essence adding a contingency. This method may cause problems, however, because the estimator can never be quite sure of what is the actual cost and what is the "padding," or safety margin, for each item. At the summary, the estimator cannot determine exactly how much has been included as a contingency factor for the project as a whole. A much more accurate and controllable approach is to price the estimate precisely and then to add one contingency amount at the bottom line.

Also shown in Figure 6.1 is a section entitled "Factors." In planning and estimating repair and remodeling projects, there are many factors that may affect the project cost beyond the basic material and labor. The economy of scale usually associated with new construction often has no influence on the cost of repair and remodeling. Small quantities of components may have to be custom-fabricated at great expense. Work-schedule coordination between trades frequently becomes difficult, and work-area restrictions can lead to subcontractor quotations with start-up and shut-down costs exceeding the cost of the specified work.

01100 | Summary

01103 | Models & Renderings

			CREW	DAILY OUTPUT	LABOR-HOURS	UNIT	MAT.	LABOR	EQUIP.	TOTAL	TOTAL INCL O&P	
500	0010	RENDERINGS Color, matted, 20" x 30", eye level,										500
	0020	1 building, minimum				Ea.	1,700			1,700	1,875	
	0050	Average					2,850			2,850	3,125	
	0100	Maximum					4,000			4,000	4,400	
	1000	5 buildings, minimum					3,500			3,500	3,850	
	1100	Maximum					6,800			6,800	7,475	

01107 | Professional Consultant

			CREW	DAILY OUTPUT	LABOR-HOURS	UNIT	MAT.	LABOR	EQUIP.	TOTAL	TOTAL INCL O&P	
100	0011	ARCHITECTURAL FEES R01107-010										100
	0020	For new construction										
	0060	Minimum				Project					4.90%	
	0090	Maximum									16%	
	0100	For alteration work, to $500,000, add to fee									50%	
	0150	Over $500,000, add to fee									25%	
200	0011	CONSTRUCTION MANAGEMENT FEES										200
	0060	For work to $10,000				Project					10%	
	0070	To $25,000									9%	
	0090	To $100,000									6%	
	0100	To $500,000									5%	
	0110	To $1,000,000									4%	
300	0010	ENGINEERING FEES R01107-030										300
	0020	Educational planning consultant, minimum				Project					.50%	
	0100	Maximum				"					2.50%	
	0400	Elevator & conveying systems, minimum				Contrct					2.50%	
	0500	Maximum									5%	
	1000	Mechanical (plumbing & HVAC), minimum									4.10%	
	1100	Maximum									10.10%	
	1200	Structural, minimum				Project					1%	
	1300	Maximum				"					2.50%	
700	0010	SURVEYING Conventional, topographical, minimum	A-7	3.30	7.273	Acre	16	236		252	405	700
	0100	Maximum	A-8	.60	53.333		48	1,700		1,748	2,800	
	0300	Lot location and lines, minimum, for large quantities	A-7	2	12		25	390		415	660	
	0320	Average	"	1.25	19.200		45	620		665	1,050	
	0400	Maximum, for small quantities	A-8	1	32		72	1,025		1,097	1,725	
	1100	Crew for building layout, 2 person crew	A-6	1	16	Day		540		540	875	
	1200	3 person crew	A-7	1	24			780		780	1,275	
	1300	4 person crew	A-8	1	32			1,025		1,025	1,650	

01200 | Price & Payment Procedures

01250 | Contract Modification Procedures

			CREW	DAILY OUTPUT	LABOR-HOURS	UNIT	MAT.	LABOR	EQUIP.	TOTAL	TOTAL INCL O&P	
200	0010	CONTINGENCIES for estimate at conceptual stage				Project					25%	200
	0050	Schematic stage									20%	
	0100	Preliminary working drawing stage (Design Dev.)									15%	
	0150	Final working drawing stage									8%	
400	0010	FACTORS Cost adjustments R01250-010										400
	0100	Add to construction costs for particular job requirements										
	0500	Cut & patch to match existing construction, add, minimum				Costs	2%	3%				
	0550	Maximum					5%	9%				

Figure 6.1

The factors explained below, and shown in Figure 6.1, are those normally associated with the loss of productivity encountered in repair and remodeling projects.

1. Cutting and patching to match the existing construction can often lead to an economical trade-off, for example, of removing entire walls rather than creating new door and window openings. Substitutions for materials that are no longer manufactured may be expensive. Piping and ductwork runs may not be as straight as in new construction, and wiring may have to be snaked through walls and floors.

2. Dust and noise protection for adjoining, nonconstruction areas may alter usual construction methods.

3. Equipment usage curtailment resulting from the physical limitations of the project may force workers to use slow, hand-operated equipment instead of power tools.

4. The confines of an enclosed building may have a costly influence on movement and material handling. Low capacity elevators and stairwells may be the only access to the upper floors of a multistory building.

5. On some repair or remodeling projects, existing construction and completed work must be secured or otherwise protected from possible damage during ongoing construction. In certain areas, completed work must be guarded to prevent theft and vandalism.

6. Work may have to be done during other than normal shifts and/or around an existing production facility that is in operation during the repair and remodeling project.

7. Shoring and bracing may be required to support the building while structural changes are being made. Another expense is the allowance for temporary storage of construction materials on above-grade floors.

The exact percentages used to modify the estimate are left to the estimator's judgment based upon the job conditions. The figures shown are for guidance only and should be used accordingly.

Chapter 7

The Estimate Summary

At the pricing stage of the estimate, there is typically a large amount of paperwork that must be assembled, analyzed, and tabulated. This documentation generally consists of the following items.

- Quantity takeoff sheets for all work items (Figure 3.1)
- Material suppliers' written quotations
- Equipment or material suppliers' or subcontractors' telephone quotations (Figure 4.1)
- Subcontractors' written quotations
- Equipment suppliers' quotations
- Pricing sheets (Figures 3.2 and 3.3)
- Estimate summary sheets

Organizing the Estimate Data

In the "real world" of estimating, many quotations, especially for large material purchases and for subcontracts, are not received until the last minute before the bidding deadline. Therefore, a system is necessary to efficiently handle the paperwork and to ensure that all relevant information will be transferred once (and only once) from the quantity takeoff to the cost analysis sheets. Some general rules for this process are as follows:

- Write on only one side of any document, when possible.
- Use Telephone Quotation forms for uniformity in recording prices received orally.
- Document the source of every quantity and price.
- Keep the entire estimate in one or more compartmentalized folder(s).
- If you are pricing your own materials, number and code each takeoff sheet and each pricing extension sheet as it is created. At the same time, keep an index list of each sheet by number. If a sheet is to be abandoned, write "VOID" on it, but do not discard it. Keep it until the bid is accepted; it will serve as a reference in accounting for all pages and sheets.

All subcontract costs should be properly noted and listed separately. These costs contain the subcontractors' mark-ups and may be treated differently from other direct costs when the estimator calculates the prime contractor's overhead and profit.

After all the unit prices and allowances have been entered on the pricing sheets, the costs are extended. In making the extensions, ignore the cents column and round all totals to the nearest dollar. In a column of figures, the cents will average out and will not be of consequence. Finally, each subdivision is added and the results checked, preferably by someone other than the person doing the extensions.

It is important to check the larger items for order of magnitude errors. If the total costs are divided by the floor area, the resulting square foot cost figures can be used to quickly pinpoint areas that are out of line with expected square foot costs. These cost figures should be recorded for comparison to past projects and as a resource for future estimating.

The takeoff and pricing method has been described as follows: Utilize a Quantity Sheet for the material takeoff, and transfer the data to a Cost Analysis form for pricing the material, labor, and equipment items.

An alternative to this method is a consolidation of the takeoff task and pricing on a single form. This approach works well for smaller bids and for change orders. For example: The Consolidated Estimate form is shown in Figure 7.1. The same sequences and recommendations used to complete the Quantity Sheet and Cost Analysis form should be followed when using the Consolidated Estimate form to price the estimate.

Final Adjustments

When the pricing of all direct costs is complete, the estimator has two choices: to make all further price changes and adjustments on the Cost Analysis or Consolidated Estimate sheets, or to transfer the total costs for each subdivision to an Estimate Summary sheet, so that all further price changes, until bid time, will be done on one sheet. Any indirect cost markups and burdens will be figured on this sheet. An example of an Estimate Summary is shown in Figure 7.2.

Unless the estimate has a limited number of items, it is recommended that direct costs be transferred to an Estimate Summary sheet. This step should be double-checked, because an error of transposition may easily occur. Preprinted forms may be useful, although a plain columnar form may suffice. This summary, with the page numbers referencing from each extension sheet, may also serve as an index.

A company that repeatedly uses certain standard listings may save valuable time by creating and printing a custom Estimate Summary sheet. The printed CSI division and subdivision headings (shown in Figure 7.2) serve as another type of checklist that may be used to ensure that all required costs are included. The following are possible column headings or categories that might be appropriate for any estimate summary form:

- Material
- Labor
- Equipment
- Subcontractor
- Total

When the items have been listed in the proper columns, the categories are added and appropriate markups applied to the total dollar values. Different percentages may be added to the sum of each column at the estimate summary. These percentages may include the following items, as discussed in Chapter 6:

- Taxes and Insurance
- Overhead
- Profit
- Contingencies

CONSOLIDATED ESTIMATE

PROJECT		CLASSIFICATION		SHEET NO.
LOCATION		ARCHITECT		ESTIMATE NO.
TAKE OFF BY	QUANTITIES BY	PRICES BY	EXTENSIONS BY	DATE
			CHECKED BY	

DESCRIPTION	NO.	DIMENSIONS	QUANTITIES		UNIT	MATERIAL		LABOR		EQUIPMENT		TOTAL	
			UNIT		UNIT COST	UNIT COST	TOTAL	UNIT COST	TOTAL	UNIT COST	TOTAL	UNIT COST	TOTAL

Figure 7.1

CONDENSED ESTIMATE SUMMARY

	SHEET NO.
PROJECT	ESTIMATE NO.
LOCATION — TOTAL AREA/VOLUME	DATE
ARCHITECT COST PER S.F./C.F.	NO. OF STORIES
PRICES BY: EXTENSIONS BY:	CHECKED BY:

DIV.	DESCRIPTION	MATERIAL	LABOR	EQUIPMENT	SUBCONTRACT	TOTAL
1.0	General Requirements					
2.0	Site Construction					
3.0	Concrete					
4.0	Masonry					
5.0	Metals					
6.0	Wood & Plastics					
7.0	Thermal & Moisture Protection					
8.0	Doors & Windows					
9.0	Finishes					
10.0	Specialties					
11.0	Equipment					
12.0	Furnishings					
13.0	Special Construction					
14.0	Conveying Systems					
15.0	Mechanical					
16.0	Electrical					
	Subtotals					
	Sales Tax %					
	Overhead %					
	Subtotal					
	Profit %					
	Contingency %					
	Adjustments					
	TOTAL BID					

Figure 7.2

Chapter 8

Pre-Bid Scheduling

The need for planning and scheduling is clear once the contract is signed and work commences on the project. However, some scheduling is also important during the bidding stage for the following reasons:

- To determine whether or not the project can be completed in the allotted or specified time, using normal crew sizes
- To identify potential overtime requirements
- To determine the time requirements for supervision
- To anticipate possible temporary heat and power requirements
- To price certain general requirement items and overhead costs
- To budget for equipment usage
- To anticipate and justify material and equipment delivery requirements

The schedule produced prior to bidding may be a simple bar chart or network diagram that includes overall quantities, probable delivery times, and available manpower. Network scheduling methods, such as the Critical Path Method (CPM) and the Precedence Chart simplify pre-bid scheduling because they do not require time-scaled line diagrams.

CPM Scheduling

In the CPM diagram, each activity is represented by an arrow. Nodes indicate start/stop between activities. The Precedence Diagram, on the other hand, shows the activity as a node, with arrows denoting precedence relationships between the activities. Different configurations of precedence arrows may represent the sequential relationships between activities. Simple examples of CPM and Precedence diagrams are shown in Figures 8.1 and 8.2, respectively. In both systems, duration times are indicated along each path. The sequence (path) of activities requiring the most total time represents the shortest possible time (critical path) in which those activities may be completed.

For example, in both Figures 8.1 and 8.2, activities A, B, and C require 20 successive days for completion before activity G may begin. Activity paths for D and E (15 days) are shorter and can easily be completed during the 20-day sequence. Therefore, this 20-day sequence is the shortest possible time (i.e., the critical path) for the completion of these activities—before activity G may begin.

Past experience or a prepared rough schedule may suggest that the time specified in the bidding documents is insufficient to complete the required work. In such cases, a more comprehensive schedule should be produced prior to bidding; this schedule will help to determine the added overtime or premium-time work costs required to meet the completion date.

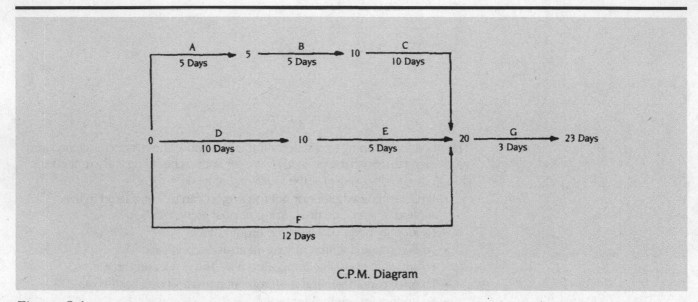

C.P.M. Diagram

Figure 8.1

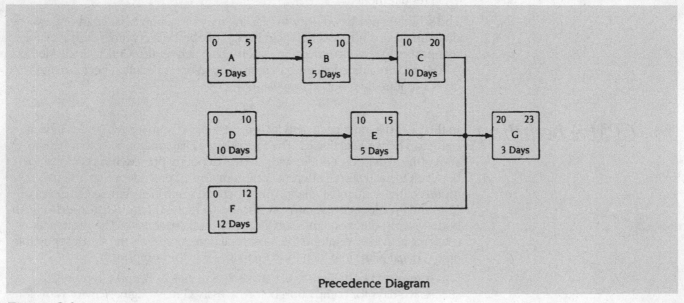

Precedence Diagram

Figure 8.2

For example, a preliminary schedule is necessary to determine the supervision and manning requirements of a small office renovation. A rough schedule for the interior work might be produced as shown in Figure 8.3. The man-days used to develop this schedule are derived from output figures determined in the estimate. Output can be determined based on the figures in *Means Repair and Remodeling Cost Data*. Man-days can also be figured by dividing the total labor cost shown on the estimate by the cost per man-day for each appropriate tradesperson.

As shown, the preliminary schedule may be used to determine supervision requirements, to develop appropriate crew sizes, and as a basis for ordering materials. All of these factors must be considered at this preliminary stage in order to determine how to meet the required completion date.

Summary

A pre-bid schedule may provide much more information than simple job duration. It may also be used to refine the estimate by introducing realistic manpower projections. Furthermore, the schedule may help the contractor to adjust the structure and size of the company based on projected requirements for months, even years, ahead. It is important to note that a schedule may become an effective tool for negotiating contracts.

For more detailed information on construction scheduling methods, see *Means Scheduling Manual, Third Edition,* by F. William Horsley.

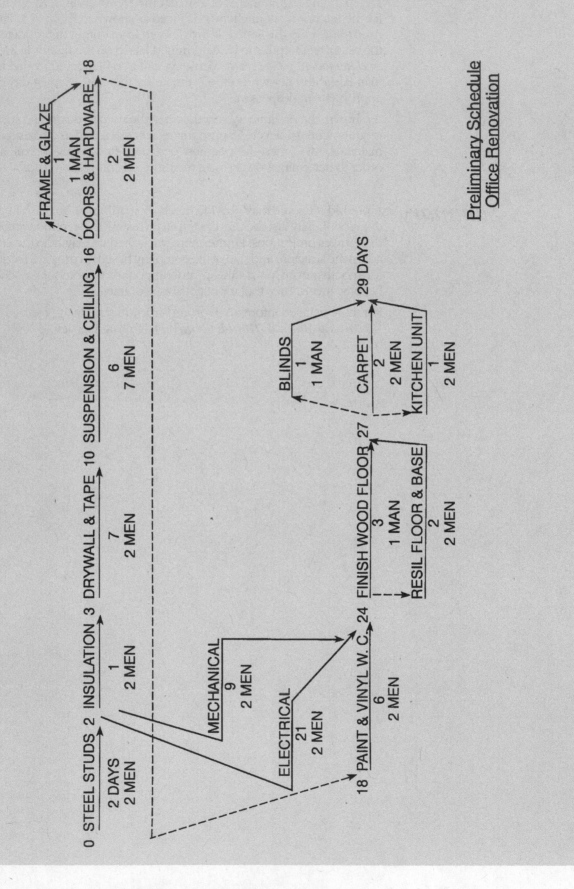

Preliminary Schedule
Office Renovation

Figure 8.3

Part II

Estimating By CSI Division

Part II

Estimating By CSI Division

There are two basic approaches to creating an estimate. One method is to proceed with the quantity takeoff and pricing in a sequence similar to the order in which a building is constructed. An Assemblies estimate follows this approach. The advantage of this method is that it enables experienced estimators to visualize the construction process. As a result, omissions are less likely. The basic disadvantage of this approach is the difficulty in determining and tracking costs for each item by division (or subcontract). Costs for a particular trade may be spread throughout the estimate.

The second approach to creating an estimate involves use of the CSI MasterFormat division (see Chapter 1). Most architectural specifications today are written according to this format, using the 16 divisions. Trades and subcontracts are generally limited to work within one division (e.g., Mechanical or Masonry). It makes sense, therefore, that the estimate should also be organized in the CSI format. Using this method, the estimator may have to exercise more care in order to ensure the inclusion of all required work. Nevertheless, the advantages outweigh potential drawbacks.

Most specifications contain references to related work for all divisions. These references serve as an aid to ensure completeness, but the estimator for the general contractor still has the responsibility of making sure that subcontractors include all that is specified and required for the appropriate portions of the project. The estimator must also decide what work will be subcontracted and what will be performed by the work force of the general contractor. These decisions may have a significant effect on the final cost of a project. Traditionally, work has been done at a lower cost, using the general contractor's own labor force. With the specialization of trades, however, subcontractors may often perform work faster, and thus at a lower cost.

Cost accounting and control is another area affected by the choice of estimating method. The chosen method should be used consistently, from estimating to field reporting to final analysis. The first approach, Assemblies estimating in the sequence of construction, brings the same challenge to cost accounting as it does to estimating—the difficulty of keeping similar items grouped together as the work of single trades.

The second method of estimating and cost accounting, by CSI MasterFormat division, may take longer when it comes to compiling all costs for each division (e.g., until all drywall or electrical work is complete). However, each trade will be separated and the records will be in accordance with the specifications.

Because material purchases and scheduling of manpower and subcontractors are based on the construction sequence, these items should be calculated based on the project schedule. This schedule is established after and formulated from the unit price estimate; it is the basis of project coordination.

Project specifications may contain, or owners may request, a list of alternates that must be included with a submitted bid. These alternates become a series of mini-estimates within the total project estimate. Each alternate may include deductions of some items from the project, originally specified, as well as the addition of other items. Mistakenly regarded as incidentals, alternates are often not addressed until the project estimate is complete. In order to efficiently determine alternate costs, without performing a completely new estimate for each, the project estimate must be organized with the alternates in mind. For example, if items are to be deducted as part of an alternate, they must be separated in the project estimate. Similarly, when measuring for quantities, pertinent dimensions and quantities for the "adds" of an alternate should be listed separately. If alternates are considered in this way, they can be estimated quickly and accurately.

Part II (Chapters 9 through 25) presents specific guidelines for estimating each of the CSI MasterFormat divisions. Two complete sample estimates follow in Part III of this book. The first estimate uses the unit price CSI MasterFormat division method. The second one is an Assemblies estimate that is carried out in the sequence of the actual construction.

Chapter 9

General Requirements

When estimating by CSI division, the estimator must be careful to include all items which, while not attributable to one trade only or directly to the physical construction of the building, are nevertheless required in order to successfully complete the project. These items are included in Division 1—General Requirements.

Often referred to as the "General Conditions" or "Project Overhead," such items are usually set forth in the first part of the specifications. Some requirements may not be directly specified, even though they are required to perform the work. Standardized sets of General Conditions have been developed by various segments of the construction industry, such as those by the American Institute of Architects, the Consulting Engineers Council/U.S., National Society of Professional Engineers, and others. These standardized documents usually include:

- Definitions
- Contract document descriptions
- Contractor's rights and responsibilities
- Architect/Engineer's authority and responsibilities
- Owner's rights and responsibilities
- Variation from contract provisions
- Payment requirements and restrictions
- Requirements for the performance of the work
- Insurance and bond requirements
- Job conditions and operation

Since these documents are generic, additions, deletions, and modifications unique to specific projects are usually included in Supplementary General Conditions.

Estimated costs for Division 1 are often recorded on a standardized form or checklist. Use of preprinted forms or checklists ensure that all requirements are included and priced. Many of the costs are dependent upon work in other divisions, or on the total cost and/or duration of the job. Project overhead costs should be determined throughout the estimating process and finalized when all other divisions have been estimated and a preliminary schedule established.

The following are brief discussions of various items that may be included as project overhead. Depending on the size and type of project, costs for these items may easily represent a significant portion of a repair or remodeling project's total costs (20% to 50%, or more). The estimator's goal is the development of an approach to estimating that will ensure that all project requirements are included.

Personnel

Job site personnel may be included as either *project overhead* or *office overhead*. This distinction often depends on the size of the project and on the contractor's accounting methods. For example, if a project is large enough to require a full-time superintendent, all time-related costs for that person may be considered project overhead. If, on the other hand, the superintendent is responsible for a number of smaller jobs, the expense may be either included in office overhead, or proportioned for each job.

Depending on the size of the project, a carpenter and/or laborer may be assigned to the job on a full-time basis for miscellaneous work. This individual would be directly responsible to the job superintendent for various tasks. The costs of this work would, therefore, not be attributable to any specific division and might appropriately be included as project overhead.

Temporary Services

Required temporary services may or may not be included in the specifications. The following is a typical statement in the specifications or scope of work.

"Contractor shall supply all material, labor, equipment, tools, utilities, and other items and services required for the proper and timely performance of the work and completion of the project."

The owner and designer may feel that such a statement eliminates ambiguity. To the estimator, this statement means that many items must be estimated. Temporary utilities, such as heat, light, power, water, and fire protection are major considerations. The estimator must anticipate monthly (or time-related) costs, and be sure that installation and removal costs are included, by the appropriate subcontractor or by the general contractor.

Storage
If there is inadequate space inside the building, an office trailer and/or storage trailers or containers may be required. Even if these facilities are owned by the contractor, depreciation and other related costs should still be allocated to the job as part of project overhead. Telephone, utility, and temporary toilet facilities are other costs that must be included when required.

Security
Depending on the location and type of project, some security services may be required. In addition to security personnel or guard dogs, fences, gates, special lighting, and alarms may also be needed.

Temporary Construction

Temporary construction may also involve many items that are not specified in the construction documents. Partitions, doors, fences and barricades may be required to delineate or isolate portions of the building or site during construction.

Worker Protection
In addition to the above items, some temporary construction may be required solely for the protection of workers. Depending on the project size, an OSHA representative may visit the site to ensure that all safety precautions are being observed.

Protection of Finished Work
Workers will almost always use a new or existing permanent elevator for access throughout the building. Even though this use is almost always restricted, precautionary measures should always be taken to protect the doors and

cabs. Invariably, some damage occurs. Protection of any and all finished surfaces (throughout the project) should be priced and included in the estimate.

Types of temporary services and construction, with costs, from *Means Repair and Remodeling Cost Data*, are shown in Figure 9.1.

Job Cleanup

An amount should always be carried in the estimate for cleanup of the construction area, both during the construction process and upon completion. The cleanup can be broken down into three basic categories, which can be estimated separately:

- Continuous (daily or otherwise) cleaning of the project area
- Rubbish handling and removal
- Final cleanup

Costs for continuous cleaning can be included as an allowance, or estimated by required labor-hours (in some cases, a full-time laborer is appropriate). Rubbish handling should include barrels; wheeled carts; a trash chute, if necessary; dumpster rental; and disposal fees. Disposal fees vary depending upon the project and location. A permit may also be required.

Costs for final cleanup should be based on past projects and may include subcontract costs for items such as the cleaning of windows and waxing of floors. Included in the costs for final cleanup may be an allowance for repair of minor damage to finished work.

Bonds

Bonding requirements for a project are specified in the Division 1—General Requirements section of the specifications. They are also included in the construction contract. Various types of bonds may be required for a given project. Some common types are listed below.

Bid Bond

This type of bond is a form of bid security executed by the bidder or principal and by a surety (bonding company) to guarantee that the bidder will enter into a contract within a specified time and furnish any required Performance of Labor and Material Payment Bonds.

Completion Bond

Also known as a "Construction" or "Contract" bond, the Completion Bond is a guarantee by a surety to the owner that the contractor will pay for all labor and materials used in the performance of the contract, as per the construction documents. The claimants under the bond have direct contracts with the contractor or any subcontractor.

Performance Bond

A Performance Bond is (1) a guarantee that a contractor will perform a job according to the terms of the contracts, and (2) a bond of the contractor in which a surety guarantees to the owner that the work will be performed in accordance with the contract documents. Except where prohibited by statute, the Performance Bond is frequently combined with the Labor and Material Payment Bond.

Surety Bond

The Surety Bond is a legal instrument under which one party agrees to answer to another party for the debt, default, or failure to perform of a third party.

			DAILY OUTPUT	LABOR-HOURS	UNIT	2002 BARE COSTS				TOTAL INCL O&P	
	01520	**Construction Facilities**	CREW			MAT.	LABOR	EQUIP.	TOTAL		
500	0010	**OFFICE** Trailer, furnished, no hookups, 20' x 8', buy	2 Skwk	1	16	Ea.	5,450	495		5,945	6,800	**500**
	0250	Rent per month					153			153	169	
	0300	32' x 8', buy	2 Skwk	.70	22.857		8,000	705		8,705	9,950	
	0350	Rent per month					168			168	185	
	0400	50' x 10', buy	2 Skwk	.60	26.667		13,600	825		14,425	16,400	
	0450	Rent per month					300			300	330	
	0500	50' x 12', buy	2 Skwk	.50	32		16,100	990		17,090	19,300	
	0550	Rent per month					350			350	385	
	0700	For air conditioning, rent per month, add					36.50			36.50	40	
	0800	For delivery, add per mile				Mile	1.53			1.53	1.68	
	1200	Storage boxes, 20' x 8', buy	2 Skwk	1.80	8.889	Ea.	3,375	275		3,650	4,150	
	1250	Rent per month					74			74	81.50	
	1300	40' x 8', buy	2 Skwk	1.40	11.429		4,025	355		4,380	5,000	
	1350	Rent per month					105			105	115	
	01530	**Temporary Construction**										
700	0010	**PROTECTION** Stair tread, 2" x 12" planks, 1 use	1 Carp	75	.107	Tread	5.10	3.20		8.30	10.90	**700**
	0100	Exterior plywood, 1/2" thick, 1 use		65	.123		1.47	3.69		5.16	7.70	
	0200	3/4" thick, 1 use		60	.133		1.99	4		5.99	8.85	
900	0010	**WINTER PROTECTION** Reinforced plastic on wood										**900**
	0100	framing to close openings	2 Clab	750	.021	S.F.	.37	.50		.87	1.24	
	0200	Tarpaulins hung over scaffolding, 8 uses, not incl. scaffolding		1,500	.011		.18	.25		.43	.61	
	0250	Tarpaulin polyester reinf. w/ integral fastening system 11 mils thick		1,600	.010		.76	.23		.99	1.23	
	0300	Prefab fiberglass panels, steel frame, 8 uses		1,200	.013		.72	.31		1.03	1.31	
	01540	**Construction Aids**										
550	0010	**PUMP STAGING**, Aluminum R01540-200										**550**
	0200	24' long pole section, buy				Ea.	345			345	380	
	0300	18' long pole section, buy					269			269	296	
	0400	12' long pole section, buy					181			181	199	
	0500	6' long pole section, buy					95.50			95.50	105	
	0600	6' long splice joint section, buy					71.50			71.50	78.50	
	0700	Pump jack					121			121	133	
	0900	Foldable brace					50			50	55	
	1000	Workbench/back safety rail support					61.50			61.50	67.50	
	1100	Scaffolding planks/workbench, 14" wide x 24' long					565			565	620	
	1200	Plank end safety rail					191			191	210	
	1250	Safety net, 22' long					277			277	305	
	1300	System in place, 50' working height, per use based on 50 uses	2 Carp	84.80	.189	C.S.F.	5.15	5.65		10.80	15.05	
	1400	100 uses		84.80	.189		2.58	5.65		8.23	12.25	
	1500	150 uses		84.80	.189		1.73	5.65		7.38	11.30	
750	0010	**SCAFFOLDING** R01540-100										**750**
	0015	Steel tubular, reg, rent/mo, no plank, incl erect & dismantle										
	0090	Building exterior, wall face, 1 to 5 stories, 6'-4" x 5' frames	3 Carp	24	1	C.S.F.	24.50	30		54.50	76	
	0200	6 to 12 stories	4 Carp	21.20	1.509		24.50	45.50		70	102	
	0310	13 to 20 stories	5 Carp	20	2		24.50	60		84.50	126	
	0460	Building interior, wall face area, up to 16' high	3 Carp	25	.960		24.50	29		53.50	74	
	0560	16' to 40' high		23	1.043		24.50	31.50		56	78.50	
	0800	Building interior floor area, up to 30' high		312	.077	C.C.F.	2.57	2.31		4.88	6.65	
	0900	Over 30' high	4 Carp	275	.116	"	2.57	3.49		6.06	8.65	
	0910	Steel tubular, heavy duty shoring, buy										
	0920	Frames 5' high 2' wide				Ea.	75			75	82.50	
	0925	5' high 4' wide					85			85	93.50	

Figure 9.1

Miscellaneous General Requirements

Many other items must be taken into account when costs are being determined for project overhead. The following items are among the major considerations:

Scaffolding or Rolling Platforms

It is important to determine who is responsible for the rental, erection, and dismantling of scaffolding, because it will often be used by more than one trade. If a subcontractor is responsible, it may be necessary to leave the scaffolding in place long enough for use by other trades. Different types of scaffolding are illustrated in Figure 9.2.

Small Tools

An allowance should be carried for small tools; the amount should be based on past experience. This sum should cover the cost of hand tools as well as small power tools supplied by the contractor. Small tools often break down or "walk," and a certain amount of replacement is inevitable.

Permits

Various types of permits may be required, depending on local codes and regulations. Both the necessity of the permit and the responsibility for acquiring it must be determined before bidding. If the work is being done in an unfamiliar location, local building officials should be consulted regarding unusual or unknown requirements.

Insurance

Insurance coverage for each project and locality—above and beyond the normal required operating insurance—should be reviewed to ensure that coverage is adequate. The contract documents will often specify certain required policy limits. The estimator should anticipate the need for specific policies or riders.

Other Considerations

Other items commonly included in Project Overhead are the following: photographs, models, job signs, costs for sample panels and materials for owner/architect approval, and an allowance for replacement of broken glass. For some materials, such as imported goods or custom-fabricated items, both shipping costs and off-site storage fees can be expected. An allowance should also be included for anticipated costs pertaining to punch list items. Determination of these costs is likely to be based on past experience.

Some project overhead costs can be calculated at the beginning of the estimate. Others are included when the estimating process is under way. Still other costs are estimated last, because they depend on the total cost and duration of the project. Many of the overhead items are not directly specified, so that the estimator must use experience and must visualize the construction process to ensure that all requirements are met. *It is not important when or where these items are included, but that they are included at all.* One contractor may list certain costs as project overhead, while another contractor may allocate the same costs (and responsibility) to a subcontractor. Either way, the costs must be, and are, recorded in the estimate.

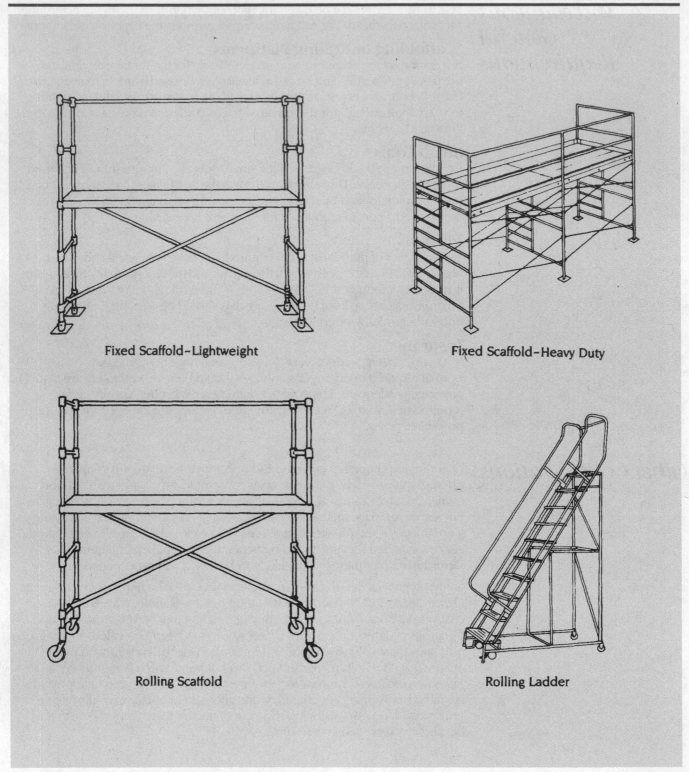

Fixed Scaffold–Lightweight

Fixed Scaffold–Heavy Duty

Rolling Scaffold

Rolling Ladder

Figure 9.2

Chapter 10

Site Work/Demolition

Site Work
In most remodeling and renovation projects, any site work involved is usually small, and relative to the amount of interior work. However small, the analysis of prospective site work is an important part of the quantity takeoff. Because most site work requires special tools or heavy equipment, its role in commercial renovation requires special attention. The estimator must determine not only the appropriate equipment options but also the cost effectiveness.

Minimum equipment charges may be in effect if the quantities of work are small. For example, Figure 10.1, from *Means Repair and Remodeling Cost Data*, shows costs for core drilling. The estimator has determined that only one six-inch hole is required through a four-inch concrete slab. Line 02210-320-0700 would suggest the cost to be 1 x $68.50, or $68.50. However, line 02210-320-2050 shows the *minimum charge* to be $137. Clearly, the minimum charge must be carried in the estimate. The estimator must determine the exact quantities involved, and he must employ careful judgment and experience in order to anticipate the consequences of actual conditions.

Demolition
In contrast to conventional site work (such as excavation, site utilities, and paving), demolition is, in most cases, a major component of remodeling work. For takeoff and estimating purposes, demolition should be broken down into individual components. Often, demolition is not directly included in the plans and specifications. A review of Division 02200, Site Preparation , in *Means Repair and Remodeling Cost Data* may help the estimator to be sure that all required work is included in the estimate. Unless the estimator has had extensive experience with selective demolition, or unless a local subcontractor can provide a bid, each item to be demolished should be listed separately. When performing the quantity takeoff, requirements for handling, hauling, and dumping the debris must also be included as separate items.

When the contractor needs only a preliminary or budget cost for demolition, the estimator may have to determine only the size of the job (square feet or cubic feet of the proposed demolition area), and then base costs on previous jobs or use cost data such as that shown in Figure 10.2. The estimator, however, must observe caution when using this technique. As emphasized, every remodeling project is different, and the amount and difficulty of demolition—whether selective or complete—will vary from job to job.

			DAILY	LABOR-		*2002 BARE COSTS*				TOTAL
02210	**Subsurface Investigation**	CREW	OUTPUT	HOURS	UNIT	MAT.	LABOR	EQUIP.	TOTAL	INCL O&P
310 1450	With crawler type drill	B-56	1	16	Day		425	810	1,235	1,575
1500	For inner city borings add, minimum									10%
1510	Maximum									20%
320 0010	**DRILLING, CORE** Reinforced concrete slab, up to 6″ thick slab									
0020	Including bit, layout and set up									
0100	1″ diameter core	B-89A	28	.571	Ea.	2.35	15.55	3.33	21.23	32
0150	Each added inch thick, add		300	.053		.42	1.45	.31	2.18	3.19
0300	3″ diameter core		23	.696		5.25	18.90	4.05	28.20	41
0350	Each added inch thick, add		186	.086		.94	2.34	.50	3.78	5.45
0500	4″ diameter core		19	.842		5.25	23	4.90	33.15	49
0550	Each added inch thick, add		170	.094		1.19	2.56	.55	4.30	6.15
0700	6″ diameter core		14	1.143		8.60	31	6.65	46.25	68.50
0750	Each added inch thick, add		140	.114		1.46	3.11	.67	5.24	7.50
0900	8″ diameter core		11	1.455		11.75	39.50	8.45	59.70	87.50
0950	Each added inch thick, add		95	.168		1.98	4.58	.98	7.54	10.80
1100	10″ diameter core		10	1.600		15.70	43.50	9.30	68.50	99.50
1150	Each added inch thick, add		80	.200		2.60	5.45	1.16	9.21	13.10
1300	12″ diameter core		9	1.778		18.85	48.50	10.35	77.70	111
1350	Each added inch thick, add		68	.235		3.12	6.40	1.37	10.89	15.50
1500	14″ diameter core		7	2.286		23	62	13.30	98.30	143
1550	Each added inch thick, add		55	.291		3.96	7.90	1.69	13.55	19.25
1700	18″ diameter core		4	4		29.50	109	23.50	162	237
1750	Each added inch thick, add	▼	28	.571		5.20	15.55	3.33	24.08	35
1760	For horizontal holes, add to above				▼	.			30%	30%
1770	Prestressed hollow core plank, 6″ thick									
1780	1″ diameter core	B-89A	52	.308	Ea.	1.56	8.35	1.79	11.70	17.50
1790	Each added inch thick, add		350	.046		.27	1.24	.27	1.78	2.64
1800	3″ diameter core		50	.320		3.44	8.70	1.86	14	20
1810	Each added inch thick, add		240	.067		.57	1.81	.39	2.77	4.05
1820	4″ diameter core		48	.333		4.58	9.05	1.94	15.57	22
1830	Each added inch thick, add		216	.074		.79	2.01	.43	3.23	4.66
1840	6″ diameter core		44	.364		5.65	9.90	2.12	17.67	25
1850	Each added inch thick, add		175	.091		.94	2.49	.53	3.96	5.70
1860	8″ diameter core		32	.500		7.60	13.60	2.91	24.11	34
1870	Each added inch thick, add		118	.136		1.31	3.69	.79	5.79	8.40
1880	10″ diameter core		28	.571		10.25	15.55	3.33	29.13	40.50
1890	Each added inch thick, add		99	.162		1.41	4.40	.94	6.75	9.85
1900	12″ diameter core		22	.727		12.50	19.80	4.23	36.53	51
1910	Each added inch thick, add		85	.188	▼	2.08	5.10	1.10	8.28	11.95
1950	Minimum charge for above, 3″ diameter core		7	2.286	Total		62	13.30	75.30	118
2000	4″ diameter core		6.80	2.353			64	13.70	77.70	121
2050	6″ diameter core		6	2.667			72.50	15.50	88	137
2100	8″ diameter core		5.50	2.909			79	16.95	95.95	149
2150	10″ diameter core		4.75	3.368			91.50	19.60	111.10	173
2200	12″ diameter core		3.90	4.103			112	24	136	211
2250	14″ diameter core		3.38	4.734			129	27.50	156.50	243
2300	18″ diameter core	▼	3.15	5.079	▼		138	29.50	167.50	261
3010	Bits for core drill, diamond, premium, 1″ diameter				Ea.	93.50			93.50	103
3020	3″ diameter					233			233	256
3040	4″ diameter					259			259	285
3050	6″ diameter					415			415	455
3080	8″ diameter					565			565	625
3120	12″ diameter					900			900	990
3180	18″ diameter					1,900			1,900	2,100
3240	24″ diameter				▼	2,550			2,550	2,800

Figure 10.1

		02220	Site Demolition	CREW	DAILY OUTPUT	LABOR-HOURS	UNIT	2002 BARE COSTS				TOTAL INCL O&P	
								MAT.	LABOR	EQUIP.	TOTAL		
100	0010	**BUILDING DEMOLITION** Large urban projects, incl. 20 Mi. haul											100
	0012	No foundation or dump fees, C.F. is volume of building standing, steel		B-8	21,500	.003	C.F.		.08	.11	.19	.25	
	0050	Concrete			15,300	.004			.11	.15	.26	.35	
	0080	Masonry			20,100	.003			.08	.12	.20	.27	
	0100	Mixture of types, average		▼	20,100	.003			.08	.12	.20	.27	
	0500	Small bldgs, or single bldgs, no salvage included, steel		B-3	14,800	.003			.08	.11	.19	.26	
	0600	Concrete			11,300	.004			.11	.14	.25	.34	
	0650	Masonry			14,800	.003			.08	.11	.19	.26	
	0700	Wood		▼	14,800	.003	▼		.08	.11	.19	.26	
	1000	Single family, one story house, wood, minimum					Ea.				2,300	2,700	
	1020	Maximum									4,000	4,800	
	1200	Two family, two story house, wood, minimum									3,000	3,600	
	1220	Maximum									5,800	7,000	
	1300	Three family, three story house, wood, minimum									4,000	4,800	
	1320	Maximum					▼				7,000	8,400	
	1400	Gutting building, see division 02225-400											
	5000	For buildings with no interior walls, deduct					Ea.				50%		
550	0010	**FOOTINGS AND FOUNDATIONS DEMOLITION**											550
	0200	Floors, concrete slab on grade,											
	0240	4" thick, plain concrete		B-9C	500	.080	S.F.		1.91	.36	2.27	3.55	
	0280	Reinforced, wire mesh			470	.085			2.03	.38	2.41	3.78	
	0300	Rods			400	.100			2.39	.44	2.83	4.44	
	0400	6" thick, plain concrete			375	.107			2.54	.47	3.01	4.73	
	0420	Reinforced, wire mesh			340	.118			2.81	.52	3.33	5.20	
	0440	Rods		▼	300	.133	▼		3.18	.59	3.77	5.90	
	1000	Footings, concrete, 1' thick, 2' wide		B-5	300	.187	L.F.		4.84	2.89	7.73	11.10	
	1080	1'-6" thick, 2' wide			250	.224			5.80	3.47	9.27	13.30	
	1120	3' wide			200	.280			7.25	4.34	11.59	16.60	
	1140	2' thick, 3' wide		▼	175	.320			8.30	4.96	13.26	19	
	1200	Average reinforcing, add									10%	10%	
	1220	Heavy reinforcing, add					▼				20%	20%	
	2000	Walls, block, 4" thick		1 Clab	180	.044	S.F.		1.04		1.04	1.73	
	2040	6" thick			170	.047			1.10		1.10	1.83	
	2080	8" thick			150	.053			1.25		1.25	2.07	
	2100	12" thick		▼	150	.053			1.25		1.25	2.07	
	2200	For horizontal reinforcing, add									10%	10%	
	2220	For vertical reinforcing, add									20%	20%	
	2400	Concrete, plain concrete, 6" thick		B-9	160	.250			5.95	1.11	7.06	11.10	
	2420	8" thick			140	.286			6.80	1.27	8.07	12.70	
	2440	10" thick			120	.333			7.95	1.48	9.43	14.80	
	2500	12" thick		▼	100	.400			9.55	1.78	11.33	17.75	
	2600	For average reinforcing, add									10%	10%	
	2620	For heavy reinforcing, add					▼				20%	20%	
	9000	Minimum labor/equipment charge		A-1	2	4	Job		94	30.50	124.50	189	
575	0010	**HYDRODEMOLITION**, concrete pavement, 4000 PSI, 2" depth		B-5	500	.112	S.F.		2.91	1.74	4.65	6.65	575
	0120	4" depth			450	.124			3.23	1.93	5.16	7.35	
	0130	6" depth			400	.140			3.63	2.17	5.80	8.30	
	0410	6000 PSI, 2" depth			410	.137			3.54	2.12	5.66	8.15	
	0420	4" depth			350	.160			4.15	2.48	6.63	9.50	
	0430	6" depth			300	.187			4.84	2.89	7.73	11.10	
	0510	8000 PSI, 2" depth			330	.170			4.40	2.63	7.03	10.10	
	0520	4" depth			280	.200			5.20	3.10	8.30	11.85	
	0530	6" depth		▼	240	.233	▼		6.05	3.62	9.67	13.85	

Figure 10.2

Determining the Appropriate Equipment

When tabulating the quantities involved in all aspects of site work, the estimator must keep in mind the methods that will be used and the restrictions involved in order to determine the appropriate equipment. For example, the plans may specify a trench for a buried tank. The estimator, having determined the size of the trench, must decide upon the best and least expensive method of work. See Figure 10.3, derived from costs in *Means Repair and Remodeling Cost Data*.

When the estimator looks only at the actual cost of excavating and backfilling, digging by hand is almost ten times more expensive than using a backhoe. The daily rental, however (usually the minimum charge), makes the backhoe the more costly alternative. In certain cases, when quantities are small, it can be more economical to work by hand. However, if other work on that job can be done by the same piece of equipment during the same day, then the equipment may become the less expensive alternative. When determining quantities involving expensive equipment, the estimator must be aware of potential savings due to economy of scale.

Trench Excavation		
Line Item	**Unit Price**	**Cost**
02315-900-1400 Excavation by hand	$39.00/CY	$138.45
02315-100-0010 Backfill by hand	$22.00/CY	$ 78.10
Actual cost by hand		$216.55
02315-440-2035 Backhoe excavation	$13.90/CY	$ 49.35
02315-100-1300 Backhoe backfill	$ 1.21/CY	$ 4.29
Actual cost by backhoe		$ 53.64
01590-200-0450 Minimum daily rental of backhoe		$260.60

Trench: 8′ length x 3′ width x 4′ depth = 96 cubic feet
3.55 cubic yards
(to be excavated)

Figure 10.3

Chapter 11
Concrete

Concrete work in remodeling may entail placing a floor slab, placing a new topping on an existing floor, constructing new column foundations, or providing equipment foundations. Figure 11.1 illustrates some different types of concrete work commonly used in remodeling projects.

Estimating concrete work may be performed using two basic methods. In the first method, estimates are made of all the individual components—formwork, reinforcing, concrete, placement, and finish—individually. This is the most accurate, but also the most time-consuming, method. It is not always practical when the concrete work is only a small portion of the total project.

In the second method, an estimate is prepared based on costs that incorporate all of the components into a *system*, or *assembly*. This type of inclusive pricing is shown in Figure 11.2. When using this method, the estimator must be sure that the system to be estimated is, in fact, the same as that for which the assembly has been developed. Slight variations in the design of a system may significantly affect costs.

Regardless of the scope of the work required, all concrete construction involves the following basic components:

- Formwork
- Reinforcing steel or welded wire fabric
- Concrete
- Placement
- Finishing

The following sections describe various concrete design details frequently found in remodeling projects. Reinforcing, strength, placement, and finish requirements are included where applicable.

Floor Slabs

Floor slabs (on grade) should be placed on compacted granular fill, such as gravel or crushed stone, which has been covered by a polyethylene vapor barrier. Slabs may be reinforced or unreinforced. Reinforcing is usually provided by welded wire fabric, which prevents any cracks from expanding.

Large concrete floors are usually placed in strips that extend across the building at column lines or at 20′ to 30′ widths. Construction joints may be keyed or straight and may contain smooth or deformed reinforcing, which in turn may be wrapped or greased on one side to allow horizontal movement and to control cracking.

Control joints are intended to limit cracking to designated lines. They may be established by saw cutting the partially cured concrete slab to a specified depth or by applying a preformed metal strip to create a crack line. Many specifications require a boxed out section around columns, with that area to be concreted after the slab has been placed and cured. This helps to prevent future cracks due to settling.

Expansion joints are generally used against confining walls, foundations, and existing construction. They are commonly constructed from a preformed expansion material. The finish of the slabs is usually dictated by use and varies between a screed, or rough, finish (associated with two-course floors) and a

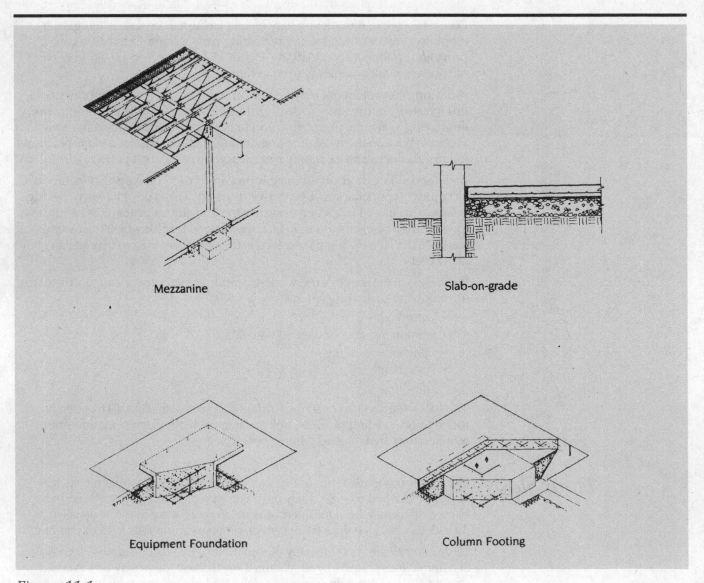

Mezzanine

Slab-on-grade

Equipment Foundation

Column Footing

Figure 11.1

			CREW	DAILY OUTPUT	LABOR-HOURS	UNIT	2002 BARE COSTS				TOTAL INCL O&P	
	03310	**Structural Concrete**					MAT.	LABOR	EQUIP.	TOTAL		
240	2300	Waffle const., 30″ domes, 125 psf Sup. Load, 20′ span R03310 -010	C-14B	37.07	5.611	C.Y.	197	169	19.35	385.35	520	**240**
	2350	30′ span		44.07	4.720		180	142	16.30	338.30	455	
	2500	One way joists, 30″ pans, 125 psf Sup. Load, 15′ span R03310 -100		27.38	7.597		231	229	26	486	665	
	2550	25′ span		31.15	6.677		216	201	23	440	600	
	2700	One way beam & slab, 125 psf Sup. Load, 15′ span R04210 -055		20.59	10.102		186	305	35	526	755	
	2750	25′ span		28.36	7.334		169	221	25.50	415.50	585	
	2900	Two way beam & slab, 125 psf Sup. Load, 15′ span		24.04	8.652		174	261	30	465	660	
	2950	25′ span	▼	35.87	5.799	▼	148	175	20	343	475	
	3100	Elevated slabs including finish, not										
	3110	including forms or reinforcing										
	3150	Regular concrete, 4″ slab	C-8	2,613	.021	S.F.	.94	.56	.26	1.76	2.23	
	3200	6″ slab		2,585	.022		1.45	.57	.26	2.28	2.81	
	3250	2-1/2″ thick floor fill		2,685	.021		.64	.55	.25	1.44	1.87	
	3300	Lightweight, 110# per C.F., 2-1/2″ thick floor fill		2,585	.022		.74	.57	.26	1.57	2.02	
	3400	Cellular concrete, 1-5/8″ fill, under 5000 S.F.		2,000	.028		.49	.74	.34	1.57	2.11	
	3450	Over 10,000 S.F.		2,200	.025		.39	.67	.31	1.37	1.86	
	3500	Add per floor for 3 to 6 stories high		31,800	.002			.05	.02	.07	.10	
	3520	For 7 to 20 stories high	▼	21,200	.003	▼		.07	.03	.10	.15	
	3800	Footings, spread under 1 C.Y.	C-14C	38.07	2.942	C.Y.	100	84.50	.94	185.44	252	
	3850	Over 5 C.Y.		81.04	1.382		92	40	.44	132.44	168	
	3900	Footings, strip, 18″ x 9″, plain		41.04	2.729		91	78.50	.87	170.37	232	
	3950	36″ x 12″, reinforced		61.55	1.820		92.50	52.50	.58	145.58	190	
	4000	Foundation mat, under 10 C.Y.		38.67	2.896		124	83.50	.93	208.43	276	
	4050	Over 20 C.Y.	▼	56.40	1.986		110	57	.64	167.64	217	
	4200	Grade walls, 8″ thick, 8′ high	C-14D	45.83	4.364		126	130	15.65	271.65	370	
	4250	14′ high		27.26	7.337		149	219	26.50	394.50	560	
	4260	12″ thick, 8′ high		64.32	3.109		111	93	11.15	215.15	288	
	4270	14′ high		40.01	4.999		119	149	17.95	285.95	400	
	4300	15″ thick, 8′ high		80.02	2.499		104	74.50	8.95	187.45	248	
	4350	12′ high		51.26	3.902		106	117	14	237	325	
	4500	18′ high	▼	48.85	4.094	▼	116	122	14.70	252.70	345	
	4520	Handicap access ramp, railing both sides, 3′ wide	C-14H	14.58	3.292	L.F.	139	98	2.44	239.44	320	
	4525	5′ wide		12.22	3.928		146	117	2.91	265.91	360	
	4530	With 6″ curb and rails both sides, 3′ wide		8.55	5.614		146	167	4.15	317.15	445	
	4535	5′ wide	▼	7.31	6.566	▼	149	195	4.86	348.86	495	
	4650	Slab on grade, not including finish, 4″ thick	C-14E	60.75	1.449	C.Y.	84.50	43	.58	128.08	167	
	4700	6″ thick	″	92	.957	″	81	28.50	.38	109.88	138	
	4751	Slab on grade, incl. troweled finish, not incl. forms										
	4760	or reinforcing, over 10,000 S.F., 4″ thick slab	C-14F	3,425	.021	S.F.	.92	.57	.01	1.50	1.95	
	4820	6″ thick slab		3,350	.021		1.35	.58	.01	1.94	2.43	
	4840	8″ thick slab		3,184	.023		1.85	.61	.01	2.47	3.02	
	4900	12″ thick slab		2,734	.026		2.77	.72	.01	3.50	4.21	
	4950	15″ thick slab	▼	2,505	.029	▼	3.48	.78	.01	4.27	5.10	
	5000	Slab on grade, incl. textured finish, not incl. forms										
	5001	or reinforcing, 4″ thick slab	C-14G	2,873	.019	S.F.	.90	.52	.01	1.43	1.84	
	5010	6″ thick		2,590	.022		1.40	.58	.01	1.99	2.49	
	5020	8″ thick	▼	2,320	.024	▼	1.83	.65	.02	2.50	3.07	
	5200	Lift slab in place above the foundation, incl. forms,										
	5210	reinforcing, concrete and columns, minimum	C-14B	2,113	.098	S.F.	4.96	2.97	.34	8.27	10.75	
	5250	Average		1,650	.126		5.60	3.80	.43	9.83	13	
	5300	Maximum	▼	1,500	.139	▼	6.35	4.18	.48	11.01	14.50	
	5500	Lightweight, ready mix, including screed finish only,										
	5510	not including forms or reinforcing										
	5550	1:4 for structural roof decks	C-14B	260	.800	C.Y.	102	24	2.76	128.76	155	
	5600	1:6 for ground slab with radiant heat	C-14F	92	.783		96.50	21.50	.38	118.38	140	
	5650	1:3:2 with sand aggregate, roof deck	C-14B	260	.800	▼	101	24	2.76	127.76	154	

Figure 11.2

steel, trowel-treated finish (common for exposed concrete floors). Dropped areas or deeper slabs may be used under concentrated loads, such as masonry walls. Formed depressions to receive other floor materials, such as mud-set ceramic tile or terrazzo, may also be necessary, at an additional cost to the slab-on-grade system.

Concrete floors on slab form, or centering, is a widely used structural system because of its light weight, fast erection time, and flexibility in bay sizes. Figure 11.3 illustrates typical concrete floor slab construction details. Concrete toppings may include new concrete slabs placed over existing floors, with colors, hardeners, integral toppings, or abrasive finishes. Granolithic topping 1/2″ to 2″ in thickness may be placed over existing slabs with the proper surface preparation.

Other Concrete Items

Other required concrete should be estimated separately. Such work may include locker bases, closures around pipes in pipe chases, and grouting around door sills, frames, and recessed hardware.

For detailed coverage of concrete problem analysis, repair and maintenance requirements and methods, refer to *Concrete Repair & Maintenance Illustrated* by Peter Emmons (R.S. Means Co., Inc.). This publication covers concrete behavior in response to corrosion, moisture and thermal and other effects; condition assessment methods; surface repair methods; strengthening and stabilization; and protection strategies.

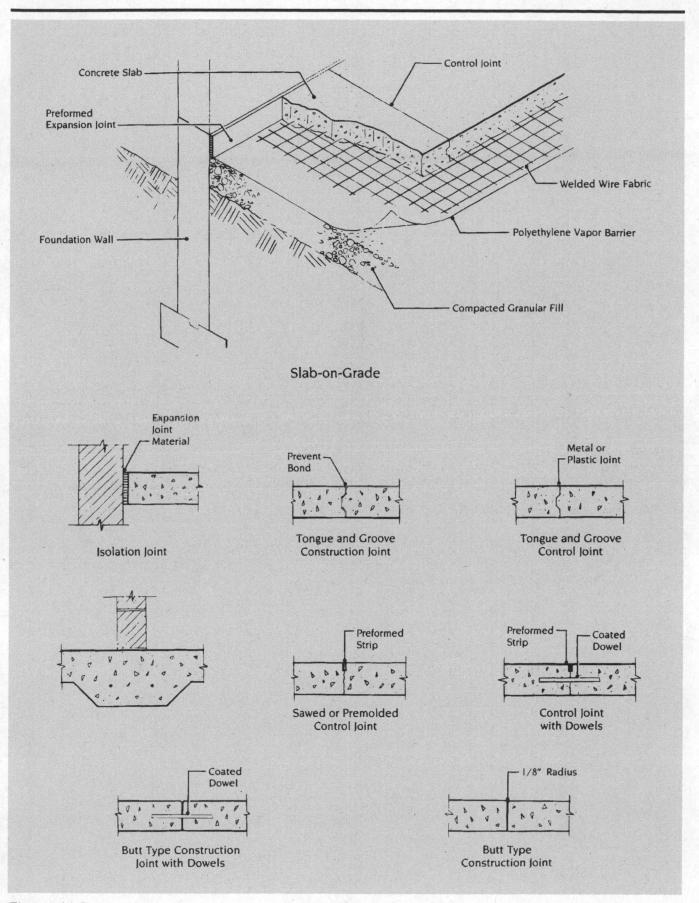

Concrete Slab

Control Joint

Preformed Expansion Joint

Foundation Wall

Welded Wire Fabric

Polyethylene Vapor Barrier

Compacted Granular Fill

Slab-on-Grade

Expansion Joint Material

Isolation Joint

Prevent Bond

Tongue and Groove Construction Joint

Metal or Plastic Joint

Tongue and Groove Control Joint

Preformed Strip

Sawed or Premolded Control Joint

Preformed Strip

Coated Dowel

Control Joint with Dowels

Coated Dowel

Butt Type Construction Joint with Dowels

1/8" Radius

Butt Type Construction Joint

Figure 11.3

Chapter 12

Masonry

A variety of masonry materials and methods are used in remodeling and renovation. The methods of installation and types of work also vary, even within the same project. The estimator can categorize masonry work according to *exterior walls, interior partitions, masonry restoration,* and *repair.* Depending on the structural design of masonry units, the reinforcement and mortar requirements also vary. Other items to be identified and listed are *insulation, embedded items, special finishes, scaffolding-related work accessories,* and *shoring requirements.*

After the estimator has identified and analyzed the walls and partitions, each kind or combination of masonry unit is then measured and entered on the takeoff sheets. The items might be grouped and listed in the order in which they would be constructed on the job:

1. Walls Below Grade
2. External Walls
3. Chimney
4. Interior Partitions
5. Cleaning and Repair
6. Miscellaneous Brickwork

This system of organization allows the estimator to tackle one part of the work at a time. Once a consistent takeoff pattern is established and followed, there is less chance that an item will be overlooked or duplicated.

The number of units per square foot is multiplied by the area of the wall to determine the total quantity involved. For items installed in a course, the quantity per linear foot is calculated and multiplied by the length involved (such as a cove base).

Before starting the quantity takeoff, the estimator should make notes based on the masonry specifications of all items that must be estimated. These notes will include the kind of masonry units, the bonds, the mortar type, joint reinforcement, grout, miscellaneous installed items, scaffolding, and cleaning.

Brick Masonry, especially brick, is generally priced by the unit, by the piece, or per thousand units. The quantity of units is determined from area measurements of walls and partitions. The areas are converted from square feet into number of units by appropriate multipliers. The multipliers are a function of the size of the masonry unit, the pattern or coursing of the masonry unit, the thickness of mortar joints, and the thickness of the wall. Figure 12.1 illustrates various types and sizes of contemporary brick, and provides a chart of brick and mortar quantities. Cement bonding and coursing patterns for brick are shown in Figures 12.2 and 12.3.

Concrete Block

For reasons of strength, versatility, and economy, concrete blocks are frequently used for constructing masonry walls and partitions in remodeling projects. They may be used for interior bearing walls, infill panels, interior partitions, backup for brick, shaft enclosures, and fire walls.

Concrete blocks are manufactured in two basic types, solid and hollow, and in various strength ratings. If the cross-sectional area, exclusive of voids, is 75% or greater than the gross area of the block, it is classified as "solid block." If the specified area is below the 75% figure, the block is classified as "hollow block." The strength of concrete block is determined by the compressive strength of the type of concrete used in its manufacture, or by the equivalent compressive strength, which is based on the gross area of the block, including voids. Applications of various types of concrete block are shown in Figures 12.4a and 12.4b. Typical block shapes are shown in Figure 12.5.

There are several special aggregates that can be used to manufacture lightweight blocks. These blocks can be identified by the weight of the concrete mixture used in their manufacture. Regular weight block is made from 125 lb. per cubic foot (P.C.F.) concrete, and lightweight block from 105 to 85 P.C.F. concrete. Costs for lightweight concrete as compared to regular weight range from 20% more for 105 P.C.F. block to 50% more for 85 P.C.F. block. Blocks of various strengths, grades, finishes, and weights should be taken off separately, as costs may vary considerably.

Concrete block partitions may be erected of concrete masonry units with a nominal thickness of 4", 6", 8", 10", or 12". The units may be regular weight or lightweight and solid or 75% solid in horizontal profile. Normal units are nominal 16" long, but some manufacturers produce a nominal 24" unit. Partitions may be left exposed, painted, epoxy-coated, or furred and then covered with gypsum board or paneling. Partitions may also be plastered (directly on the block) or covered with self-furring lath and then plastered. Figure 12.6 shows various types of concrete block specialties: lintels, corners, bond beam, and joint reinforcing.

Joint Reinforcing

Joint reinforcing and individual ties serve as important components of the various types of brick and concrete block walls. These elements must be included in the estimate. Two types of joint reinforcing are available for block walls: the *truss-type* and the *ladder-type* (shown in Figure 12.6). Because the truss-type provides better load distribution, it is normally used in bearing walls. The ladder-type is usually installed in light-duty walls that serve nonbearing functions. Both types of joint reinforcing may also be used to tie together the inner and outer wythes of composite or cavity-design walls. Corrugated strips, as well as Z-type, rectangular, and adjustable wall ties, may also be used for this purpose. Generally, one metal wall tie should be installed for each 4-1/2 square feet of brick or stone veneer. Although both types of joint reinforcing may be used as ties, individual ties should not be used as joint reinforcing to control cracking.

Structural reinforcement is commonly required in concrete block walls, especially in those that are load-bearing. Deformed steel bars may be used as vertical reinforcement when grouted into the block voids, and as horizontal reinforcement when installed above openings and in bond beams. Horizontal

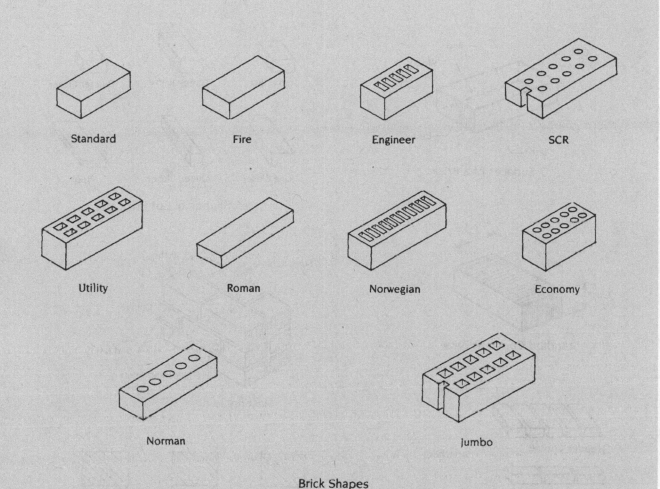

Standard · Fire · Engineer · SCR

Utility · Roman · Norwegian · Economy

Norman · Jumbo

Brick Shapes

Brick & Mortar Quantities

	Running Bond					For Other Bonds Standard Size Add to S.F. Quantities in Table to Left		
	Number of Brick per S.F. of Wall-Single Wythe with 1/2" Joints			CF of Mortar per M Bricks, Waste Included				
Type Brick	Nominal Size (Incl. Mortar) L H W	Modular Coursing	Number of Brick per S.F.	1 Wythe	2 Wythe	Bond Type	Description	Factor
Standard	8 x 2-2/3 x 4	3C = 8"	6.75	12.9	16.5	Common	full header every fifth course	+20%
Economy	8 x 4 x 4	1C = 4"	4.50	14.6	19.6		full header every sixth course	+16.7%
Engineer	8 x 3-1/5 x 4	5C = 16"	5.63	13.6	17.6	English	full header every second course	+50%
Fire	9 x 2-1/2 x 4 1/2	2C = 5"	6.40	550# Fireclay	–	Flemish	alternate headers every course	+33%
Jumbo	12 x 4 x 6 or 8	1C = 4"	3.00	34.0	41.4		every sixth course	+5.6%
Norman	12 x 2-2/3 x 4	3C = 8"	4.50	17.8	22.8	Header = W x H exposed		+100%
Norwegian	12 x 3-1/5 x 4	5C = 16"	3.75	18.5	24.4	Rowlock = H x W exposed		+100%
Roman	12 x 2 x 4	2C = 4"	6.00	17.0	20.7	Rowlock stretcher = L x W exposed		+33.3%
SCR	12 x 2-2/3 x 6	3C = 8"	4.50	26.7	31.7	Soldier = H x L exposed		—
Utility	12 x 4 x 4	1C = 4"	3.00	19.4	26.8	Sailor = W x L exposed		-33.3%

Figure 12.1

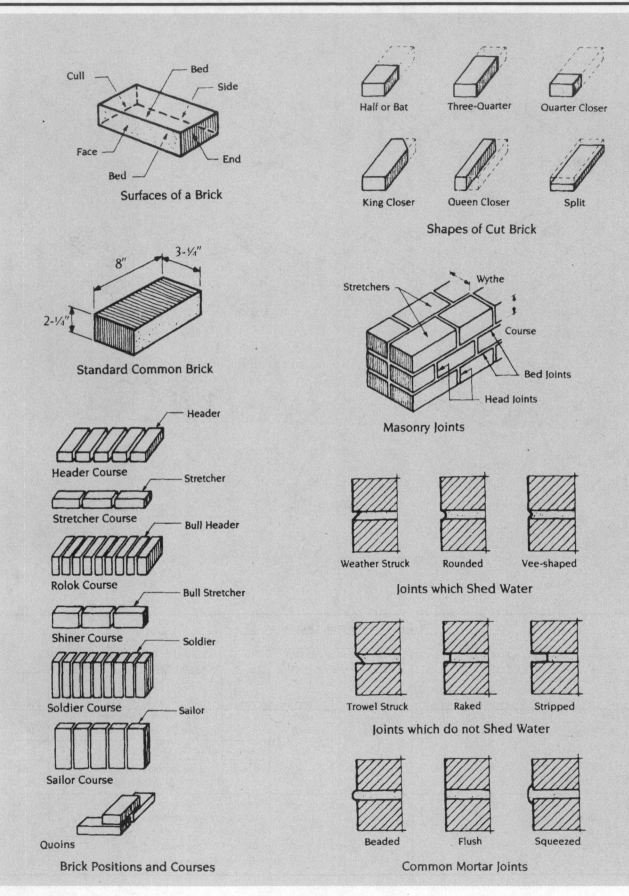

Surfaces of a Brick

Shapes of Cut Brick

Half or Bat · Three-Quarter · Quarter Closer

King Closer · Queen Closer · Split

Standard Common Brick

8″ · 3-¼″ · 2-¼″

Masonry Joints

Stretchers · Wythe · Course · Bed Joints · Head Joints

Brick Positions and Courses

Header · Header Course

Stretcher · Stretcher Course

Bull Header · Rolok Course

Bull Stretcher · Shiner Course

Soldier · Soldier Course

Sailor · Sailor Course

Quoins

Joints which Shed Water

Weather Struck · Rounded · Vee-shaped

Joints which do not Shed Water

Trowel Struck · Raked · Stripped

Common Mortar Joints

Beaded · Flush · Squeezed

Figure 12.2

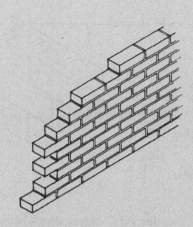

Standard Running Bond

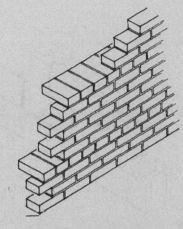

Common, Full Header Every 6th Course

English, Full Header Every 2nd Course

Flemish, Alternate Header Every Course

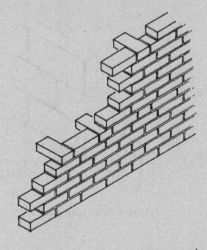

Flemish, Alternate Header Every 6th Course

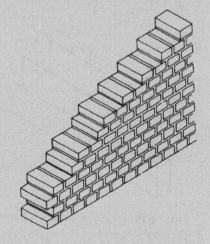

Brick Veneer, Full Header Throughout

Brick Bond Patterns

Figure 12.3

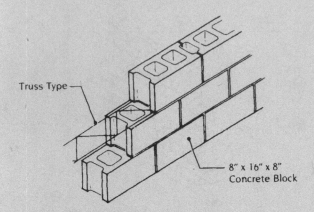

Wire Strip Joint Reinforcing

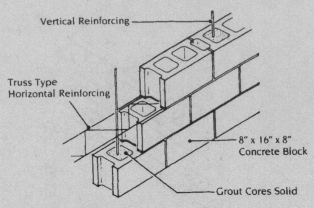

Reinforced Concrete Block Wall

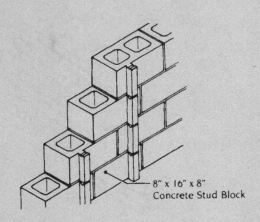

Self-furring Concrete Block

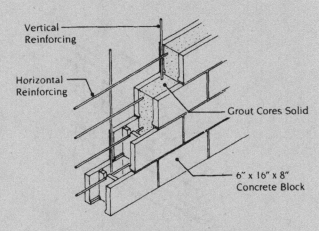

Interlocking Concrete Block

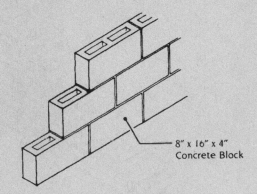

Nonbearing Concrete Block Partition

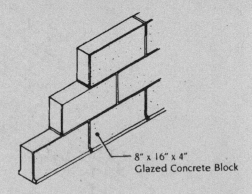

Glazed Concrete Block

Figure 12.4a

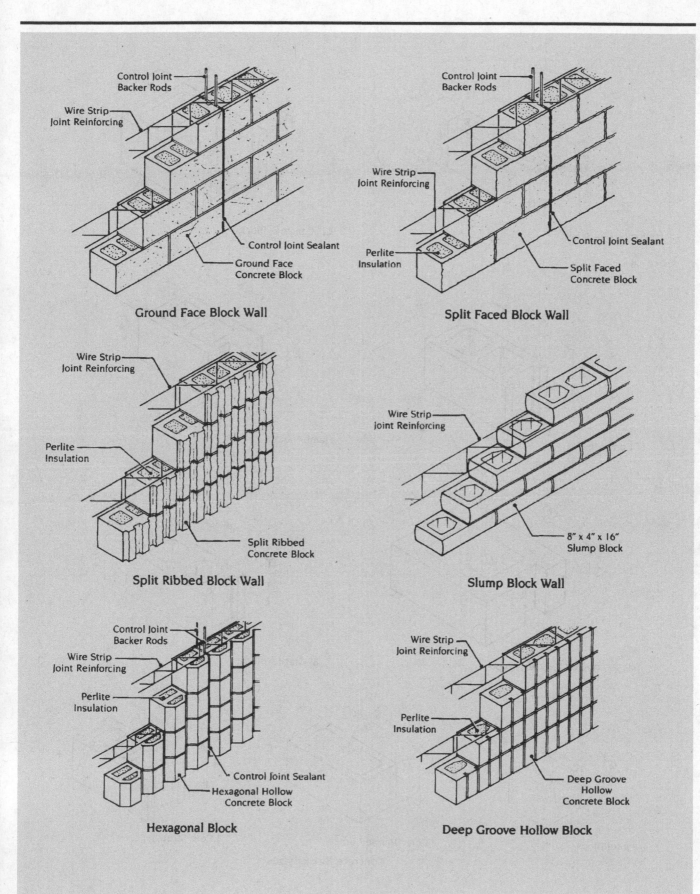

Ground Face Block Wall

Control Joint Backer Rods
Wire Strip Joint Reinforcing
Control Joint Sealant
Ground Face Concrete Block

Split Faced Block Wall

Control Joint Backer Rods
Wire Strip Joint Reinforcing
Perlite Insulation
Control Joint Sealant
Split Faced Concrete Block

Split Ribbed Block Wall

Wire Strip Joint Reinforcing
Perlite Insulation
Split Ribbed Concrete Block

Slump Block Wall

Wire Strip Joint Reinforcing
8" x 4" x 16" Slump Block

Hexagonal Block

Control Joint Backer Rods
Wire Strip Joint Reinforcing
Perlite Insulation
Control Joint Sealant
Hexagonal Hollow Concrete Block

Deep Groove Hollow Block

Wire Strip Joint Reinforcing
Perlite Insulation
Deep Groove Hollow Concrete Block

Figure 12.4b

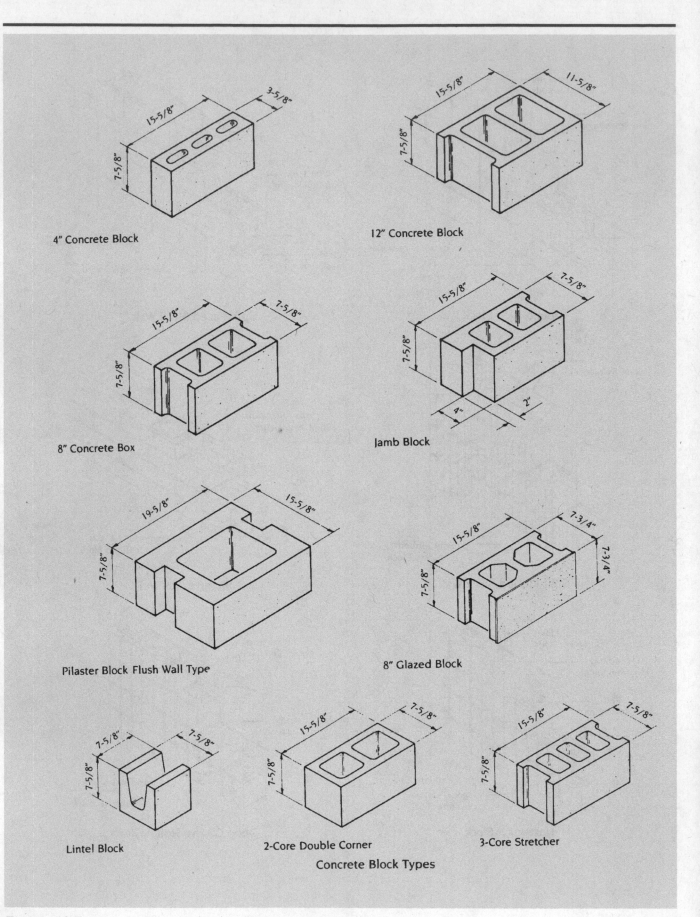

4" Concrete Block

12" Concrete Block

8" Concrete Box

Jamb Block

Pilaster Block Flush Wall Type

8" Glazed Block

Lintel Block

2-Core Double Corner

3-Core Stretcher

Concrete Block Types

Figure 12.5

114

and vertical bars may be grouted into the void that is normally used as the collar joint in a composite wall. Lintels should be installed to carry the weight of the wall above openings. Steel angles, built-up steel members, bond beams filled with steel bars and grout, and precast shapes may function as lintels.

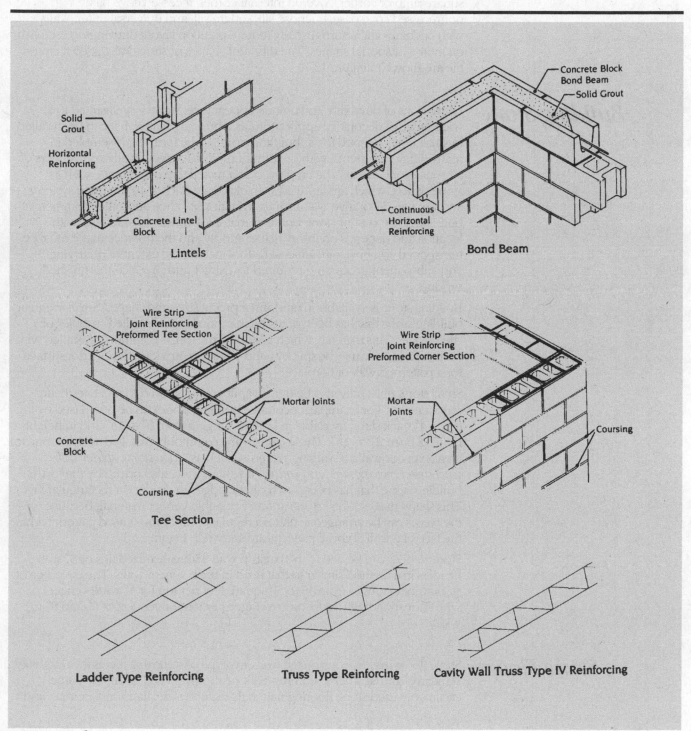

Figure 12.6

Structural Facing Tile

Structural glazed facing tile (S.G.F.T.) is kiln-fired structural clay with an integral impervious ceramic face. It may also be manufactured with an acoustical perforated face. Because glazed tile resists stains, marks, impact, abrasion, fading, and cracking, it is ideally suited for use in school corridors, locker rooms, rest rooms, kitchens, and other places where cleanliness and indestructibility are primary considerations.

Structural glazed facing tile is commonly available in a large selection of colors and color combinations in the 6T series, with 5 1/3″ x 12″ nominal face and in the 8W series, with 8″ x 16″ nominal face, both in 2″, 4″, 6″, and 8″ widths. Some manufacturers produce a 4W series with 8″ x 8″ nominal face in 2″ and 4″ widths. Some available tile shapes include stretchers, bullnose jamb or corner, square jamb or corner, covered internal corner, recessed cove base, nonrecessed cove base, bullnose sill, square sill, and universal miter. Walls with openings and returns usually require partition layout drawings to establish quantities of special shapes. The different shapes of structural glazed facing tile are shown in Figure 12.7.

Building Stone

For reasons of durability and unique appearance, stone is well suited to a wide range of decorative applications as a building material. It can be installed in small units, referred to as "building stone," which can be assembled in many different systems, with or without mortar, to create walls and veneers of all sorts. Stone can also be employed as a material in larger units, such as stone panels, which are installed with elaborate anchor and framing systems and used as decorative wall facings in high-rise office and other commercial buildings. The cost of stone building materials varies considerably with location and depends on the available supply and the distance that it must be transported. Other significant cost factors include the extent of quarrying and subsequent processing required to extract and produce the finished product.

Building stone is available in random or precut sizes and shapes. Small irregular building stone that has been quarried in random sizes is called "rubble" or "fieldstone." This material, which is sold by the ton, is commonly installed with mortar. Fieldstone may be split by hand on-site to provide a flat exterior surface for a patterned wall or fireplace.

Small stone units may also be quarry-split and processed to meet aesthetic requirements. For example, decorative building stone can be purchased by the ton in 4″ thick slabs, available in lengths ranging from 6″ to 14″ and in heights ranging from 2″ to 16″. These pieces are commonly installed with mortar to create veneer walls of varying patterns, such as *ledge stone, spider web, uncoursed rectangular*, and *squared*. Ashlar stone, also priced by the ton, is building stone that has been sawn on the edges to produce a rectangular face. This shape makes ashlar stone another possible veneer material, because the pieces can be arranged in either a regular or random-coursed pattern within the face of a wall. Typical patterns are shown in Figure 12.8.

Stone veneer can be tied to the backup wall with galvanized ties or 8″ stone headers in a method similar to that used in brick veneer walls. The coverage of stone veneer ranges from 35 to 50 square feet per ton for 4″ wide veneer, with correspondingly reduced coverages per ton for veneers of 6″ and 8″ in width.

Stone Floors

Stone floors may be constructed from any type of stone that meets the durability standards and aesthetic requirements of the proposed location. Some commonly used stone flooring materials include slate, flagstone, granite, and

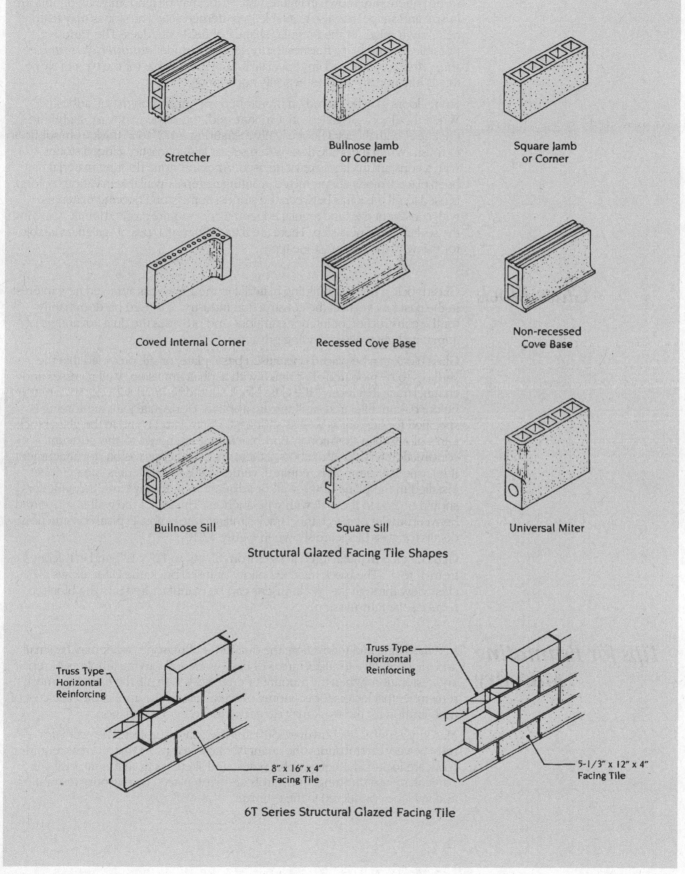

Stretcher

Bullnose Jamb
or Corner

Square Jamb
or Corner

Coved Internal Corner

Recessed Cove Base

Non-recessed
Cove Base

Bullnose Sill

Square Sill

Universal Miter

Structural Glazed Facing Tile Shapes

Truss Type
Horizontal
Reinforcing

Truss Type
Horizontal
Reinforcing

8" x 16" x 4"
Facing Tile

5-1/3" x 12" x 4"
Facing Tile

6T Series Structural Glazed Facing Tile

Figure 12.7

marble. Stone may be laid in a patterned or random design. Some popular stone patterns are shown in Figure 12.9. Stone may be randomly cut, or uniform in size and shape. It is also available in patterned sets. The stones may feature neat, sawn edges or the irregular shapes of field-cut edges. The various possible exposed-face finishes for the stones include: *natural cleft, sawn, sawn and polished*, and any specialized finish available for the type of stone that is functionally and aesthetically appropriate.

Stone floors may be placed on mortar beds or applied on mastic adhesive. When the floor is installed on a mortar bed, the stone may vary slightly in thickness, with the total floor usually measuring 1-1/2″ to 2″ thick from subfloor to finish. When the stone floor is thin set, or laid in mastic, gauged stones with a constant thickness must be used. After the stone flooring material has been placed, mortar, or premixed grouting material (available in various colors), is used to fill the joints between the stones. Some stone flooring materials with consistent sizes and regular edges may not require grouted joints. Applying the sealant is the next step. There are many different types of sealant available for use with the various stone types.

Glass Block

Glass block, a popular building material in the 1930s, has attracted new interest in the past few years in the construction industry. It is used predominantly for the construction of interior partitions and provides the dual advantage of admitting light while providing privacy.

Glass block can be placed on a raised base, plate, or sill, provided that the surfaces to be mortared are primed with asphalt emulsion. Wall recesses and channel track that receive the glass block should be lined with expansion strips before oakum filler and caulking are applied. Horizontal joint reinforcing is specified for flexural as well as shrinkage control and is laid in the glass block joints along with the mortar. End blocks are anchored to the adjacent construction with metal anchors, if there are no other provisions for attachment. If intermediate support is required, vertical I-shaped stiffeners can be either installed in the plane of the wall or adjacent to it. In any case, the stiffeners should be tied to the wall with wire anchors. The top of the wall is supported between angles or in a channel track similar to the jambs. Typical construction details for glass block are shown in Figure 12.10.

Glass block is manufactured in sizes from 6″ x 6″ to 12″ x 12″, and in thicknesses from 3″ to 4″. The block may be hollow or fused brick; the latter allows a clear view through the block. Inserts can be manufactured into the block to reduce solar transmission.

Tips for Estimating Masonry

In remodeling and renovation, the quantities of masonry work may be small and might involve localized areas of new work and varying amounts of repair and restoration. When the amount of new work is small, the estimator must remember that factors for economy of scale may not be applicable. The cost of mobilization for the work may exceed actual cost of installation.

Masonry *restoration*, however, can often be a large project. The estimator must be very careful during the quantity takeoff to ensure that all areas requiring work are included. Photographs, notes, and sketches from the site visit are particularly useful. Stringent controls are often placed on masonry restoration and must be considered in the estimate.

Rubble Stone—Random

Rubble Stone—Coursed

Ashlar Stone—Roughly Squared, Random

Ashlar Stone—Trimmed, Random

Ashlar Stone—Coursed, Narrow

Ashlar Stone—Coursed

Ashlar Stone—Cut, Stacked Joints

Ashlar Stone—Cut, Broken Joints

Patterns of Stonework

Figure 12.8

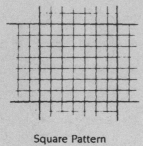

Square Pattern

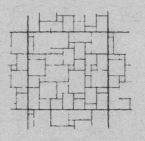

Random Rectangular Pattern

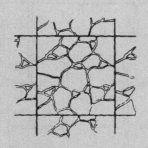

Random Irregular Pattern

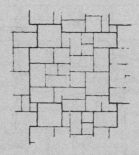

Random Repeat Pattern

Random Repeat Pattern

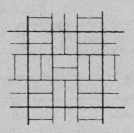

Ashlar Pattern

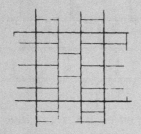

Basket Weave Pattern

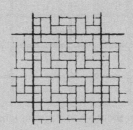

Herringbone Pattern

Figure 12.9

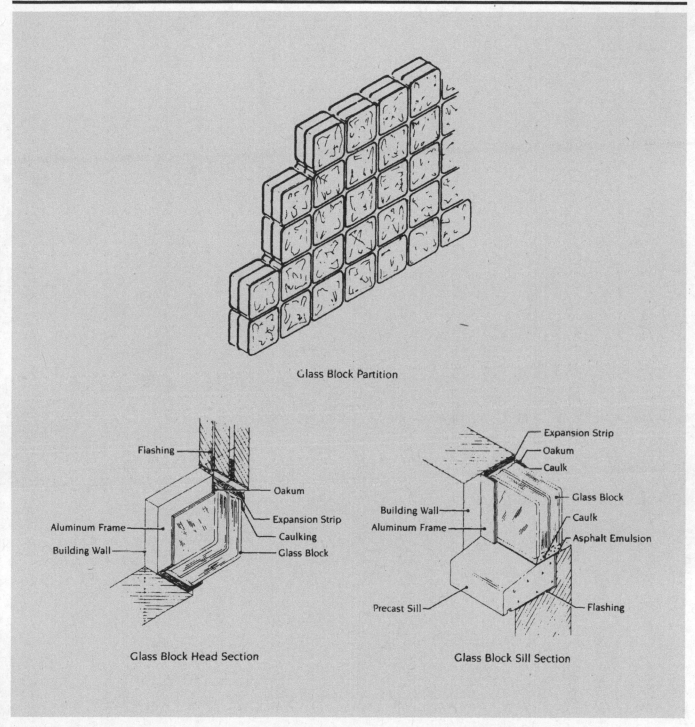

Glass Block Partition

Glass Block Head Section

Glass Block Sill Section

Figure 12.10

Chapter 13

Metals

The amount of structural steel used in remodeling projects is usually limited. Beams or plates may be employed to reinforce an existing floor system. Beams, channels, or angles may be used to frame a new opening or stairwell. The addition of a mezzanine floor might include more structural steel, requiring columns, beams, open-web joists, and steel deck. However, the most extensive use of metals in remodeling work is for miscellaneous support and ornamental purposes.

Structural Steel

The installation of structural steel into an existing building may be labor intensive. The steel may have to be moved into the building with dollies or handcarts, hoisted into position with chain falls or come-alongs, raised with a fork lift, or jacked from a scaffold. New connections are either welded (if fire codes permit) or drilled and bolted. Typical bolted-steel connections are shown in Figure 13.1. The installation process must be visualized and planned in order to properly estimate the costs. A "complexity factor" may be included for each piece of steel.

The structural steel required to repair, modify, and remodel buildings is often limited to small quantities of columns and beams purchased from local warehouses. Because warehouses sell structural steel shapes, such as wide flange, channel, square, and rectangular tubing, and round pipe sections in mill lengths, it is necessary to include the waste of the cut-off material in the estimate, as well as fabrication and cutting costs.

Temporary shoring, jacking, safety nets, work platforms, and needling of the steel into position are additional items to consider. The estimator must also keep in mind the requirements for fire lookout personnel, extinguishers, and fire curtains when cutting and welding torches are used for the dismantling or erection of structural steel.

When drawings are available for a proposed renovation project, they may include: special notes; lintel schedules; size, weight and location of each piece of steel; anchor bolt locations; special fabrication details; and special weldment. These must all be included on the Quantity Sheet.

When performing a structural steel takeoff, the estimator must consider the following: the grade of steel, method of connection, cleaning and painting (type and number of coats), and the number of pieces.

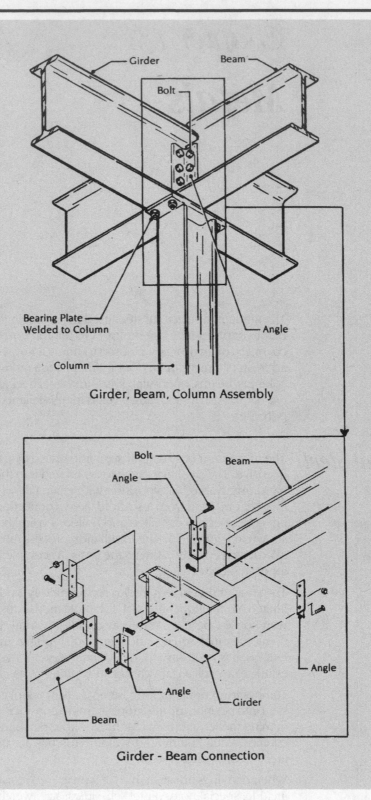

Girder, Beam, Column Assembly

Girder - Beam Connection

Figure 13.1

Remodeling and renovation often require "bringing a building up to code" to meet Building and Fire Code requirements. Such work most often includes fireproofing both the existing and new structural elements. Investigation of required fireproofing is essential. Figures 13.2a and 13.2b show surface and box areas of structural steel shapes to be used for estimating fireproofing, whether sprayed or drywall.

The estimator must also give consideration to warehouse availability, the type of fabrication required, and fabricator capability and availability. Of equal importance are delivery dates, sequence of deliveries, size of pieces to be delivered, storage area, and access.

Steel Joists and Deck

The estimator often must determine the size and type of required steel joists based on a visual inspection of existing joists. If specifications and drawings are available from the original construction, they will include type of joists, size and length, end bearing conditions, tie joist locations, bridging type and size, number and size of headers required, top and bottom chord extensions, and ceiling extensions.

The estimator must consider the various types and sizes of joists, special attachments, special paint and surface preparation, deliveries, site storage and access, and other variables that must be included in the estimate, but are not called out.

Existing floor and roof systems framed with steel joists are designed to support specified live and dead loads. When a given space is remodeled, the specified loads may be exceeded. Therefore, additional joists may be needed and must be taken off. The joists must be identified by type, size, and length. Special end bearings may also be required. Jacking or temporary shoring may be required to erect the joists.

In renovation, metal decking is usually found in floor or roof systems added to an existing building. Steel-framed mezzanines are an example of a system that normally would include a metal deck. Figure 13.3, from *Means Interior Cost Data* illustrates such a system. The type, depth, gauge, closures, finish, and attachment method of steel deck must be determined and listed.

Again, it is important to consider the quantity, availability, and delivery schedule of the types of deck specified, as well as site storage and access, special requirements, and miscellaneous items necessary to complete the work.

Miscellaneous and Ornamental Metals

Miscellaneous metals may be difficult to identify. The estimator should, therefore, carefully examine the existing conditions and add those items that are neither specified nor shown in sketches and drawings, but are necessary to complete the work. It is also important to determine the type of finish required and whether the metal is to be furnished and erected, or furnished only. When required for installation, the field measurements, shop drawings, method of installation, and equipment required to erect must all be noted and given consideration.

Miscellaneous and ornamental metals may be used extensively in remodeling projects for both decorative and functional purposes. The purchase, fabrication, and erection of these items is a specialized trade and a reliable sub-bid should be obtained for this portion of the work.

Stairs (type of construction, rails, nosing, number of risers and landings), railings (materials, type of construction, finishes, protection), gratings (types, sizes, finishes), and attachments must be thought of in the same way as miscellaneous metals.

Surface Areas and Box Areas
W Shapes
Square feet per foot of length

Designation	Case A	Case B	Case C	Case D
W 36x300	9.99	11.40	7.51	8.90
x280	9.95	11.30	7.47	8.85
x260	9.90	11.30	7.42	8.80
x245	9.87	11.20	7.39	8.77
x230	9.84	11.20	7.36	8.73
x210	8.91	9.93	7.13	8.15
x194	8.88	9.89	7.09	8.10
x182	8.85	9.85	7.06	8.07
x170	8.82	9.82	7.03	8.03
x160	8.79	9.79	7.00	8.00
x150	8.76	9.76	6.97	7.97
x135	8.71	9.70	6.92	7.92
W 33x241	9.42	10.70	7.02	8.34
x221	9.38	10.70	6.97	8.29
x201	9.33	10.60	6.93	8.24
x152	8.27	9.23	6.55	7.51
x141	8.23	9.19	6.51	7.47
x130	8.20	9.15	6.47	7.43
x118	8.15	9.11	6.43	7.39
W 30x211	8.71	9.97	6.42	7.67
x191	8.66	9.92	6.37	7.62
x173	8.62	9.87	6.32	7.57
x132	7.49	8.37	5.93	6.81
x124	7.47	8.34	5.90	6.78
x116	7.44	8.31	5.88	6.75
x108	7.41	8.28	5.84	6.72
x 99	7.37	8.25	5.81	6.68
W 27x178	7.95	9.12	5.81	6.98
x161	7.91	9.08	5.77	6.94
x146	7.87	9.03	5.73	6.89
x114	6.88	7.72	5.39	6.23
x102	6.85	7.68	5.35	6.18
x 94	6.82	7.65	5.32	6.15
x 84	6.78	7.61	5.28	6.11
W 24x162	7.22	8.30	5.25	6.33
x146	7.17	8.24	5.20	6.27
x131	7.12	8.19	5.15	6.22
x117	7.08	8.15	5.11	6.18
x104	7.04	8.11	5.07	6.14
x 94	6.16	6.92	4.81	5.56
x 84	6.12	6.87	4.77	5.52
x 76	6.09	6.84	4.74	5.49
x 68	6.06	6.80	4.70	5.45
x 62	5.57	6.16	4.54	5.13
x 55	5.54	6.13	4.51	5.10
W 21x147	6.61	7.66	4.72	5.76
x132	6.57	7.61	4.68	5.71
x122	6.54	7.57	4.65	5.68
x111	6.51	7.54	4.61	5.64
x101	6.48	7.50	4.58	5.61
x 93	5.54	6.24	4.31	5.01
x 83	5.50	6.20	4.27	4.96
x 73	5.47	6.16	4.23	4.92
x 68	5.45	6.14	4.21	4.90
x 62	5.42	6.11	4.19	4.87
x 57	5.01	5.56	4.06	4.60
x 50	4.97	5.51	4.02	4.56
x 44	4.94	5.48	3.99	4.53

Surface Areas and Box Areas
W Shapes
Square feet per foot of length

Designation	Case A	Case B	Case C	Case D
W 18x119	5.81	6.75	4.10	5.04
x106	5.77	6.70	4.06	4.99
x 97	5.74	6.67	4.03	4.96
x 86	5.70	6.62	3.99	4.91
x 76	5.67	6.59	3.95	4.87
x 71	4.85	5.48	3.71	4.35
x 65	4.82	5.46	3.69	4.32
x 60	4.80	5.43	3.67	4.30
x 55	4.78	5.41	3.65	4.27
x 50	4.76	5.38	3.62	4.25
x 46	4.41	4.91	3.52	4.02
x 40	4.38	4.88	3.48	3.99
x 35	4.34	4.84	3.45	3.95
W 16x100	5.28	6.15	3.70	4.57
x 89	5.24	6.10	3.66	4.52
x 77	5.19	6.05	3.61	4.47
x 67	5.16	6.01	3.57	4.43
x 57	4.39	4.98	3.33	3.93
x 50	4.36	4.95	3.30	3.89
x 45	4.33	4.92	3.27	3.86
x 40	4.31	4.89	3.25	3.83
x 36	4.28	4.87	3.23	3.81
x 31	3.92	4.39	3.11	3.57
x 26	3.89	4.35	3.07	3.53
W 14x730	7.61	9.10	5.23	6.72
x665	7.46	8.93	5.08	6.55
x605	7.32	8.77	4.94	6.39
x550	7.19	8.62	4.81	6.24
x500	7.07	8.49	4.68	6.10
x455	6.96	8.36	4.57	5.98
x426	6.89	8.28	4.50	5.89
x398	6.81	8.20	4.43	5.81
x370	6.74	8.12	4.36	5.73
x342	6.67	8.03	4.29	5.65
x311	6.59	7.94	4.21	5.56
x283	6.52	7.86	4.13	5.48
x257	6.45	7.78	4.06	5.40
x233	6.38	7.71	4.00	5.32
x211	6.32	7.64	3.94	5.25
x193	6.27	7.58	3.89	5.20
x176	6.22	7.53	3.84	5.15
x159	6.18	7.47	3.79	5.09
x145	6.14	7.43	3.76	5.05
x132	5.93	7.16	3.67	4.90
x120	5.90	7.12	3.64	4.86
x109	5.86	7.08	3.60	4.82
x 99	5.83	7.05	3.57	4.79
x 90	5.81	7.02	3.55	4.76
x 82	4.75	5.59	3.23	4.07
x 74	4.72	5.56	3.20	4.04
x 68	4.69	5.53	3.18	4.01
x 61	4.67	5.50	3.15	3.98
x 53	4.19	4.86	2.99	3.66
x 48	4.16	4.83	2.97	3.64
x 43	4.14	4.80	2.94	3.61
x 38	3.93	4.50	2.91	3.48
x 34	3.91	4.47	2.89	3.45
x 30	3.89	4.45	2.87	3.43
x 26	3.47	3.89	2.74	3.16
x 22	3.44	3.86	2.71	3.12

Figure 13.2a

Surface Areas and Box Areas W Shapes Square feet per foot of length				
Designation	Case A	Case B	Case C	Case D
W 12x336	5.77	6.88	3.92	5.03
x305	5.67	6.77	3.82	4.93
x279	5.59	6.68	3.74	4.83
x252	5.50	6.58	3.65	4.74
x230	5.43	6.51	3.58	4.66
x210	5.37	6.43	3.52	4.58
x190	5.30	6.36	3.45	4.51
x170	5.23	6.28	3.39	4.43
x152	5.17	6.21	3.33	4.37
x136	5.12	6.15	3.27	4.30
x120	5.06	6.09	3.21	4.24
x106	5.02	6.03	3.17	4.19
x 96	4.98	5.99	3.13	4.15
x 87	4.95	5.96	3.10	4.11
x 79	4.92	5.93	3.07	4.08
x 72	4.89	5.90	3.05	4.05
x 65	4.87	5.87	3.02	4.02
x 58	4.39	5.22	2.87	3.70
x 53	4.37	5.20	2.84	3.68
x 50	3.90	4.58	2.71	3.38
x 45	3.88	4.55	2.68	3.35
x 40	3.86	4.52	2.66	3.32
x 35	3.63	4.18	2.63	3.18
x 30	3.60	4.14	2.60	3.14
x 26	3.58	4.12	2.58	3.12
x 22	2.97	3.31	2.39	2.72
x 19	2.95	3.28	2.36	2.69
x 16	2.92	3.25	2.33	2.66
x 14	2.90	3.23	2.32	2.65
W 10x112	4.30	5.17	2.76	3.63
x100	4.25	5.11	2.71	3.57
x 88	4.20	5.06	2.66	3.52
x 77	4.15	5.00	2.62	3.47
x 68	4.12	4.96	2.58	3.42
x 60	4.08	4.92	2.54	3.38
x 54	4.06	4.89	2.52	3.35
x 49	4.04	4.87	2.50	3.33
x 45	3.56	4.23	2.35	3.02
x 39	3.53	4.19	2.32	2.98
x 33	3.49	4.16	2.29	2.95
x 30	3.10	3.59	2.23	2.71
x 26	3.08	3.56	2.20	2.68
x 22	3.05	3.53	2.17	2.65
x 19	2.63	2.96	2.04	2.38
x 17	2.60	2.94	2.02	2.35
x 15	2.58	2.92	2.00	2.33
x 12	2.56	2.89	1.98	2.31

Surface Areas and Box Areas W Shapes Square feet per foot of length				
Designation	Case A	Case B	Case C	Case D
W 8x 67	3.42	4.11	2.19	2.88
x 58	3.37	4.06	2.14	2.83
x 48	3.32	4.00	2.09	2.77
x 40	3.28	3.95	2.05	2.72
x 35	3.25	3.92	2.02	2.69
x 31	3.23	3.89	2.00	2.67
x 28	2.87	3.42	1.89	2.43
x 24	2.85	3.39	1.86	2.40
x 21	2.61	3.05	1.82	2.26
x 18	2.59	3.03	1.79	2.23
x 15	2.27	2.61	1.69	2.02
x 13	2.25	2.58	1.67	2.00
x 10	2.23	2.56	1.64	1.97
W 6x 25	2.49	3.00	1.57	2.08
x 20	2.46	2.96	1.54	2.04
x 15	2.42	2.92	1.50	2.00
x 16	1.98	2.31	1.38	1.72
x 12	1.93	2.26	1.34	1.67
x 9	1.90	2.23	1.31	1.64
W 5x 19	2.04	2.45	1.28	1.70
x 16	2.01	2.43	1.25	1.67
W 4x 13	1.63	1.96	1.03	1.37

Figure 13.2b

B10 Superstructure

B1010 Floor Construction

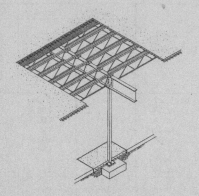

Mezzanine addition to existing building includes: Column footings; steel columns; structural steel; open web steel joists; uncoated 28 ga. steel slab forms; 2-1/2" concrete slab reinforced with welded wire fabric; steel trowel finish.

Design assumptions:

Structural steel is A36, high strength bolted. Slab form is 28 gauge, galvanized.
WWF 6 x 6 #10/#10
Conc. slab f'c = 3 ksi

System Components	QUANTITY	UNIT	COST EACH MAT.	COST EACH INST.	COST EACH TOTAL
SYSTEM B1010 310 0142					
MEZZANINE ADDITION TO EXISTING BUILDING; 100 PSF SUPERIMPOSED LOAD,					
SLAB FORM DECK, MTL. JOISTS, 2.5" CONC. SLAB, 3,000 PSI, 3,000 S.F.					
Saw cut existing slab	4.000	L.F.	426.24	1,163.52	1,589.76
Cut out existing slab	192.000	C.Y.		1,847.04	1,847.04
Remove concrete debris	9.060	C.Y.		221.97	221.97
Excavate by hand	7.104	C.Y.		518.59	518.59
Backfill by hand	1.000	C.Y.		21	21
Compaction of backfill	3.732	C.Y.		52.81	52.81
Forms for footing	144.000	C.Y.	414.72	1,965.60	2,380.32
Reinforcing bar for footing	180.000	C.Y.	57.60	79.20	136.80
Anchor bolts, 12"	24.000	Ea.	49.68	128.40	178.08
Concrete, 3000 psi, footing	5.592	C.Y.	424.99		424.99
Place concrete in footing, direct chute	5.592	C.Y.		192.48	192.48
Premolded, bituminous fiber, 1/2" x 6"	192.000	L.F.	80.64	192	272.64
Reinforcing mesh for slab	1.920	Cwt.	14.78	50.88	65.66
Concrete, 3500 psi, slab	2.784	C.Y.	217.15		217.15
Place concrete slab, direct chute	2.784	C.Y.		32.07	32.07
Machine trowel slab	192.000	S.F.		119.04	119.04
Grout for leveling plate	12.000	C.Y.	402	116.40	518.40
Column, 4" x 4" x 1/4" x12'	12.000	Ea.	1,332	990	2,322
Structural steel W 21x50	80.000	L.F.	4,440	720	5,160
W 16x31	80.000	L.F.	2,760	637.20	3,397.20
Open web joists, H or K series	6.800	Ton	10,200	5,528.40	15,728.40
Slab form, steel 28 gauge, galvanized	30.000	C.S.F.	1,980	1,560	3,540
Concrete 3,000 psi, 2 1/2" slab, incl. premium delv. chg.	23.500	C.Y.	2,679		2,679
Welded wire fabric 6x6 #10/#10 (w1.4/w1.4)	30.000	C.S.F.	231	795	1,026
Place concrete	23.500	C.Y.		835.43	835.43
Monolithic steel trowel finish	30.000	C.S.F.		1,860	1,860
Curing with sprayed membrane curing compound	30.000	C.S.F.	160.50	184.50	345
TOTAL			25,870.30	19,811.53	45,681.83
COST PER S.F.			8.62	6.60	15.22

Figure 13.3

While making the takeoff, list equipment that can be used to erect fabricated materials. Erection costs cannot be determined without giving consideration to site storage, site access, type and size of equipment required, delivery schedules, and erection sequence. This is one area where the estimator must pay particular attention to minimum labor and equipment charges.

The following is a list of some of the items normally furnished and delivered by a miscellaneous metals supplier.

- Elevator shaft beam separators
- Angle sills with welded-on anchors
- Angle corner-guards with welded-on anchors
- Pipe bollards with welded-on base-plate anchors
- Cast iron drain grates with frames
- Individual aluminum or steel sleeves for pipe or tube rails
- Cast or extruded abrasive metal nosings for concrete steps
- Templated sleeves welded on a steel flat for continuous pipe or tube-guardrails at balconies or roof
- Transformer vault door frames
- Malleable iron wedge inserts for attached or hung lintels
- Angle frames with welded-on anchors
- Elevator machine-room double-leaf aluminum floor hatches, with compensating hinges
- Elevator machine-room ceiling hoist monorail beams
- Roof scuttles
- Stainless steel sleeves for swimming pool ladders and guardrails
- Slab inserts for toilet partitions, operating room lights, and x-ray room ceiling supports
- Loose lintels, 12″ longer than the net opening
- Steel stairs, complete as shown on drawing, including abrasive extruded nosings, if any, but excluding concrete fill or terrazzo
- Elevator shaft sill angles mounted on inserts, built-in channel door jambs
- All gratings, including any support angles bolted to masonry or concrete or connected to inserts built in by others
- All open-riser ship's or engineer's ladders, with diamond plate or grating treads
- Interior ladders to roof scuttles and under hatches
- Exterior ladders with goosenecks from low to high roofs, with or without safety cages as per drawings
- Steel bench supports
- Tube or pipe guardrails in steel, nonferrous or stainless steel at balconies, roofs, and elsewhere as per drawings, including at interior concrete stairs
- Hung or attached angle lintels for brick supports connected to inserts built in by others
- Catwalks, strutted or suspended, complete with grating walkways, guardrails, and ladder accesses
- Spiral metal angle bases at wooden floor accesses, auditorium stage, book-stack accesses, mezzanines, etc.
- Sheet-metal angle bases in wooden floor rooms, such as gyms, interior racquetball, squash, wrestling stages, and prosceniums
- Wall handrails at ramps, places other than stairs, hospital hallways, etc.

- Stainless steel swimming pool ladders and pipe guardrails at bleachers
- Toilet partition supports in "Unistrut"
- X-ray machine supports in "Unistrut" or angles
- Operating and autopsy room light "spider-leg" supports
- Monitor supports throughout a hospital
- Hospital linear accelerator supports
- Rolling and fire partitions supports
- Computer room floor supports
- Proscenium grillages
- Acoustic baffle cloud supports
- Motor supports for overhead doors
- Entrance door and other door supports
- Welding of inserts on decking sheets for uses underneath, for ceiling supports
- Projection booth counterweighted port doors for fire protection
- Exterior door saddles, with or without Rixsons
- Interior floor transition door slip-saddles
- Exterior door combination slip-saddles
- Floor, wall, and ceiling nonferrous expansion joints
- Folding partition supports
- Banquet hall movable partition supports
- Steel or aluminum louvres not in contact with any ductwork
- Ornamental metals for glass railings
- Ornamental metals for composite acrylic/wood railings
- Ornamental metals for combination panels and railings
- Ornamental metals for balusters, posts, trillage, and scroll railings
- Nonferrous expansion joint covers
- Nonferrous door saddles

Some of the items listed above are illustrated in Figure 13.4. The list of ornamental items included in any interior estimate may be extensive, and the takeoff and pricing may be time consuming. A reliable subcontractor should be contacted and a proper quotation requested.

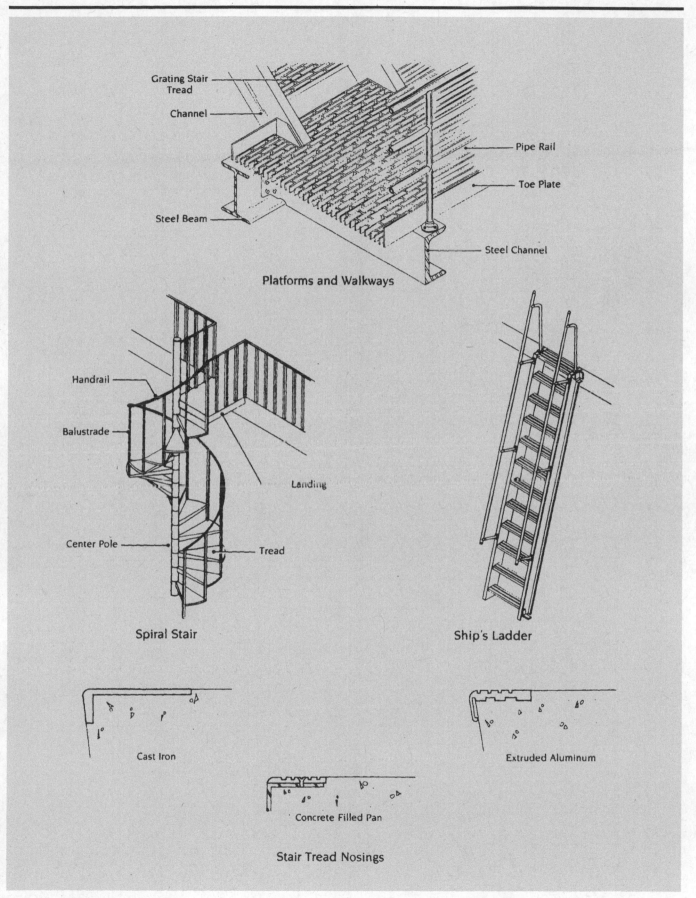

Platforms and Walkways

Grating Stair Tread
Channel
Pipe Rail
Toe Plate
Steel Beam
Steel Channel

Spiral Stair

Handrail
Balustrade
Center Pole
Landing
Tread

Ship's Ladder

Cast Iron

Extruded Aluminum

Concrete Filled Pan

Stair Tread Nosings

Figure 13.4

Chapter 14

Wood and Plastics

Carpentry work can be broken down into the following categories: rough carpentry, finish carpentry and millwork, and laminated framing and decking. The material prices for carpentry work fluctuate more widely and with greater frequency than is the case with most other building materials. For this reason, when the material list is complete, it is important to obtain *current, local* prices for the lumber. Installation costs greatly depend on productivity. Accurate cost records from past jobs may be most helpful. Since lumber tends to be used extensively in remodeling projects, a careful estimate of this component is essential.

Rough Carpentry

Rough carpentry is most often required in projects where there is extensive renovation, or where the location of partitions, windows, or doors is significantly changed. Lumber is commonly estimated in board feet and purchased in 1,000 board foot quantities. A board foot is the equivalent of 1″ x 12″ x 12″ (nominal) or 3/4″ x 11-1/2″ x 12″ milled (actual). To determine the board feet of a piece of framing, the nominal dimensions can be multiplied, and the result divided by 12. The final result represents the number of board feet per linear foot of that framing size.

Example:

> 2 x 10 joists
>
> 2 x 10 = 20
>
> 20/12 = 1.67 B.F./L.F.

The Quantity Sheet for lumber should indicate species, grade, and any type of wood preservative or fire retardant treatment specified or required by code. Floor joists, shown or specified by size and spacing, should be taken off by nominal length and the quantity required. Add for double joists under partitions, headers and cripple joists at openings, overhangs, laps at bearings, and blocking or bridging.

Studs required are noted on the drawings by spacing, usually 16″ on center (O.C.) or 24″ O.C., with the stud size given. The linear feet of like partitions (having the same stud size, height, and spacing) divided by the spacing will give the estimator the approximate number of studs required. Additional studs for openings, corners, double top plates, sole plates, and intersecting partitions must be taken off separately. An allowance for waste should be included, or heights should be recorded as a standard purchased length, for example, 8', 10', 12', etc. One "rule of thumb" is to allow one stud for every linear foot of wall, for 16" O.C. spacing.

Number and size of openings are important takeoff information. Even though there are no studs in these areas, the estimator must take off headers, subsills, king studs, trimmers, cripples, and knee studs. Where bracing and fire blocking are noted, both the type and quantity should be indicated.

Tongue-and-groove decks of various woods, solid planks, or laminated construction are nominally 2" to 4" thick, and are often used with glued laminated beams or heavy timber framing. The square foot method is used to determine quantities and consideration given to nonmodular areas for the amount of waste involved. The materials are purchased by board-foot measurement. The conversion from square feet to board feet must allow for *net* sizes, as opposed to *board measure*. In this way, loss of coverage due to the tongue and available mill lengths can be taken into account.

Sheathing on walls can be plywood of different grades and thicknesses, particle board, or solid boards nailed directly to the studs. Plywood may be applied with the grain vertical, horizontal, or rarely, diagonal to the studding. Solid boards are usually nailed diagonally, but can be applied horizontally when lateral forces are not present. For solid-board sheathing, add 15% to 20% more material to the takeoff when using tongue-and-groove (as opposed to square-edge) sheathing. Plywood or particle-board sheathing can be installed either horizontally or vertically, depending upon wall height and Fire Code restrictions. When estimating quantities of panel sheathing, the estimator calculates the number of sheets required by measuring the square feet of area to be covered, adding waste, and then dividing by sheet size. Applying these materials diagonally or on nonmodular areas creates additional waste. This waste factor must be included in the estimate. For diagonal application of boards, plywood, or wallboard, include an additional 10% to 15% material waste factor.

Subfloors can be CDX-type plywood (with the thickness dependent on the load and span), solid boards laid diagonally or perpendicular to the joists, or tongue-and-groove planks. The quantity takeoff for subfloors is similar to that for sheathing (previously noted).

Grounds are normally 1" x 2" wood strips used for case work or plaster; the quantities are estimated in L.F.

Furring (1" x 2" or 3") wood strips are fastened to wood, masonry, or concrete walls, so that wall coverings may be attached thereto. Furring may also be used on the underside of ceiling joists to fasten ceiling finishes. Quantities are estimated by L.F.

Framing members (studs or joists) are measured in linear feet. The quantity required is based on square feet of surface area (wall, floor, ceiling). Use the table below to estimate rough quantities of basic framing members. Supplemental framing, such as plates, sills, headers and bands, must be added separately.

Spacing of Framing Members	Linear Feet per Square Foot Surface
12" O.C.	1.2 L.F./S.F.
16" O.C.	1.0 L.F./S.F.
24" O.C.	0.8 L.F./S.F

Additional requirements for rough carpentry, especially those for temporary construction, may not all be directly stated in the plans and specifications. Such items may include blocking, temporary stairs, wood inserts for metal-pan stairs, and railings, along with various other requirements for different trades. Temporary construction may also be included in Division 1–General Requirements.

Typical framing installations are shown in Figures 14.1a, 14.1b and 14.1c.

Laminated Construction

Laminated construction should be listed separately, as it is frequently supplied by a specialty subcontractor. Sometimes the beams are supplied and erected by one subcontractor, and the decking installed by the general contractor or another subcontractor. The takeoff units must be adapted to the system: *square foot* for floors, *linear foot* for members, or *board foot* for lumber. Since the members are factory-fabricated, the plans and specifications must be submitted to a fabricator for takeoff and pricing. Some examples of laminated construction are shown in Figure 14.2.

Finish Carpentry and Millwork

Finish carpentry and millwork—wood rails, paneling, shelves, casements, and cabinetry—are common features, even in buildings that might have no other wood. Upon examination of the plans and specifications, the estimator must determine which items will be built on-site, and which will be fabricated off-site by a millwork subcontractor. Shop drawings are often required for architectural woodwork and are usually included in the subcontract price.

Moldings and door trim may be taken off and priced by the "set" or by the linear foot. The common use of prehung doors makes it convenient to take off this trim with the doors. Exterior trim, other than door and window trim, should be taken off with the siding, because the details and dimensions are interrelated.

Paneling is taken off by type, finish, and square foot (converted to full sheets). Be sure to list any millwork that would show up on the details. Panel siding and associated trim are taken off by the square foot and linear foot, respectively. Be sure to provide an allowance for waste.

Decorative beams and columns that are nonstructural should be estimated separately. Decorative trim may be used to wrap exposed structural elements. Particular attention should be paid to the joinery. Long, precise joints are difficult (and expensive) to construct in the field.

Cabinets, counters, and shelves are most often priced by the linear foot or by the unit. Job fabricated, prefabricated, and subcontracted work should be estimated separately.

Stairs should be estimated by individual components unless accurate, complete system costs have been developed from previous projects. Typical components and units for estimating are shown in Figure 14.3.

General Rule

The following is a general rule for budgeting millwork: Total costs may be two to three times the cost of the materials. Millwork is often ordered and purchased directly by the owner. When installation is the responsibility of the contractor, costs for handling, storage, and protection, as well as those for installation, should be included. Typical finish carpentry and millwork items are illustrated in Figures 14.4a and 14.4b.

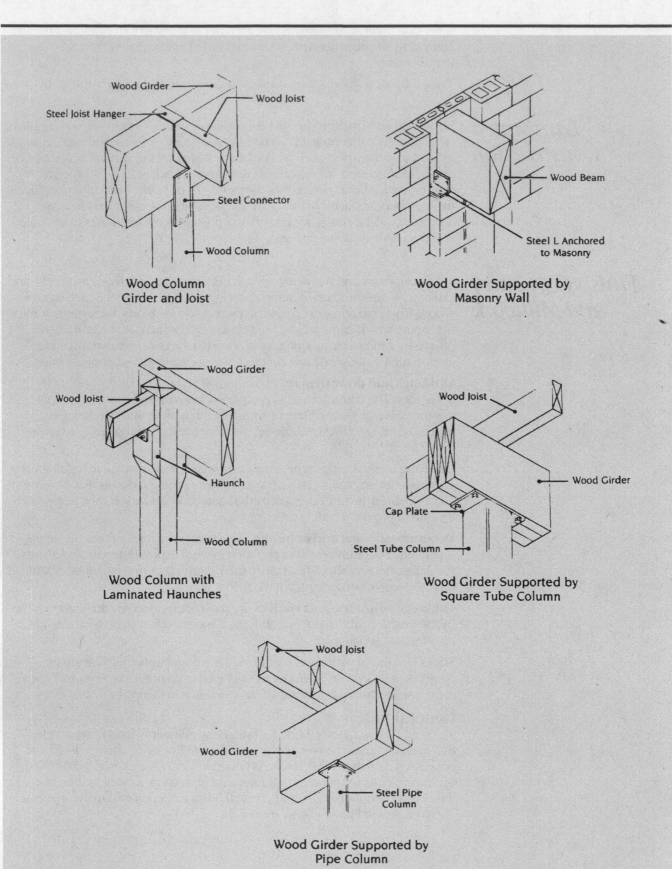

Wood Girder

Wood Joist

Steel Joist Hanger

Steel Connector

Wood Column

**Wood Column
Girder and Joist**

Wood Beam

Steel L Anchored
to Masonry

**Wood Girder Supported by
Masonry Wall**

Wood Girder

Wood Joist

Haunch

Wood Column

**Wood Column with
Laminated Haunches**

Wood Joist

Wood Girder

Cap Plate

Steel Tube Column

**Wood Girder Supported by
Square Tube Column**

Wood Joist

Wood Girder

Steel Pipe
Column

**Wood Girder Supported by
Pipe Column**

Figure 14.1a

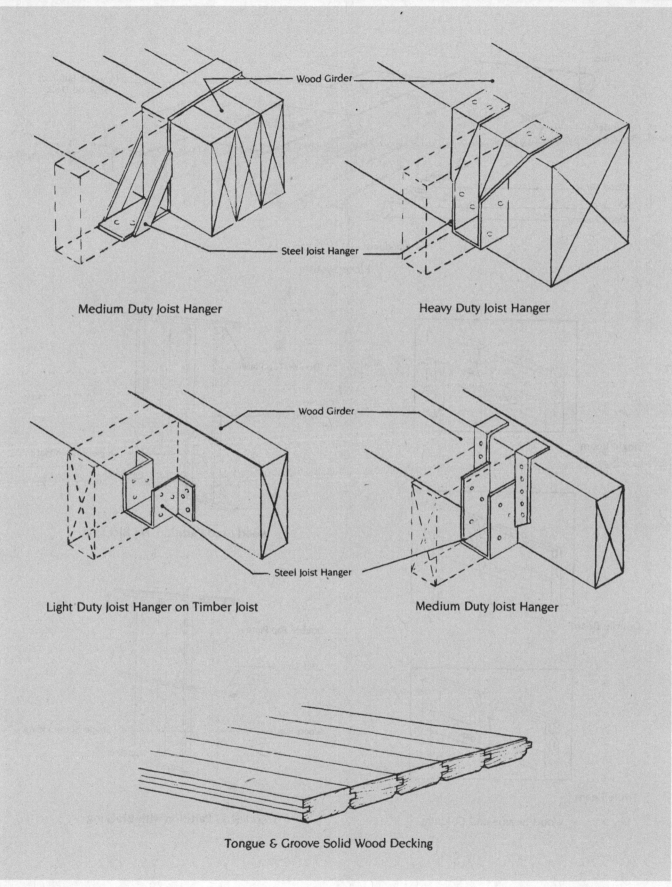

Medium Duty Joist Hanger

Heavy Duty Joist Hanger

Light Duty Joist Hanger on Timber Joist

Medium Duty Joist Hanger

Wood Girder

Steel Joist Hanger

Wood Girder

Steel Joist Hanger

Tongue & Groove Solid Wood Decking

Figure 14.1b

137

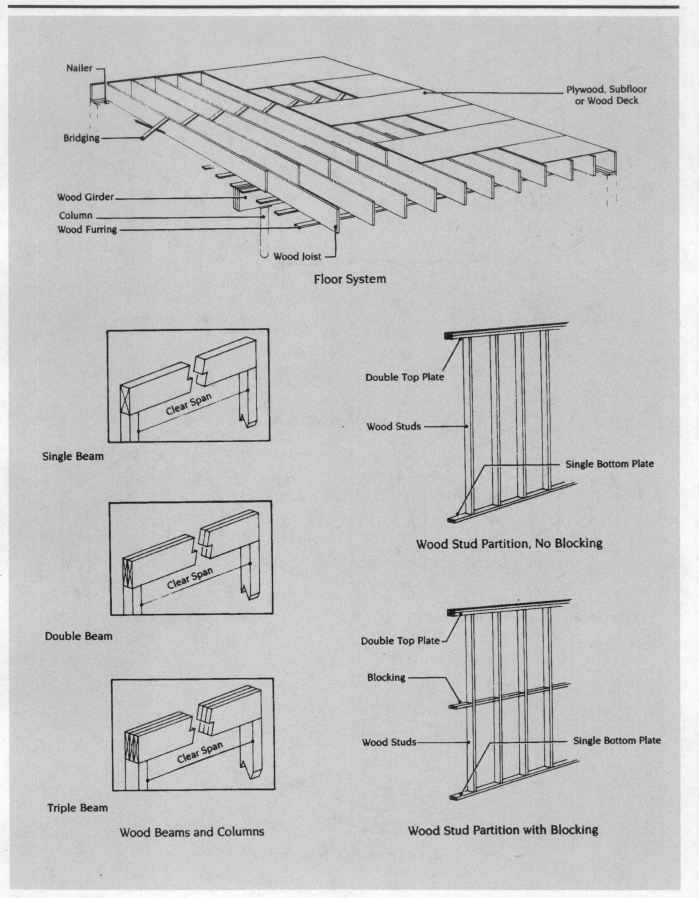

Floor System

Nailer

Bridging

Wood Girder

Column

Wood Furring

Wood Joist

Plywood, Subfloor or Wood Deck

Single Beam

Clear Span

Double Beam

Clear Span

Triple Beam

Clear Span

Wood Beams and Columns

Double Top Plate

Wood Studs

Single Bottom Plate

Wood Stud Partition, No Blocking

Double Top Plate

Blocking

Wood Studs

Single Bottom Plate

Wood Stud Partition with Blocking

Figure 14.1c

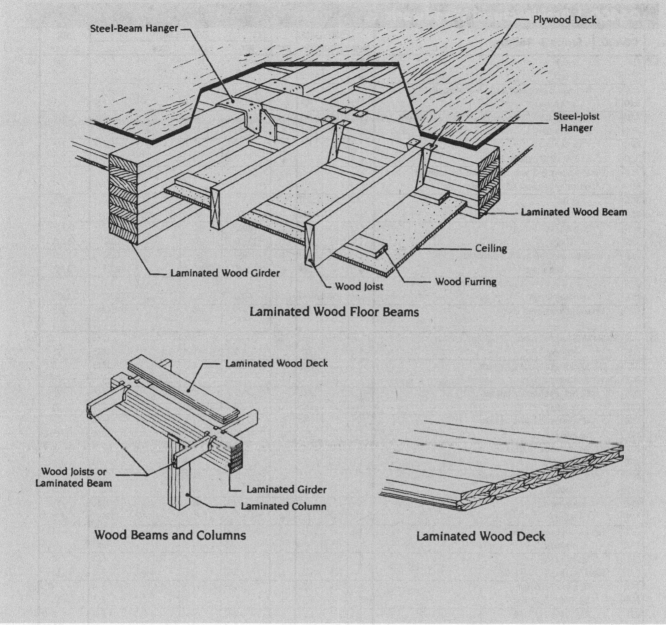

Laminated Wood Floor Beams

Steel-Beam Hanger

Plywood Deck

Steel-Joist Hanger

Laminated Wood Beam

Ceiling

Wood Furring

Wood Joist

Laminated Wood Girder

Wood Beams and Columns

Laminated Wood Deck

Wood Joists or Laminated Beam

Laminated Girder

Laminated Column

Laminated Wood Deck

Figure 14.2

		06430	**Stairs & Railings**	CREW	DAILY OUTPUT	LABOR-HOURS	UNIT	2002 BARE COSTS				TOTAL INCL O&P	
								MAT.	LABOR	EQUIP.	TOTAL		
620	2200		6' high	2 Carp	3.50	4.571	Flight	555	137		692	840	620
	2300		8' high	↓	3	5.333	↓	730	160		890	1,075	
	2500		For prefab. 3 piece wood railings & balusters, add for										
	2600		3' high stairs	2 Carp	15	1.067	Ea.	31.50	32		63.50	87.50	
	2700		4' high stairs		14	1.143		51	34.50		85.50	114	
	2800		6' high stairs		13	1.231		63	37		100	131	
	2900		8' high stairs		12	1.333		96.50	40		136.50	173	
	3100		For 3'-6" x 3'-6" platform, add	↓	4	4	↓	72.50	120		192.50	279	
	3300		Curved stairways, 3'-3" wide, prefabricated, oak, unfinished,										
	3310		incl. curved balustrade system, open one side										
	3400		9' high	2 Carp	.70	22.857	Flight	6,375	685		7,060	8,150	
	3500		10' high		.70	22.857		7,200	685		7,885	9,050	
	3700		Open two sides, 9' high		.50	32		10,000	960		10,960	12,600	
	3800		10' high		.50	32		10,800	960		11,760	13,500	
	4000		Residential, wood, oak treads, prefabricated		1.50	10.667		930	320		1,250	1,550	
	4200		Built in place	↓	.44	36.364	↓	1,325	1,100		2,425	3,250	
	4400		Spiral, oak, 4'-6" diameter, unfinished, prefabricated,										
	4500		incl. railing, 9' high	2 Carp	1.50	10.667	Flight	4,000	320		4,320	4,925	
	9000		Minimum labor/equipment charge	"	3	5.333	Job		160		160	265	
630	0010		**STAIR PARTS** Balusters, turned, 30" high, pine, minimum	1 Carp	28	.286	Ea.	4.11	8.55		12.66	18.70	630
	0100		Maximum		26	.308	"	9	9.25		18.25	25	
	0300		30" high birch balusters, minimum		28	.286	Ea.	6.30	8.55		14.85	21	
	0400		Maximum		26	.308		10.20	9.25		19.45	26.50	
	0600		42" high, pine balusters, minimum		27	.296		6	8.90		14.90	21.50	
	0700		Maximum		25	.320		13	9.60		22.60	30	
	0900		42" high birch balusters, minimum		27	.296		7.65	8.90		16.55	23	
	1000		Maximum		25	.320	↓	27	9.60		36.60	45.50	
	1050		Baluster, stock pine, 1-1/16" x 1-1/16"		240	.033	L.F.	1.98	1		2.98	3.84	
	1100		1-5/8" x 1-5/8"		220	.036	"	2.19	1.09		3.28	4.22	
	1200		Newels, 3-1/4" wide, starting, minimum		7	1.143	Ea.	33.50	34.50		68	94	
	1300		Maximum		6	1.333		126	40		166	206	
	1500		Landing, minimum		5	1.600		73.50	48		121.50	161	
	1600		Maximum		4	2	↓	179	60		239	296	
	1800		Railings, oak, built-up, minimum		60	.133	L.F.	6.50	4		10.50	13.80	
	1900		Maximum		55	.145		15.75	4.36		20.11	24.50	
	2100		Add for sub rail		110	.073		4.21	2.18		6.39	8.25	
	2300		Risers, beech, 3/4" x 7-1/2" high		64	.125		5.80	3.75		9.55	12.55	
	2400		Fir, 3/4" x 7-1/2" high		64	.125		1.59	3.75		5.34	7.95	
	2600		Oak, 3/4" x 7-1/2" high		64	.125		5	3.75		8.75	11.70	
	2800		Pine, 3/4" x 7-1/2" high		66	.121		2.80	3.64		6.44	9.10	
	2850		Skirt board, pine, 1" x 10"		55	.145		3.15	4.36		7.51	10.70	
	2900		1" x 12"		52	.154	↓	3.85	4.62		8.47	11.90	
	3000		Treads, 1-1/16" x 9-1/2" wide, 3' long, oak		18	.444	Ea.	23	13.35		36.35	47.50	
	3100		4' long, oak		17	.471		29	14.10		43.10	55.50	
	3300		1-1/16" x 11-1/2" wide, 3' long, oak		18	.444		23.50	13.35		36.85	48	
	3400		6' long, oak	↓	14	.571	↓	56	17.15		73.15	90	
	3600		Beech treads, add					40%					
	3800		For mitered return nosings, add				L.F.	8.40			8.40	9.25	
	9000		Minimum labor/equipment charge	1 Carp	3	2.667	Job		80		80	133	

				CREW	DAILY OUTPUT	LABOR-HOURS	UNIT	MAT.	LABOR	EQUIP.	TOTAL	TOTAL INCL O&P	
150	0010		**BEAMS, DECORATIVE** Rough sawn cedar, non-load bearing, 4" x 4"	2 Carp	180	.089	L.F.	1.30	2.67		3.97	5.85	150
	0100		4" x 6"		170	.094		2.50	2.82		5.32	7.45	
	0200		4" x 8"		160	.100		3.21	3		6.21	8.50	
	0300		4" x 10"	↓	150	.107	↓	4.46	3.20		7.66	10.20	

Figure 14.3

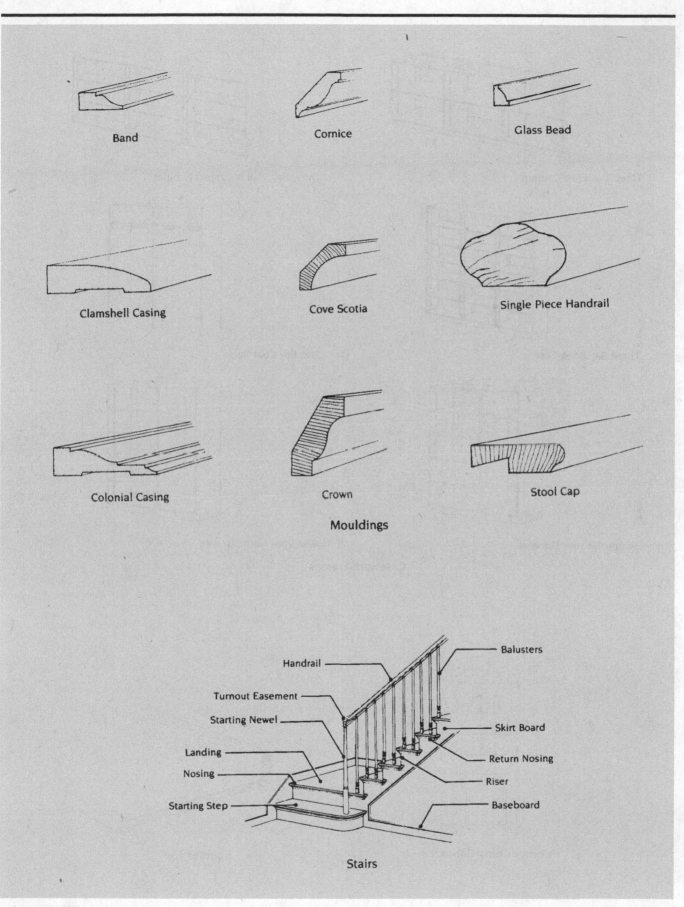

Band

Cornice

Glass Bead

Clamshell Casing

Cove Scotia

Single Piece Handrail

Colonial Casing

Crown

Stool Cap

Mouldings

Handrail

Balusters

Turnout Easement

Starting Newel

Skirt Board

Landing

Return Nosing

Nosing

Riser

Starting Step

Baseboard

Stairs

Figure 14.4a

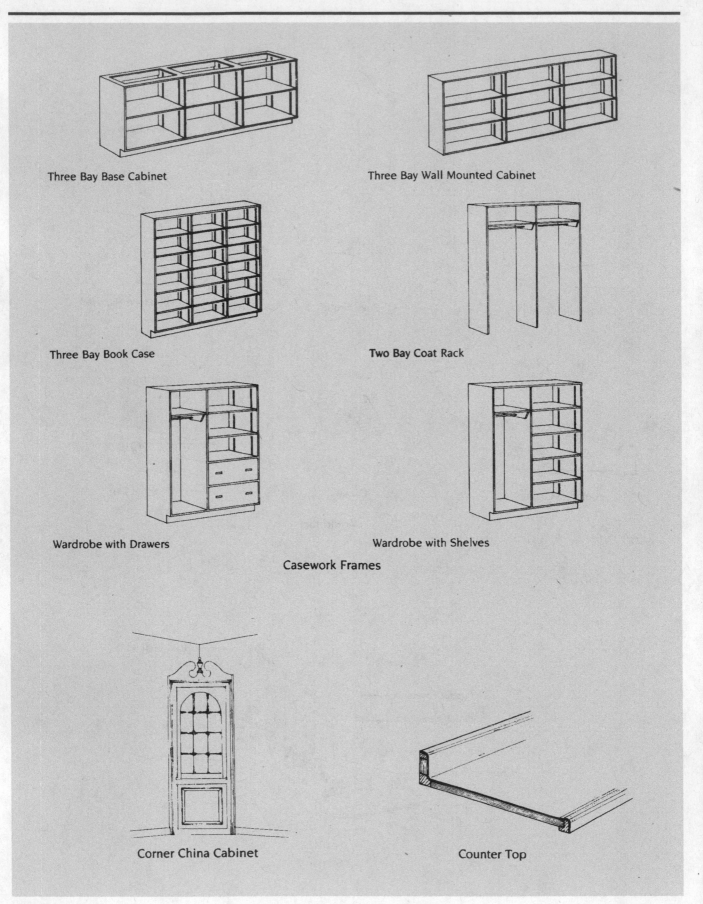

Three Bay Base Cabinet

Three Bay Wall Mounted Cabinet

Three Bay Book Case

Two Bay Coat Rack

Wardrobe with Drawers

Wardrobe with Shelves

Casework Frames

Corner China Cabinet

Counter Top

Figure 14.4b

Chapter 15

Thermal and Moisture Protection

This division includes materials for sealing the outside of a building for protection against moisture and air infiltration, as well as for insulation and associated accessories. When reviewing the plans and specifications and inspecting the site, the estimator should visualize the construction process, and in this way determine all probable areas where these materials will be required. The technique used for quantity takeoff depends on the specific materials and installation methods.

Waterproofing

- Dampproofing
- Caulking and Sealants
- Sheet and Membrane
- Integral Cement Coatings

A distinction should be made between dampproofing and waterproofing. Dampproofing is used to inhibit the migration of moisture or water vapor. In most cases, dampproofing will not stop the flow of water (even at minimal pressures). Waterproofing, on the other hand, consists of a continuous, impermeable membrane and is used to prevent or stop the flow of water.

Dampproofing

Dampproofing usually consists of one or two bituminous coatings applied to foundation walls from the bottom of the footings to approximately the finished grade line. The areas involved are calculated based on the total height of the dampproofing and the length of the wall. After separate areas are figured and added together to provide a total square foot area, a unit cost per square foot can be selected for the type of material, the number of coats, and the method of application specified for the building.

Waterproofing Below Grade

Waterproofing at or below grade with elastomeric sheets or membranes is estimated on the same basis as dampproofing, with two basic exceptions. First, the installed unit costs for the elastomeric sheets do not include bonding adhesive or splicing tape, which must be figured as an additional cost. Second, the membrane waterproofing under slabs must be estimated separately from the higher cost installation on walls. In all cases, unit costs are per square foot of covered surface.

For walls below grade, protection board is often specified to prevent damage to the waterproofing membrane when the excavation is backfilled. Rigid foam insulation installed outside of the barrier may also have a protective function. Metallic coating material may be applied to floors or walls, usually on the interior or dry side, after the masonry surface has been prepared (usually by chipping) for bonding to the new material. The unit cost per square foot for these materials depends on the thickness of the material, the orientation of the area to be covered, and the preparation required. In many cases, these materials must be applied in locations where access is difficult and under the control of others. The estimator should make an allowance for delays caused by this problem.

Caulking and Sealants

Caulking and sealants are most often required on the exterior of the building and for certain conditions on the interior. In most cases, caulking and sealing are done to prevent water and/or air from entering a building. Caulking and sealing are usually specified at joints, expansion joints, control joints, door and window frames, and in places where dissimilar materials meet over the surface of the building exterior. To estimate the installed cost of this type of material, two things must be determined. First, the estimator must note (from the specifications) the kind of material to be used for each caulking or sealing job. Second, the size of the joints to be caulked or sealed must be measured, with attention given to any requirements for backer rods. With this information, the estimator can select the applicable cost per linear foot and multiply it by the total length in feet. The result is an estimated cost for each kind of caulking or sealing on the job. Caulking and sealing may often be overlooked as incidental items. They may, in fact, represent a significant cost, depending on the type of construction and the quantity.

Insulation

- Batt or Roll
- Board (Rigid and Semi–rigid)
- Cavity Masonry
- Perimeter Foundation
- Poured in Place
- Reflective
- Roof
- Sprayed

Insulation is primarily used to reduce heat transfer through the exterior enclosure of the building. The choice of insulation type and form varies according to its location in the structure and the size of the space it occupies. Major categories of insulation include mineral granules, fibers and foams, vegetable fibers and solids, and plastic foams. These materials may be required around foundations, or on inside walls, and under roofing. Many different details of the drawings must be examined in order to determine types, methods, and quantities of insulation. The cost of insulation depends on the type of material, its form (loose, granular, batt or boards), its thickness in inches, the method of installation, and the total area in square feet.

Specifying by "R" Value

It is becoming popular to specify insulation by "R" value only. The estimator may have a choice of materials, given a required "R" value and a certain cavity space (which may dictate insulation thickness). For example, the specifications may require an "R" value of 11, and only 2″ of wall cavity is available for the thickness of the insulation. From Figure 15.1, it is seen that only isocyanurate (line 07210-900-1660) meets the design criteria. Note that if more cavity space were available, 3-1/2″ nonrigid fiberglass (line 07210-950-0420) would be a much less expensive alternative. The estimator may have to do some comparison

07210 | Building Insulation

		CREW	DAILY OUTPUT	LABOR-HOURS	UNIT	MAT.	LABOR	EQUIP.	TOTAL	TOTAL INCL O&P	
900											**900**
1610	1/2" thick, R3.9	1 Carp	800	.010	S.F.	.29	.30		.59	.82	
1620	5/8" thick, R4.5		800	.010		.30	.30		.60	.83	
1630	3/4" thick, R5.4		800	.010		.31	.30		.61	.84	
1640	1" thick, R7.2		800	.010		.34	.30		.64	.87	
1650	1-1/2" thick, R10.8		730	.011		.38	.33		.71	.96	
1660	2" thick, R14.4		730	.011		.47	.33		.80	1.06	
1670	3" thick, R21.6		730	.011		1.11	.33		1.44	1.76	
1680	4" thick, R28.8		730	.011		1.37	.33		1.70	2.05	
1700	Perlite, 1" thick, R2.77		800	.010		.26	.30		.56	.79	
1750	2" thick, R5.55		730	.011		.50	.33		.83	1.09	
1900	Extruded polystyrene, 25 PSI compressive strength, 1" thick, R5		800	.010		.34	.30		.64	.87	
1940	2" thick R10		730	.011		.67	.33		1	1.28	
1960	3" thick, R15		730	.011		.96	.33		1.29	1.60	
2100	Expanded polystyrene, 1" thick, R3.85		800	.010		.14	.30		.44	.65	
2120	2" thick, R7.69		730	.011		.38	.33		.71	.96	
2140	3" thick, R11.49		730	.011		.52	.33		.85	1.11	
9000	Minimum labor/equipment charge		4	2	Job		60		60	99.50	
950	0010	**WALL OR CEILING INSUL., NON-RIGID**									**950**
	0040	Fiberglass, kraft faced, batts or blankets									
	0060	3-1/2" thick, R11, 11" wide	1 Carp	1,150	.007	S.F.	.23	.21		.44	.60
	0080	15" wide		1,600	.005		.23	.15		.38	.50
	0100	23" wide		1,600	.005		.23	.15		.38	.50
	0140	6" thick, R19, 11" wide		1,000	.008		.33	.24		.57	.76
	0160	15" wide		1,350	.006		.33	.18		.51	.65
	0180	23" wide		1,600	.005		.33	.15		.48	.61
	0200	9" thick, R30, 15" wide		1,150	.007		.60	.21		.81	1.01
	0220	23" wide		1,350	.006		.60	.18		.78	.95
	0240	12" thick, R38, 15" wide		1,000	.008		.76	.24		1	1.24
	0260	23" wide		1,350	.006		.76	.18		.94	1.13
	0400	Fiberglass, foil faced, batts or blankets									
	0420	3-1/2" thick, R11, 15" wide	1 Carp	1,600	.005	S.F.	.34	.15		.49	.62
	0440	23" wide		1,600	.005		.34	.15		.49	.62
	0460	6" thick, R19, 15" wide		1,350	.006		.41	.18		.59	.74
	0480	23" wide		1,600	.005		.41	.15		.56	.70
	0500	9" thick, R30, 15" wide		1,150	.007		.71	.21		.92	1.13
	0550	23" wide		1,350	.006		.71	.18		.89	1.07
	0800	Fiberglass, unfaced, batts or blankets									
	0820	3-1/2" thick, R11, 15" wide	1 Carp	1,350	.006	S.F.	.21	.18		.39	.52
	0830	23" wide		1,600	.005		.21	.15		.36	.48
	0860	6" thick, R19, 15" wide		1,150	.007		.34	.21		.55	.72
	0880	23" wide		1,350	.006		.34	.18		.52	.66
	0900	9" thick, R30, 15" wide		1,000	.008		.60	.24		.84	1.06
	0920	23" wide		1,150	.007		.60	.21		.81	1.01
	0940	12" thick, R38, 15" wide		1,000	.008		.76	.24		1	1.24
	0960	23" wide		1,150	.007		.76	.21		.97	1.19
	1300	Mineral fiber batts, kraft faced									
	1320	3-1/2" thick, R12	1 Carp	1,600	.005	S.F.	.26	.15		.41	.54
	1340	6" thick, R19		1,600	.005		.39	.15		.54	.68
	1380	10" thick, R30		1,350	.006		.62	.18		.80	.97
	1850	Friction fit wire insulation supports, 16" O.C.		960	.008	Ea.	.05	.25		.30	.47
	1900	For foil backing, add				S.F.	.04			.04	.04
	9000	Minimum labor/equipment charge	1 Carp	4	2	Job		60		60	99.50

Figure 15.1

shopping to find the least expensive material for the specified "R" value and thickness. Installation costs may vary from one material to another. Also, wood blocking, furring, and/or nailers are often required to match the insulation thickness in some instances.

Associated Costs

Working with the above data, the estimator can accurately select the installed cost per square foot and estimate the total cost. The estimate for insulation should also include associated costs, such as cutting and patching for difficult installation, or requirements for air vents and other accessories.

Insulating for Sound

Insulation is not just used for controlling heat transfer. It is also specified for use in internal walls and ceilings for controlling sound transfer. Although the noise reduction coefficient of batt insulation is not as great as specialized sound attenuation blankets, the costs are considerably less.

Shingles

Most residences and many smaller types of commercial buildings have sloping roofs that are covered with some form of shingle or watershed material. The materials used in shingles vary from the more common granular-covered asphalt and fiberglass units to wood, metal, clay, concrete, or slate.

The first step in estimating the cost of a shingle roof is to determine the material specified, the shingle size and weight, and the installation method. With this information, the estimator can select the accurate installed cost of the roofing material.

Determining Roof Area

In a sloping roof deck, the ridge and eave lengths, as well as the ridge to eaves dimension, must be known or measured before the actual roof area can be calculated. When the plan dimensions of the roof are known and the sloping dimensions are not known, the actual roof area can still be estimated, providing the slope of the roof is known. Figure 15.2 is a table of multipliers that can be used for this purpose. The roof slope is given in both the inches of rise per foot of horizontal run and in the degree of slope, which allows direct conversion of the horizontal plan dimension into the dimension on the slope.

Converting Area to Units

After the roof area has been estimated in square feet, it must be divided by 100 to convert it into roofing squares, the conventional "unit" for roofing (one square equals 100 square feet). To determine the quantity of shingles required for hips or ridges, add one square for each 100 linear feet of hips and/or ridges.

Allowance for Waste

When the total squares of roofing have been calculated, the estimator should make an allowance for waste based on the design of the roof. A minimum allowance of 3% to 5% is needed if the roof has two straight sides with two gable ends and no breaks. At the other extreme, any roof with several valleys, hips, and land ridges may need a waste allowance of 15% or more to cover the excess cutting required.

Related Work

Accessories that are part of a shingle roof include drip edges and flashings at chimneys, dormers, skylights, vents, valleys, and walls. These are items necessary to complete the roof and should be included in the estimate for the shingles.

Other Roofing Materials and Siding

In addition to shingles, many types of roofing and siding are used on commercial and industrial buildings. These roofing and siding materials are provided in several different kinds of substances, and they are available in many forms, including panels, sheets, membranes, and boards.

The materials used in roofing and siding panels include: aluminum, mineral fiber-cement, epoxy, fibrous glass, steel, vinyl, many types of synthetic sheets and membranes, coal tar, asphalt, tar felt and asphalt felt. Most of the latter materials are used in job-fabricated, built-up roofs and as backing for other materials, such as shingles. The basic data required for estimating either roofing or siding includes the specification of the material, the supporting structure, the method of installation, and the area to be covered. When selecting the current unit price for these materials, the estimator must remember that basic installed unit costs are *per square foot* for siding and *per square* for roofing. The major exceptions to this general rule are prefabricated roofing panels and single-ply roofing, which are priced per square foot.

Single-ply Roofs

Since the early 1970s the use of single-ply roofing (SPR) membranes in the construction industry has been on the rise. Market surveys have recently shown that of all the single-ply systems being installed, about one in three is on new construction. Use of SPR is also increasing in renovation and remodeling. Materially, these roofs are more expensive than other, more conventional roofs.

Factors for Converting Inclined to Horizontal					
Roof Slope	Approx. Angle	Factor	Roof Slope	Approx. Angle	Factor
Flat	0	1.000	12 in 12	45.0	1.414
1 in 12	4.8	1.003	13 in 12	47.3	1.474
2 in 12	9.5	1.014	14 in 12	49.4	1.537
3 in 12	14.0	1.031	15 in 12	51.3	1.601
4 in 12	18.4	1.054	16 in 12	53.1	1.667
5 in 12	22.6	1.083	17 in 12	54.8	1.734
6 in 12	26.6	1.118	18 in 12	56.3	1.803
7 in 12	30.3	1.158	19 in 12	57.7	1.873
8 in 12	33.7	1.202	20 in 12	59.0	1.943
9 in 12	36.9	1.250	21 in 12	60.3	2.015
10 in 12	39.8	1.302	22 in 12	61.4	2.088
11 in 12	42.5	1.357	23 in 12	62.4	2.162

Example:
[20′ (1.302) 90′] 2 = 4,687.2 S.F. = 46.9 Sq.
OR
[40′ (1.302) 90′] = 4,687.2 S.F. = 46.9 Sq.

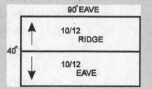

Figure 15.2

However, labor costs are much lower because of faster installation. Reroofing represents the largest market for single-ply roofing today. Single-ply roof systems are normally installed in one of the following ways.

Loose-laid and Ballasted

Generally, this is the easiest type of single-ply roof to install. Some special consideration must be given, however, when flashing is attached to the roof. The membrane is typically fused together at the seams, stretched out flat and ballasted with stone (1-1/2″ @ 10-12 PSF) to prevent wind blow-off. This extra load must be considered during design stages. It is particularly important if reroofing over an existing built-up roof that already weighs 10-15 PSF. A slip-sheet or vapor barrier is sometimes required to separate the new roof from the old.

Partially Adhered

This method of installation uses a series of bar or point attachments that adhere the membrane to a substrate. The membrane manufacturer typically specifies the method to be used based on the material and substrate. Partially adhered systems do not use ballast material. A slip-sheet may be required.

Fully Adhered

This is generally the most time-consuming of the single-plies to install, because these roofs employ a contact cement, cold adhesive, or hot bitumen to adhere the membrane uniformly to the substrate. Only manufacturer-approved insulation board or substrate should be used to receive the membrane. No ballast is required.

The most common single-ply materials available can be classified in three basic categories:

- Thermo-Setting: EPDM, Neoprene, and PIB
- Thermo-Plastic: Hypalon, PVC, and CPE
- Composites: Modified Bitumen

Each has its own requirements and performance characteristics. Most are available for all three installation methods.

Single-ply roof systems are available from many sources. However, most, if not all, manufacturers sell their materials only to franchised installers. As a result, there may be only one source for a price in any given area. Read the specifications carefully. Estimate the system required, exactly as specified; substitutes are usually not allowed.

Sheet Metal

- Copper and Stainless Steel
- Gutters and Downspouts
- Edge Cleats and Gravel Stops
- Flashings
- Trim
- Miscellaneous

Sheet-metal work included in this division is limited to that used on roofs or sidewalls of buildings, usually on the exterior exposed to the weather. Many of the items covered are wholly or partially prefabricated, with labor added for installation. Several are materials that require labor added for on-site fabrication; this cost must be estimated separately.

Pricing shop-made items such as downspouts, drip edges, expansion joints, gravel stops, gutters, regrets, and termite shields requires that the estimator determine the material, size, and shape of the fabricated section, and the linear feet of the item.

The cost of items like copper roofing and metal flashing is estimated in a similar manner, except that unit costs are per square foot. Some roofing systems, particularly single-ply, require flashing materials that are unique to that roofing system.

Roofing materials like monel, stainless steel, and zinc copper alloy are also estimated by the same method, except that the unit costs are per square (100 square feet). Prefabricated items like strainers and louvers are priced on a cost-per-unit basis. Adhesives are priced by the gallon. The installed cost of roofing adhesives depends on the cost per gallon and the coverage per gallon. With trowel-grade adhesive, the coverage varies from a light coating at 25 S.F. per gallon to a heavy coating at 10 S.F. per gallon. With most flashing work, the asphalt adhesive covers an average of 15 S.F. per gallon for each layer or course. Many specifications state coverage of special materials like adhesives. This information should be used as the basis for the estimate.

Roof Accessories

- Hatches
- Skylights
- Vents
- Snow Guards

Roof accessories must be considered as part of the complete weatherproofing system. Standard-size accessories, such as ceiling, roof, and smoke vents or hatches, and snow guards are priced per installed unit. Accessories that must be fabricated to meet project specifications may be priced per square foot, per linear foot, or per unit. Skylight costs, for example, are listed by the square foot, with unit costs decreasing in steps as the nominal size of individual units increases. Some types of skylights and hatches are shown in Figure 15.3.

Skyroofs are priced on the same basis. Due to the many variations in the shape and construction of these units, however, costs are per square foot of surface area. These costs will vary with the size and type of unit. In many cases, maximum and minimum costs give the estimator a range of prices for different design variations. Because there are many types and styles, the estimator must determine the exact specifications for the skyroof being priced. The accuracy of the total cost figure will depend entirely on the selection of the proper unit cost and calculation of the skyroof area. Skyroofs are becoming widely used in the industry, and the work is growing more and more specialized. Often a particular manufacturer is specified. Specialty installing subcontractors are often factory-authorized and required to perform the installation to maintain warranties and waterproof integrity.

Accessories such as roof drains, plumbing vents, and duct penetrations are usually installed by other appropriate subcontractors. However, costs for flashing and sealing these items are often included by the roofing subcontractor.

When estimating Division 7, associated costs must be included for items that may not be directly stated in the specifications. Placement of materials, for example, may require the use of a crane or conveyors. Pitch pockets, sleepers, pads, and walkways may be required for rooftop equipment. Blocking and cant strips, and items associated with different trades must also be coordinated. Once again, the estimator must visualize the construction process.

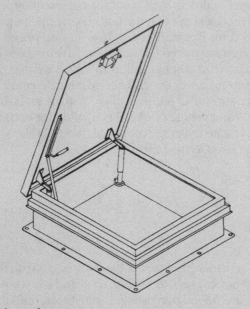

Roof Hatch

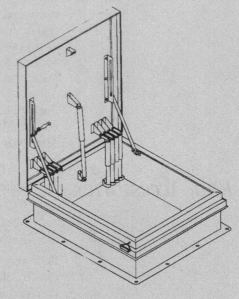

Smoke Hatch

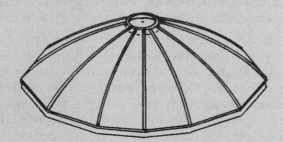

Domed Skylight

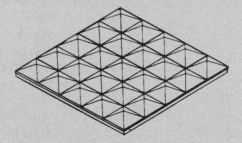

Pyramid Skylights in Grid Form

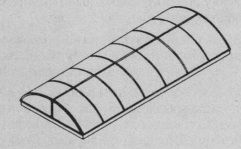

Vaulted Skylight

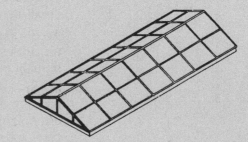

Double Pitch Skylight

Figure 15.3

Chapter 16

Doors, Windows, and Glass

The cost of doors, windows, and glass may represent a sizeable portion of a total remodeling project. Windows and doors require hardware, an additional material cost. Any one door assembly (door, frame, hardware) can be a combination of the variable features listed below:

Door	Frame	Hardware
Size	Size	Lockset
Thickness	Throat	Passage set
Wood-type	Wood-type	Panic bar
Metal gauge	Metal gauge	Closer
Laminate	Casing	Hinges
Hollow-core type	Stops	Stops
Solid-core material	Fire rating	Bolts
Fire rating	Knock-down	Finish
Finish	Welded	Plates

When available, most architectural plans and specifications usually include door, window, and hardware schedules, listing the items needed for the required combinations. The estimator should use these schedules and details in conjunction with the plans to avoid duplication or omission of units when determining the quantities. The schedules should identify the location, size, and type of each unit. Schedules should also include information regarding the frame, fire rating, hardware, and special notes. If no such schedules are included, the estimator should prepare them in order to provide an accurate quantity takeoff. Figure 16.1 is an example of a schedule that may be prepared by the estimator. Most suppliers prepare separate schedules; each must be approved by the architect or owner.

A proper door schedule on the architectural drawings identifies each opening in detail. The estimator should define each opening using the schedule and any other pertinent data. Installation information should be carefully reviewed in the specifications.

For the quantity takeoff, the numbers of all similar doors and frames should be combined. Each should be checked off on the plans to ensure that none have been left out. An easy and obvious double-check is to count the total number of openings, making certain that two doors and only one frame have been included where double doors are used. Important details to check for both door and frame are the following:

DOOR AND FRAME SCHEDULE

PROJECT ______ ARCHITECT ______ PAGE ____ OF ____

LOCATION ______ OWNER ______ DATE ______

BY ______

DOOR NO.	DOOR							FRAME					FIRE RATING		HARDWARE		REMARKS
	SIZE			MAT.	TYPE	GLASS	LOUVER	MAT.	TYPE	DETAILS			LAB	CON	SET NO.	KEYSIDE ROOM NO.	
	W	H	T							JAMB	HEAD	SILL					

Figure 16.1

- Material
- Gauge
- Size
- Core Material
- Fire-Rating Label
- Finish
- Style

Wood Doors

Wood doors are manufactured in either flush or paneled designs, and are separated into three grades: *architectural/commercial, residential*, and *decorator*. A wide variety of frames are available—in metal, pine, hardwood, and for various partition thicknesses—for remodeling installations. Some doors are available prehung in frames for quick installation.

Architectural or Commercial Wood Doors

This type of door is most often specified in commercial renovation. The stiles are made of hardwood, and the core is dense and of hot-bonded construction. These doors feature thick-face veneers that are exterior-glued and matched in their grain patterns. Because of its durability, this grade of door often carries a lifetime warranty.

Residential Wood Doors

These doors are chosen for low-frequency use where economy is a primary consideration. The stiles are manufactured from soft wood; the core, from low-density materials. The face veneers are thin, interior-glued, and broken in their grain patterns.

Decorator Wood Doors

These doors are manufactured from solid wood and are usually hand carved. Because of the choice woods used and the special craftsmanship required in their production, the cost of these doors is several times that of similar-sized, architectural wood doors.

Flush Doors

The cores of these doors are produced in varying densities: hollow, particle board, or veneer core. Lauan mahogany, birch, oak, or other hardwood veneers are used for facings. Synthetic veneers, created from a medium-density overlay or high-pressure plastic laminate, may serve as an alternate choice to natural wood veneers. Flush wood doors may be fire-rated. Door swings and wood-door details are shown in Figure 16.2.

Fire Doors

The estimator must pay particular attention to fire-door specifications when performing the quantity takeoff. It is important to determine the exact type of door required. Figure 16.3 is a table describing various types of fire doors. Please note that a "B" label door may be one of four types. If the plans or door schedule do not specify exactly which temperature rise is required, the estimator should consult the architect or local building inspector. Many Building and Fire Codes also require that frames and hardware at fire doors be fire-rated, and labeled as such. When determining quantities, the estimator must also include any glass (usually wired) or special inserts to be installed in fire doors (or in any doors).

Metal Doors and Frames

Hollow Metal Doors

Hollow metal doors are available in stock or custom fabrication, flush or embossed, with glazing or louvers, labeled or unlabeled, and in various steel

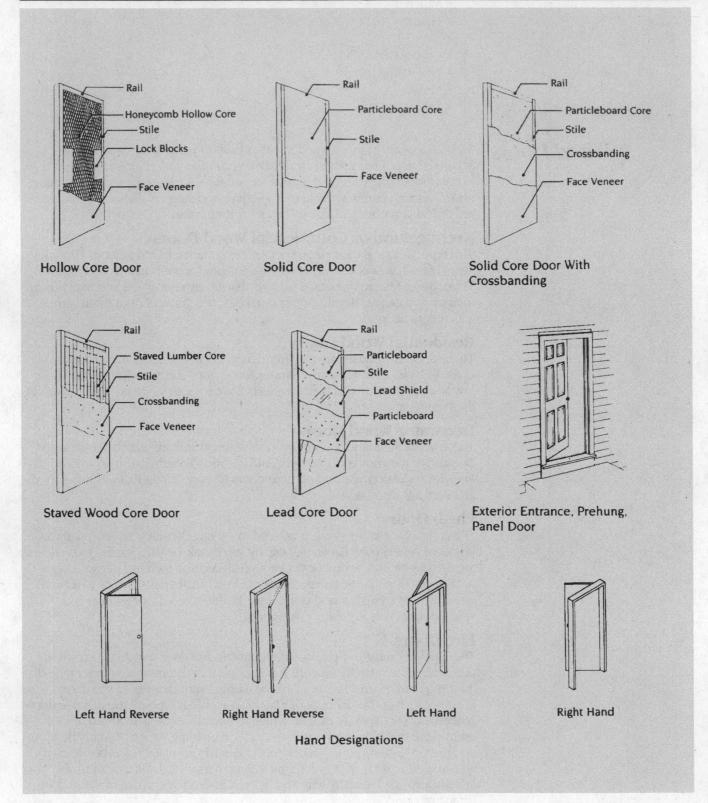

Hollow Core Door
- Rail
- Honeycomb Hollow Core
- Stile
- Lock Blocks
- Face Veneer

Solid Core Door
- Rail
- Particleboard Core
- Stile
- Face Veneer

Solid Core Door With Crossbanding
- Rail
- Particleboard Core
- Stile
- Crossbanding
- Face Veneer

Staved Wood Core Door
- Rail
- Staved Lumber Core
- Stile
- Crossbanding
- Face Veneer

Lead Core Door
- Rail
- Particleboard
- Stile
- Lead Shield
- Particleboard
- Face Veneer

Exterior Entrance, Prehung, Panel Door

Left Hand Reverse

Right Hand Reverse

Left Hand

Right Hand

Hand Designations

Figure 16.2

gauges and core fills. Stock doors may be supplied by some manufacturers for low-, moderate-, or high-frequency use. The doors are available in widths of 2′ to 4′ and heights varying from 6′6″ to 10′ or more. They may be used as single doors, in pairs with both leaves active, and in pairs with one active leaf, including an astragal. Bifold hollow metal doors are available for specified applications.

Hollow metal doors are reinforced at the stress points and premortised for the hardware required for the door application. Hollow metal, labelled fire doors can be supplied stock or custom-manufactured with A, B, C, D, E labels, with 3/4-hour to 3-hour ratings, depending on the glass area, height and width restrictions, and maximum expected temperature rise (shown in Figure 16.3). The door types shown in Figure 16.4 are examples of typical metal fire doors. Code requirements for fire doors and ratings vary from state to state and often from city to city.

Fire Door				
Classification	Time Rating (as Shown on Label)		Temperature Rise (as Shown on Label)	Maximum Glass Area
3 Hour fire doors (A) are for use in openings in walls separating buildings or dividing a single building into the areas.	3 Hr.	(A)	30 Min. 250°F Max	None
	3 Hr.	(A)	30 Min. 450°F Max	
	3 Hr.	(A)	30 Min. 650°F Max	
	3 Hr.	(A)	*	
1–1/2 Hour fire doors (B) and (D) are for use in openings in 2 Hour enclosures of vertical communication through buildings (stairs, elevators, etc.) or in exterior walls which are subject to severe fire exposure from outside of the building. 1 Hour fire doors (B) are for use in openings in 1 Hour enclosures of vertical communication through buildings (stairs, elevators, etc.)	1–1/2 Hr.	(B)	30 Min. 250°F Max	100 square inches per door
	1–1/2 Hr.	(B)	30 Min. 450°F Max	
	1–1/2 Hr.	(B)	30 Min. 650°F Max	
	1–1/2 Hr.	(B)	*	
	1 Hr.		30 Min. 250°F Max	
	1–1/2 Hr.	(D)	30 Min. 250°F Max	None
	1–1/2 Hr.	(D)	30 Min. 450°F Max	
	1–1/2 Hr.	(D)	30 Min. 650°F Max	
	1–1/2 Hr.	(D)	*	
3/4 Hour fire doors (C) and (E) are for use in openings in corridor and room partitions or in exterior walls which are subject to moderate fire exposure from outside of the building.	3/4 Hr.	(C)	**	1296 Square
	3/4 Hr.	(E)	**	720 square inches per light No limit
1/2 Hour fire doors and 1/3 Hour fire doors are for use where smoke control is a primary consideration and are for the protection of openings in partitions between a habitable room and a corridor when the wall has a fire–resistance rating of not more than one hour.	1/2 Hr.		**	
	1/3 Hr.		**	

*The labels do not record any temperature rise limits. This means that the temperature rise on the unexposed face of the door at the end of 30 minutes of test is in excess of 650°F.
**Temperature rise is not recorded.

Figure 16.3

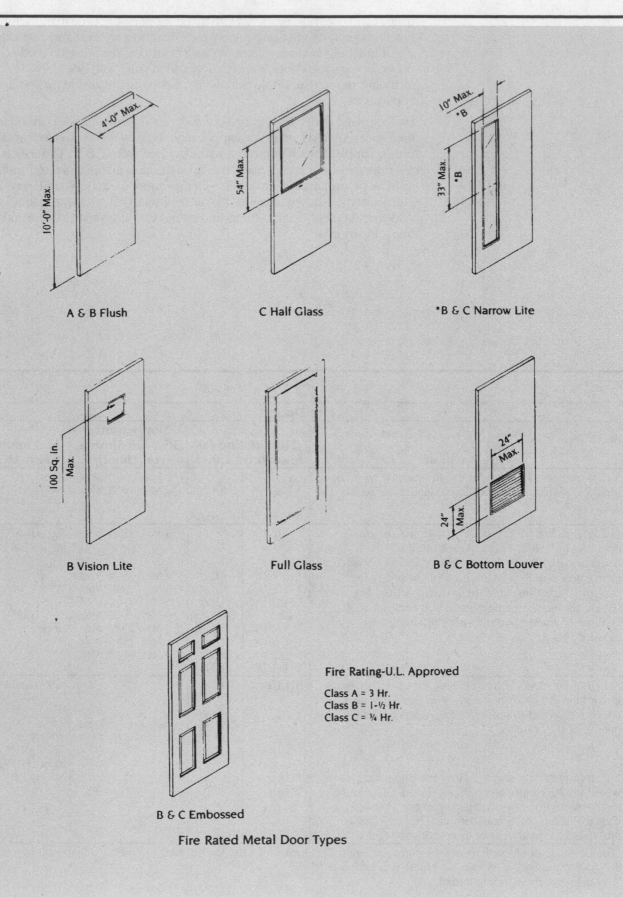

A & B Flush

C Half Glass

*B & C Narrow Lite

B Vision Lite

Full Glass

B & C Bottom Louver

Fire Rating-U.L. Approved

Class A = 3 Hr.
Class B = 1-½ Hr.
Class C = ¾ Hr.

B & C Embossed

Fire Rated Metal Door Types

Figure 16.4

Hollow Metal Frames

Hollow metal frames may be supplied in 14-, 16-, or 18-gauge galvanized or plain steel in knock-down standard frames or welded customized frames that can be fabricated to satisfy most design conditions. Frames with borrowed lights, transoms, or cased openings are available in stick components from some manufacturers.

Frames may be wraparound (enclosing the wall), or they may butt up against the opening. A wraparound frame may terminate into the enclosed wall when it is covered by a finish such as plaster, or the frame may return along the enclosed wall when it is exposed, as in drywall construction. Frames are sometimes supplied in two pieces to suit varied wall thicknesses, or in one piece to satisfy standard wall thicknesses. Frames are normally reinforced at stress points and are prepared for hinges and strikes. Anchors to attach the frame to the wall are supplied to suit wall construction requirements. Custom frames normally require a hardware schedule and templates to produce required shop drawings and to accomplish fabrication. Typical hollow metal frames for various applications are shown in Figure 16.5.

Special Doors

Metal Access Panels and Doors

These doors are available in steel or stainless steel for fire-rated or nonfire-rated applications. Panels are fabricated for flush installations in drywall (both skim-coated or taped), for masonry and tile applications, for plastered walls and ceilings, and for acoustical ceilings. These doors are available in stock sizes and types to suit most applications.

Blast Doors

These doors are available in standard designs, and may also be custom-designed to withstand specified pressures and to resist penetration.

Cold-Storage Doors

This type of door is available in standard designs in wood, steel, fiberglass, plastic, and stainless steel—for all types of cold-storage requirements. These doors are manufactured to provide insulation for cool zones, coolers, and freezers for manual, air, electric, or hydraulic operations. These doors may operate in any of the following ways: sliding, vertical lift, biparting overhead, and single- and double-swing. Pass windows (vertical and horizontal sliding), rotating shelf windows, ticket windows, cashier doors, and coin and cash trays are available in aluminum, steel, or stainless steel. Pass windows, roll-up shutters, and projection booth shutters are available, labelled or nonlabelled, with fusible links to suit most applications and requirements. A roll-up shutter and a roll-up gate are shown in Figure 16.6.

Glass and Glazing

Glass and glazing in remodeling construction most often includes interior glazed partitions, entrances, and storefronts. Interior glazed partitions are commonly constructed of tubular aluminum framing. Glazing subcontractors may estimate such partitions by measuring and pricing the length of each component of the frame separately. Examples are shown in Figures 16.7 and 16.8. The glass can be plate, tempered, safety, tinted, insulated, or combinations thereof, depending on project and code requirements. Glass is estimated by the square foot or by united length (length plus width). Glass doors and hardware may also be estimated separately. Another method of estimating is to use complete system prices, whether by square foot (Figure 16.9) or by the opening.

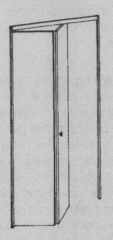

Bi-folding Closet Door and Frame

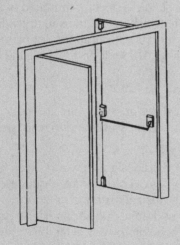

Double Egress Door and Frame

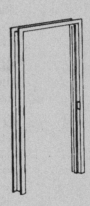

Hollow Metal Frame–Butt or Wraparound

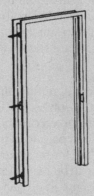

Hollow Metal Frame with Anchors

Figure 16.5

Entrances and storefronts are almost all special designs and combinations of unit items to fit a unique situation. The estimator should submit the plans and specifications to a specialty installer for takeoff and pricing. Typical entrances are shown in Figure 16.10. The general procedure for the installer's takeoff is:

For stationary units:
- Determine the height and width of each like unit.
- Determine the linear feet of intermediate, horizontal, and vertical members, rounded to the next higher foot.
- Determine the number of joints.

For entrance units:
- Determine the number of joints.
- Determine special frame hardware per unit.
- Determine special door hardware per unit.
- Determine thresholds and closers.

Curtain Walls

A curtain wall is a nonstructural facade consisting of panels in a wide variety of materials and constructions. A curtain wall is held in place in a metal frame by caulking, gaskets, and sealants. The curtain wall can be prefabricated to the following degree depending on the type.

- Stick, in which all components are field assembled
- Panel and mullion, in which the panels are prefabricated into frames and field-connected to mullions
- Total panel systems, in which the mullions are pre-assembled into the panels

Figure 16.11 shows different types of wall systems.

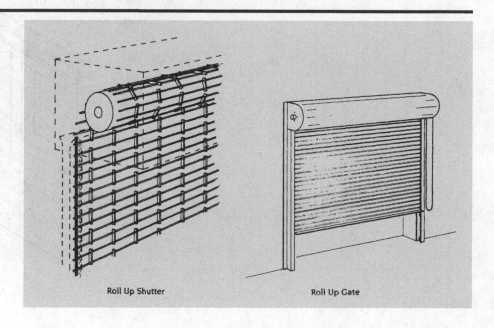

Roll Up Shutter   Roll Up Gate

Figure 16.6

Finish Hardware

Finish hardware is the construction industry term for the devices used to operate doors, windows, drawers, shutters, closets, and cabinets. This category includes such items as hinges, latches, locks, panic devices, security and detection systems, astragals, and weatherstripping. Some examples are shown in Figure 16.12a and 16.12b. In a typical building, the finish hardware accounts for between 2% and 3% of the total job cost. Consequently, the difference between economy and quality hardware can mean a 1% difference in the total building cost.

A hardware specialist will often prepare a schedule and specify the hardware that is to be used for each opening. There are two general classifications of hardware: *Builder's*, and *Commercial*. Builder's hardware is generally used for residential construction.

Another hardware specification is based on frequency of use: *heavy, light*, or *medium*. For example, the size, weight, and material of a door and its frame dictate the size and number of hinges. Building Codes and security requirements determine the selection of the proper fire barrier and electronic hardware in different remodeling situations.

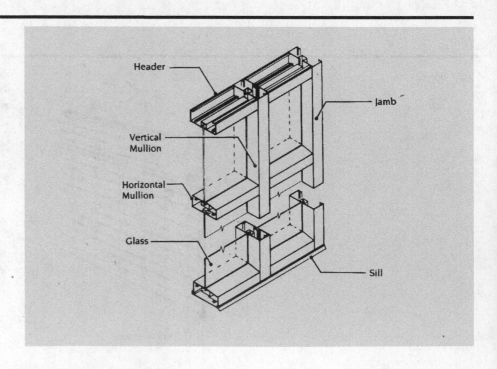

Figure 16.7

08900 | Glazed Curtain Wall

08911 | Glazed Aluminum Curtain Wall

		CREW	DAILY OUTPUT	LABOR-HOURS	UNIT	MAT.	LABOR	EQUIP.	TOTAL	TOTAL INCL O&P	
700							2002 BARE COSTS					700
0010	**TUBE FRAMING** For window walls and store fronts, aluminum, stock										
0050	Plain tube frame, mill finish, 1-3/4" x 1-3/4"	2 Glaz	103	.155	L.F.	6.15	4.66		10.81	14.30	
0150	1-3/4" x 4"		98	.163		7.95	4.90		12.85	16.60	
0200	1-3/4" x 4-1/2"		95	.168		8.85	5.05		13.90	17.90	
0250	2" x 6"		89	.180		13.30	5.40		18.70	23.50	
0350	4" x 4"		87	.184		13.05	5.50		18.55	23.50	
0400	4-1/2" x 4-1/2"		85	.188		14.50	5.65		20.15	25	
0450	Glass bead		240	.067		1.68	2		3.68	5.10	
1000	Flush tube frame, mill finish, 1/4" glass, 1-3/4" x 4", open header	2 Glaz	80	.200	L.F.	7.80	6		13.80	18.25	700
1050	Open sill		82	.195		6.80	5.85		12.65	16.90	
1100	Closed back header		83	.193		10.90	5.80		16.70	21.50	
1150	Closed back sill		85	.188		10.35	5.65		16	20.50	
1200	Vertical mullion, one piece		75	.213		11.55	6.40		17.95	23	
1250	Two piece		73	.219		12.35	6.60		18.95	24	
1300	90° or 180° vertical corner post		75	.213		19.45	6.40		25.85	32	
1400	1-3/4" x 4-1/2", open header		80	.200		9.50	6		15.50	20	
1450	Open sill		82	.195		7.85	5.85		13.70	18.05	
1500	Closed back header		83	.193		11.90	5.80		17.70	22.50	
1550	Closed back sill		85	.188		11.15	5.65		16.80	21.50	
1600	Vertical mullion, one piece		75	.213		12.50	6.40		18.90	24	
1650	Two piece		73	.219		13.20	6.60		19.80	25	
1700	90° or 180° vertical corner post		75	.213		13.50	6.40		19.90	25	
2000	Flush tube frame, mill fin. for ins. glass, 2" x 4-1/2", open header		75	.213		10.80	6.40		17.20	22.50	
2050	Open sill		77	.208		9.50	6.25		15.75	20.50	
2100	Closed back header		78	.205		11.80	6.15		17.95	23	
2150	Closed back sill		80	.200		11.65	6		17.65	22.50	
2200	Vertical mullion, one piece		70	.229		13.05	6.85		19.90	25.50	
2250	Two piece		68	.235		13.95	7.05		21	27	
2300	90° or 180° vertical corner post		70	.229		13.20	6.85		20.05	25.50	
5000	Flush tube frame, mill fin., thermal brk., 2-1/4"x 4-1/2", open header		74	.216		11.90	6.50		18.40	23.50	
5050	Open sill		75	.213		10.40	6.40		16.80	22	
5100	Vertical mullion, one piece		69	.232		14.35	6.95		21.30	27	
5150	Two piece		67	.239		15.35	7.15		22.50	28.50	
5200	90° or 180° vertical corner post		69	.232		13.80	6.95		20.75	26.50	
6980	Door stop (snap in)		380	.042		2.27	1.26		3.53	4.54	
7000	For joints, 90°, clip type, add				Ea.	19			19	21	
7050	Screw spline joint, add					14.60			14.60	16.05	
7100	For joint other than 90°, add					30.50			30.50	33.50	
8000	For bronze anodized aluminum, add					15%					
8020	For black finish, add					27%					
8050	For stainless steel materials, add					350%					
8100	For monumental grade, add					50%					
8150	For steel stiffener, add	2 Glaz	200	.080	L.F.	7.45	2.40		9.85	12.05	
8200	For 2 to 5 stories, add per story				Story		5%				
9000	Minimum labor/equipment charge	2 Glaz	2	8	Job		240		240	385	

Figure 16.8

08400 | Entrances & Storefronts

08411 | Aluminum Framed Storefront

			CREW	DAILY OUTPUT	LABOR-HOURS	UNIT	MAT.	LABOR	EQUIP.	TOTAL	TOTAL INCL O&P	
140	0500	Wide stile, 2'-6" x 7'-0" opening	2 Sswk	2	8	Ea.	665	274		939	1,250	140
	0520	3'-0" x 7'-0" opening		2	8		655	274		929	1,225	
	0540	3'-6" x 7'-0" opening		2	8		685	274		959	1,275	
	0560	5'-0" x 7'-0" opening		2	8		1,050	274		1,324	1,675	
	0580	6'-0" x 7'-0" opening		1.30	12.308	Pr.	1,000	420		1,420	1,900	
	0600	7'-0" x 7'-0" opening		1	16	"	1,150	550		1,700	2,275	
	1100	For full vision doors, with 1/2" glass, add				Leaf	55%					
	1200	For non-standard size, add					67%					
	1300	Light bronze finish, add					36%					
	1400	Dark bronze finish, add					18%					
	1500	For black finish, add					36%					
	1600	Concealed panic device, add					930			930	1,025	
	1700	Electric striker release, add				Opng.	239			239	263	
	1800	Floor check, add				Leaf	710			710	780	
	1900	Concealed closer, add				"	475			475	520	
	2000	Flush 3' x 7' Insulated, 12"x 12" lite, clear finish	2 Sswk	2	8	Ea.	900	274		1,174	1,500	
	9000	Minimum labor/equipment charge	2 Carp	4	4	Job		120		120	199	
600	0010	STAINLESS STEEL AND GLASS Entrance unit, narrow stiles										600
	0020	3' x 7' opening, including hardware, minimum	2 Sswk	1.60	10	Opng.	4,600	345		4,945	5,700	
	0050	Average		1.40	11.429		4,975	390		5,365	6,200	
	0100	Maximum		1.20	13.333		5,325	455		5,780	6,725	
	1000	For solid bronze entrance units, statuary finish, add					60%					
	1100	Without statuary finish, add					45%					
	2000	Balanced doors, 3' x 7', economy	2 Sswk	.90	17.778	Ea.	6,225	610		6,835	8,000	
	2100	Premium		.70	22.857	"	10,700	785		11,485	13,300	
	9000	Minimum labor/equipment charge		2	8	Job		274		274	515	
650	0010	STOREFRONT SYSTEMS Aluminum frame, clear 3/8" plate glass,										650
	0020	incl. 3' x 7' door with hardware (400 sq. ft. max. wall)										
	0500	Wall height to 12' high, commercial grade	2 Glaz	150	.107	S.F.	12	3.20		15.20	18.35	
	0600	Institutional grade		130	.123		16	3.69		19.69	23.50	
	0700	Monumental grade		115	.139		23	4.17		27.17	32	
	1000	6' x 7' door with hardware, commercial grade		135	.119		12.25	3.56		15.81	19.25	
	1100	Institutional grade		115	.139		16.80	4.17		20.97	25.50	
	1200	Monumental grade		100	.160		31	4.80		35.80	42	
	1500	For bronze anodized finish, add					15%					
	1600	For black anodized finish, add					30%					
	1700	For stainless steel framing, add to monumental					75%					
	9000	Minimum labor/equipment charge	2 Glaz	1	16	Job		480		480	775	

08460 | Automatic Entrance Doors

			CREW	DAILY OUTPUT	LABOR-HOURS	UNIT	MAT.	LABOR	EQUIP.	TOTAL	TOTAL INCL O&P	
600	0010	SLIDING ENTRANCE 12' x 7'-6" opng., 5' x 7' door, 2 way traf.,										600
	0020	mat activated, panic pushout, incl. operator & hardware,										
	0030	not including glass or glazing	2 Glaz	.70	22.857	Opng.	5,900	685		6,585	7,600	
	9000	Minimum labor/equipment charge	"	.70	22.857	Job		685		685	1,100	
650	0010	SLIDING PANELS										650
	0020	Mall fronts, aluminum & glass, 15' x 9' high	2 Glaz	1.30	12.308	Opng.	2,325	370		2,695	3,150	
	0100	24' x 9' high		.70	22.857		3,350	685		4,035	4,800	
	0200	48' x 9' high, with fixed panels		.90	17.778		6,250	535		6,785	7,725	
	0500	For bronze finish, add					17%					
	9000	Minimum labor/equipment charge	2 Glaz	1	16	Job		480		480	775	

08480 | Balanced Entrance Doors

			CREW	DAILY OUTPUT	LABOR-HOURS	UNIT	MAT.	LABOR	EQUIP.	TOTAL	TOTAL INCL O&P	
150	0010	BALANCED DOORS										150
	0020	Hardware & frame, alum. & glass, 3' x 7', econ.	2 Sswk	.90	17.778	Ea.	4,725	610		5,335	6,350	

Figure 16.9

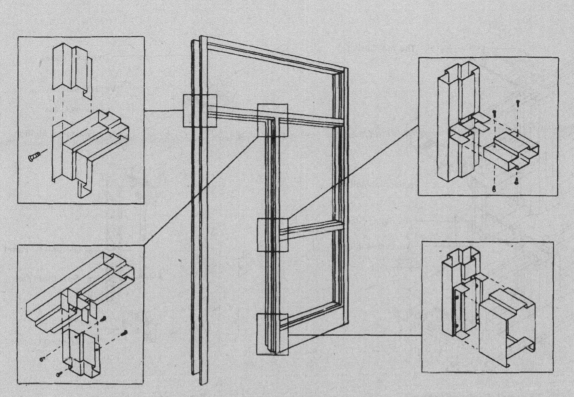

Pre-Engineered, "Stick" System Entrance

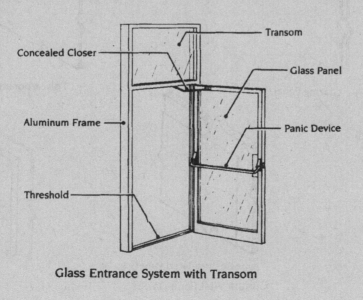

Transom

Concealed Closer

Glass Panel

Aluminum Frame

Panic Device

Threshold

Glass Entrance System with Transom

Figure 16.10

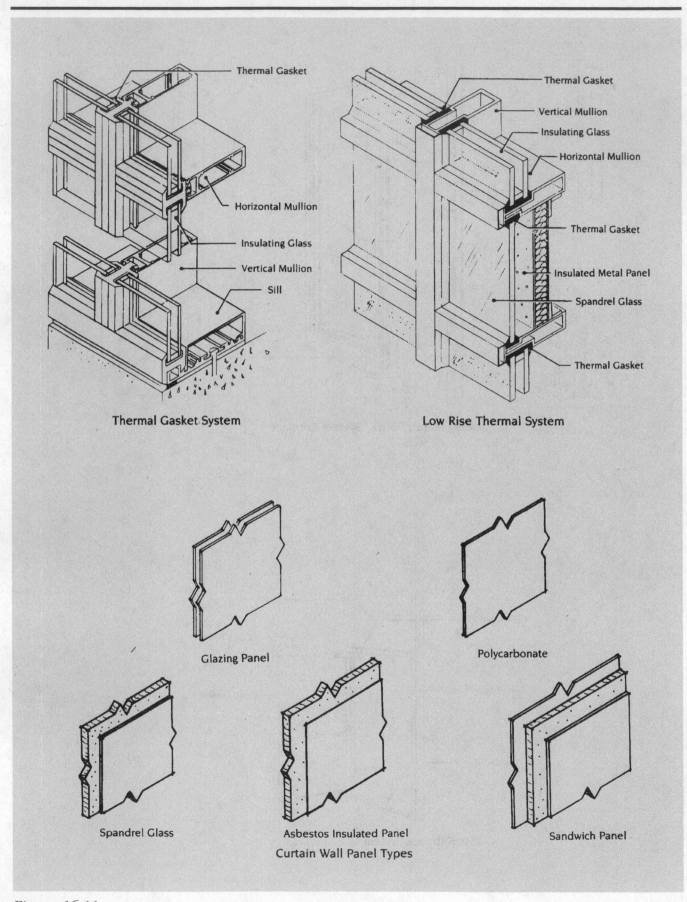

Thermal Gasket System

- Thermal Gasket
- Horizontal Mullion
- Insulating Glass
- Vertical Mullion
- Sill

Low Rise Thermal System

- Thermal Gasket
- Vertical Mullion
- Insulating Glass
- Horizontal Mullion
- Thermal Gasket
- Insulated Metal Panel
- Spandrel Glass
- Thermal Gasket

Glazing Panel

Polycarbonate

Spandrel Glass

Asbestos Insulated Panel

Sandwich Panel

Curtain Wall Panel Types

Figure 16.11

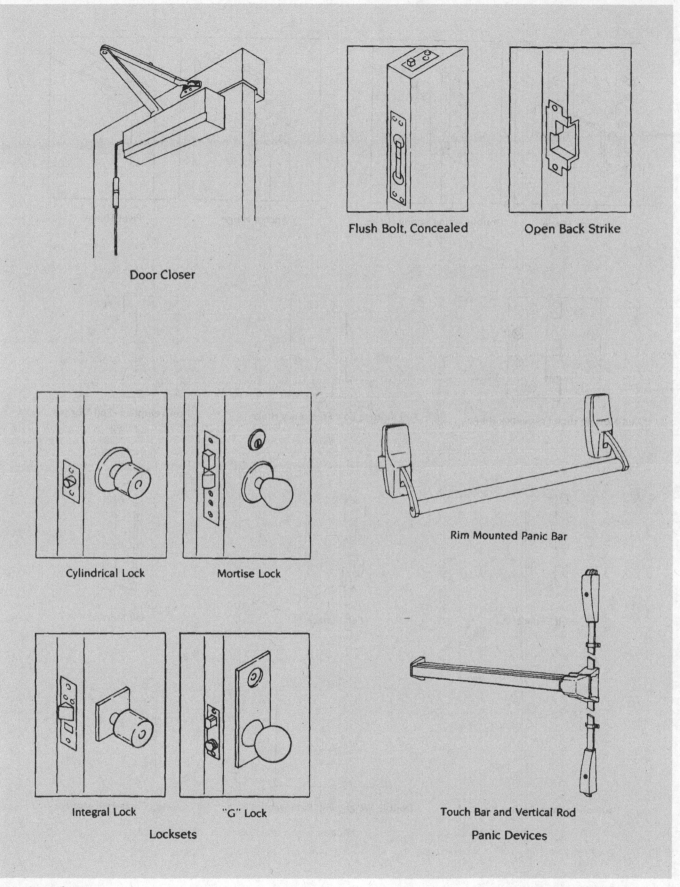

Door Closer

Flush Bolt, Concealed

Open Back Strike

Cylindrical Lock

Mortise Lock

Rim Mounted Panic Bar

Integral Lock

"G" Lock

Locksets

Touch Bar and Vertical Rod

Panic Devices

Figure 16.12a

165

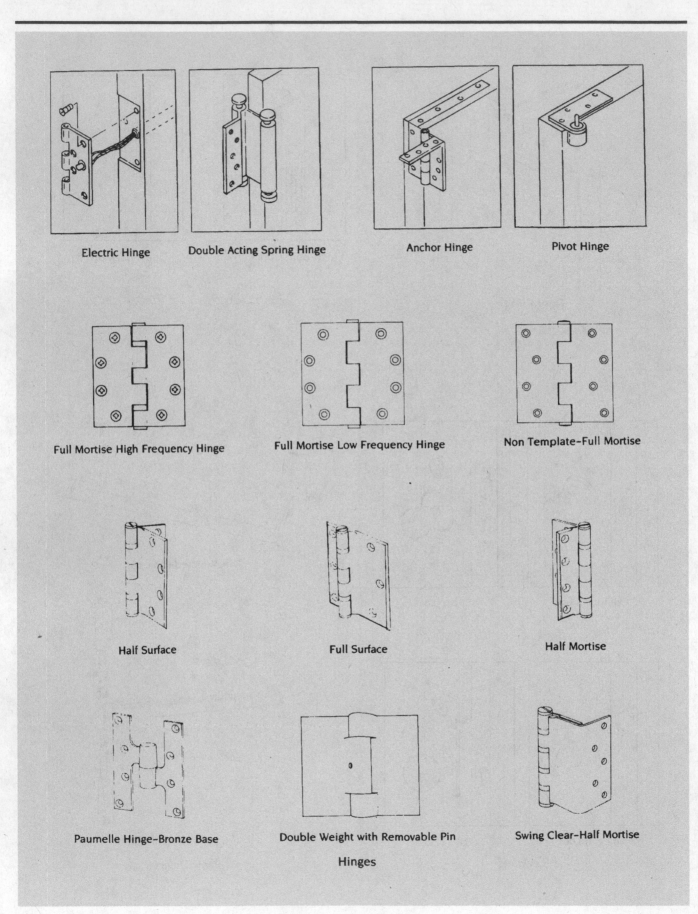

Electric Hinge

Double Acting Spring Hinge

Anchor Hinge

Pivot Hinge

Full Mortise High Frequency Hinge

Full Mortise Low Frequency Hinge

Non Template–Full Mortise

Half Surface

Full Surface

Half Mortise

Paumelle Hinge–Bronze Base

Double Weight with Removable Pin

Swing Clear–Half Mortise

Hinges

Figure 16.12b

166

Chapter 17

Finishes

This chapter contains descriptions of the basic types of finishes, with appropriate estimating methods. Materials used to finish walls, ceilings, doors, windows, and trim work may include any of a number of available products. Most building codes (and specifications) require strict adherence to standards for protection against maximum fire, flame spread, and smoke generation characteristics. In today's fireproof and fire-resistant types of construction, some finish materials may be the only combustibles used in a building project. These combustible materials may have to be treated for fire retardancy, at an additional cost. The estimator must be sure that all materials meet the specified requirements and that any necessary fireproofing be included in the estimate for interior finishes.

Lath and Plaster

The different types of plaster work require different pricing strategies. Large open areas of continuous walls or ceilings involve considerably less labor per unit of area than small areas, repair work, or intricate work such as archways, curved walls, cornices, and window returns. Gypsum and metal lath are most often used as sub-bases. However, plaster may be applied directly on masonry, concrete, and, in some restoration work, wood. In the latter cases, a bonding agent may be specified. Illustrations of plaster partitions and metal lath are shown in Figure 17.1.

Number and Type of Coatings

The number of coats of plaster may also vary. Traditionally, a scratch coat is applied to the substrate. A brown coat is then applied two days later, and the finish, smooth coat, seven days after the brown coat. Currently, the systems most often used are two-coat and one-coat (imperial plaster on "blueboard"). Textured surfaces, with and without patterns, may be required. The many variables in plaster work make it difficult to develop "system" prices. Each project, and even areas within each project, must be examined individually.

Sequence of Work

The quantity takeoff should proceed in the normal construction sequence: furring (or studs), lath, plaster, and accessories. Studs, furring, and/or ceiling suspension systems, whether wood or steel, should be taken off separately. Responsibility for the installation of these items should be clearly noted. Depending on local work practices, lathers may or may not install studs or furring. These materials are usually estimated by the piece or linear foot, and sometimes by the square foot.

Determining Quantities

Lath is traditionally estimated by the square yard for both gypsum and metal lath and, more recently, by the square foot. Usually, a 5% allowance for waste is included. Casing bead, corner bead, and other accessories are measured by the linear foot. An extra foot of surface area should be allowed for each linear foot of corner or stop.

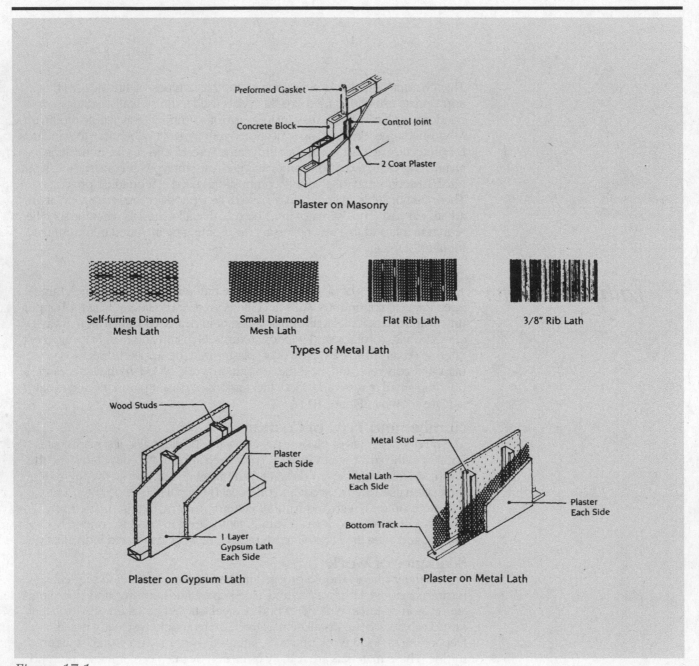

Preformed Gasket

Concrete Block

Control Joint

2 Coat Plaster

Plaster on Masonry

Self-furring Diamond Mesh Lath

Small Diamond Mesh Lath

Flat Rib Lath

3/8" Rib Lath

Types of Metal Lath

Wood Studs

Plaster Each Side

1 Layer Gypsum Lath Each Side

Plaster on Gypsum Lath

Metal Stud

Metal Lath Each Side

Bottom Track

Plaster Each Side

Plaster on Metal Lath

Figure 17.1

Although wood plaster grounds are usually installed by carpenters, they should be measured when taking off the plaster requirements.

Plastering is also traditionally measured by the square yard. Deductions for openings vary by preference—from a zero deduction to 50% of all openings over two feet in width. Some estimators deduct a percentage of the total yardage for openings. One extra square foot of wall area should be allowed for each linear foot of inside or outside corner located below the ceiling level. The areas of small radius work should be doubled.

Plaster quantities are determined by measuring surface area (walls, ceilings, etc.). The estimator must consider the complexity and the intricacy of the work, as well as the quality. There are two basic quality categories:

- **Ordinary quality** is used for commercial purposes. Waves 1/8″ to 3/16″ in 10 feet are acceptable. Angles and corners must be fairly true.
- **First quality** requires that variations be less than 1/16″ in 10 feet. Labor costs for first quality work are approximately 20% more than those for ordinary plastering.

Drywall

Framing

With the advent of light-gauge metal framing, tin snips are becoming as important a tool to the carpenter as the circular saw. Metal studs and framing are usually installed by the drywall subcontractor. The estimator should make sure that studs and other framing, whether metal or wood, are not included twice by different subcontractors. In some drywall systems, such as shaftwall, the framing is integral and installed simultaneously with the drywall panels. Typical drywall partition systems are shown in Figure 17.2.

Metal studs are manufactured in various widths (1–5/8″, 2–1/2″, 3–5/8″, 4″, and 6″) and in various gauges, or metal thicknesses. They may be used for both load-bearing and nonload-bearing partitions, depending on design criteria and code requirements. Metal framing is particularly useful as a replacement for structural wood, since the latter material is often prohibited due to its combustible quality. Metal studs, track, and accessories are purchased by the linear foot, and usually stocked in 8′ to 16′ lengths, by 2′ increments. For large orders, metal studs can be purchased in any length up to 20′.

For estimating, light-gauge metal framing is taken off by the linear foot or by the square foot of wall area of each type. Different wall types—(with different stud widths, stud spacing, or drywall requirements)—should each be taken off separately, especially if estimating by the square foot.

Fasteners

Metal studs can be installed very quickly. Depending on the specification, they may have to be fastened to the track with self-tapping screws, tack welds or clips, or they may not have to be prefastened. Each requirement will affect the labor costs. Fasteners such as self-tapping screws, clips and powder-actuated studs are very expensive, though labor-saving. These costs must be included.

Drywall

Drywall may be purchased in various thicknesses (1/4″ to 1″) and in various sizes (2′ x 8′ to 4′ x 20′). Different types include *standard, fire-resistant, water-resistant, blueboard, coreboard*, and *prefinished*. There are many variables and possible combinations of sizes and types. While the installation cost of 5/8″ standard drywall may be the same as that of 5/8″ fire-resistant drywall, these two types (and all others) should be taken off separately. The takeoff will be used for purchasing; material costs will vary.

Because drywall is used in such large quantities, current, local prices should always be checked. A variation of a few cents per square foot may amount to many thousands of dollars over a whole project.

Firewalls

Fire-resistant drywall provides an excellent design advantage in creating relatively lightweight, easy-to-install firewalls (as opposed to masonry walls). As with any type of drywall partition, the variations are numerous. The estimator must be very careful to take off the appropriate firewalls exactly as specified.

Even more important, the contractor must build the firewalls exactly as specified. There is a tremendous potential for liability in the event the system fails. For example, a metal stud partition with two layers of 1/2″ fire-resistant drywall on each side may constitute a *two-hour* partition (when all other requirements such as staggered joints, taping, sealing openings, and so forth,

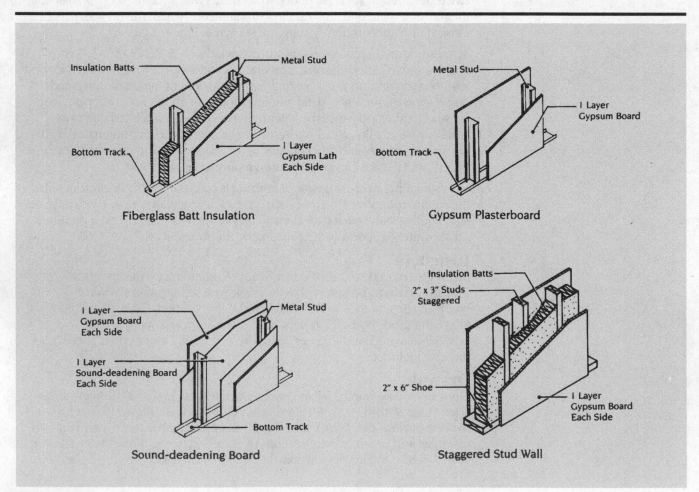

Figure 17.2

are met). If a *one-hour* partition is called for, the estimator cannot assume that one layer of 1/2″ fire-resistant drywall on each side of a metal stud partition will suffice. Alone, it does not.

When left to choose the appropriate assembly (given the rating required), the estimator must be sure that the system has been tested and approved for use—by Underwriters Laboratory as well as local Building and Fire Codes and responsible authorities. In all cases, the drywall (and studs) for firewalls must extend completely from the deck below to the underside of the deck above, covering the area above and around any and all obstructions. All penetrations must be protected.

In the past, structural members to be fireproofed, such as beams or columns, had to be "wrapped" with a specified number of layers of fire-resistant drywall. This is a labor-intensive and expensive task. With the advent of spray-on fireproofing, structural members can be much more easily protected. This type of work is usually performed by a specialty subcontractor. Takeoff and pricing are done by square foot of surface area. (See Figure 13.2 for surface and boxed areas of structural steel members.)

Soundproofing

When walls are specified for minimal sound transfer, the same continuous, unbroken construction used for firewalls is required. Soundproofing specifications may include additional accessories and related work. Resilient channels attached to studs, mineral fiber batts, and staggered studs may all be used. In order to develop high, noise-reduction coefficients, double stud walls may be required with sheet lead between double or triple layers of drywall. (Sheet lead may also be required for X-ray installations.) Caulking is required at all joints and seams. All openings must be specially framed with double, "broken" door and window jambs.

Shaftwall

Cavity shaftwall, developed for a distinct design advantage, is another drywall assembly that should be estimated separately. While firewalls require equal protection from both sides, shaftwall (used at vertical openings such as elevators and utility chases) can be installed completely from one side. Special track, studs (C-H or double E type), and drywall (usually 1″ thick and 2′ wide coreboard) are used. These items should be priced separately from other drywall partition components. Figure 17.3 illustrates several types of cavity shaftwall.

Material Handling

Because of the size and weight of drywall, costs for material handling and loading should be included with installation and material costs. Using larger sheets (manufactured up to 4′ x 20′) may require less taping and finishing, but these sheets may each weigh well in excess of 100 pounds and are awkward to handle. So that allowable floor loads are not exceeded, the weight of drywall is also a factor to be taken into account (for distribution) when loading a job in existing buildings. All material handling involves costs that must be included.

Pricing

As with plaster work, open spans of drywall should be priced differently from small, intricate areas that require much cutting and patching. Similarly, areas with many corners or curves require higher finishing costs than open walls or ceilings. Corners, both inside and outside, should be estimated by the linear foot, in addition to the square feet of surface area.

Although difficult because of variations, the estimator may be able to develop his own historical systems or assemblies prices for metal studs, drywall, and taping and finishing, similar to those shown in Figure 17.4. When using any system, whether his own or one developed by Means, the estimator must be sure that it is identical to the materials specified for the job. For example, it can be seen in Figure 17.4 that the cost difference between a single layer of 1/2″ regular and 1/2″ fire-resistant drywall is $.02/S.F. To develop the cost for a system of two layers of 1/2″ fire-resistant drywall on both faces, this system can be modified by removing the line of 1/2″ regular drywall and substituting two lines of 1/2″ fire-resistant drywall, each at $.02/S.F. added material cost.

Tile   Ceramic tile and tile manufactured from other special-use materials provide an almost endless source of interior wall, floor, and countertop coverings. Some of the more common applications of ceramic and other tile include: toilet

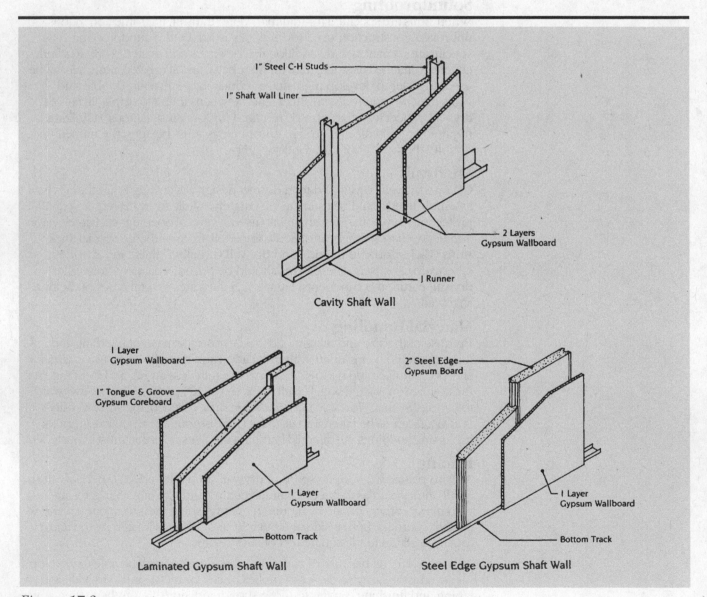

Cavity Shaft Wall

Laminated Gypsum Shaft Wall   Steel Edge Gypsum Shaft Wall

Figure 17.3

C1010 Partitions

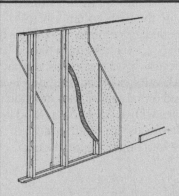

This page illustrates and describes a drywall system including gypsum plasterboard, taped and finished, metal studs with runners, insulation, baseboard and painting. Lines within System Components give the unit price and total price per square foot for this system. Prices for alternate drywall systems are on Line Items C1010 136 1300 thru 1900. Both material quantities and labor costs have been adjusted for the system listed.

Factors: To adjust for job conditions other than normal working situations use Lines C1010 136 2900 thru 4000.

Example: You are to install the system and control dust in the work area. Go to Line C1010 136 3100 and apply these percentages to the appropriate MAT. and INST. costs.

System Components			COST PER S.F.		
	QUANTITY	UNIT	MAT.	INST.	TOTAL
Gypsum drywall, taped, finished and painted 2 faces, galvanized metal studs					
Including top & bottom runners, insulation, painted baseboard, wall 10'high					
Gypsum drywall, 5/8" thick, standard	2.000	S.F.	.48	.80	1.28
Taping and finishing	2.000	S.F.	.08	.80	.88
Metal studs, 20 ga., 3-5/8" wide, 24" O.C.	1.000	S.F.	.16	.54	.70
Insulation, 3-1/2" fiberglass batts	1.000	S.F.	.37	.25	.62
Baseboard	.200	L.F.	.25	.33	.58
Paint baseboard, primer + 2 coats	.200	L.F.	.02	.09	.11
Painting, roller 2 coats	2.000	S.F.	.20	.88	1.08
TOTAL		S.F.	1.56	3.69	5.25

C1010 136	Partitions, Drywall	COST PER S.F.		
		MAT.	INST.	TOTAL
1200	For alternate drywall systems:			
1300	Gypsum drywall, 5/8" thick, fire resistant	1.56	3.69	5.25
1400	Water resistant	1.68	3.69	5.37
1500	1/2" thick, standard	1.54	3.69	5.23
1600	Fire resistant	1.56	3.69	5.25
1700	Water resistant	1.56	3.69	5.25
1800	3/8" thick, vinyl faced, standard	2.32	4.65	6.97
1900	5/8" thick, vinyl faced, fire resistant	2.52	4.65	7.17
2000				
2100				
2200				
2300				
2400				
2500				
2600				
2700				
2900	Cut & patch to match existing construction, add, minimum	2%	3%	
3000	Maximum	5%	9%	
3100	Dust protection, add, minimum	1%	2%	
3200	Maximum	4%	11%	
3300	Material handling & storage limitation, add, minimum	1%	1%	
3400	Maximum	6%	7%	
3500	Protection of existing work, add, minimum	2%	2%	
3600	Maximum	5%	7%	
3700	Shift work requirements, add, minimum		5%	
3800	Maximum		30%	
3900	Temporary shoring and bracing, add, minimum	2%	5%	
4000	Maximum	5%	12%	

Figure 17.4

rooms, tubs, steam rooms, swimming pools, and other related installations where easily cleaned, water-repellent, and durable surfaces are required.

Available Styles and Materials

Ceramic and other tiles are manufactured from several basic materials in a wide range of shapes, sizes, and finishes. Ceramic tiles are manufactured from clay, porcelain, or cement. Metal and plastic are the most commonly used materials for hard surface tiles that are not labelled as "ceramic." Floor tiles may be ceramic or they may be brick pavers, quarry tiles, or terra cotta tiles. Tile shapes and sizes vary from 1″ squares to variously sized rectangles, mosaics, patterned combinations, hexagons, octagons, valencia, wedges, and circles. They are available in multicolored designs and mural sets or with individual pieces embossed with designs or pictures. Tiles with special designs and logos may also be custom manufactured. Standard and custom ceramic tiles are produced with glazed and unglazed surfaces with bright, matte, nonslip, and textured finishes.

Installation

Proper installation methods for ceramic tiles are determined by the location and the type of backing surface, which may vary from stud or masonry walls to wood or concrete floors. The tiling material may be placed in individual pieces, or installed in factory-prepared, back-mounted and ungrouted sections or sheets that cover two square feet or more as a unit. Tiling material is also available in factory-prepared sections and patterns, in which the tiles have been preset and pregrouted with silicone rubber.

The two recognized installation methods are the thick-set, or mud-set method and the thin-set method, in which the tile is directly adhered to the base, subbase, or wall material. In thick-set *floor* installations, Portland cement and sand are placed and screeded to a thickness of 3/4″ to 1-1/4″. For thick-set *wall* installations, Portland cement, sand, and lime are placed on the backing surface and troweled to a thickness of 3/4″ to 1-1/4″. With both floor and wall placement, the mortar may be reinforced with metal lath or mesh, and can be backed with impervious membranes. The tiles may be placed on and adhered to the mortar bed while it remains plastic, or they may be placed after the bed has cured, and adhered with a thin bond coat of Portland cement with sand additives. The thin-set method of tile installation requires specially prepared mortars and adhesives on properly prepared surfaces. Latex Portland cement, epoxy mortar, epoxy emulsion mortar, epoxy adhesive, and organic adhesives are some of these specialized thin-set preparations.

The grouting of the placed and adhered tile material is the final critical step in the installation process, because it ensures the sealing of the joints between the tiles and affects the appearance of the installation. The choice of grouting material to be used depends on the type of tile material and the conditions of its exposure. Some of the available grouting mixtures include Portland cement grouts with additives, mastic grout, furan resin grout, epoxy grout, and silicone rubber grout. The recommendations of the manufacturer should be carefully followed in all aspects of the grouting operation. A typical ceramic tile installation is shown in Figure 17.5.

Pricing

Tile is usually taken off and priced per square foot. Linear features such as bullnose trim and cove base are estimated per linear foot. Specialties (accessories) are taken off as *each*.

Ceilings

In addition to exposed structural systems that are painted or stained, there are three basic types of ceilings:

- Directly applied acoustical tile
- Gypsum board, or plaster
- Suspended systems, hung below the supporting superstructure to allow space for ductwork, piping, and lighting systems

Suspension Systems

Suspension systems usually consist of aluminum or steel main runners of *light, intermediate*, or *heavy-duty* classification, spaced two, three, four, or five feet on center, with snap-in cross tees usually available in one- to five-foot lengths. Main runners are usually hung from the supporting structure with tie wire or metal straps spaced as required by the system. The runners and cross tees may be exposed, natural, or painted; or they may be concealed (enclosed by the ceiling tiles). Drywall suspension systems consist of main and cross-tee members, with flanges to allow the gypsum board to be screwed to the supports.

Acoustical Tiles

Acoustical ceiling tiles are available in mineral fiber with many patterns and textures. The most common sizes are 1' x 1', 2' x 2', and 2' x 4'. The face may be perforated, fissured, textured, or plastic-covered. In addition to mineral fiber, ceiling tiles are also available in perforated steel, stainless steel, or aluminum to allow for easy cleaning in high-humidity or kitchen areas. To retain the acoustical characteristics, these panels are filled with sound-absorbing pads. Specialty metal systems are also available as linear ceilings in varying widths in many finishes and configurations. Integrated ceiling systems combine a suspension grid, air handling, lighting, and ceiling tiles into one modular system.

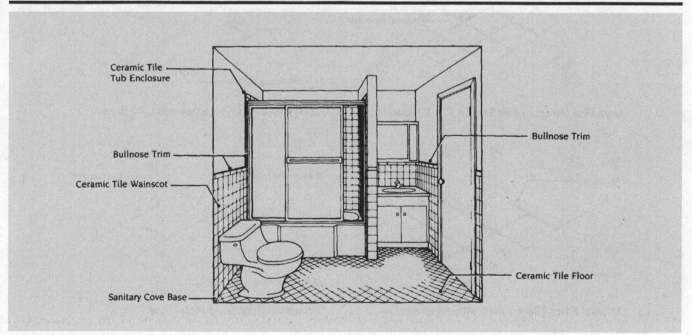

Figure 17.5

Gypsum Board or Plaster

Drywall ceilings may be applied directly to furring strips attached to the supporting structure, or a compatible suspension system. They may be painted, sprayed with acoustical material, or covered with acoustical tiles. Plaster ceilings may be applied directly on self-furring or gypsum lath. Suspended plaster ceilings attach to cold-rolled channels hung from the supporting structure and covered with metal or gypsum lath. Figures 17.6a and b show typical ceiling applications.

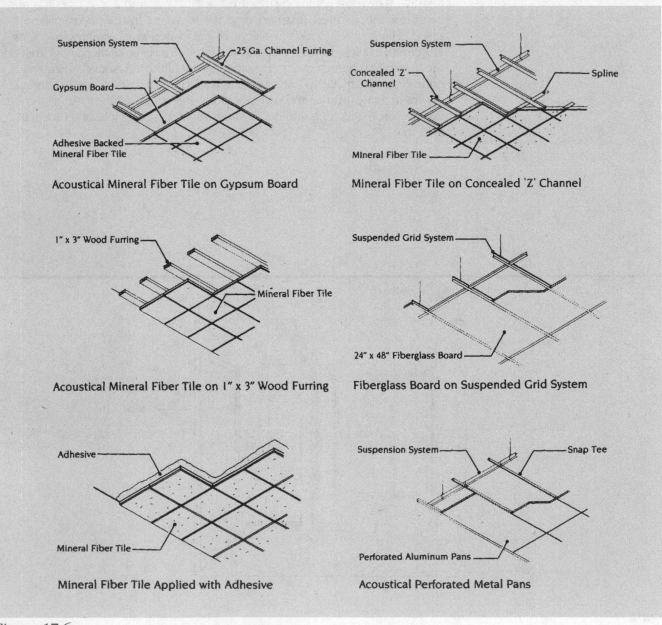

Acoustical Mineral Fiber Tile on Gypsum Board

Mineral Fiber Tile on Concealed 'Z' Channel

Acoustical Mineral Fiber Tile on 1" x 3" Wood Furring

Fiberglass Board on Suspended Grid System

Mineral Fiber Tile Applied with Adhesive

Acoustical Perforated Metal Pans

Figure 17.6a

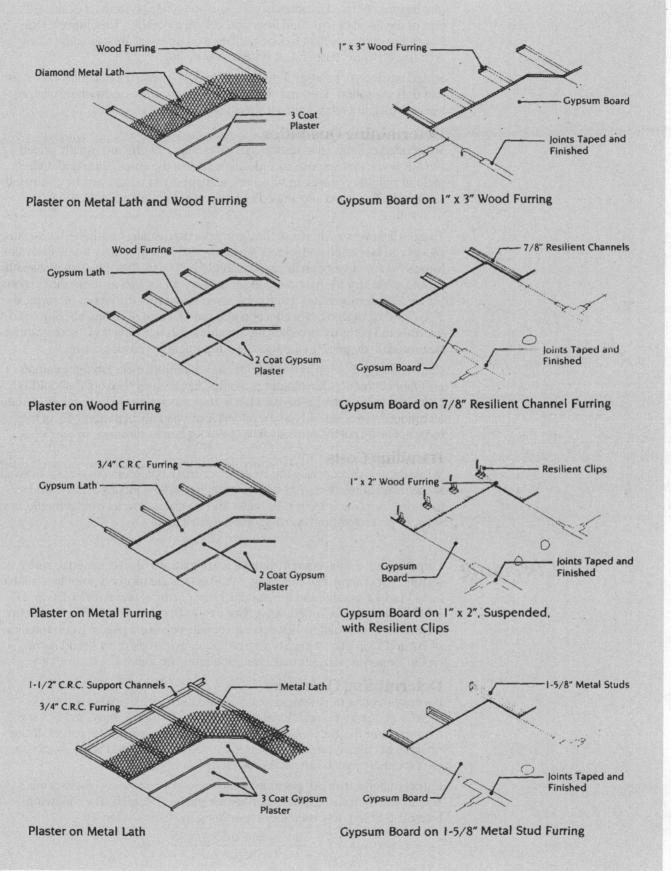

Plaster on Metal Lath and Wood Furring

Wood Furring
Diamond Metal Lath
3 Coat Plaster

Gypsum Board on 1" x 3" Wood Furring

1" x 3" Wood Furring
Gypsum Board
Joints Taped and Finished

Plaster on Wood Furring

Wood Furring
Gypsum Lath
2 Coat Gypsum Plaster

Gypsum Board on 7/8" Resilient Channel Furring

7/8" Resilient Channels
Gypsum Board
Joints Taped and Finished

Plaster on Metal Furring

3/4" C.R.C. Furring
Gypsum Lath
2 Coat Gypsum Plaster

Gypsum Board on 1" x 2", Suspended, with Resilient Clips

1" x 2" Wood Furring
Resilient Clips
Gypsum Board
Joints Taped and Finished

Plaster on Metal Lath

1-1/2" C.R.C. Support Channels
3/4" C.R.C. Furring
Metal Lath
3 Coat Gypsum Plaster

Gypsum Board on 1-5/8" Metal Stud Furring

1-5/8" Metal Studs
Gypsum Board
Joints Taped and Finished

Figure 17.6b

Fire Resistance

Fire resistance (rated in hours) for the various ceiling systems is evaluated as part of the floor, or structural floor and ceiling assembly. To achieve a rating in hours, fire dampers, light fixtures, grilles, and diffusers are evaluated along with the structure, suspension system, and ceiling.

Sound barriers for acoustical deadening usually consist of materials with mass that deflects sound. They may be masonry, metal with sandwich insulation, leaded vinyl, fiberglass batts, or sheet lead.

Determining Quantities

Suspended ceilings of acoustical tiles, drywall, or plaster are usually priced per S.F. Some system costs include all or part of the suspension grid. Others include only the panels, in which case suspension costs must be estimated separately. Patterned and angled ceilings require more material and labor to install.

There is little waste when installing grid systems (usually less than 5%), because pieces can be butted and joined. Waste for tile, however, may range from as low as 5% for large open areas, to as high as 30% to 40% for small areas and rooms. Waste for tile may depend on grid layout as well as room dimensions. Figure 17.7 demonstrates that for the same size room, the layout of a typical 2' x 4' grid has a significant effect on generated waste of ceiling tile. Since most textures and patterns on ceiling tile are aligned in one direction, pieces cannot be turned 90 degrees (to the specified alignment) to reduce waste.

Certain tile types, such as tegular (recessed), require extra labor for cutting and fabrication at edge moldings. Soffits, fascias and "boxouts" should be estimated separately due to extra labor, material waste, and special attachment techniques. Costs should also be added for unusually high numbers of tiles to be specially cut for items such as sprinkler heads, diffusers, or telepoles.

Handling Costs

While the weight of ceiling tile is not the primary consideration, unlike drywall, some material handling and storage costs will still be incurred and must be included. Acoustical tile is very bulky and cumbersome, as well as fragile, and must be protected from damage before installation.

Carpeting

Carpeting is manufactured in various combinations of size, material, and texture. For commercial carpeting, rolled goods are most commonly available in 54" and 12' widths, and the standard size of carpet tiles is 18" x 18" or 24" x 24". Carpet may be tufted, woven, or fusion-bonded. The surface texture may be level loop or multilevel loop, cut level pile, velvet cut pile, or a combination of cut and loop pile. Materials may be wool, nylon (different brand name types), or acrylic. Carpets may be of one fabric or a blend.

Determining Quantities

If a custom color or pattern is required and the quantity exceeds 100 S.Y., there is generally no additional cost for manufacturing, provided that the manufacturer has the appropriate equipment to construct the carpet. If the quantity of custom carpet required is less than 100 S.Y. and is a custom pattern or color, there may be an additional charge of up to 20%.

Carpet material may be specified or measured by its face weight in ounces per square yard, its pile height, its density, or stitches per inch. The following formula is commonly used for determining the carpet's density.

$$\text{Density (oz./S.Y.)} = \frac{\text{Face weight, oz./S.Y. x 36}}{\text{Pile height, in.}}$$

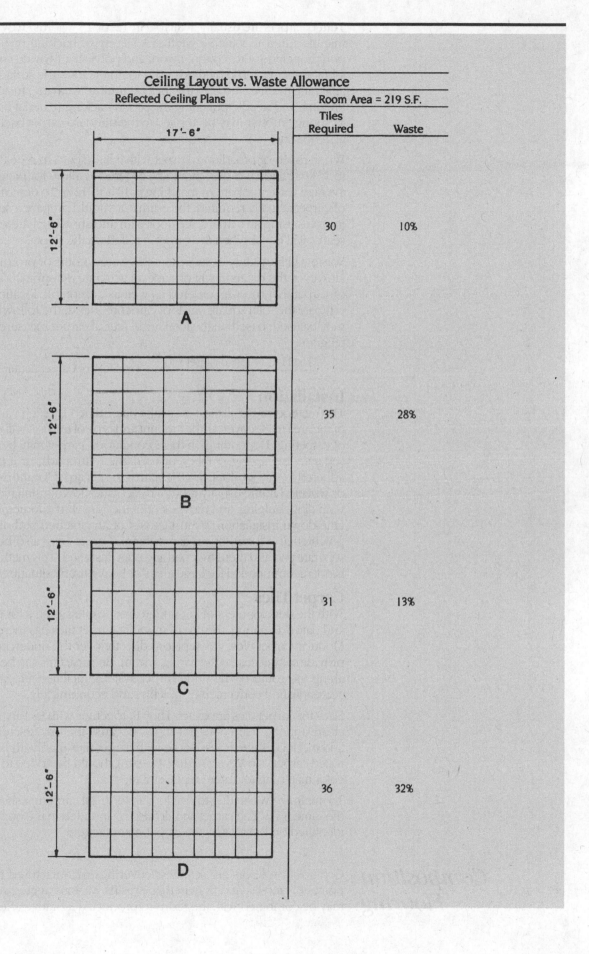

Figure 17.7

Tufted carpets are usually available in 12' or 15' widths. Woven carpets are manufactured in variable widths. A variety of backing materials, including polypropelene, jute, cotton, rayon, and polyester, provide support for the surface material. The methods used for integrating the surface materials and attaching them to the backing include tufting, weaving, fusion bonding, knitting, and needle punching. A second backing material of foam, urethane, or sponge rubber may be applied to the standard carpet backing to serve as an attached pad.

When rolled or broadloom carpet with a large pattern repeat is specified, up to 15% additional carpet should be ordered to allow for pattern matching. The average waste for pattern matching is 10% to 12%. To determine the quantity of carpet actually required, the estimator should prepare a seam diagram. This involves using an actual scale floor plan and showing where all the carpet seams will occur when the carpet is rolled on the floor.

Waste will inevitably occur where there are cutouts or protruding corners. However, the diagram will provide an accurate determination of the amount of carpet that must be ordered as well as acceptable locations for seams. To estimate the total square yards of carpet required, the following equation may be used, based on the total lineal feet of carpet measured on the seam diagram:

$$\frac{\text{Total L.F. of rolled carpet} \times 12}{9} = \text{Total S.Y. of Carpet Required}$$

Installation

The method of installation is usually determined by such variables as the size of the area to be covered, the amount and type of traffic it will carry, and the style of carpeting. Depending on these conditions, carpet may be installed over separate felt, sponge rubber, or urethane foam pads, or it may be directly adhered to the subfloor. Wide expanses and carpet locations that carry heavy or wheeled traffic usually require direct glue-down installation to prevent wrinkling, bulging, and movement of the carpet. If padding is required in a glue-down installation, it must consist of an attached pad of synthetic latex, polyurethane foam, or similar material. Carpet may also be installed over a separate pad and then stretched and edge fastened. This method of installation is not recommended for large areas or heavy traffic situations.

Carpet Tiles

With the development of flat undercarpet cable systems for power, telephone, and data line wiring, the use of modular carpet tiles has increased greatly. Modular carpet tiles, when placed directly over the undercarpet cable run, provide easy access to the wiring system. Because taps can be made at any point along the route of the undercarpet system, facilities relocation can be accomplished conveniently, quickly, and economically.

Sizes for carpet tiles are exact. They fit together with factory edges, which are clean, and the seam does not show. Because the manufactured sizes are small and modular, there is less waste with carpet tiles than with broadloom or rolled carpet. However, 7% to 10% extra should be added to ensure proper color matching for future replacement.

To minimize waste due to dye lot variation, all carpeting should be ordered at the same time. The carpet should also be installed in the numerical sequence identified on the rolls by the carpet manufacturer.

Composition Flooring

Composition floors are seamless coverings manufactured from epoxy, polyester, acrylic, and polyurethane resins. Various aggregates or color chips may be combined with the resins to produce decorative effects, or special coatings may be applied to enhance the functional effectiveness of a floor's surface. Composition floors are normally installed in special-use situations, where other flooring materials do not meet sanitary, safety, or durability standards.

In most cases, the materials that are combined to form composition flooring are determined by and designed for the specific application. Some of the available composition flooring systems include: terrazzo (for decorative floors), wood toppings and sealers, concrete toppings and sealers, waterproof and chemical-resistant floors, gymnasium sports surfaces, interior and exterior sports surfaces, and conductive floors. The materials employed in these systems are applied by trowel, roller, squeegee, notched trowel, spray gun, brush, or similar tools. After the flooring materials have been applied and allowed to cure, they may be sanded and/or sealed to produce the desired texture and surface finish. Composition floors are normally estimated and installed by trained and approved specialty contractors.

Resilient Floors

Resilient floors are designed for situations where durability and low maintenance are primary considerations. The various resilient flooring materials and their installation formats include: asphalt tiles, cork tiles, polyethylene in rolls, polyvinyl chloride in sheets, rubber in tiles and sheets, vinyl composition tiles, and vinyl in sheets and tiles. All of these materials may be manufactured with or without resilient backing and, except for the polyethylene rolls, they are available in a wide range of colors, designs, textures, compositions, and styles. Rubber and vinyl accessories, which are designed to complement any type of flooring material, include the following: bases, beveled edges, thresholds, corner guards, stair treads, risers, nosings, and stringer covers. Some of these materials are illustrated in Figure 17.8.

Manufacturers' recommendations should be carefully followed for the installation details of any resilient floor. Generally, any concrete floor surface should be dry, clean, and free from depressions and adhered droppings. Curing and separating compounds should also be thoroughly removed from the surface before the resilient floor is placed. Special consideration is required for installations on slabs or wood surfaces that are located below grade level or above low crawl spaces.

The floor thickness specified is important to note, because there are different thickness grades for commercial and residential use. Resilient floors are usually estimated and priced per square foot; base and stair stringers are priced per linear foot.

Due to the thinness and flexibility of resilient goods, defects in the subfloor easily "telegraph" through the material. Consequently, the subfloor material and the quality of surface preparation are very important. Subcontractors will often make contracts conditional on a smooth, level subfloor. Surface preparation, which can involve chipping and patching, grinding or washing, is often an "extra." Some surface preparation will invariably be required and an allowance for this work should be included in the estimate. This is especially true in renovation where, in extreme cases, the floor may have to be leveled with a complete application of special lightweight leveling concrete.

Terrazzo

Terrazzo flooring materials provide many options to produce colorful, durable, and easily cleaned floors. Conventional ground and polished terrazzo floors employ granite, marble, glass, and onyx chips in a choice of specialized matrices. Well-graded gravel and other stone materials may be added to the mix to create different textures. Precast terrazzo tiles and bases, which are

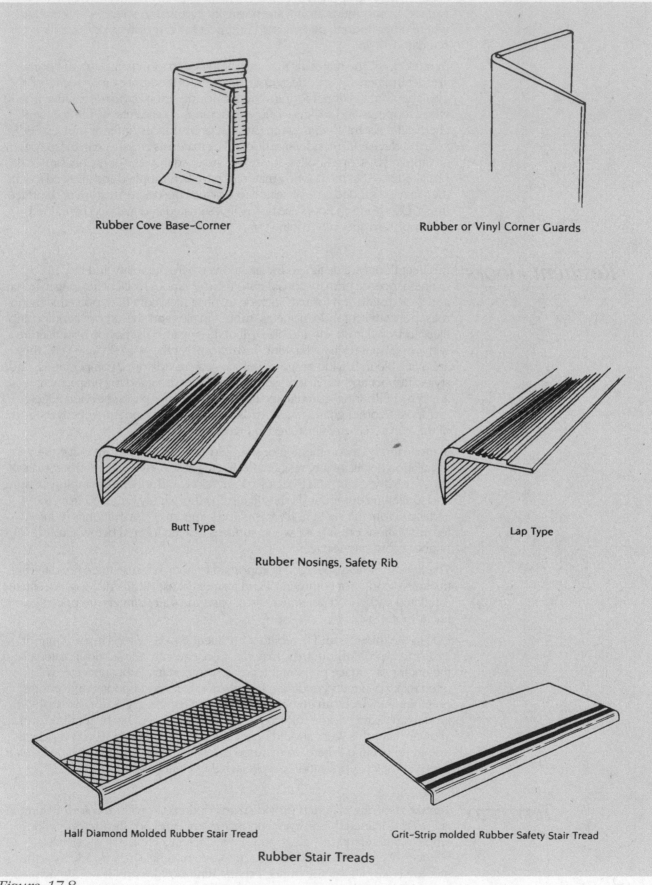

Rubber Cove Base-Corner

Rubber or Vinyl Corner Guards

Butt Type

Lap Type

Rubber Nosings, Safety Rib

Half Diamond Molded Rubber Stair Tread

Grit-Strip molded Rubber Safety Stair Tread

Rubber Stair Treads

Figure 17.8

normally installed finished and polished on a cement sand base, are also available in many color combinations and aggregates.

Conventional installation of terrazzo flooring involves mixing the aggregate with one of three commonly used bonding matrices and placing the mix in sections defined by divider strips. The matrices employed to bond the aggregate include the following: cementitious matrices (which consist of natural or colored Portland cement), cement with an acrylic additive, and resinous matrices (which consist of epoxy and polyester). The divider strips, which are manufactured from zinc, brass, or colored plastic, provide for expansion and are, therefore, positioned over breaks in the substrate and at critical locations where movement is expected in large sections of flooring. The divider strips are also used to terminate pours, to act as leveling guides, and to permit changes in the aggregate mix, section-to-section, to create designs or patterns.

The terrazzo flooring materials may be installed on a cement sand underbed, or they may be applied in the thin-set method directly to the concrete slab. Sand-cushioned terrazzo employs three layers of material, as follows: a 1/4"-thick sand cushion placed on the slab and covered by an isolation membrane; a mesh-reinforced underbed of approximately 1-3/4" thickness; and a 1/2"-thick terrazzo topping (see Figure 17.9). Bonded terrazzo consists of two layers of material: a 1-3/4"-thick underbed placed on the slab and a 1/2"-thick terrazzo topping. Monolithic terrazzo flooring is comprised of a single 1/2"-thick layer of terrazzo topping applied directly to the slab after control joints have been saw cut into the slab and the divider strips grouted into place. With the thin-set method, the terrazzo mix is applied in a single 1/4"-thick layer directly on the concrete slab. The surface of stair treads and risers may be poured in place with a 1/2"-thick layer of terrazzo topping on a 3/4" underbed, or they may be installed as precast terrazzo pieces on a 3/4"-thick underbed.

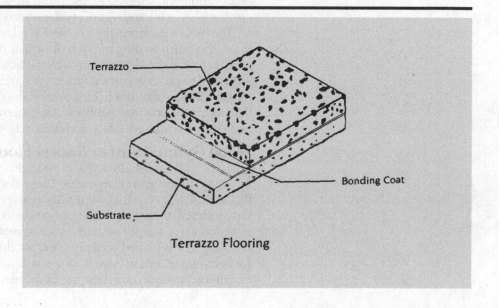

Terrazzo Flooring

Figure 17.9

Wood Floors

Wood flooring material is manufactured in the following three common formats: solid or laminated planks, solid or laminated parquet, and end-grain blocks, which are available in individual pieces or in preassembled strips. Commonly employed hardwood flooring materials include ash, beech, cherry, mahogany, oak, pecan, teak, and walnut, as well as exotic hardwood species, such as ebony, karpa wood, rosewood, and zebra wood. Cedar, fir, pine, and spruce are among the popular softwood flooring materials. End-grain block flooring is manufactured from alder, fir, hemlock, mesquite, oak, and yellow pine.

Strip Flooring

This material is supplied in several different milling formats and combinations, including tongue-and-groove and matched, square-edged, and jointed with square edges and splines. The installation of tongue-and-groove flooring usually requires blind nailing or fastening by means of metal attachment clips, while square-edged flooring is usually face-fastened. Strip floors may be installed over wood-framed subfloors of planks or plywood sheets that have been fastened to the floor joists. A layer of building paper should be laid between the subfloor and the finish strip floor to provide a comfort cushion and to reduce noise. If the strip floor is placed over a concrete surface, a subfloor of exterior plywood should be fastened directly to the slab and protected from moisture by a polyethylene vapor barrier between the concrete and the subfloor. The plywood subfloor may also be fastened to lapped sleepers that have been imbedded in asphalt floor mastic on the slab and draped with the polyethylene vapor barrier before the subfloor is placed.

Parquet Floors

These floors are prefabricated in panels of various sizes that are milled with square edges, tongue-and-groove edges, or splines. The panels may also contain optional adhered backings that protect the flooring material from moisture, add comfort to the walking area, provide insulation, and deaden sound. Adhesives are normally employed to attach the parquet flooring panels to firm, level subfloors of concrete, wood, plywood, particle board, resilient tile, or terrazzo. Some parquet floor manufacturers also supply feature strips and factory-fashioned moldings to cover the required expansion space at the edges of the floor.

Wood Block Floors

Wood block floors for industrial application are manufactured in individual blocks and preassembled strips. The surface dimensions of individual blocks are 3″ x 6″, 4″ x 6″, and 4″ x 8″, with nominal thicknesses ranging from 1–1/2″ to 4″. The blocks are normally installed in a layer of pitch applied to a concrete floor. The finish coating of pitch or similar material is then squeezed into the joints between the blocks to provide additional fastening strength. A sealer is applied to provide surface finish. End-grain strip block flooring is placed in mastic adhesive that has been troweled into a dampproofing membrane that covers the concrete subfloor. After the strip block flooring material has been laid, it is sanded, filled, and finished with penetrating oil.

Wood Gymnasium or Sports Floors

This type of floor usually requires specialized installation methods because of its unique function and large size. These floors may be placed over sleepers that are installed on cushions or pads, or they may be laid directly over a resilient base material or plywood sub-base on cushioned pads. If the flooring material is placed on a sleeper support system, metal clips may be applied for fastening. If it is placed on a plywood subfloor, direct nailing is normally used for fastening. Because large floor areas, such as gymnasiums, require a wide expansion space at the edges, the placement and type of closure strip installed to cover the space deserve special consideration. Wood floors are usually priced per square foot of coverage required.

Painting and Finishing

Painting and finishing are required to protect interior and exterior surfaces against wear and corrosion and to provide a coordinated finish appearance on the protected material. Normally, paints are classified by their binders or vehicles. **Alkyds** are oil-modified resin used to manufacture fast-drying enamels. **Chlorinated rubber** produces coatings that are resistant to alkalies, acids, chemicals, and water. **Catalyzed epoxies** are two-part coatings that produce a hard film resistant to abrasion, traffic, chemicals, and cleaning. **Epoxy esters** are epoxies (modified with oil) that dry by oxidation, but are not as hard as catalyzed epoxies. **Latex binders** (commonly polyvinylacetate, acrylics, or vinyl acetate-acrylics) are binders mixed with a water base. **Silicone alkyd binders** are oil-modified to produce coatings with heat resistance and high-gloss retention. **Urethanes** are isocynate polymers that are modified with drying oils or alkyds. **Vinyl coating solutions** are plasticized copolymers of vinyl chloride and vinyl acetate dissolved in strong solvents. **Zinc coatings** are primers high in zinc dust content dispersed in various vehicles to provide coatings to protect steel from oxidation.

Surface Preparation
In order to achieve successful painting, the surface to be painted must first be cleaned. Preparation methods for painting include solvent cleaning, hand-tool cleaning, (including wire brushing, scraping, chipping, and sanding), power-tool cleaning, white metal blasting (including removal of all rust and scale), commercial blasting (to remove oil, grease, dirt, rust), and scaling or brushing off blast (to remove oil, grease, dirt, and loose rust). These surface preparation processes may be costly and must be included in the estimate, where required.

Primers, Sealers, and Topcoats
Fillers, primers, and sealers are manufactured for use with paint or clear coatings for metal, wood plaster, drywall, and masonry. Sanding fillers are available for clear coatings and block fillers for masonry construction.

Topcoats or finish coats should be specified and applied to meet service requirements. Finishes may be supplied for *normal* use and atmospheric conditions, *hard service* areas, and *critical* areas. Finishes can be furnished in gloss, semigloss, low lustre, eggshell, or flat. Intumescent coatings (expandable paints) are available to meet fire-retardant requirements.

Meeting Specified Requirements
Architects and designers will be particularly insistent about adherence to specifications and specified colors for painting, as with most finishes. This is not an area in which to cut corners in estimating or performance of the work. Usually, the specifications clearly define acceptable materials, manufacturers, preparation, and application methods for each different type of surface to be painted. Many samples of various colors and/or finishes may be required for approval by the architect or owner before final decisions are made.

Determining Quantities
The materials and methods for painting should be included in the specifications. Areas to be painted are usually defined on a Room Finish Schedule and taken off from the plans and elevations in square feet of surface area. Odd-shaped and special items may be converted to an equivalent wall area. The table in Figure 17.10 includes suggested conversion factors for various types of

Balustrades:		1 Side × 4
Blinds:	Plain	Actual area × 2
	Slotted	Actual area × 4
Cabinets:	Including interior	Front area × 5
Downspouts and Gutters:		Actual area × 2
Drop Siding:		Actual area × 1.1
Cornices:	1 Story	Actual area × 2
	2 Story	Actual area × 3
	1 Story Ornamental	Actual area × 4
	2 Story Ornamental	Actual area × 6
Doors:	Flush	Actual area × 1.5
	Two Panel	Actual area × 1.75
	Four Panel	Actual area × 2.0
	Six Panel	Actual area × 2.25
Door Trim:		LF × 0.5
Fences:	Chain Link	1 side × 3 for both sides
	Picket	1 side × 4 for both sides
Gratings:		1 side × 0.66
Grilles:	Plain	1 side × 2.0
	Lattice	Actual area × 2.0
Mouldings:	Under 12″ Wide	1 SF/LF
Open Trusses:		Length × Depth × 2.5
Pipes:	Up to 4″	1 SF per LF
	4″ to 8″	2 SF per LF
	8″ to 12″	3 SF per LF
	12″ to 16″	4 SF per LF
	Hangers Extra	
Radiators:		Face area × 7
Sanding and Puttying:	Quality Work	Actual area × 2
	Average Work	Actual area × 0.5%
	Industrial	Actual area × 0.25%
Shingle Siding:		Actual area × 1.5
Stairs:		No. of risers × 8 widths
Tie Rods:		2 SF per LF
Wainscotting, Paneled:		Actual area × 2
Walls and Ceilings:		Length × Width no deductions for less than 100 SF
Window Sash:		1 LF of part = 1 SF

Figure 17.10

surfaces. Although the factors in Figure 17.10 are used to determine quantities of equivalent wall surface areas, the appropriate, specified application method must be used for pricing.

Pricing

The choice of application method will have a significant impact on the final cost. Spraying is very fast, but the costs of masking the areas to be protected may offset the savings. Oversized rollers may be used to increase production. Brushwork, on the other hand, is labor intensive. The specifications often include (or restrict) certain application methods. Typical coverage and labor-hour rates are shown in Figure 17.11, from *Means Building Construction Cost Data*, for different application methods. These figures include masking and protection of adjacent surfaces.

Depending on local work rules, some unions require that painters be paid higher rates for spraying, and even for roller work. Higher rates also tend to apply for structural steel painting, for high work and for the application of fire-retardant paints. The estimator should determine which restrictions may apply.

Wall Coverings

Wall coverings are manufactured, printed, or woven in burlaps, jutes, weaves, grasses, paper, leather, vinyl, silks, stipples, corks, foils, sheets, cork tiles, flexible wood veneers, and flexible mirrors. Wallpaper, vinyl wall coverings, and woven coverings are usually available in different weights, backings, and quality. Surface preparation and adhesive selection are important considerations in placing wall coverings.

R09910-220   Painting

Item	Coat	One Gallon Covers			In 8 Hours a Laborer Covers			Labor-Hours per 100 S.F.		
		Brush	Roller	Spray	Brush	Roller	Spray	Brush	Roller	Spray
Paint wood siding	prime	250 S.F.	225 S.F.	290 S.F.	1150 S.F.	1300 S.F.	2275 S.F.	.695	.615	.351
	others	270	250	290	1300	1625	2600	.615	.492	.307
Paint exterior trim	prime	400	—	—	650	—	—	1.230	—	—
	1st	475	—	—	800	—	—	1.000	—	—
	2nd	520	—	—	975	—	—	.820	—	—
Paint shingle siding	prime	270	255	300	650	975	1950	1.230	.820	.410
	others	360	340	380	800	1150	2275	1.000	.695	.351
Stain shingle siding	1st	180	170	200	750	1125	2250	1.068	.711	.355
	2nd	270	250	290	900	1325	2600	.888	.603	.307
Paint brick masonry	prime	180	135	160	750	800	1800	1.066	1.000	.444
	1st	270	225	290	815	975	2275	.981	.820	.351
	2nd	340	305	360	815	1150	2925	.981	.695	.273
Paint interior plaster or drywall	prime	400	380	495	1150	2000	3250	.695	.400	.246
	others	450	425	495	1300	2300	4000	.615	.347	.200
Paint interior doors and windows	prime	400	—	—	650	—	—	1.230	—	—
	1st	425	—	—	800	—	—	1.000	—	—
	2nd	450	—	—	975	—	—	.820	—	—

Figure 17.11

Determining Quantities

Wall coverings are usually estimated by the number of rolls. Single rolls contain approximately 36 S.F. This figure is used to determine the number of rolls required. Wall coverings are, however, usually sold in double- or triple-roll bolts.

The area to be covered is measured by multiplying the length times height of wall above the baseboards in order to get the square footage of each wall. This figure is divided by 30 to obtain the number of single rolls, allowing 6 S.F. of waste per roll. One roll should be deducted for every two door openings. Two pounds of dry paste make about three gallons of ready-to-use adhesive, which will cover about 36 single rolls of light- to medium-weight paper, or 14 rolls of heavyweight paper. Application labor costs vary with the quality, pattern, and type of joint required.

With vinyls and grass cloths requiring no pattern match, a waste allowance of 10% is normal (approximately 3.5 S.F. per roll). Wall coverings that require a pattern match may have about 25% to 30% waste, or 9 S.F. to 11 S.F. per roll. Waste can run as high as 50% to 60% on wall coverings with a large, bold, or intricate pattern repeat.

Commercial wallcoverings are available in widths from 21" to 54" and in lengths from 5-1/3 yards (single roll) to 100-yard bolts. To determine quantities, independent of width, the linear (perimeter) footage of walls to be covered should be measured. The linear footage should then be divided by the width of the goods, to determine the number of "strips" or drops. The number of strips per bolt or package can be determined by dividing the length per bolt by the ceiling height.

$$\frac{\text{Linear Footage of Walls}}{\text{Width of Goods}} = \text{No. of Strips Required}$$

$$\frac{\text{Length of Bolt (roll)}}{\text{Ceiling Height}} = \text{No. of Strips (whole no.) per Bolt (roll) Height}$$

Finally, divide the quantity of strips required by the number of strips per bolt (roll) in order to determine the required amount of material, using the same waste allowances as above.

$$\frac{\text{No. of Strips Required}}{\text{No. of Strips per Bolt (roll)}} = \text{No. of Bolts (rolls)}$$

Surface preparation costs for wall covering must also be included. If the wall covering is to be installed over new surfaces, the walls must be treated with a wall sizing, shellac, or primer coat for proper adhesion. For existing surfaces, scraping, patching, and sanding may be necessary. Requirements will be included in the specifications.

Chapter 18

Specialties

Specialties include prefinished, manufactured items that are usually installed at the end of a project when other finish work is complete. Following is a partial list of items that may be included in the specifications under Division 10–Specialties.

- Bathroom Accessories
- Bulletin Boards
- Chutes
- Control Boards
- Directory Boards
- Display Cases
- Key Cabinets
- Lockers
- Mailboxes
- Medicine Cabinets
- Partitions:
  - folding accordion
  - folding leaf
  - hospital
  - moveable office
  - operable
  - portable
  - shower
  - toilet
  - woven wire
- Part Bins
- Projection Screens
- Security Gates
- Shelving
- Signs
- Telephone Enclosures
- Turnstiles

A thorough review of the drawings and specifications is necessary to be sure that all items are accounted for. The estimator should list each type of item and the recommended manufacturers. Often, no substitutes are allowed. Each type of item is then counted. Takeoff units will vary with different items.

Quotations and bids should be solicited from local suppliers and specialty subcontractors. The estimator must include all appropriate shipping and handling costs. When a specialty item is particularly large, job-site equipment may be needed for placement or installation.

The estimator should pay particular attention to the construction requirements that are necessary to Division 10 work but are included in other divisions. Almost all items in Division 10 require some form of base, or backing, or preparation work for proper installation. These requirements may or may not be included in the construction documents, but they are usually listed in the manufacturers' recommendations for installation. The General Conditions of the specifications often state that "the contractor shall install all products according to manufacturers' recommendations." This is a catch-all phrase that places responsibility on the contractor (and estimator).

Preparation costs prior to the installation of specialty items may, in some cases, exceed the costs of the items themselves. The estimator must visualize the installation in order to anticipate all of the work and costs.

Partitions

Folding partitions, operable walls, and relocatable partitions are manufactured in a variety of sizes, shapes, and finishes. Operating partitions include folding accordion, folding leaf (both shown in Figure 18.1), or individual panel systems. These units may be operated by hand or power. Relocatable partitions include the portable type, which are designed for frequent relocation, and the demountable type, for infrequent relocation.

Operating partitions are supported by aluminum, steel, or wood-framing members. The panels are usually filled with sound-insulation material, because most partitions are rated by their sound-reduction qualities. Panel skin materials include aluminum, composition board, fabric, or wood. The panels may be painted or covered with carpeting, fabric, plastic laminate, vinyl, or wood paneling. Large operating partitions are generally installed by factory specialists after the supporting members and framing have been supplied and erected by the building contractor.

Toilet Partitions, Dressing Compartments, and Screens

Toilet partitions, dressing compartments, and privacy screens are manufactured in a variety of materials, finishes, and colors. They are available in many stock sizes for both regular and handicapped-equipped installations. These partitions may be custom-fabricated to fit special size or use requirements, and they may be supported from the floor, braced overhead, or hung from the ceiling or wall. Available finish materials include marble, painted metal, plastic laminate, porcelain enamel, and stainless steel. Various types of toilet partitions are shown in Figure 18.2.

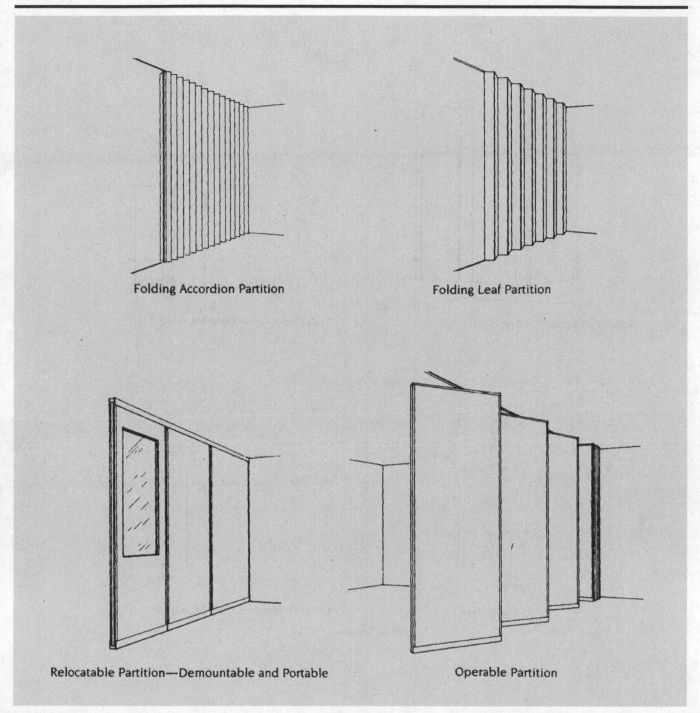

Folding Accordion Partition

Folding Leaf Partition

Relocatable Partition—Demountable and Portable

Operable Partition

Figure 18.1

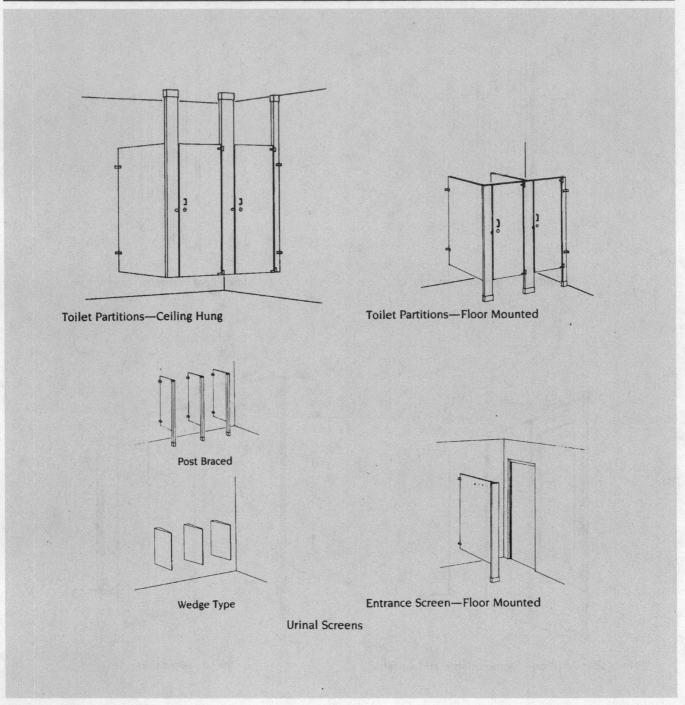

Toilet Partitions—Ceiling Hung

Toilet Partitions—Floor Mounted

Post Braced

Wedge Type

Entrance Screen—Floor Mounted

Urinal Screens

Figure 18.2

Chapter 19

Architectural Equipment

Architectural equipment includes the permanent fixtures that cause the space to function as designed, such as book stacks for libraries and vaults for banks. The construction documents may specify that the owner will purchase architectural equipment directly and that the contractor will install it. In such cases, the estimator must include the installation cost of the equipment. If architectural equipment is furnished by the owner, it is common practice to build approximately 10% of the materials cost into the estimate. This allowance covers handling costs associated with the materials. Often the contractor is responsible for receipt, storage, and protection of these owner-purchased items until they are installed and accepted.

The following items might be included in the architectural equipment category:

- Appliances
- Bank Equipment
- Baking Equipment
- Church Equipment
- Checkout Counter
- Darkroom Equipment
- Dental Equipment
- Detention Equipment
- Health Club Equipment
- Kitchen Equipment
- Laboratory Equipment
- Laundry Equipment
- Medical Equipment
- Movie Equipment
- Refrigerator Food Cases
- Safe
- Sauna
- School Equipment
- Stage Equipment
- Steam Bath
- Vocational Shop Equipment
- Waste Handling Equipment
- Wine Vault

Like specialties, which are described in Chapter 18, architectural equipment must also be evaluated to determine what is required from other divisions for successful installation. Following are some possibilities:

- Concrete Work
- Miscellaneous Metal Supports
- Rough Carpentry and Backing
- Mechanical Coordination
- Electrical Requirements

Division 11−Architectural Equipment includes equipment that can be packaged and delivered complete or partially assembled by the factory. Also, some items can or must be purchased and installed by an authorized factory representative. The estimator must investigate these variables in order to include adequate costs.

Chapter 20

Furnishings

Numerous furnishings may be specified in the plans for remodeling projects. Some may be mass produced, while others may be custom-made, one-of-a-kind items. Furnishings may be supplied by the owner and installed by the vendor, supplied by the owner and installed by the contractor, or carried under a specified allowance.

Before beginning a furniture estimate, the estimator should carefully review both the furnishings drawings and the complete set of construction documents and specifications. Items such as casework, wall art, or banquette seating in a restaurant may not be specified in the furnishings drawings, but they will be found in the construction documents. The estimator must devise a systematic method to ensure that all furnishings items are accounted for. The full set of drawings might be scanned and any item that may affect the furnishings costs might be marked with a colored pencil. Use of a standardized form may also aid in accounting for all items in the project.

A system for labeling each type or style component is also helpful. The following coding system is offered as a quick reference and for accounting purposes. Label all tables of a particular type or style with the symbol "T-1," another style, "T-2," and yet another, "T-3." This system enables the estimator to take off each style table specified and to record each one by category and quantity. A designation such as "T-1:43" might then indicate that there were 43 tables of the specific type and style denoted by "T-1."

Furniture can also be estimated using the *systems estimating method*. This method involves doing the takeoff by work stations or groupings, rather than by individual items.

Furniture ## Office Furniture

Office materials are generally made of wood, metal, or plastic laminate. Differences in the construction of drawer glides, detailing, and the configuration of drawers and compartments distinguish the various products available. Because the construction may differ greatly, thus affecting the cost appreciably, the estimator should verify that apparently identical products (from different manufacturers) are actually the same.

Office seating is usually ergonomic in design. However, each style may offer different features, and chairs vary in style and quality. The estimator should review the specifications carefully, because each style chair within a given price range may have unique options. For example, lounge chairs vary in price based on construction quality and covering fabric.

Tables for office use may range from generic tables to specific-name designer tables. The competitiveness of bidding will be determined by design and style. For example, designer furniture, items that are patented by a specific designer, may only be available through one specific dealership. Generic items, on the other hand, can be obtained through several furniture dealerships, and bidding for these items may be competitive.

Conference tables are usually fabricated by specialty manufacturers. They may be made of wood, plastic laminate, or stone (marble or granite). Prices should be obtained from conference table manufacturers. There may be added costs for shipping, crating, and delivery, due to special size or special handling requirements. Bases for conference tables may vary greatly in type and style. Before adding the cost of shipping, delivery, and installation, the base specifications should be noted, because size may affect these costs significantly.

The quality, style, and price of available office furnishings may vary greatly. Therefore, the estimator should determine exactly what is being specified by technical description, if not by specific manufacturer. Any "grey areas" should be clarified by the specifier, because bidding can be competitive.

Hotel Furniture

Hotel furnishings include bed frames, headboards, mattresses, benches, desks, chairs, dressers, guest tables, mirrors, sleep sofas, and cocktail and end tables. The estimator should review the specifications to determine the requirements for each of these items. Large quantities of hotel furniture may be ordered from a hotel furniture manufacturer, with quantity discounts if no special tooling is required. However, the estimator should verify availability of the specified items, so that bids are not based on discontinued items. Substitutions might otherwise incur additional costs.

Restaurant Furniture

The manufacture and sale of restaurant furnishings may often be a competitive business. Because the cost of chairs and tables of all types, styles, sizes and finishes may vary greatly, the estimator should first determine the amount of detail specified. For example, the specified quality of upholstered seating and finishes may affect costs significantly, especially if the fabric or finish is available from only a single manufacturer. Fabric may be COM (Customer's Own Material), ordered from a fabric manufacturer and delivered to a furniture manufacturer, who then covers a specific style of chair with the material.

The cost of tables varies considerably, depending on the quality specified. Restaurant tabletops and bases are generally sold separately, and, therefore, should be priced separately. The estimator should determine the requirements for the type of top and base from the specifications. Top materials may be wood, glass, metal, or stone, with various edge and top finishes. Bases may be made of wood, metal, or stone. Tops and bases are usually shipped separately, the top having predrilled holes for installation of the base. The estimator should consider the additional cost for attaching the table to the base when figuring installation cost.

Dormitory Furniture

Dormitory furniture includes all bookcases, beds, chairs, bureaus, desks, dressing units, mattresses, mirrors, nightstands, and wardrobes. These units are usually made of wood, plastic laminate, and/or metal, or combinations thereof.

In the calculation of the cost of dormitory furniture, material and installation requirements should be estimated separately. Additional material, such as hardware, or additional labor may be necessary to install these units.

Library Furniture

Library furniture includes book stacks, attendant desks, book display items, book trucks, card catalogs, study carrels (both single- and double-face), chairs, charge desks, dictionary stands, exhibit cases, globe stands, newspaper and magazine racks, and tables for card catalog reference, study use, computers, printers, and word processors. Bids for these types of furnishings should be obtained from a specialty contractor, since special detailing may be required in many of the pieces of casework to meet the library's functional requirements.

School Furniture

School furniture includes chairs and desks made of molded plastic or wood and metal. Estimates should be obtained from local furniture dealers specializing in school furniture.

Furnishings Items

The following section describes items often encountered in furniture estimates. These items are often ordered through a commercial furniture dealership, an interior designer, or an architectural office with purchasing services. Since costs may vary greatly, quotations for specific items and quantities should be obtained. These quotes are often only good for a limited period of time, and any time limit should be noted.

Stack and Folding Chairs

Chairs may be metal, wood, or plastic. The estimator should check the specifications for all finishes prior to determining cost. Before estimating the number of chairs required, the estimator should find out from the manufacturer how many chairs can be stored/stacked on each dolly, or carrier. This information can then be used to determine the number of dollies or carriers that might be required. If upholstered chairs are specified, the fabric should be determined. This cost may be significant, especially for stacking chairs used for hotel ballroom projects, because a large quantity of these chairs is usually required. Floor glides for chairs differ depending on whether they are intended for use with carpet or hard floors. The cost of the glides might affect the cost of the chairs. Another factor is the requirement for tablet arms, and whether these devices, if called for, will be operable or fixed. Chair finishes also vary, from lacquered wood to painted metal.

Booths

Booths, also known as *fixed seating*, are available in a variety of sizes and shapes. The cost varies according to the type of seating specified. For example, banquette upholstered seating for restaurant use differs in price from other types of fixed seating. When banquette seating is called for, the estimator should check the specifications and drawings to verify the exact requirements. In many cases banquette seating should be estimated by a specialty contractor, especially where curved units are required, because these are not standard manufactured items. Other types of fixed seating, however, consist of one-piece plastic chairs and plastic-laminated tabletop units, such as those in fast-food restaurants.

Multiple Seating

Multiple seating is used in airports, hotel lobbies, and reception areas. Costs vary depending on the construction specifications. Numerous types, styles, and materials are available. Therefore, the estimator should carefully review the specifications and drawings to determine what is required.

Lecture Hall Seating

Lecture hall seating requires pedestal-mounted (floor-mounted) seats. The cost for this kind of seating is affected by the following factors: the amount and quality of upholstery required, whether a veneer surface is called for, whether the tablet arm is to be retractable or fixed, and the type of material the shell is to be made of (wood, metal, or plastic). Other considerations that may affect the cost of lecture hall seating are design and style.

Upholstery

Interior projects may call for upholstery on chairs or as wall covering. Upholstery materials include various blends of nylons, polyester, silk, or wool. To estimate the cost of upholstery, the estimator should carefully review the specifications to determine the type of weave, weight, color, pattern requirements, and fabric width. There may be dye-lot variations in some rolls. Therefore, the estimator must carefully calculate the quantity of material needed to ensure that an adequate amount of material from the same dye lot is available for the job.

Fabric treatments may include flame-retardant, acrylic backing, paper backing, and stain protection. When estimating the cost for fabric treatments, the estimator should also include costs for testing of fabric/treatment compatibility.

Folding Tables

Folding tables are used in conference centers, hotels, schools, and wherever large assemblies of people are expected for dining or conference purposes. Shapes may be rectangular or circular. Sizes vary from conventional depths of 18", 30", and 36", and lengths of 48", 60", 72" and 96". Rounds are usually 60" and 72" in diameter. Carts for storage and moving these items should be included in the estimate.

Files and File Systems

There are grades of filing systems for different uses, ranging from *light residential* to *heavy administrative* use. Files may be two- to four-drawer, vertical or lateral. The specifications should be carefully checked, because the specifier has determined the appropriate type of file for each function. Floor loading capacities must also be considered.

Where large, mobile filing systems are used with floor-mounted tracks in filing banks, or for file systems on rollers or rotating banks, a specialty supplier and or contractor should be consulted for accurate pricing. The price of these systems may vary. The specialty contractor is experienced in installing these file systems and, therefore, is familiar with the costs. A structural engineer should be consulted whenever excessive loads are to be placed on a floor not specifically designed for such a purpose.

Cabinets and Countertops

Cabinets for kitchen, hospital, and residential use are generally prefabricated in fixed sizes, ready for installation. Plastic-laminated countertops are available in a variety of widths and may be stock or custom-made. Laminated tops are usually cut and installed by a specialty contractor skilled in assembling laminated products, since special post-forming machines are required to make plastic-laminated, curved surfaces. Costs should be adjusted, however, when small quantities are ordered.

Blinds

Blinds may be either vertical or horizontal, in a variety of finishes including vinyl, metal, fabric, and wood. Slat sizes and the type of operation may vary. The estimator should verify all options in the specifications. The finish for both sides should be checked, since the two sides may differ and this may affect the cost. Unique installations, such as door-hung blinds that require "hold down clips," or blinds in sloped skylights that require special side rails to keep the blinds in place, should be verified with a local specialty contractor to ensure accurate estimates.

Panels and Dividers
Panels and dividers for office use vary in type, style, and size. They may be powered or nonpowered, with a fabric edge or a hard edge of metal, plastic, or wood. The acoustical rating of panels varies with each manufacturer. Available colors and types of connectors also vary.

Coat Racks and Wardrobes
Coat racks and wardrobes are usually wood or metal, with components that vary from project to project. The specifications should be checked to determine the height, width, number of hangers required, and the amount of hat storage required overhead. (Hat storage is indicated by the number and width of shelves at the top of the unit.) Finishes should be verified. If custom colors are required, a specialty manufacturer should be consulted, as there may be a surcharge for custom colors.

Floor Mats
Floor mats are usually made-to-order. Size, material, and installation should be considered when estimating. Material for floor mats may be recessed units made of rubber, aluminum, or synthetics. Floor mat sizes should be confirmed and field-measured prior to ordering. Tile units, for example, come in boxes with a certain amount per box. The specific manufacturer should be consulted to determine the number of boxes required. Installation procedures may vary, from mechanical fasteners to glue, and should be confirmed with the manufacturer.

Ash/Trash Receivers
Trash receivers come in a variety of shapes and sizes with various options. Ash/Trash receivers may be simple sand urns, or units with mechanical self-cleaning tops. The estimator should check the furnishings specifications for the type of operation required and other options, as well as finishes, as these have an impact on the cost of the unit.

Packing, Crating, Insurance, Shipping, and Delivery

There is often an added cost for packing, crating, insurance, shipping, and delivery of furnishings. This added cost for insurance and shipping should be anticipated for each furnishings item, where applicable. The specifications dictate how the furniture is to be packed and crated. Otherwise, large pieces of furniture might be blanket-wrapped, at no additional cost, and put on a truck without the extra protection of a crate or proper packing. Large pieces of furniture, such as hotel and office items such as chairs and desks, should be packed and crated. The extra charge varies depending on the size of the piece of furniture. Since crating and additional insurance charges are extra, they apply to items that are of "high value." The specifications should be checked for this requirement.

Shipping costs for furnishings can generally be estimated at 10% to 15% of the furniture cost. Furnishings may be delivered with an FOB (Freight on Board) designation, and are drop-shipped at the receiving dock.

Delivery time can vary from project to project. Delivery time for small, stock orders will be 30 to 120 days. For medium-sized office and hotel projects, or for custom orders, delivery can be from 60 to 120 days. Therefore, for each furniture item, delivery time should be anticipated. Many manufacturers dealing in office furnishings offer a quick ship/delivery program, but this is usually only for certain basic stock items offered in limited quantities, since they do not usually store large inventories of office furniture.

Delivery charges are generally extra. For example, if furnishings are for the fourth floor of an office building, the contract may be only for delivery to the loading dock of the address specified. Provisions for storage and handling should be made ahead of time for large quantities of furnishings, and the costs appropriately included.

Chapter 21

Special Construction

A partial list of items in Division 13–Special Construction of the specifications and drawings appears below. Subcontractor quotations should be carefully evaluated to make sure that all required items are included. Some of the materials may have to be supplied and/or installation performed by other suppliers or subcontractors.

- Acoustical Enclosures
- Bulk Asbestos Removal
- Building Automation & Control
- Greenhouses
- Fire Suppression Systems
- Lead Paint Removal
- Refrigerators–Walk-in
- Shielding
- Solar Energy
- Sports Courts
- Swimming Pools
- Storage Tanks

It is a good idea to review this portion of the project with the subcontractor(s) to determine both the exact scope of the work and those items that are not covered by the quotation. If the subcontractor requires services such as excavation, unloading, or other temporary work, these otherwise excluded items must be included elsewhere in the estimate.

Often, manufacturers of these products require that only trained personnel install the product. Otherwise, material and performance warranties may be voided.

The specialty subcontractor may provide more detailed information concerning the system. The more detailed the estimator's knowledge of a system, the easier it may be to subdivide the costs into material, labor, and equipment to fully identify the direct cost of the specialty item for future purposes.

Chapter 22

Conveying Systems

Conveying systems used in remodeling may include, but are not limited to, the following:

- Correspondence Lifts
- Dumbwaiters
- Elevators
- Escalators
- Material Handling Conveyers
- Motorized Car Distribution Systems
- Moving Ramps and Walks
- Parcel Lifts
- Pneumatic Tube Systems
- Vertical Conveyers

Current quotations from competent contractors should be obtained for all of the items listed above, if specified and shown on the plans. Budget costs are available and should, when used properly, allow sufficient money to cover material and installation costs for new buildings. Installation in existing buildings must be priced for each individual project. It should be noted that new hydraulic elevators with telescoping shafts have been developed to combat some of the problems encountered when installing elevators in existing buildings. A checklist for the various systems is shown in Figure 22.1.

The following illustrations show typical conveying systems used in renovation projects. Figures 22.2a and 22.2b are schematic drawings of typical conveying systems. Figures 22.3a and 22.3b illustrate an elevator selective cost sheet from *Means Building Construction Cost Data*. This chart may be useful in the development of budget costs for various types of elevators, based on specific project requirements. Note the number of variables that may significantly affect the cost of elevators.

Conveying Systems

Dumbwaiters:
- capacity
- floors
- speed

- size
- stops
- finish

Elevators:
- hydraulic or electric
- capacity
- floors
- stops
- finish
- door type
- special requirements

- geared or gearless
- size
- number required
- speed
- machinery location
- signals

Material handling systems:
- automated

- non-automated

Moving stairs and walks:
- capacity
- floors
- story height
- finish
- incline angle

- size
- number required
- speed
- machinery location
- special requirements

Pneumatic tube systems:
- automatic
- size
- length

- manual
- stations
- special requirements

Vertical conveyer:
- automatic

- non-automatic

Figure 22.1

204

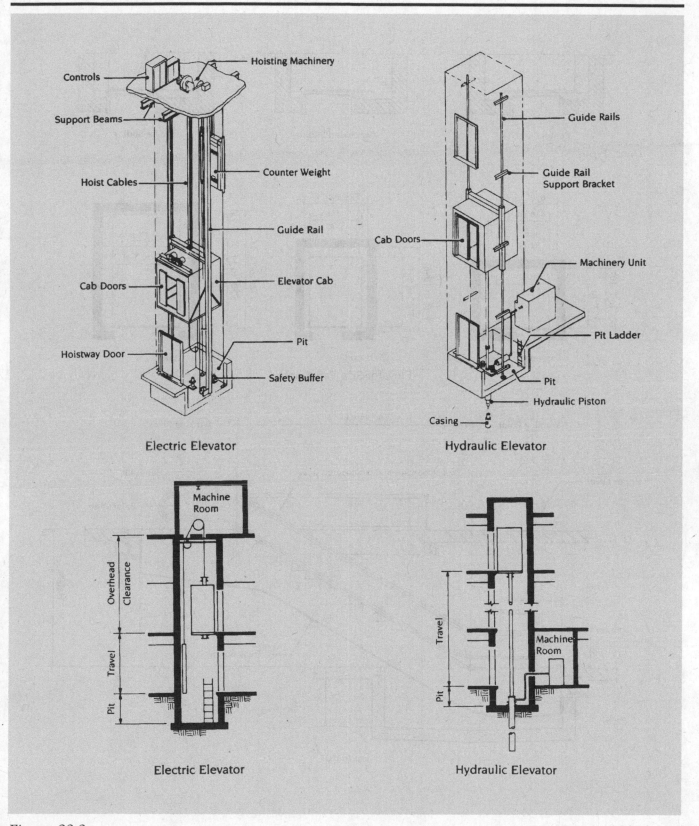

Electric Elevator

Hydraulic Elevator

Electric Elevator

Hydraulic Elevator

Figure 22.2a

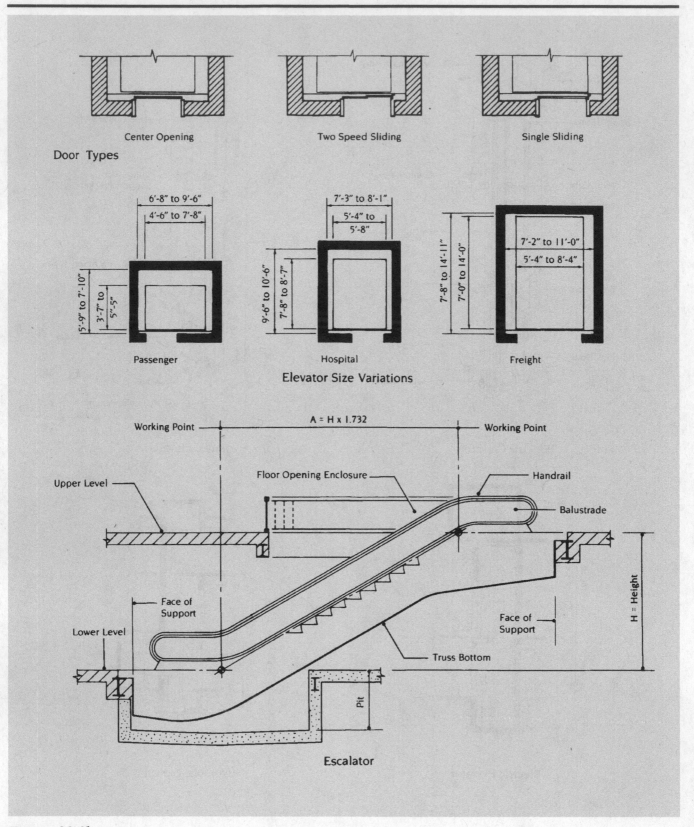

Door Types

Center Opening   Two Speed Sliding   Single Sliding

Elevator Size Variations

Passenger   Hospital   Freight

Escalator

Figure 22.2b

R14200-200 Elevator Selective Costs See R14200-400 for cost development.

A. Base Unit	Passenger		Freight		Hospital	
	Hydraulic	**Electric**	**Hydraulic**	**Electric**	**Hydraulic**	**Electric**
Capacity	1,500 lb.	2,000 lb.	2,000 lb.	4,000 lb.	4,000 lb.	4,000 lb.
Speed	100 F.P.M.	200 F.P.M.	100 F.P.M.	200 F.P.M.	100 F.P.M.	200 F.P.M.
#Stops/Travel Ft.	2/12	4/40	2/20	4/40	2/20	4/40
Push Button Oper.	Yes	Yes	Yes	Yes	Yes	Yes
Telephone Box & Wire	"	"	"	"	"	"
Emergency Lighting	"	"	No	No	"	"
Cab	Plastic Lam. Walls	Plastic Lam. Walls	Painted Steel	Painted Steel	Plastic Lam. Walls	Plastic Lam. Walls
Cove Lighting	Yes	Yes	No	No	Yes	Yes
Floor	V.C.T.	V.C.T.	Wood w/Safety Treads	Wood w/Safety Treads	V.C.T.	V.C.T.
Doors, & Speedside Slide	Yes	Yes	Yes	Yes	Yes	Yes
Gates, Manual	No	No	No	No	No	No
Signals, Lighted Buttons	Car and Hall	Car and Hall	Car and Hall	Car and Hall	Car and Hall	Car and Hall
O.H. Geared Machine	N.A.	Yes	N.A.	Yes	N.A.	Yes
Variable Voltage Contr.	"	"	N.A.	"	"	"
Emergency Alarm	Yes	"	Yes	"	Yes	"
Class "A" Loading	N.A.	N.A.	"	"	N.A.	N.A.

Figure 22.3

Chapter 23

Mechanical

It is doubtful that the remodeling estimator can spend the time or is capable of preparing a realistic unit price estimate for the mechanical trades. A reliable sub-bid should be sought for the mechanical portion of the interior work as soon as specific requirements are known. As a check, or for preliminary budgets, costs can be quickly calculated using the systems approach to mechanical estimating. A systems estimate is basically accomplished by counting the fixtures in the plumbing portion, establishing the class of fire protection required, and determining the heat and air conditioning load, source, and method of distribution. Appropriate systems costs are applied based on determined quantities and/or the size and occupancy use of the project.

For renovation projects, costs for the mechanical portion of the estimate must be estimated by using good judgment and past experience. The "cutting and patching" may be extensive, and existing conditions may severely restrict normal work procedures; both factors can significantly affect cost. Mechanical work is usually broken down into three basic systems:

- Plumbing
- Fire Protection
- Heating, Ventilating, and Air Conditioning

Plumbing

In order to determine a good budget estimate for plumbing, the estimator must count the different fixture types and also determine other specific requirements. Systems costs can be applied per bathroom, as shown in Figure 23.1, from *Means Repair & Remodeling Cost Data*. Note that quantities for rough-in, partitions, and accessories are listed in the "systems components" section, as well as the fixture itself.

Prior to or during design development, the estimator may be required to determine or anticipate plumbing fixture requirements for the proposed project. Figure 23.2 (from *Means Plumbing Cost Data*), adapted from the BOCA Plumbing Code, can be used to determine fixture requirements at this stage. For final design purposes, local building officials must be consulted to ensure compliance with applicable codes and requirements.

For many remodeling projects, plumbing stacks already exist in the space. Toilet rooms and other fixtures are usually located close to this piping. However, when the plumbing fixtures are located away from the stacks or if the stacks are to be part of the work being estimated, costs must be determined and included. For small quantities, costs for pipe (linear feet) and fittings

D2010 Plumbing Fixtures

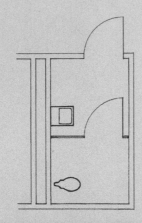

This page illustrates and describes a women's public restroom system including a water closet, lavatory, accessories, and service piping. Lines within System Components give the unit price and total price on a cost each basis for this system. Prices for alternate women's public restroom systems are on Line Items D2010 956 1400 thru 1700. Both material quantities and labor costs have been adjusted for the system listed.

Factors: To adjust for job conditions other than normal working situations use Lines D2010 956 2900 thru 4000.

Example: You are to install the system and protect surrounding area from dust. Go to Line D2010 956 3100 and apply these percentages to the MAT. and INST. costs.

System Components	QUANTITY	UNIT	COST EACH MAT.	COST EACH INST.	COST EACH TOTAL
Public women's restroom incl. water closet, lavatory, accessories and Necessary service piping to install this system in one wall.					
Water closet, wall mounted, one piece	1.000	Ea.	375	139	514
Rough-in waste and vent for water closet	1.000	Set	298	315	613
Lavatory, 20" x 18" P.E. cast iron with accessories	1.000	Ea.	218	101	319
Rough-in waste and vent for lavatory	1.000	Set	240	485	725
Toilet partition, painted metal between walls, floor mounted	1.000	Ea.	405	133	538
For handicap unit, add	1.000	Ea.	286		286
Grab bar, 36" long	1.000	Ea.	58	39.80	97.80
Mirror, 18" x 24", with stainless steel shelf	1.000	Ea.	109	19.90	128.90
Napkin/tampon dispenser, recessed	1.000	Ea.	430	26.50	456.50
Soap Dispenser, chrome, surface mounted, liquid	1.000	Ea.	47	19.90	66.90
Toilet tissue dispenser, surface mounted, stainless steel	1.000	Ea.	12.95	13.25	26.20
Towel dispenser, surface mounted, stainless steel	1.000	Ea.	42	25	67
TOTAL		System	2,520.95	1,317.35	3,838.30

D2010 956	Plumbing - Public Restroom	COST EACH MAT.	COST EACH INST.	COST EACH TOTAL
1200				
1300	For alternate size restrooms:			
1400	Two water closets, two lavatories	4,200	2,500	6,700
1500				
1600	For each additional water closet over 2, add	1,075	580	1,655
1700	For each additional lavatory over 2, add	615	625	1,240
1800				
1900				
2400	NOTE: PLUMBING APPROXIMATIONS			
2500	WATER CONTROL: water meter, backflow preventer,			
2600	Shock absorbers, vacuum breakers, mixer....10 to 15% of fixtures			
2700	PIPE AND FITTINGS: .... 30 to 60% of fixtures			
2800				
2900	Cut & patch to match existing construction, add, minimum	2%	3%	
3000	Maximum	5%	9%	
3100	Dust protection, add, minimum	1%	2%	
3200	Maximum	4%	11%	
3300	Equipment usage curtailment, add, minimum	1%	1%	
3400	Maximum	3%	10%	
3500	Material handling & storage limitation, add, minimum	1%	1%	

Figure 23.1

R15100-410   Minimum Plumbing Fixture Requirements

Minimum Plumbing Fixture Requirements

Type of Building or Occupancy (2)	Water Closets (14) (Fixtures per Person)		Urinals (5,10) (Fixtures per Person)		Lavatories (Fixtures per Person)		Bathtubs or Showers (Fixtures per Person)	Drinking Fountains (Fixtures per Person) (3, 13)
	Male	Female	Male	Female	Male	Female		
Assembly Places-Theatres, Auditoriums, Convention Halls, etc.-for permanent employee use	1: 1 - 15 2: 16 - 35 3: 36 - 55 Over 55, add 1 fixture for each additional 40 persons	1: 1 - 15 2: 16 - 35 3: 36 - 55	0: 1 - 9 1: 10 - 50 Add one fixture for each additional 50 males		1 per 40	1 per 40		
Assembly Places-Theatres, Auditoriums, Convention Halls, etc. - for public use	1: 1 - 100 2: 101 - 200 3: 201 - 400 Over 400, add 1 fixture for each additional 500 males and 1 for each additional 125 females	3: 1 - 50 4: 51 - 100 8: 101 - 200 11: 201 - 400	1: 1 - 100 2: 101 - 200 3: 201 - 400 4: 401 - 600 Over 600, add 1 fixture for each additional 300 males		1: 1 - 200 2: 201 - 400 3: 401 - 750 Over 750, add 1 fixture for each additional 500 persons	1: 1 - 200 2: 201 - 400 3: 401 - 750		1: 1 - 150 2: 151 - 400 3: 401 - 750 Over 750, add one fixture for each additional 500 persons
Dormitories (9) School or Labor	1 per 10 Add 1 fixture for each additional 25 males (over 10) and 1 for each additional 20 females (over 8)	1 per 8	1 per 25 Over 150, add 1 fixture for each additional 50 males		1 per 12 Over 12 add 1 fixture for each additional 20 males and 1 for each 15 additional females	1 per 12	1 per 8 For females add 1 bathtub per 30. Over 150, add 1 per 20	1 per 150 (12)
Dormitories-for Staff Use	1: 1 - 15 2: 16 - 35 3: 36 - 55 Over 55, add 1 fixture for each additional 40 persons	1: 1 - 15 3: 16 - 35 4: 36 - 55	1 per 50		1 per 40	1 per 40	1 per 8	
Dwellings: Single Dwelling Multiple Dwelling or Apartment House	1 per dwelling 1 per dwelling or apartment unit				1 per dwelling 1 per dwelling or apartment unit		1 per dwelling 1 per dwelling or apartment unit	
Hospital Waiting rooms	1 per room				1 per room			1 per 150 (12)
Hospitals-for employee use	1: 1 - 15 2: 16 - 35 3: 36 - 55 Over 55, add 1 fixture for each additional 40 persons	1: 1 - 15 3: 16 - 35 4: 36 - 55	0: 1 - 9 1: 10 - 50 Add 1 fixture for each additional 50 males		1 per 40	1 per 40		
Hospitals: Individual Room Ward Room	1 per room 1 per 8 patients				1 per room 1 per 10 patients		1 per room 1 per 20 patients	1 per 150 (12)
Industrial (6) Warehouses Workshops, Foundries and similar establishments-for employee use	1: 1 -10 2: 11 - 25 3: 26 - 50 4: 51 - 75 5: 76 - 100 Over 100, add 1 fixture for each additional 30 persons	1: 1 -10 2: 11 - 25 3: 26 - 50 4: 51 - 75 5: 76 - 100			Up to 100, per 10 persons Over 100, 1 per 15 persons (7, 8)		1 shower for each 15 persons exposed to excessive heat or to skin contamination with poisonous, infectious or irritating material	1 per 150 (12)
Institutional - Other than Hospitals or Penal Institutions (on each occupied floor)	1 per 25	1 per 20	0: 1 - 9 1: 10 - 50 Add 1 fixture for each additional 50 males		1 per 10	1 per 10	1 per 8	1 per 150 (12)
Institutional - Other than Hospitals or Penal Institutions (on each occupied floor)-for employee use	1: 1 - 15 2: 16 - 35 3: 36 - 55 Over 55, add 1 fixture for each additional 40 persons	1: 1 - 15 3: 16 - 35 4: 36 - 55	0: 1 - 9 1: 10 - 50 Add 1 fixture for each additional 50 males		1 per 40	1 per 40	1 per 8	1 per 150 (12)
Office or Public Buildings	1: 1 - 100 2: 101 - 200 3: 201 - 400 Over 400, add 1 fixture for each additional 500 males and 1 for each additional 150 females	3: 1 - 50 4: 51 - 100 8: 101 - 200 11: 201 - 400	1: 1 - 100 2: 101 - 200 3: 201 - 400 4: 401 - 600 Over 600, add 1 fixture for each additional 300 males		1: 1 - 200 2: 201 - 400 3: 401 - 750 Over 750, add 1 fixture for each additional 500 persons	1: 1 - 200 2: 201 - 400 3: 401 - 750		1 per 150 (12)
Office or Public Buildings - for employee use	1: 1 - 15 2: 16 - 35 3: 36 - 55 Over 55, add 1 fixture for each additional 40 persons	1: 1 - 15 3: 16 - 35 4: 36 - 55	0: 1 - 9 1: 10 - 50 Add 1 fixture for each additional 50 males		1 per 40	1 per 40		

Figure 23.2

211

R15100-410 Minimum Plumbing Fixture Requirements

Minimum Plumbing Fixture Requirements

Type of Building or Occupancy	Water Closets (14) (Fixtures per Person)		Urinals (5, 10) (Fixtures per Person)		Lavatories (Fixtures per Person)		Bathtubs or Showers (Fixtures per Person)	Drinking Fountains (Fixtures per Person) (3, 13)
	Male	Female	Male	Female	Male	Female		
Penal Institutions - for employee use	1: 1 - 15 2: 16 - 35 3: 36 - 55 Over 55, add 1 fixture for each additional 40 persons	1: 1 - 15 3: 16 - 35 4: 36 - 55	0: 1 - 9 1: 10 - 50 Add 1 fixture for each additional 50 males		1 per 40	1 per 40		1 per 150 (12)
Penal Institutions - for prison use Cell	1 per cell				1 per cell			1 per cellblock floor
Exercise room	1 per exercise room		1 per exercise room		1 per exercise room			1 per exercise room
Restaurants, Pubs and Lounges (11)	1: 1 - 50 2: 51 - 150 3: 151 - 300 Over 300, add 1 fixture for each additional 200 persons	1: 1 - 50 2: 51 - 150 4: 151 - 300	1: 1 - 150 Over 150, add 1 fixture for each additional 150 males		1: 1 - 150 2: 151 - 200 3: 201 - 400 Over 400, add 1 fixture for each additional 400 persons	1: 1 - 150 2: 151 - 200 3: 201 - 400		
Schools - for staff use All Schools	1: 1 - 15 2: 16 - 35 3: 36 - 55 Over 55, add 1 fixture for each additional 40 persons	1: 1 - 15 3: 16 - 35 4: 36 - 55	1 per 50		1 per 40	1 per 40		
Schools - for student use: Nursery	1: 1 - 20 2: 21 - 50 Over 50, add 1 fixture for each additional 50 persons	1: 1 - 20 2: 21 - 50			1: 1 - 25 2: 26 - 50 Over 50, add 1 fixture for each additional 50 persons	1: 1 - 25 2: 26 - 50		1 per 150 (12)
Elementary	1 per 30	1 per 25	1 per 75		1 per 35	1 per 35		1 per 150 (12)
Secondary	1 per 40	1 per 30	1 per 35		1 per 40	1 per 40		1 per 150 (12)
Others (Colleges, Universities, Adult Centers, etc.	1 per 40	1 per 30	1 per 35		1 per 40	1 per 40		1 per 150 (12)
Worship Places Educational and Activities Unit	1 per 150	1 per 75	1 per 150		1 per 2 water closets			1 per 150 (12)
Worship Places Principal Assembly Place	1 per 150	1 per 75	1 per 150		1 per 2 water closets			1 per 150 (12)

Notes:

1. The figures shown are based upon one (1) fixture being the minimum required for the number of persons indicated or any fraction thereof.
2. Building categories not shown on this table shall be considered separately by the Administrative Authority.
3. Drinking fountains shall not be installed in toilet rooms.
4. Laundry trays. One (1) laundry tray or one (1) automatic washer standpipe for each dwelling unit or one (1) laundry trays or one (1) automatic washer standpipes, or combination thereof, for each twelve (12) apartments. Kitchen sinks, one (1) for each dwelling or apartment unit.
5. For each urinal added in excess of the minimum required, one water closet may be deducted. The number of water closets shall not be reduced to less than two-thirds (2/3) of the minimum requirement.
6. As required by ANSI Z4.1-1968, Sanitation in Places of Employment.
7. Where there is exposure to skin contamination with poisonous, infectious, or irritating materials, provide one (1) lavatory for each five (5) persons.
8. Twenty-four (24) lineal inches of wash sink or eighteen (18) inches of a circular basin, when provided with water outlets for such space shall be considered equivalent to one (1) lavatory.
9. Laundry trays, one (1) for each fifty (50) persons. Service sinks, one (1) for each hundred (100) persons.
10. General. In applying this schedule of facilities, consideration shall be given to the accessibility of the fixtures. Conformity purely on a numerical basis may not result in an installation suited to the need of the individual establishment. For example, schools should be provided with toilet facilities on each floor having classrooms.
    a. Surrounding materials, wall and floor space to a point two (2) feet in front of urinal lip and four (4) feet above the floor, and at least two (2) feet to each side of the urinal shall be lined with non-absorbent materials.
    b. Trough urinals shall be prohibited.
11. A restaurant is defined as a business which sells food to be consumed on the premises.
    a. The number of occupants for a drive-in restaurant shall be considered as equal to the number of parking stalls.
    b. Employee toilet facilities shall not to be included in the above restaurant requirements. Hand washing facilities shall be available in the kitchen for employees.
12. Where food is consumed indoors, water stations may be substituted for drinking fountains. Offices, or public buildings for use by more than six (6) persons shall have one (1) drinking fountain for the first one hundred fifty (150) persons and one additional fountain for each three hundred (300) persons thereafter.
13. There shall be a minimum of one (1) drinking fountain per occupied floor in schools, theaters, auditoriums, dormitories, offices of public building.
14. The total number of water closets for females shall be at least equal to the total number of water closets and urinals required for males.

Figure 23.2b

(each) can be estimated and can be included separately. For budgeting large projects, percentage multipliers, and rules of thumb, such as those shown in Figures 23.3 and 23.4 (from *Means Plumbing Cost Data*), may be useful.

Fire Protection

Square foot costs for fire protection systems should be developed from past projects for budget purposes. These costs may be based on the relative hazards of occupancy—*light, ordinary*, and *extra*—and on the type of system. Some requirements of the different level hazards are shown in Figure 23.5, from *Means Plumbing Cost Data*.

Consideration must also be given to special or unusual requirements. For example, many architects specify that sprinkler heads must be located in the center of ceiling tiles. Each head may require extra elbows and nipples for precise location. Recessed heads are more expensive. Special dry pendent heads are required in areas subject to freezing. When installing a sprinkler system in an existing structure, a completely new water service may be required in addition to the existing domestic water service. These are just a few examples of requirements that may necessitate an adjustment of square foot costs.

In addition to the hazard, the *type* of sprinkler system is the most significant factor affecting cost. The size of the system (square footage) and the number of floors served by one system also affect the cost. This cost variation is illustrated along with the components of a typical wet pipe sprinkler system in Figure 23.6.

Wet Pipe Systems

Wet pipe systems employ automatic sprinklers attached to a piping system. The pipes contain water and are connected to a water supply so that water discharges immediately from sprinklers activated by a fire.

R15100-040 Plumbing Approximations for Quick Estimating

Water Control
Water Meter; Backflow Preventer, ... 10 to 15% of Fixtures
Shock Absorbers; Vacuum Breakers;
Mixer.

Pipe And Fittings .... 30 to 60% of Fixtures

> **Note:** Lower percentage for compact buildings or larger buildings with plumbing in one area.
> Larger percentage for large buildings with plumbing spread out.
> In extreme cases pipe may be more than 100% of fixtures.
> Percentages **do not** include special purpose or process piping.

Plumbing Labor
1 & 2 Story Residential .... Rough-in Labor = 80% of Materials
Apartment Buildings .... Rough-in Labor = 90 to 100% of Materials
Labor for handling and placing fixtures is approximately 25 to 30% of fixtures

Quality/Complexity Multiplier (for all installations)
Economy installation, add. .... 0 to 5%
Good quality, medium complexity, add .... 5 to 15%
Above average quality and complexity, add .... 15 to 25%

Figure 23.3

Dry Pipe Systems

These systems employ automatic sprinklers attached to piping that contains air under pressure. As the air is released by the opening of sprinklers, the water pressure opens a valve known as a dry pipe valve. The water then flows into the piping system and out the opened sprinklers.

Pre-Action Systems

Like the dry and wet pipe systems, automatic sprinklers are again used for the pre-action systems. The sprinklers are attached to piping containing air that may or may not be under pressure. There is also a supplemental heat-responsive system of generally more sensitive characteristics than the automatic sprinklers themselves, installed in the same areas as the sprinklers. Actuation of the heat-responsive system, as from a fire, opens a valve which permits water to fill the sprinkler piping system and to be discharged from any sprinklers that may open.

Deluge Systems

These systems employ open sprinklers attached to piping that is connected to a water supply through a valve. The valve is opened by the operation of a heat-responsive system installed in the same areas as the sprinklers. When this valve opens, water flows into the piping system and discharges from all sprinklers at once.

Combined Dry Pipe and Pre-Action Sprinkler Systems

In these systems, automatic sprinklers are attached to piping containing air under pressure. There is also a supplemental heat-responsive system that is more sensitive than the automatic sprinklers themselves, installed in the same areas as the sprinklers. Activation of the heat-responsive system also opens approved air-exhaust valves at the end of the feed main, which facilitates the

R15100-420   Plumbing Fixture Installation Time

Item	Rough-In	Set	Total Hours	Item	Rough-In	Set	Total Hours
Bathtub	5	5	10	Shower head only	2	1	3
Bathtub and shower, cast iron	6	6	12	Shower drain	3	1	4
Fire hose reel and cabinet	4	2	6	Shower stall, slate		15	15
Floor drain to 4 inch diameter	3	1	4	Slop sink	5	3	8
Grease trap, single, cast iron	5	3	8	Test 6 fixtures			14
Kitchen gas range		4	4	Urinal, wall	6	2	8
Kitchen sink, single	4	4	8	Urinal, pedestal or floor	6	4	10
Kitchen sink, double	6	6	12	Water closet and tank	4	3	7
Laundry tubs	4	2	6	Water closet and tank, wall hung	5	3	8
Lavatory wall hung	5	3	8	Water heater, 45 gals. gas, automatic	5	2	7
Lavatory pedestal	5	3	8	Water heaters, 65 gals. gas, automatic	5	2	7
Shower and stall	6	4	10	Water heaters, electric, plumbing only	4	2	6

Figure 23.4

filling of the system with water. This process usually precedes the opening of sprinklers. The heat-responsive system also serves as an automatic fire alarm system.

Limited Water Supply Systems

These systems employ automatic sprinklers. They conform to standards, but they are supplied by a pressure tank of limited capacity.

Chemical Systems

Chemical systems use FM200, carbon dioxide, dry chemicals, or high-expansion foam as selected for special requirements. The chemical agent may extinguish flames by excluding oxygen, interrupting chemical action of the oxygen uniting with fuel, or sealing and cooling the combustion center.

Firecycle Systems

Firecycle systems are fixed fire protection sprinkler systems utilizing water as the extinguishing agent. These are time-delayed, recycling, preaction-type

---

R10520-120 System Classification

System Classification

Rules for installation of sprinkler systems vary depending on the classification of occupancy falling into one of three categories as follows:

Light Hazard Occupancy

The protection area allotted per sprinkler should not exceed 200 S.F. with the maximum distance between lines and sprinklers on lines being 15'. The sprinklers do not need to be staggered. Branch lines should not exceed eight sprinklers on either side of a cross main. Each large area requiring more than 100 sprinklers and without a sub-dividing partition should be supplied by feed mains or risers sized for ordinary hazard occupancy.

Included in this group are:

Auditoriums	Museums
Churches	Nursing Homes
Clubs	Offices
Educational	Residential
Hospitals	Restaurants
Institutional	Schools
Libraries	Theaters
(except large stack rooms)	

Ordinary Hazard Occupancy

The protection area allotted per sprinkler shall not exceed 130 S.F. of noncombustible ceiling and 120 S.F. of combustible ceiling. The maximum allowable distance between sprinkler lines and sprinklers on line is 15'. Sprinklers shall be staggered if the distance between heads exceeds 12'. Branch lines should not exceed eight sprinklers on either side of a cross main.

Included in this group are:

Automotive garages	Electric generating stations
Bakeries	Feed mills
Beverage manufacturing	Grain elevators
Bleacheries	Ice manufacturing
Boiler houses	Laundries
Canneries	Machine shops
Cement plants	Mercantiles
Clothing factories	Paper mills
Cold storage warehouses	Printing and Publishing
Dairy products manufacturing	Shoe factories
Distilleries	Warehouses
Dry cleaning	Wood product assembly

Extra Hazard Occupancy

The protection area allotted per sprinkler shall not exceed 90 S.F. of noncombustible ceiling and 80 S.F. of combustible ceiling. The maximum allowable distance between lines and between sprinklers on lines is 12'. Sprinklers on alternate lines shall be staggered if the distance between sprinklers on lines exceeds 8'. Branch lines should not exceed six sprinklers on either side of a cross main.

Included in this group are:

Aircraft hangars	Paint shops
Chemical works	Shade cloth manufacturing
Explosives manufacturing	Solvent extracting
Linoleum manufacturing	Varnish works
Linseed oil mills	Volatile flammable
Oil refineries	liquid manufacturing & use

Figure 23.5

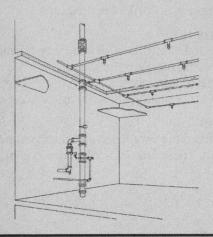

This page illustrates and describes a wet type fire sprinkler system. Lines within System Components give the unit cost of a system for a 2,000 square foot building. Lines D4010 315 1900 thru 2600 give the square foot costs for alternate systems. Both material quantities and labor costs have been adjusted for the system listed.

Factors: To adjust for conditions other than normal working conditions, use Lines D4010 315 2900 thru 4000.

System Components	QUANTITY	UNIT	COST EACH MAT.	COST EACH INST.	COST EACH TOTAL
Wet pipe fire sprinkler system, ordinary hazard, open area to 2000 S.F. On one floor.					
4" OS & Y valve	1.000	Ea.	390	269	659
Wet pipe alarm valve	1.000	Ea.	835	405	1,240
Water motor alarm	1.000	Ea.	183	113	296
3" check valve	1.000	Ea.	155	271	426
Pipe riser, 4" diameter	10.000	L.F.	99	243.80	342.80
Water gauges and trim	1.000	Set	1,300	815	2,115
Electric fire horn	1.000	Ea.	40	65.50	105.50
Sprinkler head supply piping, 1" diameter	168.000	L.F.	155.25	633.75	789
Sprinkler head supply piping, 1-1/2" diameter	168.000	L.F.	151.80	555.50	707.30
Sprinkler head supply piping, 2-1/2" diameter	168.000	L.F.	210.90	613.70	824.60
Pipe fittings, 1"	8.000	Ea.	33.60	448	481.60
Pipe fittings, 1-1/2"	8.000	Ea.	68	496	564
Pipe fittings, 2-1/2"	4.000	Ea.	124	358	482
Pipe fittings, 4"	2.000	Ea.	222	404	626
Sprinkler heads	16.000	Ea.	72.64	456	528.64
Fire department connection	1.000	Ea.	297	163	460
TOTAL		System	4,337.19	6,310.25	10,647.44

D4010 315	Fire Sprinkler Systems, Wet	COST PER S.F. MAT.	COST PER S.F. INST.	COST PER S.F. TOTAL
1900	Ordinary hazard, one floor, area to 2000 S.F./floor	2.19	3.16	5.35
2000	For each additional floor, add per floor	.58	2.11	2.69
2100	Area to 3200 S.F./floor	1.56	2.86	4.42
2200	For each additional floor, add per floor	.56	2.21	2.77
2300	Area to 5000 S.F./ floor	1.37	2.70	4.07
2400	For each additional floor, add per floor	.72	2.30	3.02
2500	Area to 8000 S.F./ floor	.97	2.23	3.20
2600	For each additional floor, add per floor	.57	1.98	2.55
2700				
2900	Cut & patch to match existing construction, add, minimum	2%	3%	
3000	Maximum	5%	9%	
3100	Dust protection, add, minimum	1%	2%	
3200	Maximum	4%	11%	
3300	Equipment usage curtailment, add, minimum	1%	1%	
3400	Maximum	3%	10%	
3500	Material handling & storage limitation, add, minimum	1%	1%	

Figure 23.6

systems that automatically shut the water off when heat is reduced below the detector-operating temperature. When that temperature is exceeded, they turn the water back on. The system senses a fire condition through a closed circuit electrical detector system which automatically controls water flow to the fire. Batteries supply up to a 90-hour emergency power supply for system operation. The piping system is dry (until water is required) and is monitored with pressurized air. Should any leak in the system piping occur, an alarm will sound. However, water will not enter the system until a firecycle detector senses heat.

Heating, Ventilating, and Air Conditioning

As with fire protection, square foot (or systems) costs can be developed for HVAC (heating, ventilating, and air conditioning) by keeping records from past projects. The estimator should obtain quotations or detailed estimates from experienced engineers or subcontractors whenever possible. HVAC is a specialized trade; specific knowledge is required for a proper estimate. However, budgets can be developed based on square foot, cubic foot, or systems costs. Such costs will vary based on the type of system and the use of the occupied space as well as the size of the space. Different types of systems are illustrated in Figures 23.7a and 23.7b.

For preliminary budgets, prior to or during design development, the estimator can calculate rough heating and cooling requirements in order to determine costs. The tables in Figure 23.8, from *Means Mechanical Cost Data*, may be used to determine heat loss (in BTUs per hour). This figure, for all practical purposes, is the capacity required of the heat source, for example, a boiler or a furnace. Figure 23.9, from *Means Mechanical Cost Data*, may be used to determine cooling requirements in tons for 45 types of building uses. (One ton of cooling equals 12,000 BTUs per hour.) When heating and cooling systems are combined, as with rooftop systems, the cooling capacity is used to determine the size of the system. After the system capacity has been calculated, budget costs can be determined.

For most remodeling projects, the main heating and cooling systems will already be in place. However, distribution (ductwork, hot water baseboard, fan coil units, etc.) usually depends on the final design details and is often executed as part of the renovation. Computer rooms usually require complete, new, and often independent cooling systems. (Computer rooms rarely require supplemental heating.)

Computer Rooms

Computer rooms impose special requirements on air conditioning systems. A prime requirement is reliability, due to the potential monetary loss that could be incurred by a system failure. A second basic requirement is the tolerance of control with which temperature and humidity are regulated and dust is eliminated. Because the air conditioning system's reliability is so vital, the additional cost of reserve capacity and redundant components is often justified.

Computer areas may be environmentally controlled by one of three methods as follows:

- **Self-contained Units**. These are units built to high standards of performance and reliability, and usually contain alarms and controls to indicate component operation failure, the need to change filters, etc. It is important to remember the following: (1) that these units occupy interior space that is relatively expensive to build, and (2) that all alterations and service of the equipment will also have to be accomplished within the computer area.
- **Decentralized Air-Handling Units**. In operation, these units are similar to the self-contained units, except that cooling capability

comes from remotely located refrigeration equipment, such as refrigerant or chilled water. Since no compressors or refrigerating equipment are needed in the air units, they are smaller and require less service than self-contained units.

- **Central System Supply**. In this system, the cooling comes from a central source. Because it is not located within the computer room, this central source may have excess capacity and permit greater flexibility without interfering with the computer components. System performance criteria must still be met. This type of system may provide less control than independent units.

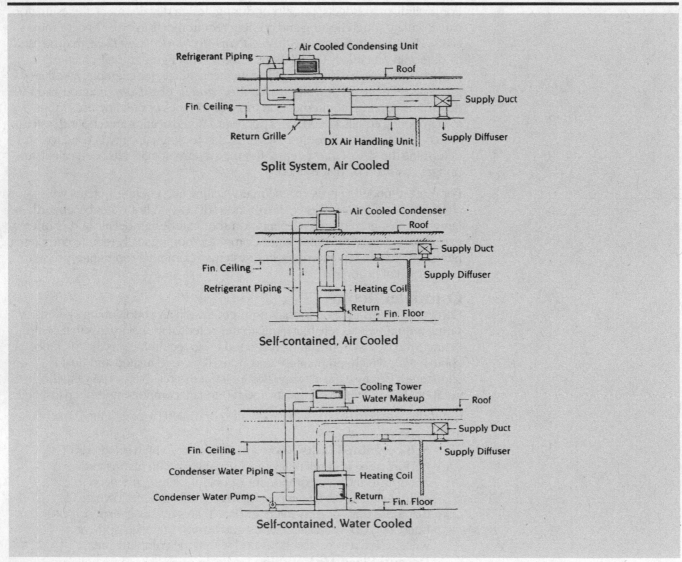

Split System, Air Cooled

Self-contained, Air Cooled

Self-contained, Water Cooled

Figure 23.7a

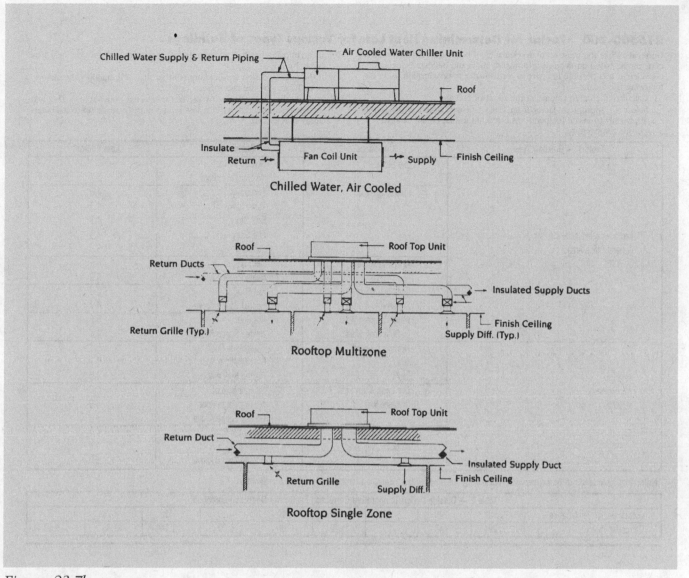

Figure 23.7b

R15500-050   Factor for Determining Heat Loss for Various Types of Buildings

General: While the most accurate estimates of heating requirements would naturally be based on detailed information about the building being considered, it is possible to arrive at a reasonable approximation using the following procedure:

1. Calculate the cubic volume of the room or building.
2. Select the appropriate factor from Table 1 below. Note that the factors apply only to inside temperatures listed in the first column and to 0°F outside temperature.

3. If the building has bad north and west exposures, multiply the heat loss factor by 1.1.
4. If the outside design temperature is other than 0°F, multiply the factor from Table 1 by the factor from Table 2.
5. Multiply the cubic volume by the factor selected from Table 1. This will give the estimated BTUH heat loss which must be made up to maintain inside temperature.

Table 1 — Building Type	Conditions	Qualifications	Loss Factor*
Factories & Industrial Plants General Office Areas at 70°F	One Story	Skylight in Roof	6.2
		No Skylight in Roof	5.7
	Multiple Story	Two Story	4.6
		Three Story	4.3
		Four Story	4.1
		Five Story	3.9
		Six Story	3.6
	All Walls Exposed	Flat Roof	6.9
		Heated Space Above	5.2
	One Long Warm Common Wall	Flat Roof	6.3
		Heated Space Above	4.7
	Warm Common Walls on Both Long Sides	Flat Roof	5.8
		Heated Space Above	4.1
Warehouses at 60°F	All Walls Exposed	Skylights in Roof	5.5
		No Skylight in Roof	5.1
		Heated Space Above	4.0
	One Long Warm Common Wall	Skylight in Roof	5.0
		No Skylight in Roof	4.9
		Heated Space Above	3.4
	Warm Common Walls on Both Long Sides	Skylight in Roof	4.7
		No Skylight in Roof	4.4
		Heated Space Above	3.0

*Note: This table tends to be conservative particularly for new buildings designed for minimum energy consumption.

Table 2 — Outside Design Temperature Correction Factor (for Degrees Fahrenheit)									
Outside Design Temperature	50	40	30	20	10	0	-10	-20	-30
Correction Factor	.29	.43	.57	.72	.86	1.00	1.14	1.28	1.43

Figure 23.8

R15700-020 Air Conditioning Requirements

BTU's per hour per S.F. of floor area and S.F. per ton of air conditioning.

Type of Building	BTU per S.F.	S.F. per Ton	Type of Building	BTU per S.F.	S.F. per Ton	Type of Building	BTU per S.F.	S.F. per Ton
Apartments, Individual	26	450	Dormitory, Rooms	40	300	Libraries	50	240
Corridors	22	550	Corridors	30	400	Low Rise Office, Exterior	38	320
Auditoriums & Theaters	40	300/18*	Dress Shops	43	280	Interior	33	360
Banks	50	240	Drug Stores	80	150	Medical Centers	28	425
Barber Shops	48	250	Factories	40	300	Motels	28	425
Bars & Taverns	133	90	High Rise Office—Ext. Rms.	46	263	Office (small suite)	43	280
Beauty Parlors	66	180	Interior Rooms	37	325	Post Office, Individual Office	42	285
Bowling Alleys	68	175	Hospitals, Core	43	280	Central Area	46	260
Churches	36	330/20*	Perimeter	46	260	Residences	20	600
Cocktail Lounges	68	175	Hotel, Guest Rooms	44	275	Restaurants	60	200
Computer Rooms	141	85	Corridors	30	400	Schools & Colleges	46	260
Dental Offices	52	230	Public Spaces	55	220	Shoe Stores	55	220
Dept. Stores, Basement	34	350	Industrial Plants, Offices	38	320	Shop'g. Ctrs., Supermarkets	34	350
Main Floor	40	300	General Offices	34	350	Retail Stores	48	250
Upper Floor	30	400	Plant Areas	40	300	Specialty	60	200

*Persons per ton

12,000 BTU = 1 ton of air conditioning

Figure 23.9

Chapter 24

Electrical

For most remodeling projects, electrical power is supplied, at least to a panel at or near the space to be constructed. In some cases, a new electric service and feeder distribution system may be required. A new service may include charges to be paid to the local utility, expensive switchgear, and, in some cases, transformers (exterior or within an interior vault). A feeder distribution system involves runs of conduit and large wire to electrical panels located throughout a building (to each floor or tenant space). This type of work should be estimated by an experienced engineer or electrical contractor. The project estimator should nevertheless be familiar with the various types of distribution systems and methods.

Distribution Systems

For many projects an engineered specialty distribution network may be in place prior to the start of the remodeling work. Such networks are often modular and may be installed during initial construction of the building. In industrial or manufacturing space, these networks may be cable tray or bus duct systems (shown in Figure 24.1). These systems are most advantageous where flexibility for power usage and ease of installation are most important. In commercial office buildings, in-place distribution networks may include underfloor raceway systems, trench duct, and cellular concrete floor raceway systems. These are illustrated in Figure 24.2.

If there is no "built-in" distribution system in place within the space to be constructed, certain options exist. Conventional wiring consists primarily of the following types of raceways and/or conductors:

- Rigid galvanized steel conduit
- Aluminum conduit
- EMT (thin wall conduit)
- BX (armored cable)
- Romex (nonmetallic sheathed cable)

The choice depends on national and local Electrical, Building, and Fire Codes, and on occupant requirements. Figure 24.3 demonstrates how each of the different types of wiring may affect the project costs.

Undercarpet Systems

In recent years, a new type of distribution system has been developed that offers the advantages of the underfloor raceway and the trench duct systems, while it allows the flexibility of conventional methods—custom installation

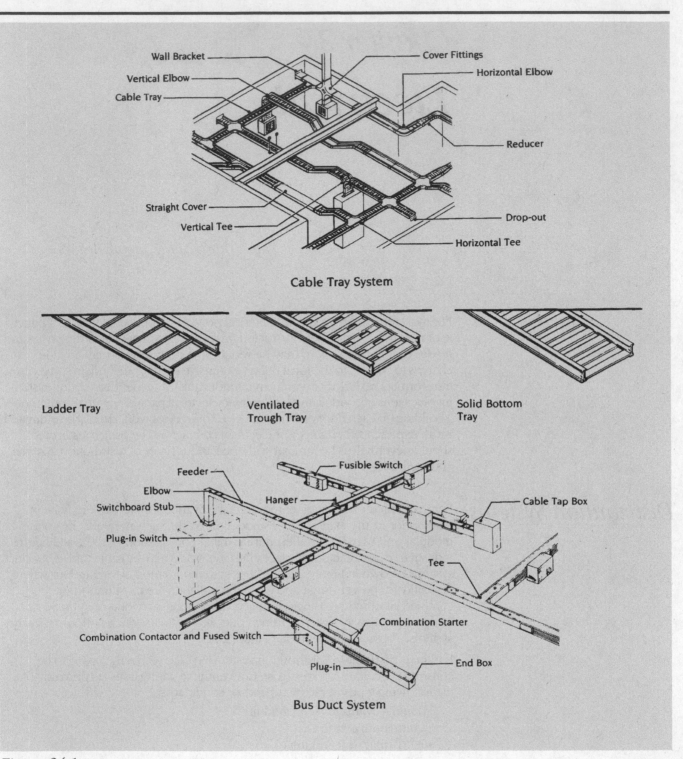

Cable Tray System

Wall Bracket
Vertical Elbow
Cable Tray
Cover Fittings
Horizontal Elbow
Reducer
Straight Cover
Vertical Tee
Drop-out
Horizontal Tee

Ladder Tray

Ventilated
Trough Tray

Solid Bottom
Tray

Bus Duct System

Feeder
Elbow
Switchboard Stub
Plug-in Switch
Fusible Switch
Hanger
Cable Tap Box
Tee
Combination Contactor and Fused Switch
Combination Starter
Plug-in
End Box

Figure 24.1

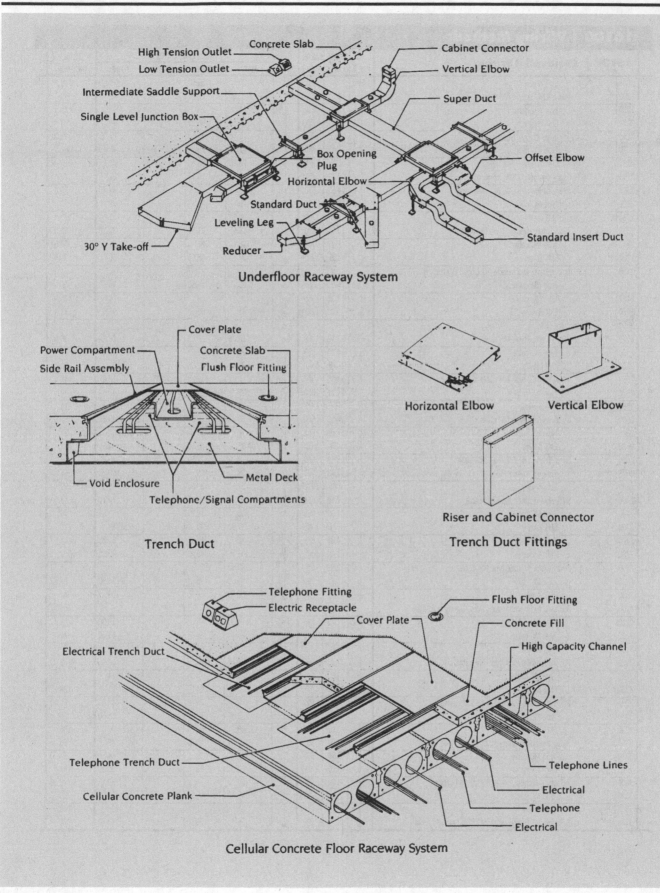

Underfloor Raceway System

High Tension Outlet
Low Tension Outlet
Concrete Slab
Cabinet Connector
Vertical Elbow
Intermediate Saddle Support
Super Duct
Single Level Junction Box
Box Opening Plug
Offset Elbow
Horizontal Elbow
Standard Duct
Leveling Leg
30° Y Take-off
Reducer
Standard Insert Duct

Trench Duct

Power Compartment
Side Rail Assembly
Cover Plate
Concrete Slab
Flush Floor Fitting
Void Enclosure
Metal Deck
Telephone/Signal Compartments

Trench Duct Fittings

Horizontal Elbow
Vertical Elbow
Riser and Cabinet Connector

Cellular Concrete Floor Raceway System

Telephone Fitting
Electric Receptacle
Cover Plate
Flush Floor Fitting
Concrete Fill
Electrical Trench Duct
High Capacity Channel
Telephone Trench Duct
Telephone Lines
Cellular Concrete Plank
Electrical
Telephone
Electrical

Figure 24.2

16100 | Wiring Methods

16139 | Residential Wiring

		CREW	DAILY OUTPUT	LABOR-HOURS	UNIT	2002 BARE COSTS				TOTAL INCL O&P
						MAT.	LABOR	EQUIP.	TOTAL	
2180	EMT & wire	1 Elec	5	1.600	Ea.	20	56.50		76.50	110
2200	4-way, #14/3, type NM cable		14.55	.550		22.50	19.50		42	55
2220	Type MC cable		12.31	.650		34	23		57	73
2230	EMT & wire		5	1.600		32.50	56.50		89	124
2250	S.P., 20 amp, #12/2, type NM cable		13.33	.600		11.55	21.50		33.05	45.50
2270	Type MC cable		11.43	.700		21	25		46	62
2280	EMT & wire		4.85	1.649		23.50	58.50		82	116
2290	S.P. rotary dimmer, 600W, no wiring		17	.471		15.75	16.70		32.45	43
2300	S.P. rotary dimmer, 600W, type NM cable		14.55	.550		18.25	19.50		37.75	50
2320	Type MC cable		12.31	.650		29	23		52	67
2330	EMT & wire		5	1.600		30.50	56.50		87	121
2350	3-way rotary dimmer, type NM cable		13.33	.600		15.65	21.50		37.15	50.50
2370	Type MC cable		11.43	.700		26	25		51	67.50
2380	EMT & wire	↓	4.85	1.649	↓	27.50	58.50		86	121
2400	Interval timer wall switch, 20 amp, 1-30 min., #12/2									
2410	Type NM cable	1 Elec	14.55	.550	Ea.	29	19.50		48.50	62
2420	Type MC cable		12.31	.650		36	23		59	75.50
2430	EMT & wire	↓	5	1.600	↓	41	56.50		97.50	133
2500	Decorator style									
2510	S.P., 15 amp, type NM cable	1 Elec	17.10	.468	Ea.	10.05	16.60		26.65	36.50
2520	Type MC cable		14.30	.559		20.50	19.85		40.35	53
2530	EMT & wire		5.71	1.401		21.50	49.50		71	100
2550	3-way, #14/3, type NM cable		14.55	.550		13.55	19.50		33.05	45
2570	Type MC cable		12.31	.650		25	23		48	63
2580	EMT & wire		5	1.600		23.50	56.50		80	114
2600	4-way, #14/3, type NM cable		14.55	.550		26	19.50		45.50	59
2620	Type MC cable		12.31	.650		38	23		61	77
2630	EMT & wire		5	1.600		36	56.50		92.50	128
2650	S.P., 20 amp, #12/2, type NM cable		13.33	.600		15.10	21.50		36.60	49.50
2670	Type MC cable		11.43	.700		25	25		50	66
2680	EMT & wire		4.85	1.649		27	58.50		85.50	120
2700	S.P., slide dimmer, type NM cable		17.10	.468		25.50	16.60		42.10	53.50
2720	Type MC cable		14.30	.559		36	19.85		55.85	70
2730	EMT & wire		5.71	1.401		37.50	49.50		87	118
2770	Type MC cable		14.30	.559		32	19.85		51.85	65.50
2780	EMT & wire		5.71	1.401		33	49.50		82.50	113
2800	3-way touch dimmer, type NM cable		13.33	.600		38.50	21.50		60	75.50
2820	Type MC cable		11.43	.700		49.50	25		74.50	92.50
2830	EMT & wire	↓	4.85	1.649	↓	50.50	58.50		109	146
3100	S.P. switch/15 amp recpt., Ivory, 1-gang box, plate									
3110	Type NM cable	1 Elec	11.43	.700	Ea.	14.30	25		39.30	54.50
3120	Type MC cable		10	.800		25	28.50		53.50	71
3130	EMT & wire		4.40	1.818		26.50	64.50		91	129
3150	S.P. switch/pilot light, type NM cable		11.43	.700		14.95	25		39.95	55
3170	Type MC cable		10	.800		25.50	28.50		54	71.50
3180	EMT & wire		4.43	1.806		27	64		91	128
3200	2-S.P. switches, 2-#14/2, type NM cables		10	.800		16.10	28.50		44.60	61.50
3220	Type MC cable		8.89	.900		33.50	32		65.50	86
3230	EMT & wire		4.10	1.951		28	69		97	138
3250	3-way switch/15 amp recpt., #14/3, type NM cable		10	.800		20.50	28.50		49	66
3270	Type MC cable		8.89	.900		32	32		64	84.50
3280	EMT & wire		4.10	1.951		30.50	69		99.50	141
3300	2-3 way switches, 2-#14/3, type NM cables		8.89	.900		27	32		59	78.50
3320	Type MC cable		8	1		46.50	35.50		82	106
3330	EMT & wire	↓	4	2	↓	35	71		106	148
3350	S.P. switch/20 amp recpt., #12/2, type NM cable		10	.800		25	28.50		53.50	71

Figure 24.3

when the remodeling work is performed, and ease of change during operation for specific occupant requirements. This new type of system is the undercarpet power, data, and communication system.

Undercarpet systems are an alternative to conventional, round cable for wiring commercial and industrial offices. They provide a method of distributing power almost anywhere on the floors without having to channel through underfloor ducts, walls, or ceilings.

The flat, low-profile design of this type of system allows for its installation directly on top of wood, concrete, composition, or ceramic floors. It is then covered with carpet tiles (18" to 30" square) that allow for change at any time.

The basic elements of undercarpet systems include three groupings of components:

- Specialized flat, low-profile cable
- Transition fittings, which house the round-to-flat conductor connections at the supply end of the cable
- Floor fittings, which house the flat-to-round connections and provide various access configurations at the other end of the cable

Undercarpet power, telephone, and data systems are illustrated in Figure 24.4.

Undercarpet systems involve specialized materials and installation methods. Specific manufacturers should be consulted for material costs. Installation costs (for other than rough budget purposes) should be estimated by experienced subcontractors.

Lighting and Power

The following types of lighting may be included in a remodeling estimate:

- Fluorescent
- Incandescent
- High-Intensity Discharge
- Emergency Lights and Power

Figure 24.5 illustrates typical fluorescent and incandescent fixtures, and Figure 24.6 shows typical high-intensity discharge fixtures. Figure 24.7 demonstrates a graphic relationship of the light output versus power usage for the different types light sources.

Fluorescent Lighting

A fluorescent lamp consists of a hot cathode in a phosphor-coated tube that contains inert gas and mercury vapor. When energized, the cathode causes a mercury arc to produce ultraviolet light and fluorescence on the phosphor coating of the tube. The color of the light varies according to the type of phosphor used in the coating. Fluorescent lamps are high in efficiency, and with limited switching on and off, they have a life in excess of 20,000 hours. A ballast is required in the lamp circuit to limit the current. Ballasts are required in various watt-saving types and can be matched with special energy-saving lamps. Special ballasts are required for dimming.

Manufacturers produce fluorescent tubes in many different wattages, sizes, and types. One manufacturer lists lamps of 4 watts to 215 watts with lengths of 6" to 96". Three basic types of fluorescent lamps are currently manufactured: *preheat, instant start*, and *rapid start*. The preheat lamp, which is the oldest type, requires a starter. The instant start lamp, or *slimline*, was developed after the preheat type. The rapid start lamp, which is most commonly used today, operates at 425 mA. High output lamps operate at 800 mA, and very high output, at 1500 mA. Because the ballasts used in high output and very high output lamps tend to be noisy, these types of fluorescent lamps are not recommended for use in quiet areas.

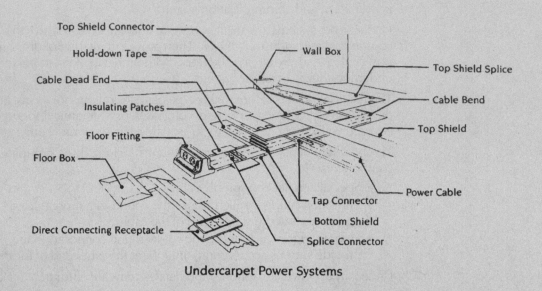

Undercarpet Power Systems

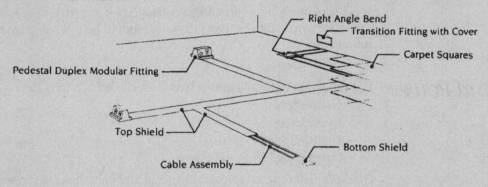

Undercarpet Telephone Systems

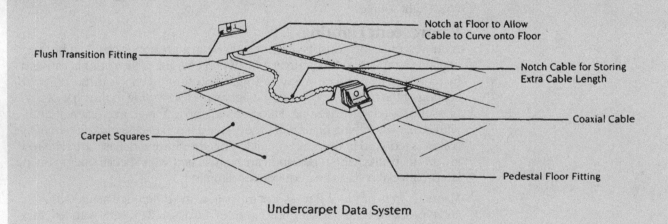

Undercarpet Data System

Figure 24.4

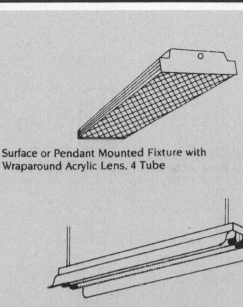

Surface or Pendant Mounted Fixture with
Wraparound Acrylic Lens, 4 Tube

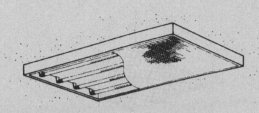

Surface Mounted Fixture with Acrylic Lens, 4 Tube

Pendant Mounted Industrial Fixture, 2 Tube

Surface Mounted Strip Fixture, 2 Tube

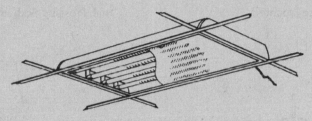

Troffer Mounted Fixture with Acrylic Lens, 4 Tube

Fluorescent Lighting

Track Lighting Spotlight

Exterior Fixture, Wall Mounted, Quartz

Round Ceiling Fixture, Recessed, with
Alzak Reflector

Round Ceiling Fixture with Concentric Louver

Round Ceiling Fixture with Reflector, No Lens

Square Ceiling Fixture, Recessed, with Glass Lens, Metal Trim

Incandescent Lighting

Figure 24.5

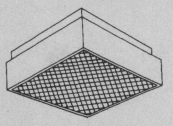

Mercury Vapor Ceiling Fixture, Recessed, Integral Ballast

High Pressure Sodium Fixture, Round, Surface

Mercury Vapor Fixture, Surface Mounted

High Pressure Sodium, Round, Wall Mounted

Mercury Vapor Fixture, Round, Pendent

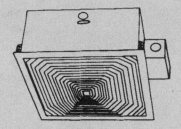

Vaporproof, High Pressure Sodium Fixture, Recessed

Mercury Vapor Fixture, Square, Pendent Mounted

Vaporproof, High Pressure Sodium Fixture, Wall Mounted

Mercury Vapor Fixture, Square, Wall Mounted

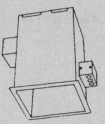

Metal Halide Fixture, Square, Recessed

H.I.D. Lighting

Figure 24.6

Incandescent Lighting

An incandescent lamp is a glass bulb that contains a tungsten filament with a mixture of argon and nitrogen gas. The base of the bulb is usually capped with a screw base made of brass or aluminum. Incandescent lamps are versatile sources of light, because they are manufactured in many different sizes, shapes, wattages, and base configurations. Some of these variations include the following: bulbs with clear, frosted, and hard glass (for weatherproof applications); aluminized reflectors; wide, narrow, and spot beam prefocused; and three-way wattage switching. For general applications, incandescent lamps are rated from 2 watts to 1500 watts, but some street lighting lamps may be rated as high as 15,000 watts.

Along with the advantage of variety in lamp sizes and special features, the relatively small size of incandescents allows them to be fit easily into the design of the fixtures that hold them. They are low in cost, soft in color, and easy to use with dimmers. The disadvantages of incandescent lamps include relatively short life, usually less than 1000 hours, and higher energy consumption than fluorescent and mercury vapor lamps (shown in Figure 24.7). If the incandescent lamp is used in a system with a higher voltage than that

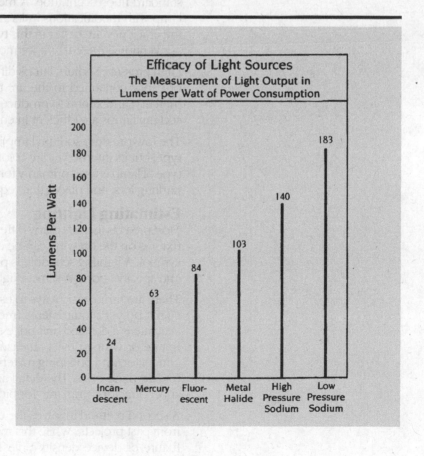

Figure 24.7

recommended by the manufacturer, the life of the lamp decreases significantly. An excess of just 10 volts above the recommended voltage can reduce the lamp's life considerably.

The quartz lamp, or *tungsten halogen lamp*, is a special type of incandescent lamp that consists of a quartz tube with various configurations. Some quartz lamps are simple, double-ended tubes, while others include a screw base or are mounted inside an R- or PAR-shaped bulb. A quartz lamp maintains maximum light output throughout its life, which varies from 2000 hours to 4000 hours, according to its type and size. Generally, quartz lamps are more energy efficient than regular incandescents, but their purchase price is higher. Quartz lamps are available in sizes from 36 watts to 1500 watts.

Fixtures for regular incandescent lamps vary widely, depending on their function, location, and desired appearance. Simple lampholders, decorative multilamp chandeliers, down lights, spotlights, accent wall lights, and track lights are just a few of the many types of fixtures. Some of the fixtures used for quartz lamps include exterior flood, track, accent, and emergency lights, to name a few.

High Intensity Discharge Lighting

High intensity discharge lighting (HID) includes *mercury vapor, metal halide*, and *high and low pressure sodium lamps*. HID lamps are usually installed to light large indoor and outdoor areas, such as factories, gymnasiums, sports complexes, parking lots, building perimeters, streets, and highways.

A mercury vapor lamp is a glass bulb containing high pressure mercury vapor. It works on the same principle as do fluorescent lamps, except that the pressure of the mercury vapor within the bulb is much higher than that of a standard fluorescent tube. A metal halide lamp is basically a mercury vapor lamp with modifications in its arc tube arrangement. Generally, the color and the efficiency are better in this type of lamp, when they are compared to conventional mercury vapor models.

High pressure sodium lamps differ from mercury lamps primarily in the type of vapor contained in the arc tube. These lamps use a mixture of sodium, mercury, and xenon to produce a slightly yellow color. Like mercury lamps, sodium lamps are efficient in energy consumption and high-rated lamp life.

The low pressure sodium lamp is more efficient than the high pressure sodium type, but its intense yellow color makes it unsuitable for indoor use. This type of lamp is used primarily for lighting large outdoor areas, such as roadways, parking lots, and places that require security lighting.

Estimating Lighting

Most projects include many different types and sizes of lighting fixtures. Lighting fixtures on the drawing can be organized if each type is assigned a letter symbol. A lighting schedule is prepared to list each type with a description and specifications. A sample lighting schedule is shown in Figure 24.8.

There are three basic ways to estimate the costs for lighting and other branch circuit power requirements (receptacles, switches, special circuits). The most accurate is a detailed unit price estimate. This method involves counting each fixture or wiring device, measuring the lengths of each type of raceway and conductor, and obtaining material prices and labor productivity rates in order to determine costs. The detail and time required limits this type of estimate to subcontractors, primarily for bidding purposes.

A second method involves the use of square foot costs that may be developed from past projects. While this method is fast, it does not allow for variation in fixture or device density unless records are kept for different types of occupancy, for example, an office or school.

In recent years, most state and local governments have adopted energy codes that, in effect, limit the maximum power usage based on the type of occupancy. These limits are usually defined as the maximum wattage per square foot of area. Besides type of building use, different types of spaces within a building (e.g., office areas, bathrooms, and hallways) may have different requirements. Figure 24.9 shows typical allowances based on type of occupancy and use. Local codes and officials should be consulted.

The estimator may use these power usage requirements to the best advantage when determining costs. Device and lighting costs may be determined based on wattage per square foot. This third method of estimating may account for fixture and device density variations based on different types of occupancy and use.

Type	Manufacturer & Catalog #	Fixture	Type	Lamps Qty Volts		Watts	Mounting	Remarks
A	Meansco #7054	2'x4' Troffer	F-40 CW	4	277	40	Recessed	Acrylic Lens
B	Meansco #7055	1'x4' Troffer	F-40 CW	2	277	40	Recessed	Acrylic Lens
C	Meansco #7709	6"x4'	F-40 CW	1	277	40	Surface	Acrylic Wrap
D	Meansco #7710	6"x8' Strip	F96T12 CW	1	277	40	Surface	
E	Meansco #7900A	6"x4'	F-40	1	277	40	Surface	Mirror Light
F	Kingston #100A	6"x4'	F-40	1	277	40	Surface	Acrylic Wrap
G	Kingston #110C	' Strip	F-40 CW	1	277	40	Surface	
H	Kingston #3752	Wallpack	HPS	1	277	150	Bracket	W/Photo Cell
J	Kingston #201-202	Floodlight	HPS	1	277	400	Surface	2' Below Fascia
K	Kingston #203		HPS	1	277	100	Wall Bracket	
L	Meansco #8100	Exit Light	1-13W 20W T6-1/2	1 2	120 6 1/2	13 20	Surface	
M	Meansco #9000	Battery Unit	Sealed Beam	2	12	18	Wall Mount	12 Volt Unit

Figure 24.8

General Lighting Loads by Occupancies

Type of Occupancy	Unit Load per S.F. (Watts)
Armories and Auditoriums	1
Banks	5
Barber Shops and Beauty Parlors	3
Churches	1
Clubs	2
Court Rooms	2
*Dwelling Units	3
Garages–Commercial (storage)	1/2
Hospitals	2
*Hotels and Motels, including apartment houses without provisions for cooking by tenents	2
Industrial Commercial (loft) Buildings	2
Lodge Rooms	1–1/2
Office Buildings	5
Restaurants	2
Schools	3
Stores	3
Warehouses (storage)	1/4
*In any of the above occupancies except one-family dwellings and individual dwelling units of multi-family dwellings:	
Assembly Halls and Auditoriums	1
Halls, Corridors, Closets	1/2
Storage Spaces	1/4

Lighting Limit (Connected Load) for Listed Occupancies: New Building Proposed Energy Conservation Guideline

Type of Use	Maximum Walls per S.F.
Interior:	
Category A: Classrooms, office areas, automotive mechanical areas, museums, conference rooms, drafting rooms, clerical areas, laboratories, merchandising areas, kitchens, examining rooms, book stacks, athletic facilities.	3.00
Category B: Auditoriums, waiting areas, spectator areas, restrooms, dining areas, transportation terminals, working corridors in prisons and hospitals, book storage areas, active inventory storage, hospital bedrooms, hotel and motel bedrooms, enclosed shopping mall concourse areas, stairways.	1.00
Category C: Corridors, lobbies, elevators, inactive storage areas.	0.50
Category D: Indoor parking	0.25
Exterior:	
Category E: Building perimeter: wall-wash, facade, canopy.	5.00 (per linear foot)
Category F: Outdoor parking.	0.10

Figure 24.9

234

Chapter 25

Using Means Repair & Remodeling Cost Data

Means Repair & Remodeling Cost Data, published annually, eliminates the guesswork when it is necessary to price unknowns by providing quick, reasonable, average prices for remodeling, renovation, and repair work. The book contains cost data, along with a wealth of information to aid the estimator, the contractor, the designer, and the owner to better plan and manage remodeling projects. Productivity data is provided in order to assist with scheduling. National labor rates are analyzed. Tables and charts for location and time adjustments are included and help the estimator tailor the prices to a specific location. The costs in *Means Repair & Remodeling Cost Data* consist of over 13,000 unit price line items, as well as prices for thousands of construction assemblies. The unit price information, organized according to the Construction Specifications Institute's MasterFormat divisions, also provides an invaluable checklist to the construction professional to ensure that all required items are included in a project. Many people who use the cost data, for bids or verification of quotations or budgets, find that knowledge of how the prices are derived enables them to use this resource to the fullest advantage.

Format and Data

The major portion of *Means Repair & Remodeling Cost Data* is the **Unit Price section.** This is the primary source of unit cost data, and it is organized according to CSI MasterFormat divisions. This index was developed by representatives of all parties concerned with the building construction industry. It has been accepted by the American Institute of Architects (AIA), the Associated General Contractors of America, Inc. (AGC), the Construction Specifications Institute, Inc. (CSI), and Construction Specifications Canada (CSC).

CSI MasterFormat Divisions:

Division 1–General Requirements

Division 2–Site Construction

Division 3–Concrete

Division 4–Masonry

Division 5–Metals

Division 6–Wood & Plastics

Division 7–Thermal & Moisture Protection

Division 8–Doors & Windows

Division 9–Finishes

Division 10–Specialties

Division 11–Equipment

Division 12–Furnishings

Division 13–Special Construction

Division 14–Conveying Systems

Division 15–Mechanical

Division 16–Electrical

The Assemblies section contains over 2,000 costs for related assemblies, or systems. Components of the assemblies are fully detailed and accompanied by illustrations. Included also are factors used to adjust costs for restrictive existing conditions. The assemblies cost data is organized according to the 7 UNIFORMAT II divisions.

A Substructure

B Shell

C Interiors

D Services

E Equipment

F Special Construction & Demolition

G Building Sitework

The Reference section contains background information that relates to the "Reference Numbers" that appear in the Unit Price section. This information is provided in the form of reference tables, charts, explanations, and estimating principles that support how the unit price data was arrived at, along with alternate pricing methods, technical data, estimating procedures and information on design and economy in construction. Also included in this section are Change Orders that contain information on pricing changes to contract documents; Crew Listings that give a full listing of all the crews, equipment, and their costs; Historical Cost Indexes for cost comparisons over time; City Cost Indexes and Location Factors for adjusting costs to 305 major U.S. and Canadian cities and over 930 three-digit zip code areas, respectively; an explanation of all abbreviations used in the book; and a comprehensive index.

Source of Costs

The prices presented in *Means Repair & Remodeling Cost Data* are national averages. Material and equipment costs are developed through annual contact with manufacturers, dealers, distributors, and contractors throughout the United States. Means' staff of engineers is constantly updating prices and keeping abreast of changes and fluctuations within the industry. Labor costs are based on the average of wage rates from 30 major U.S. cities. Means' staff of statisticians determines these wage rates by constantly monitoring negotiated labor union agreements or prevailing wages for over 35 construction trades.

Following is a list of factors and assumptions on which the costs presented in *Means Repair & Remodeling Cost Data* have been based:

- **Quality:** The costs are based on methods, materials, and workmanship in accordance with U.S. Government standards; they represent good, sound construction practice.
- **Overtime:** The costs, as presented, include *no* allowance for overtime. If overtime or premium time is anticipated, labor costs must be factored accordingly.

- **Productivity:** The daily output and labor-hour figures are based on an eight-hour workday, during daylight hours. The chart in Figure 25.1 (from *Means Building Construction Cost Data*) shows that as the number of hours worked per day (over eight) increases, and as the days per week (over five) increase, production efficiency decreases.

- **Size of Project:** Costs in *Means Repair & Remodeling Cost Data* are based on projects that cost approximately $10,000 to $1,000,000. Large residential projects are also included.

- **Local Factors:** Weather conditions, season of the year, local union restrictions, and unusual building code requirements can all have a significant impact on construction costs. The availability of a skilled labor force, sufficient materials and even adequate energy and utilities will also affect costs. These factors vary in impact and do not necessarily depend on location. They must be reviewed for each project in every area.

Days per Week	Hours per Day	Production Efficiency					Payroll Cost Factors	
		1 Week	2 Weeks	3 Weeks	4 Weeks	Average 4 Weeks	@ 1-1/2 Times	@ 2 Times
	8	100%	100%	100%	100%	100%	100%	100%
	9	100	100	95	90	96.25	105.6	111.1
5	10	100	95	90	85	91.25	110.0	120.0
	11	95	90	75	65	81.25	113.6	127.3
	12	90	85	70	60	76.25	116.7	133.3
	8	100	100	95	90	96.25	108.3	116.7
	9	100	95	90	85	92.50	113.0	125.9
6	10	95	90	85	80	87.50	116.7	133.3
	11	95	85	70	65	78.75	119.7	139.4
	12	90	80	65	60	73.75	122.2	144.4
	8	100	95	85	75	88.75	114.3	128.6
	9	95	90	80	70	83.75	118.3	136.5
7	10	90	85	75	65	78.75	121.4	142.9
	11	85	80	65	60	72.50	124.0	148.1
	12	85	75	60	55	68.75	126.2	152.4

Figure 25.1

Certain rounding rules are employed, in presenting prices in *Means Repair & Remodeling Cost Data*, to make the numbers easy to use without significantly affecting accuracy. The rules are used consistently, and they are as follows:

Prices From	To	Rounded to Nearest
0.01	5.00	0.01
5.01	20.00	0.05
20.01	100.00	1.00
100.01	1,000.00	5.00
1,000.01	10,000.00	25.00
10,000.01	50,000.00	100.00
50,000.01	up	500.00

Unit Price Section

The Unit Price section of *Means Repair & Remodeling Cost Data* contains a great deal of information in addition to the unit cost for each construction component. Figure 25.2 is a typical page, showing costs for drywall. Note that prices are included for several types of drywall, each based on the type of installation and finishing requirements. In addition, approximate crews, workers, and productivity data are indicated. The information and cost data is broken down and itemized in this way to provide for the most detailed pricing possible.

Within each individual line item, there is a description of the **construction component,** information regarding **typical crews** designated to perform the work, and **productivity** shown as daily output and as labor-hours. Costs are presented as "bare," or unburdened, as well as with markups for overhead and profit. Figures 25.3 and 25.4 are a graphic representation of how to use the Unit Price section, as presented in *Means Repair & Remodeling Cost Data*.

Line Numbers

Every construction item in the Means unit price cost data has a unique line number. This line number acts as an "address," so that each item may be quickly located and/or referenced. The numbering system is based on the CSI MasterFormat levels. In Figure 25.2, note the bold number in reverse type, "09200". This number represents Level 2, in this case "Plaster & Gypsum Board," of the CSI Division 9–Finishes. All 16 divisions are organized in this manner. Within Level 2, the data is broken down into Level 3 and Level 4 designations. Each item, or line, is further defined by an individual number. As shown in Figure 25.3, the full line number for each item consists of: a Level 3 number—a Level 4 number—and an item line number. Each full line number describes a unique construction element. For example, in Figure 25.2, the line number for 1/2″ drywall, on walls, taped and finished (Level 4 finish), is 09250-700-0350.

Line Description

Each line has a text description of the item for which costs are listed. The description may be self-contained and all-inclusive. Or, if indented, the complete description for a line depends on the information provided above. All indented items are delineations (by size, color, material, etc.) or breakdowns of previously described items. An index is provided in the back of *Means Repair & Remodeling Cost Data* to aid in locating particular items.

Crew

For each construction element (each line item), a minimum typical crew is designated as appropriate to perform the work. The crew may include one or more trades, foremen, craftsmen and helpers, and any equipment required for proper installation of the described item. If an individual trade installs the item using only hand tools, the smallest efficient number of tradesmen will be

09200 | Plaster & Gypsum Board

09250	Gypsum Board		CREW	DAILY OUTPUT	LABOR-HOURS	UNIT	2002 BARE COSTS				TOTAL INCL O&P
							MAT.	LABOR	EQUIP.	TOTAL	
700 0350	Taped and finished (level 4 finish)	R09250 -100	2 Carp	965	.017	S.F.	.25	.50		.75	1.10 **700**
0390	With compound skim coat (level 5 finish)			775	.021		.27	.62		.89	1.33
0400	Fire resistant, no finish included			2,000	.008		.22	.24		.46	.64
0450	Taped and finished (level 4 finish)			965	.017		.26	.50		.76	1.11
0490	With compound skim coat (level 5 finish)			775	.021		.28	.62		.90	1.34
0500	Water resistant, no finish included			2,000	.008		.22	.24		.46	.64
0550	Taped and finished (level 4 finish)			965	.017		.26	.50		.76	1.11
0590	With compound skim coat (level 5 finish)			775	.021		.28	.62		.90	1.34
0600	Prefinished, vinyl, clipped to studs			900	.018		.56	.53		1.09	1.50
1000	On ceilings, standard, no finish included			1,800	.009		.21	.27		.48	.67
1050	Taped and finished (level 4 finish)			765	.021		.25	.63		.88	1.32
1090	With compound skim coat (level 5 finish)			610	.026		.27	.79		1.06	1.60
1100	Fire resistant, no finish included			1,800	.009		.22	.27		.49	.68
1150	Taped and finished (level 4 finish)			765	.021		.26	.63		.89	1.33
1195	With compound skim coat (level 5 finish)			610	.026		.28	.79		1.07	1.61
1200	Water resistant, no finish included			1,800	.009		.22	.27		.49	.68
1250	Taped and finished (level 4 finish)			765	.021		.26	.63		.89	1.33
1290	With compound skim coat (level 5 finish)			610	.026		.28	.79		1.07	1.61
1500	On beams, columns, or soffits, standard, no finish included			675	.024		.24	.71		.95	1.45
1550	Taped and finished (level 4 finish)			475	.034		.29	1.01		1.30	1.99
1590	With compound skim coat (level 5 finish)			540	.030		.31	.89		1.20	1.81
1600	Fire resistant, no finish included			675	.024		.25	.71		.96	1.46
1650	Taped and finished (level 4 finish)			475	.034		.30	1.01		1.31	2
1690	With compound skim coat (level 5 finish)			540	.030		.32	.89		1.21	1.82
1700	Water resistant, no finish included			675	.024		.25	.71		.96	1.46
1750	Taped and finished (level 4 finish)			475	.034		.30	1.01		1.31	2
1790	With compound skim coat (level 5 finish)			540	.030		.32	.89		1.21	1.82
2000	5/8" thick, on walls, standard, no finish included			2,000	.008		.22	.24		.46	.64
2050	Taped and finished (level 4 finish)			965	.017		.26	.50		.76	1.11
2090	With compound skim coat (level 5 finish)			775	.021		.28	.62		.90	1.34
2100	Fire resistant, no finish included			2,000	.008		.22	.24		.46	.64
2150	Taped and finished (level 4 finish)			965	.017		.26	.50		.76	1.11
2195	With compound skim coat (level 5 finish)			775	.021		.28	.62		.90	1.34
2200	Water resistant, no finish included			2,000	.008		.27	.24		.51	.70
2250	Taped and finished (level 4 finish)			965	.017		.31	.50		.81	1.16
2290	With compound skim coat (level 5 finish)			775	.021		.33	.62		.95	1.39
2300	Prefinished, vinyl, clipped to studs			900	.018		.65	.53		1.18	1.60
3000	On ceilings, standard, no finish included			1,800	.009		.22	.27		.49	.68
3050	Taped and finished (level 4 finish)			765	.021		.26	.63		.89	1.33
3090	With compound skim coat (level 5 finish)			615	.026		.28	.78		1.06	1.60
3100	Fire resistant, no finish included			1,800	.009		.22	.27		.49	.68
3150	Taped and finished (level 4 finish)			765	.021		.26	.63		.89	1.33
3190	With compound skim coat (level 5 finish)			615	.026		.28	.78		1.06	1.60
3200	Water resistant, no finish included			1,800	.009		.27	.27		.54	.74
3250	Taped and finished (level 4 finish)			765	.021		.31	.63		.94	1.38
3290	With compound skim coat (level 5 finish)			615	.026		.33	.78		1.11	1.65
3500	On beams, columns, or soffits, no finish included			675	.024		.25	.71		.96	1.46
3550	Taped and finished (level 4 finish)			475	.034		.30	1.01		1.31	2
3590	With compound skim coat (level 5 finish)			380	.042		.32	1.26		1.58	2.44
3600	Fire resistant, no finish included			675	.024		.25	.71		.96	1.46
3650	Taped and finished (level 4 finish)			475	.034		.30	1.01		1.31	2
3690	With compound skim coat (level 5 finish)			380	.042		.32	1.26		1.58	2.44
3700	Water resistant, no finish included			675	.024		.31	.71		1.02	1.52
3750	Taped and finished (level 4 finish)			475	.034		.36	1.01		1.37	2.06
3790	With compound skim coat (level 5 finish)			380	.042		.38	1.26		1.64	2.51
4000	Fireproofing, beams or columns, 2 layers, 1/2" thick, incl finish			330	.048		.48	1.45		1.93	2.94

Figure 25.2

How to Use the Unit Price Pages

The following is a detailed explanation of a sample entry in the Unit Price Section. Next to each bold number below is the item being described with appropriate component of the sample entry following in parenthesis. Some prices are listed as bare costs, others as costs that include overhead and profit of the installing contractor. In most cases, if the work is to be subcontracted, the general contractor will need to add an additional markup (R.S. Means suggests using 10%) to the figures in the column "Total Incl. O&P."

1 Division Number/Title (03300/Cast-In-Place Concrete)

Use the Unit Price Section Table of Contents to locate specific items. The sections are classified according to the CSI MasterFormat (1995 Edition).

2 Line Numbers (03310 240 3900)

Each unit price line item has been assigned a unique 12-digit code based on the CSI MasterFormat classification.

Level One - CSI-MasterFormat Division
Level Two - CSI

03300
03310-240-3900

Means 12-digit Line Number
Level Four - Means
Level Three - CSI

3 Description (Concrete-In-Place, etc.)

Each line item is described in detail. Sub-items and additional sizes are indented beneath the appropriate line items. The first line or two after the main item (in boldface) may contain descriptive information that pertains to all line items beneath this boldface listing.

4 Reference Number Information

R03310 -010

You'll see reference numbers shown in bold rectangles at the beginning of some sections. These refer to related items in the Reference Section, visually identified by a vertical gray bar on the edge of pages.

The relation may be: (1) an estimating procedure that should be read before estimating, (2) an alternate pricing method, or (3) technical information.

The "R" designates the Reference Section. The numbers refer to the MasterFormat classification system.

It is strongly recommended that you review all reference numbers that appear within the section in which you are working.

Note: Not all reference numbers appear in all Means publications.

		03300	Cast-In-Place Concrete						2002 BARE COSTS				TOTAL	
		03310	Structural Concrete		CREW	DAILY OUTPUT	LABOR-HOURS	UNIT	MAT.	LABOR	EQUIP.	TOTAL	INCL O&P	
240	0010		CONCRETE IN PLACE (including forms (4 uses) reinforcing											240
	0050		steel, including finishing unless otherwise indicated											
	3800		Footings, spread under 1 C.Y.		C-14C	38.07	2.942	C.Y.	100	84.50	.94	185.44	252	
	3850		Over 5 C.Y.			81.04	1.382		92	40	.44	132.44	168	
	3900		Footings, strip, 18" x 9", plain			41.04	2.729		91	78.50	.87	170.37	232	
	3950		36" x 12", reinforced			61.55	1.820		92.50	52.50	.58	145.58	190	
	4000		Foundation mat, under 10 C.Y.			38.67	2.896		124	83.50	.93	208.43		
	4050		Over 20 C.Y.			56	2.036		110	57	.54	167.64		
	4200		Grade walls, 8" thick, 8' high		C-14D	45	4.154		126	130	15.65	271.65	370	
	4250		14' high			27.26	7.337		149	219	26.50	394.50	560	
	4260		12" thick, 8' high			64.32	3.109		111	93	11.15	215.15	288	
	4270		14' high			40.01	4.999		119	149	17.95	285.95	400	

Figure 25.3

5 Crew (C-14C)

The "Crew" column designates the typical trade or crew used to install the item. If an installation can be accomplished by one trade and requires no power equipment, that trade and the number of workers are listed (for example, "2 Carpenters"). If an installation requires a composite crew, a crew code designation is listed (for example, "C-14C"). You'll find full details on all composite crews in the Crew Listings.

- For a complete list of all trades utilized in this book and their abbreviations, see the inside back cover.

Crews

Crew No.	Bare Costs		Incl. Subs O & P		Cost Per Labor-Hour	
Crew C-14C	Hr.	Daily	Hr.	Daily	Bare Costs	Incl. O&P
1 Carpenter Foreman (out)	$32.00	$256.00	$53.00	$424.00	$28.79	$48.03
6 Carpenters	30.00	1440.00	49.70	2385.60		
2 Rodmen (reinf.)	34.25	548.00	60.20	963.20		
4 Laborers	23.45	750.40	38.85	1243.20		
1 Cement Finisher	28.70	229.60	45.35	362.80		
1 Gas Engine Vibrator		35.60		39.15	.32	.35
112 L.H., Daily Totals		$3259.60		$5417.95	$29.11	$48.38

6 Productivity: Daily Output (41.04)/Labor-Hours (2.729)

The "Daily Output" represents the typical number of units the designated crew will install in a normal 8-hour day. To find out the number of days the given crew would require to complete the installation, divide your quantity by the daily output. For example:

Quantity	÷	Daily Output	=	Duration
100 C.Y.	÷	41.04/ Crew Day	=	2.44 Crew Days

The "Labor-Hours" figure represents the number of labor-hours required to install one unit of work. To find out the number of labor-hours required for your particular task, multiply the quantity of the item times the number of labor-hours shown. For example:

Quantity	x	Productivity Rate	=	Duration
100 C.Y.	x	2.729 Labor-Hours/ C.Y.	=	272.9 Labor-Hours

7 Unit (C.Y.)

The abbreviated designation indicates the unit of measure upon which the price, production, and crew are based (C.Y. = Cubic Yard). For a complete listing of abbreviations refer to the Abbreviations Listing in the Reference Section of this book.

8 Bare Costs:

Mat. (Bare Material Cost) (91)

The unit material cost is the "bare" material cost with no overhead and profit included. *Costs shown reflect national average material prices for January of the current year and include delivery to the job site. No sales taxes are included.*

Labor (78.50)

The unit labor cost is derived by multiplying bare labor-hour costs for Crew C-14C by labor-hour units. The bare labor-hour cost is found in the Crew Section under C-14C. (If a trade is listed, the hourly labor cost—the wage rate—is found on the inside back cover.)

Labor-Hour Cost Crew C-14C	x	Labor-Hour Units	=	Labor
$28.79	x	2.729	=	$78.50

Equip. (Equipment) (.87)

Equipment costs for each crew are listed in the description of each crew. Tools or equipment whose value justifies purchase or ownership by a contractor are considered overhead as shown on the inside back cover. The unit equipment cost is derived by multiplying the bare equipment hourly cost by the labor-hour units.

Equipment Cost Crew C-14C	x	Labor-Hour Units	=	Equip.
.32	x	2.729	=	.87

Total (170.37)

The total of the bare costs is the arithmetic total of the three previous columns: mat., labor, and equip.

Material	+	Labor	+	Equip.	=	Total
$91	+	$78.50	+	$.87	=	$170.37

9 Total Costs Including O&P

This figure is the sum of the bare material cost plus 10% for profit; the bare labor cost plus total overhead and profit (per the inside back cover or, if a crew is listed, from the crew listings); and the bare equipment cost plus 10% for profit.

Material is Bare Material Cost + 10% = 91 + 9.10	=	$100.10
Labor for Crew C-14C = Labor-Hour Cost (48.03) x Labor-Hour Units (2.729)	=	$131.07
Equip. is Bare Equip. Cost + 10% = .87 + .09	=	$ .96
Total (Rounded)	=	$232

Figure 25.4

indicated (e.g., 1 Carp, 2 Elec, etc.). Abbreviations for trades are shown in Figure 25.5. If more than one trade is required to install the item and/or if powered equipment is needed, a crew number will be designated (e.g., B-5, D-3, etc.). A complete listing of crews is presented in the Reference section of *Means Repair & Remodeling Cost Data*. On these pages, each crew is broken down into the following components:

- Number and type of workers designated.
- Number, size, and type of any equipment required.
- Hourly labor costs listed two ways: bare (base rate including fringe benefits) and including installing contractor's overhead and profit (billing rate). (See Figure 25.5 from the inside back cover of *Means Repair & Remodeling Cost Data* for labor rate information.)
- Daily equipment costs, based on the weekly equipment rental cost divided by 5 days per week, plus the hourly operating cost, times 8 hours. This cost is listed two ways: as a bare cost, and with a 10% markup to cover handling and management costs.
- Labor and equipment are broken down further into *cost per labor-hour* for labor, and cost per labor-hour for the equipment.
- The total daily labor-hours for the crew.
- The total bare cost for the crew, including equipment.
- The total daily cost of the crew, including the installing contractor's overhead and profit.

The total daily cost of the required crew is used to calculate the unit installation cost for each item (for both bare costs and total costs including overhead and profit).

The crew designation does not mean that this is the only crew that can perform the work. Crew size and content have been developed and chosen based on practical experience and feedback from contractors. These designations represent a labor and equipment makeup commonly found in the industry.

The most appropriate crew for a given task is best determined based on particular project requirements. Unit costs may vary if crew sizes or content are significantly changed.

Unit

The unit column (see Figures 25.2 and 25.3) identifies the component for which the costs have been calculated. It is this "unit" on which unit price estimating is based. The units as used represent standard estimating and quantity takeoff procedures. However, the estimator should always check to be sure that the units taken off are the same as those priced. A list of standard abbreviations is included in the Reference section of *Means Repair & Remodeling Cost Data*.

Bare Costs

The four columns listed under *Bare Cost—Material, Labor, Equipment,* and *Total* represent the actual cost of construction items to the contractor. In other words, bare costs are those which *do not* include the overhead and profit of the installing contractor, whether for a subcontractor or a general contracting company using its own crews.

Material costs are based on the national average contractor purchase price delivered to the job site. Delivered costs are assumed to be within a 20-mile radius of metropolitan areas. No sales tax is included in the material prices because of variation from state to state.

The prices are based on quantities that would normally be purchased for remodeling projects costing $10,000 to $1,000,000. Prices for small quantities

Installing Contractor's Overhead & Profit

Below are the **average** installing contractor's percentage mark-ups applied to base labor rates to arrive at typical billing rates.

Column A: Labor rates are based on union wages averaged for 30 major U.S. cities. Base rates including fringe benefits are listed hourly and daily. These figures are the sum of the wage rate and employer-paid fringe benefits such as vacation pay, employer-paid health and welfare costs, pension costs, plus appropriate training and industry advancement funds costs.

Column B: Workers' Compensation rates are the national average of state rates established for each trade.

Column C: Column C lists average fixed overhead figures for all trades. Included are Federal and State Unemployment costs set at 7.0%; Social Security Taxes (FICA) set at 7.65%; Builder's Risk Insurance costs set at 0.34%; and Public Liability costs set at 1.55%. All the percentages except those for Social Security Taxes vary from state to state as well as from company to company.

Columns D and E: Percentages in Columns D and E are based on the presumption that the installing contractor has annual billing of $1,500,000 and up. Overhead percentages may increase with smaller annual billing. The overhead percentages for any given contractor may vary greatly and depend on a number of factors, such as the contractor's annual volume, engineering and logistical support costs, and staff requirements. The figures for overhead and profit will also vary depending on the type of job, the job location, and the prevailing economic conditions. All factors should be examined very carefully for each job.

Column F: Column F lists the total of Columns B, C, D, and E.

Column G: Column G is Column A (hourly base labor rate) multiplied by the percentage in Column F (O&P percentage).

Column H: Column H is the total of Column A (hourly base labor rate) plus Column G (Total O&P).

Column I: Column I is Column H multiplied by eight hours.

Abbr.	Trade	A Base Rate Incl. Fringes Hourly	Daily	B Workers' Comp. Ins.	C Average Fixed Overhead	D Overhead	E Profit	F Total Overhead & Profit %	Amount	H Rate with O & P Hourly	I Daily
Skwk	Skilled Workers Average (35 trades)	$30.95	$247.60	16.8%	16.5%	16.0%	15%	64.3%	$19.90	$50.85	$406.80
	Helpers Average (5 trades)	22.75	182.00	18.5				66.0	15.00	37.75	302.00
	Foreman Average, Inside ($.50 over trade)	31.45	251.60	16.8				64.3	20.20	51.65	413.20
	Foreman Average, Outside ($2.00 over trade)	32.95	263.60	16.8				64.3	21.20	54.15	433.20
Clab	Common Building Laborers	23.45	187.60	18.1				65.6	15.40	38.85	310.80
Asbe	Asbestos/Insulation Workers/Pipe Coverers	33.45	267.60	16.2				63.7	21.30	54.75	438.00
Boil	Boilermakers	36.25	290.00	14.7				62.2	22.55	58.80	470.40
Bric	Bricklayers	30.50	244.00	16.0				63.5	19.35	49.85	398.80
Brhe	Bricklayer Helpers	23.50	188.00	16.0				63.5	14.90	38.40	307.20
Carp	Carpenters	30.00	240.00	18.1				65.6	19.70	49.70	397.60
Cefi	Cement Finishers	28.70	229.60	10.6				58.1	16.65	45.35	362.80
Elec	Electricians	35.45	283.60	6.7				54.2	19.20	54.65	437.20
Elev	Elevator Constructors	37.10	296.80	7.7				55.2	20.50	57.60	460.80
Eqhv	Equipment Operators, Crane or Shovel	32.35	258.80	10.6				58.1	18.80	51.15	409.20
Eqmd	Equipment Operators, Medium Equipment	31.20	249.60	10.6				58.1	18.15	49.35	394.80
Eqlt	Equipment Operators, Light Equipment	29.80	238.40	10.6				58.1	17.30	47.10	376.80
Eqol	Equipment Operators, Oilers	26.65	213.20	10.6				58.1	15.50	42.15	337.20
Eqmm	Equipment Operators, Master Mechanics	32.80	262.40	10.6				58.1	19.05	51.85	414.80
Glaz	Glaziers	30.00	240.00	13.8				61.3	18.40	48.40	387.20
Lath	Lathers	28.75	230.00	11.1				58.6	16.85	45.60	364.80
Marb	Marble Setters	30.10	240.80	16.0				63.5	19.10	49.20	393.60
Mill	Millwrights	31.75	254.00	10.6				58.1	18.45	50.20	401.60
Mstz	Mosaic & Terrazzo Workers	29.25	234.00	9.8				57.3	16.75	46.00	368.00
Pord	Painters, Ordinary	27.15	217.20	13.8				61.3	16.65	43.80	350.40
Psst	Painters, Structural Steel	27.90	223.20	48.4				95.9	26.75	54.65	437.20
Pape	Paper Hangers	27.10	216.80	13.8				61.3	16.60	43.70	349.60
Pile	Pile Drivers	29.80	238.40	24.9				72.4	21.60	51.40	411.20
Plas	Plasterers	28.10	224.80	15.8				63.3	17.80	45.90	367.20
Plah	Plasterer Helpers	23.70	189.60	15.8				63.3	15.00	38.70	309.60
Plum	Plumbers	35.95	287.60	8.3				55.8	20.05	56.00	448.00
Rodm	Rodmen (Reinforcing)	34.25	274.00	28.3				75.8	25.95	60.20	481.60
Rofc	Roofers, Composition	26.60	212.80	32.6				80.1	21.30	47.90	383.20
Rots	Roofers, Tile & Slate	26.75	214.00	32.6				80.1	21.45	48.20	385.60
Rohe	Roofers, Helpers (Composition)	19.80	158.40	32.6				80.1	15.85	35.65	285.20
Shee	Sheet Metal Workers	35.10	280.80	11.7				59.2	20.80	55.90	447.20
Spri	Sprinkler Installers	36.20	289.60	8.7				56.2	20.35	56.55	452.40
Stpi	Steamfitters or Pipefitters	36.20	289.60	8.3				55.8	20.20	56.40	451.20
Ston	Stone Masons	30.65	245.20	16.0				63.5	19.45	50.10	400.80
Sswk	Structural Steel Workers	34.25	274.00	39.8				87.3	29.90	64.15	513.20
Tilf	Tile Layers	29.15	233.20	9.8				57.3	16.70	45.85	366.80
Tilh	Tile Layers Helpers	23.35	186.80	9.8				57.3	13.40	36.75	294.00
Trlt	Truck Drivers, Light	24.30	194.40	14.9				62.4	15.15	39.45	315.60
Trhv	Truck Drivers, Heavy	25.00	200.00	14.9				62.4	15.60	40.60	324.80
Sswl	Welders, Structural Steel	34.25	274.00	39.8				87.3	29.90	64.15	513.20
Wrck	*Wrecking	23.45	187.60	41.2	↓	↓	↓	88.7	20.80	44.25	354.00

*Not included in averages

Figure 25.5

must be adjusted accordingly. If more current costs for materials are available for the appropriate location, it is recommended that unit costs be adjusted to reflect any cost difference.

Labor costs are calculated by multiplying the bare labor cost per labor-hour times the number of labor-hours, from the *Labor-Hours* column. The *bare labor rate* is determined by adding the base rate plus fringe benefits. The base rate is the actual hourly wage of a worker, as used in figuring the payroll. It is this figure from which employee deductions are taken (e.g., federal withholding, FICA, and state withholding). Fringe benefits include all employer-paid benefits, above and beyond the payroll amount (e.g., employer-paid health, vacation pay, pension, profit sharing). The *Bare Labor Cost* is, therefore, the actual amount that the contractor must pay directly for construction workers. Figure 25.5 shows labor rates for the 35 standard construction trades plus skilled worker, helper, and foreman averages. These rates are the averages of union wage agreements effective January 1 of the year of publication, from 30 major cities in the United States. The *Bare Labor Cost* for each trade, as used in *Means Repair & Remodeling Cost Data* is shown in Column "A" as the base rate including fringes. Refer to the *Crew* column to determine the rate used to calculate the *Bare Labor Cost* for a particular line item.

Equipment costs are calculated by multiplying the *Bare Equipment Cost* per labor-hour, from the appropriate *Crew* listing, times the labor-hours in the *Labor-Hours* column.

Total Bare Costs

This column simply represents the arithmetic sum of the bare material, labor, and equipment costs. This total is the average cost to the contractor for the particular item of construction, furnished and installed, or "in place." No overhead and/or profit is included.

Total Including Overhead and Profit

The prices in the *Total Including Overhead and Profit* column represent the total cost of an item including the installing contractor's overhead and profit. The installing contractor could be either the general contractor or a subcontractor. If these costs are used for an item to be installed by a subcontractor, the general contractor should include an additional percentage (usually 10% to 20%) to cover the expenses of supervision and management.

The costs in this column are the arithmetic sum of the following three calculations:

- Bare Material Cost, plus 10%.
- Labor Cost, including fixed overhead, overhead, and profit, per labor-hour.
- Equipment Costs, plus 10%.

The following items are included in the increase for fixed overhead, overhead, and profit:

- **Worker's Compensation Insurance** rates vary from state to state and are tied into the construction trade safety records in that particular state. Rates also vary by trade according to the hazard involved. The proper authorities will most likely keep the contractor well informed of the rates and obligations.
- **State and Federal Unemployment Insurance** rates are adjusted by a merit rating system according to the number of former employees applying for benefits. Contractors who find it possible to offer a maximum of steady employment may enjoy a reduction in the unemployment tax rate.

- **Employer-Paid Social Security (FICA)** is adjusted annually by the federal government. It is a percentage of an employee's salary up to a maximum annual contribution.
- **Builder's Risk and Public Liability** insurance rates vary according to the trades involved and the state in which the work is done.

Overhead is an average percentage to be added for office or operating overhead. This is the cost of doing business. The percentages are presented as national averages by trade, as shown in Figure 25.5. [Note that the operating overhead costs are applied to labor only in *Means Repair & Remodeling Cost Data.*]

Profit is the fee (usually a percentage) added by the contractor to offer both a return on investment and an allowance to cover the risk involved in the type of construction being bid. The profit percentage may vary from 4% on large, straightforward projects to as much as 25% on smaller, high-risk jobs. Profit percentages are directly affected by economic conditions, the expected number of bidders, and the estimated risk involved in the project. For estimating purposes, *Means Repair & Remodeling Cost Data* assumes 15% on labor costs to be a reasonable average profit factor.

Assemblies Section

Means' assemblies data is divided into 7 *UNIFORMAT II* divisions, which reorganize the components of construction into logical groupings that follow the sequence of building construction. The Assemblies, or Systems, approach was devised to provide quick and easy methods for estimating, even when only preliminary design data is available.

The groupings, or systems, are presented in such a way that the estimator can easily vary components within the systems, as well as substitute one system for another. This flexibility is extremely useful when adapting to budget, design, or other considerations. Figure 25.6 shows how the data is presented in the Assemblies section.

Each assembly is illustrated and accompanied by a detailed description. The book lists the components and sizes of each system, usually in the order of construction. Alternates for the most commonly variable components are also listed. Each individual component is found in the Unit Price section. If an alternate component (not listed in the assembly) is required, it can easily be substituted.

Quantity
A unit of measure is established for each system. For example, partition systems are measured by the square foot of wall area, and doors are measured by "each." Within each system, the components are measured by industry standard, using the same units as in the Unit Price section.

Material
The cost of each component in the *Material* column is the *Bare Material Cost*, plus 10% handling, for the unit and quantity as defined in the *Quantity* column.

Installation
Installation costs, as listed in the *Systems* pages, contain both labor and equipment costs. The labor rate includes the *Bare Labor Cost* plus the installing contractor's overhead and profit (shown in Figure 25.5). The equipment rate is the *Bare Equipment Cost*, plus 10%.

Reference Tables

Throughout the Unit Price pages of *Means Repair & Remodeling Cost Data*, certain line items contain reference numbers that refer the reader to expanded

C10 Interior Construction

C1020 Interior Doors

This page illustrates and describes interior metal door systems including a metal door, metal frame and hardware. Lines within System Components give the unit price and total price on a cost each for this system. Prices for alternate interior metal door systems are on Line Items C1020 110 1100 thru 2100. Both material quantities and labor costs have been adjusted for the system listed.

Factors: To adjust for job conditions other than normal working situations use Lines C1020 110 2900 thru 4000.

Example: You are to install the system while protecting existing construction. Go to Line C1020 110 3700 and apply these percentages to the appropriate MAT. and INST. costs.

System Components	QUANTITY	UNIT	COST EACH		
			MAT.	INST.	TOTAL
Single metal door, including frame and hardware.					
Hollow metal door, 1-3/8″ thick, 2′-6″ x 6′-8″, painted	1.000	Ea.	210	53	263
Metal frame, 5-3/4″ deep	1.000	Set	78	49.50	127.50
Paint door and frame, 1 coat	1.000	Ea.	1.82	20.50	22.32
Hardware, hinges	1.000	Set	57.50		57.50
Hardware, passage lockset	1.000	Set	13.85	25	38.85
TOTAL		Ea.	361.17	148	509.17

C1020 110	Doors, Interior Flush, Metal	COST EACH		
		MAT.	INST.	TOTAL
1000	For alternate systems:			
1100	Hollow metal doors, 1-3/8″ thick, 2′-8″ x 6′-8″	365	148	513
1200	3′-0″ x 7′-0″	355	156	511
1300				
1400	Interior fire door, 1-3/8″ thick, 2′-6″ x 6′-8″	410	148	558
1500	2′-8″ x 6′-8″	410	148	558
1600	3′-0″ x 7′-0″	410	156	566
1700				
1800	Add to fire doors:			
1900	Baked enamel finish	30%	15%	
2000	Galvanizing	15%		
2200				
2300				
2400				
2900	Cut & patch to match existing construction, add, minimum	2%	3%	
3000	Maximum	5%	9%	
3100	Dust protection, add, minimum	1%	2%	
3200	Maximum	4%	11%	
3300	Equipment usage curtailment, add, minimum	1%	1%	
3400	Maximum	3%	10%	
3500	Material handling & storage limitation, add, minimum	1%	1%	
3600	Maximum	6%	7%	
3700	Protection of existing work, add, minimum	2%	2%	
3800	Maximum	5%	7%	
3900	Shift work requirements, add, minimum		5%	
4000	Maximum		30%	

Figure 25.6

data and information in the back of the book. The Reference Table section contains over 50 pages of tables, charts, definitions, and costs, all of which corroborate the unit price data. The development of many unit costs is explained and detailed in this section. This information may be particularly helpful to the estimator, especially when more information is needed about materials and considerations that have gone into a unit price line item. Figures 25.7 and 25.8 show two examples of the kinds of information provided in the Reference Tables.

City Cost Indexes

The unit prices in *Means Repair & Remodeling Cost Data* are national averages. When they are to be applied to a particular location, these prices must be adjusted to local conditions. Means has developed the City Cost Indexes for just that purpose. This section contains tables for 305 U.S. and Canadian cities, based on a 30 major city average of 100. The figures are broken down into material and installation for all CSI divisions, as shown in Figures 25.9 and 25.10. Please note that for each city there is a weighted average based on total project costs. This average is based on the relative contribution of each division to the construction process as a whole.

In addition to adjusting the figures in *Means Repair & Remodeling Cost Data* for particular locations, the City Cost Index can also be used to adjust costs from one city to another. For example, costs for a particular building type are known for City A. In order to budget the costs of the same building type in City B, the following calculation can be made:

$$\frac{\text{City B Index}}{\text{City A Index}} \times \text{City A Cost} = \text{City B Cost}$$

While City Cost Indexes provide a means to adjust prices for location, the Historical Cost Index, (also included in *Means Repair & Remodeling Cost Data* and shown in Figure 25.11) provides a method for adjusting for time. Using the same principle as the City Cost Index, a time adjustment factor can be calculated as follows:

$$\frac{\text{Index for Year X}}{\text{Index for Year Y}} = \text{Time Adjustment Factor}$$

This time adjustment factor can be used to determine the budget costs for a particular building type in Year X, based on the cost for a similar building type known from Year Y. Used together, the two indexes allow for cost adjustments from one city during a given year to another city in another year (the present or otherwise). For example, an office building built in San Francisco in 1974 originally cost $1,000,000. How much will a similar building cost in Phoenix in 2002 Adjustment factors are developed as shown above, using data from Figures 25.9, 25.10 and 25.11:

$$\frac{\text{Phoenix Index}}{\text{San Francisco Index}} = \frac{89.4}{124.4} = 0.72$$

$$\frac{\text{2002 Index}}{\text{1974 Index}} = \frac{112.8}{41.4} = 2.72$$

Original cost × location adjustment × time adjustment = Proposed new cost,

$1,000,000 × 0.72 × 2.72 = $1,958,400

Note: For a list of estimating software that offers Means Repair and Remodeling data with their estimating systems, refer to the yellow pages in *Means Repair and Remodeling Cost Data*. For more information and free demos for downloading, visit Means Web Site **http://www.rsmeans.com/demo/**

R01540-100   Steel Tubular Scaffolding

On new construction, tubular scaffolding is efficient up to 60′ high or five stories. Above this it is usually better to use a hung scaffolding if construction permits. Swing scaffolding operations may interfere with tenants. In this case, the tubular is more practical at all heights.

In repairing or cleaning the front of an existing building the cost of tubular scaffolding per S.F. of building front increases as the height increases above the first tier. The first tier cost is relatively high due to leveling and alignment.

The minimum efficient crew for erection is three workers. For heights over 50′, a crew of four is more efficient. Use two or more on top and two at the bottom for handing up or hoisting. Four workers can erect and

dismantle about nine frames per hour up to five stories. From five to eight stories they will average six frames per hour. With 7′ horizontal spacing this will run about 400 S.F. and 265 S.F. of wall surface, respectively. Time for placing planks must be added to the above. On heights above 50′, five planks can be placed per labor-hour.

The table below shows the number of pieces required to erect tubular steel scaffolding for 1000 S.F. of building frontage. This area is made up of a scaffolding system that is 12 frames (11 bays) long by 2 frames high.

For jobs under twenty-five frames, add 50% to rental cost. Rental rates will be lower for jobs over three months duration. Large quantities for long periods can reduce rental rates by 20%.

Description of Component	CSI Line Item	Number of pieces for 1000 S.F. of Building Front	Unit
5′ Wide Standard Frame, 6′-4″ High	01540-750-2200	24	Ea.
Leveling Jack & Plate	01540-750-2650	24	
Cross Brace	01540-750-2500	44	
Side Arm Bracket, 21″	01540-750-2700	12	
Guardrail Post	01540-750-2550	12	
Guardrail, 7′ section	01540-750-2600	22	
Stairway Section	01540-750-2900	2	
Stairway Starter Bar	01540-750-2910	1	
Stairway Inside Handrail	01540-750-2920	2	
Stairway Outside Handrail	01540-750-2930	2	
Walk-Thru Frame Guardrail	01540-750-2940	2	

Scaffolding is often used as falsework over 15′ high during construction of cast-in-place concrete beams and slabs. Two foot wide scaffolding is generally used for heavy beam construction. The span between frames depends upon the load to be carried with a maximum span of 5′.

Heavy duty shoring frames with a capacity of 10,000#/leg can be spaced up to 10′ O. C. depending upon form support design and loading.

Scaffolding used as horizontal shoring requires less than half the material required with conventional shoring.

On new construction, erection is done by carpenters.

Rolling towers supporting horizontal shores can reduce labor and speed the job. For maintenance work, catwalks with spans up to 70′ can be supported by the rolling towers.

R01540-200   Pump Staging

Pump staging is generally not available for rent. The table below shows the number of pieces required to erect pump staging for 2400 S.F. of building frontage. This area is made up of a pump jack system that is 3 poles (2 bays) wide by 2 poles high.

Item	CSI Line Item	Number of pieces for 2400 S.F. of Building Front	Unit
Aluminum pole section, 24′ long	01540-550-0200	6	Ea.
Aluminum splice joint, 6′ long	01540-550-0600	3	
Aluminum foldable brace	01540-550-0900	3	
Aluminum pump jack	01540-550-0700	3	
Aluminum support for workbench/back safety rail	01540-550-1000	3	
Aluminum scaffold plank/workbench, 14″ wide x 24′ long	01540-550-1100	4	
Safety net, 22′ long	01540-550-1250	2	
Aluminum plank end safety rail	01540-550-1200	2	

The cost in place for this 2400 S.F. will depend on how many uses are realized during the life of the equipment. Several options are given in Division 01540-550.

Figure 25.7

Wood & Plastics   R061 | Rough Carpentry

R06100-010   Thirty City Lumber Prices

Prices for boards are for #2 or better or sterling, whichever is in best supply. Dimension lumber is "Standard or Better" either Southern Yellow Pine (S.Y.P.), Spruce-Pine-Fir (S.P.F.), Hem-Fir (H.F.) or Douglas Fir (D.F.). The species of lumber used in a geographic area is listed by city. Plyform is 3/4" BB oil sealed fir or S.Y.P. whichever prevails locally, 3/4" CDX is S.Y.P. or Fir.

These are prices at the time of publication and should be checked against the current market price. Relative differences between cities will stay approximately constant.

City	Species	Contractor Purchases per M.B.F.								Contractor Purchases per M.S.F.	
		S4S						Boards		3/4" Ext. Plyform	3/4" Thick CDX T&G
		Dimensions									
		2"x4"	2"x6"	2"x8"	2"x10"	2"x12"	4"x4"	1"x6"	1"x12"		
Atlanta	S.P.F.	$468	$526	$504	$785	$ 878	$ 749	$1,069	$1,454	$ 905	$ 810
Baltimore	S.P.F.	523	559	618	639	1,067	1,012	1,044	1,420	746	790
Boston	S.P.F.	532	568	582	603	835	1,253	837	1,138	1,027	832
Buffalo	H.F.	631	670	698	714	766	930	1,425	1,938	635	516
Chicago	S.P.F.	517	570	634	791	698	785	926	1,259	784	774
Cincinnati	S.P.F.	511	564	763	878	1,018	896	1,275	1,734	824	904
Cleveland	S.P.F.	442	482	583	566	825	1,236	748	1,017	1,025	840
Columbus	S.P.F.	511	564	763	877	1,018	881	992	1,349	1,186	775
Dallas	S.P.F.	459	504	684	835	1,011	772	904	1,229	712	718
Denver	H.F.	681	717	808	818	899	894	719	977	1,202	898
Detroit	S.P.F.	472	458	544	573	661	1,102	1,200	1,632	804	889
Houston	S.Y.P.	648	662	684	821	1,017	1,001	1,300	1,768	985	703
Indianapolis	S.Y.F.	449	480	683	828	929	763	913	1,241	825	845
Kansas City	D.F.	441	468	544	609	636	835	1,127	1,533	785	830
Los Angeles	D.F.	598	562	598	606	615	984	795	1,081	696	790
Memphis	S.P.F.	566	582	755	887	946	808	904	1,229	948	1,040
Milwaukee	S.P.F.	484	513	629	691	855	1,528	1,431	1,945	828	859
Minneapolis	S.P.F.	449	467	471	585	624	756	1,090	1,482	1,138	753
Nashville	S.P.F.	567	582	725	888	946	1,548	1,120	1,523	916	754
New Orleans	S.Y.P.	477	488	600	717	850	792	1,165	1,584	870	720
New York City	S.P.F.	662	645	641	664	732	881	1,007	1,369	1,064	796
Philadelphia	H.F.	488	491	576	801	878	832	1,246	1,695	750	796
Phoenix	D.F.	441	441	605	621	631	806	1,243	1,690	1,310	928
Pittsburgh	H.F.	495	562	612	717	838	900	1,225	1,665	780	830
St. Louis	S.P.F.	621	652	720	858	946	959	1,010	1,374	742	691
San Antonio	S.Y.P.	492	527	634	780	943	1,289	514	699	754	810
San Diego	D.F.	674	788	769	769	759	1,604	895	1,217	1,284	904
San Francisco	D.F.	662	740	720	821	825	966	676	919	906	940
Seattle	S.P.F.	523	583	829	893	986	933	895	1,217	810	790
Washington, DC	S.P.F.	531	534	567	711	795	894	990	1,346	1,125	815
Average		$534	$565	$651	$745	$ 848	$ 986	$1,023	$1,391	$ 912	$ 811

To convert square feet of surface to board feet, 4% waste included.

S4S Size	Multiply S.F. by	T & G Size	Multiply S.F. by	Flooring Size	Multiply S.F. by
1 x 4	1.18	1 x 4	1.27	25/32" x 2-1/4"	1.37
1 x 6	1.13	1 x 6	1.18	25/32" x 3-1/4"	1.29
1 x 8	1.11	1 x 8	1.14	15/32" x 1-1/2"	1.54
1 x 10	1.09	2 x 6	2.36	1" x 3"	1.28
				1" x 4"	1.24

Figure 25.8

City Cost Indexes

UNITED STATES & ALABAMA

DIVISION		US 30 CITY AVERAGE MAT.	INST.	TOTAL	BIRMINGHAM MAT.	INST.	TOTAL	HUNTSVILLE MAT.	INST.	TOTAL	MOBILE MAT.	INST.	TOTAL	MONTGOMERY MAT.	INST.	TOTAL	TUSCALOOSA MAT.	INST.	TOTAL
01590	EQUIPMENT RENTAL	.0	100.0	100.0	.0	101.6	101.6	.0	101.5	101.5	.0	98.0	98.0	.0	98.0	98.0	.0	101.5	101.5
02	SITE CONSTRUCTION	100.0	100.0	100.0	87.3	93.8	92.2	85.3	92.9	91.1	96.4	86.7	89.0	96.9	87.2	89.6	85.8	92.6	91.0
03100	CONCRETE FORMS & ACCESSORIES	100.0	100.0	100.0	91.5	76.7	78.6	93.3	55.6	60.5	93.3	58.7	63.3	91.7	53.1	58.2	93.2	43.7	50.2
03200	CONCRETE REINFORCEMENT	100.0	100.0	100.0	92.9	85.5	88.6	92.9	85.9	88.8	95.9	63.2	76.8	95.9	84.5	89.2	92.9	84.8	88.2
03300	CAST-IN-PLACE CONCRETE	100.0	100.0	100.0	93.8	66.8	82.9	88.7	63.5	78.5	93.7	61.6	80.7	95.3	55.5	79.3	92.4	53.0	76.5
03	CONCRETE	100.0	100.0	100.0	89.2	76.1	82.7	86.8	65.8	76.3	89.7	62.2	76.0	90.4	61.7	76.1	88.6	56.8	72.8
04	MASONRY	100.0	100.0	100.0	85.6	72.8	77.8	85.5	57.7	68.5	86.1	58.2	69.1	86.6	40.4	58.4	85.8	42.9	59.6
05	METALS	100.0	100.0	100.0	98.5	95.6	97.5	98.2	94.4	96.8	96.8	85.6	92.7	96.8	93.5	95.6	97.2	94.0	96.1
06	WOOD & PLASTICS	100.0	100.0	100.0	92.1	76.9	84.3	91.3	53.9	72.1	91.3	58.8	74.6	89.5	53.9	71.2	91.3	42.8	66.4
07	THERMAL & MOISTURE PROTECTION	100.0	100.0	100.0	96.1	80.2	88.6	95.8	72.3	84.7	95.8	73.5	85.3	95.5	66.5	81.8	95.8	65.2	81.3
08	DOORS & WINDOWS	100.0	100.0	100.0	96.4	78.5	92.1	96.4	60.0	87.6	96.4	59.2	87.4	96.4	61.1	87.9	96.4	60.2	87.7
09200	PLASTER & GYPSUM BOARD	100.0	100.0	100.0	110.1	76.7	88.5	106.9	53.0	72.1	106.9	58.1	75.3	106.9	53.0	72.1	106.9	41.5	64.7
095,098	CEILINGS & ACOUSTICAL TREATMENT	100.0	100.0	100.0	95.7	76.7	82.9	95.7	53.0	67.0	95.7	58.1	70.4	95.7	53.0	67.0	95.7	41.5	59.3
09600	FLOORING	100.0	100.0	100.0	101.2	58.0	90.6	101.2	50.2	88.6	109.8	62.9	98.2	109.8	32.5	90.7	101.2	47.7	88.0
097,099	WALL FINISHES, PAINTS & COATINGS	100.0	100.0	100.0	92.6	54.7	70.4	92.6	53.7	69.8	92.6	62.1	74.7	92.6	61.0	74.1	92.6	52.5	69.1
09	FINISHES	100.0	100.0	100.0	98.4	70.6	83.9	97.8	53.2	74.6	101.8	59.4	79.7	101.9	48.9	74.3	97.8	43.9	69.7
10-14	TOTAL DIV. 10000-14000	100.0	100.0	100.0	100.0	81.5	96.1	100.0	74.9	94.8	100.0	71.1	94.0	100.0	72.6	94.3	100.0	69.7	93.7
15	MECHANICAL	100.0	100.0	100.0	99.9	69.8	86.5	99.9	52.0	78.6	99.9	65.2	84.5	99.9	46.9	76.3	99.9	41.2	73.8
16	ELECTRICAL	100.0	100.0	100.0	96.3	64.8	74.4	96.7	72.5	79.8	96.7	59.5	70.8	96.3	72.2	79.6	96.7	64.8	74.5
01-16	WEIGHTED AVERAGE	100.0	100.0	100.0	96.2	75.8	86.4	95.8	67.2	82.0	96.7	66.5	82.1	96.8	62.7	80.3	95.9	59.7	78.4

ALASKA & ARIZONA

DIVISION		ANCHORAGE MAT.	INST.	TOTAL	FAIRBANKS MAT.	INST.	TOTAL	JUNEAU MAT.	INST.	TOTAL	FLAGSTAFF MAT.	INST.	TOTAL	MESA/TEMPE MAT.	INST.	TOTAL	PHOENIX MAT.	INST.	TOTAL
01590	EQUIPMENT RENTAL	.0	118.9	118.9	.0	118.9	118.9	.0	118.9	118.9	.0	93.5	93.5	.0	94.0	94.0	.0	94.6	94.6
02	SITE CONSTRUCTION	135.4	135.4	135.4	119.2	135.4	131.5	131.2	135.4	134.4	86.4	99.4	96.3	83.1	99.0	95.2	83.5	99.8	95.9
03100	CONCRETE FORMS & ACCESSORIES	131.1	115.5	117.6	132.7	122.4	123.7	132.5	115.5	117.8	106.0	71.2	75.8	103.0	70.6	74.9	104.3	76.7	80.4
03200	CONCRETE REINFORCEMENT	141.5	102.7	118.9	118.9	102.7	109.5	105.2	102.7	103.7	103.7	78.6	89.1	103.9	71.2	84.9	102.1	79.2	88.7
03300	CAST-IN-PLACE CONCRETE	189.8	117.6	160.7	158.3	118.2	142.1	190.6	117.6	161.2	97.3	84.7	92.2	99.2	76.6	90.1	99.3	84.7	93.4
03	CONCRETE	153.7	113.1	133.5	130.0	116.3	123.2	149.8	113.1	131.6	122.5	77.0	99.9	100.1	72.6	86.4	99.7	79.6	89.7
04	MASONRY	195.4	122.8	151.0	182.4	122.8	146.0	184.6	122.8	146.9	101.4	63.1	78.0	109.3	57.6	77.7	96.2	72.6	81.8
05	METALS	130.0	98.4	118.4	130.0	98.8	118.6	130.3	98.4	118.6	98.0	70.9	88.1	98.3	68.9	87.6	99.6	73.5	90.1
06	WOOD & PLASTICS	118.0	113.5	115.7	118.3	122.3	120.3	118.0	113.5	115.7	107.0	70.7	88.3	101.0	77.5	89.0	102.2	77.7	89.6
07	THERMAL & MOISTURE PROTECTION	199.7	115.4	159.7	195.7	118.9	159.3	196.4	115.4	158.0	111.2	73.5	93.4	108.6	68.5	89.6	108.5	75.9	93.1
08	DOORS & WINDOWS	128.0	105.9	122.7	125.1	111.0	121.7	125.1	105.9	120.5	101.1	72.2	94.1	98.0	70.5	91.4	99.2	75.9	93.6
09200	PLASTER & GYPSUM BOARD	129.7	113.9	119.4	129.7	122.9	125.3	129.7	113.9	119.4	90.9	69.9	77.3	91.5	76.8	82.0	92.0	76.9	82.3
095,098	CEILINGS & ACOUSTICAL TREATMENT	129.4	113.9	118.9	129.4	122.5	125.3	129.4	113.9	118.9	102.9	69.9	80.7	105.6	76.8	86.2	105.6	76.9	86.3
09600	FLOORING	127.3	126.3	127.1	127.5	126.3	127.2	127.3	126.3	127.1	98.9	58.7	89.0	102.1	73.2	95.0	102.4	76.4	96.0
097,099	WALL FINISHES, PAINTS & COATINGS	116.2	107.8	111.3	116.2	117.9	117.2	116.2	107.8	111.3	98.7	55.7	73.5	109.9	59.9	80.6	109.9	69.0	85.9
09	FINISHES	140.2	116.9	128.1	138.0	123.2	130.3	138.7	116.9	127.4	97.1	66.6	81.2	98.5	70.0	83.6	98.7	75.6	86.7
10-14	TOTAL DIV. 10000-14000	100.0	114.1	102.9	100.0	115.2	103.2	100.0	114.1	102.9	100.0	82.1	96.3	100.0	77.5	95.3	100.0	83.2	96.5
15	MECHANICAL	108.2	108.0	108.1	108.2	117.3	112.2	108.2	108.0	108.1	100.0	83.1	92.6	100.1	73.0	88.0	100.1	83.2	92.0
16	ELECTRICAL	161.7	110.8	126.3	164.6	110.8	127.2	164.6	110.8	127.2	100.4	55.8	69.4	98.1	37.7	56.1	107.2	66.9	79.2
01-16	WEIGHTED AVERAGE	134.1	113.7	124.3	129.5	117.2	123.6	132.6	113.7	123.5	102.5	73.4	88.4	99.6	67.2	83.9	99.8	78.2	89.4

ARIZONA & ARKANSAS

DIVISION		PRESCOTT MAT.	INST.	TOTAL	TUCSON MAT.	INST.	TOTAL	FORT SMITH MAT.	INST.	TOTAL	JONESBORO MAT.	INST.	TOTAL	LITTLE ROCK MAT.	INST.	TOTAL	PINE BLUFF MAT.	INST.	TOTAL
01590	EQUIPMENT RENTAL	.0	93.5	93.5	.0	94.0	94.0	.0	85.1	85.1	.0	106.6	106.6	.0	85.1	85.1	.0	85.1	85.1
02	SITE CONSTRUCTION	73.6	99.0	92.9	80.3	99.6	94.9	77.9	83.7	82.3	101.8	98.8	99.5	77.7	83.7	82.3	79.9	83.7	82.8
03100	CONCRETE FORMS & ACCESSORIES	100.9	65.3	70.0	103.7	76.2	79.8	101.5	47.4	54.6	85.7	54.4	58.5	95.6	52.0	57.8	77.5	51.9	55.3
03200	CONCRETE REINFORCEMENT	103.7	75.2	87.1	101.0	78.6	87.9	95.6	73.7	82.8	91.5	52.8	68.9	95.8	69.7	80.6	95.8	69.7	80.6
03300	CAST-IN-PLACE CONCRETE	97.2	71.9	87.0	99.3	84.5	93.3	91.2	68.8	82.2	86.9	62.8	77.2	91.2	68.8	82.2	83.7	68.8	77.7
03	CONCRETE	107.0	69.3	88.2	99.5	79.2	89.4	87.8	60.7	74.3	84.4	58.7	71.6	87.5	61.9	74.8	85.1	61.9	73.5
04	MASONRY	101.9	68.7	81.6	96.7	63.0	76.1	96.8	58.5	73.4	90.8	53.8	68.2	95.0	58.5	72.7	114.4	58.5	80.2
05	METALS	98.1	68.2	87.2	98.9	71.6	88.9	96.7	72.8	87.9	90.9	80.6	87.2	96.3	71.5	87.2	95.4	71.4	86.6
06	WOOD & PLASTICS	102.2	64.2	82.7	102.7	77.7	89.9	104.8	46.4	74.8	89.1	55.8	72.0	101.5	52.5	76.3	81.5	52.5	66.6
07	THERMAL & MOISTURE PROTECTION	109.4	69.8	90.7	109.2	70.1	90.7	99.2	53.2	77.4	109.1	57.5	84.7	98.1	53.8	77.1	97.8	53.8	76.9
08	DOORS & WINDOWS	101.1	64.6	92.3	96.0	75.9	91.1	96.6	49.4	85.2	98.1	53.3	87.3	96.6	52.6	86.0	92.0	52.6	82.5
09200	PLASTER & GYPSUM BOARD	88.9	63.1	72.3	92.3	76.9	82.4	96.0	45.7	63.5	102.1	55.0	71.6	96.0	51.9	67.5	89.3	51.9	65.1
095,098	CEILINGS & ACOUSTICAL TREATMENT	102.9	63.1	76.2	107.0	76.9	86.8	101.9	45.7	64.1	102.5	55.0	70.5	101.9	51.9	68.3	97.6	51.9	66.9
09600	FLOORING	97.2	58.5	87.7	101.5	58.7	91.0	118.1	74.4	107.3	79.1	51.6	72.3	119.4	74.4	108.3	106.8	74.4	98.8
097,099	WALL FINISHES, PAINTS & COATINGS	98.7	55.7	73.5	107.4	55.7	77.2	95.8	65.3	78.0	84.7	58.7	69.5	95.8	50.0	69.0	95.8	50.0	69.0
09	FINISHES	95.1	62.6	78.1	98.4	70.7	84.0	101.9	54.1	77.0	92.5	54.5	72.7	102.3	55.8	78.1	96.6	55.8	75.4
10-14	TOTAL DIV. 10000-14000	100.0	80.8	96.0	100.0	83.2	96.5	100.0	70.0	93.8	100.0	63.0	92.3	100.0	70.8	93.9	100.0	70.8	93.9
15	MECHANICAL	100.2	79.8	91.1	100.0	74.5	88.7	100.1	48.8	77.3	100.2	48.8	77.4	100.1	58.3	81.5	100.1	51.6	78.5
16	ELECTRICAL	100.0	53.7	67.8	101.6	64.9	76.1	95.6	69.5	77.5	100.7	55.0	68.9	96.7	69.9	78.1	94.5	69.9	77.4
01-16	WEIGHTED AVERAGE	99.9	70.7	85.8	98.8	74.3	87.0	96.7	61.6	79.8	95.5	60.9	78.8	96.6	64.0	80.8	95.8	62.7	79.9

Figure 25.9

City Cost Indexes

DIVISION		ARKANSAS			CALIFORNIA														
		TEXARKANA			ANAHEIM			BAKERSFIELD			FRESNO			LOS ANGELES			OAKLAND		
		MAT.	INST.	TOTAL	MAT.	INST.	TOTAL	MAT.	INST.	TOTAL	MAT.	INST.	TOTAL	MAT.	INST.	TOTAL	MAT.	INST.	TOTAL
01590	EQUIPMENT RENTAL	.0	85.8	85.8	.0	102.4	102.4	.0	99.7	99.7	.0	99.7	99.7	.0	98.2	98.2	.0	101.6	101.6
02	SITE CONSTRUCTION	97.1	84.5	87.5	95.8	110.2	106.7	100.3	107.1	105.5	101.8	107.3	105.9	90.8	109.0	104.6	130.1	104.3	110.5
03100	CONCRETE FORMS & ACCESSORIES	86.6	45.1	50.5	103.7	120.2	118.1	96.1	120.0	116.8	100.4	122.8	119.8	105.4	119.9	118.0	110.2	138.7	134.9
03200	CONCRETE REINFORCEMENT	95.3	51.1	69.5	106.9	114.4	111.3	106.5	114.2	111.0	107.0	114.3	111.3	108.4	114.4	111.9	98.9	115.3	108.4
03300	CAST-IN-PLACE CONCRETE	91.2	49.0	74.2	104.0	120.3	110.6	99.3	119.1	107.3	108.7	114.4	111.0	86.2	117.6	98.8	129.6	118.3	125.0
03	CONCRETE	84.4	48.6	66.6	111.9	118.2	115.0	109.0	117.6	113.3	114.1	117.2	115.7	105.1	117.0	111.0	126.2	125.9	126.0
04	MASONRY	96.8	37.7	60.7	87.7	116.6	105.4	107.3	113.9	111.3	109.8	113.3	112.0	94.8	118.6	109.3	150.4	128.3	136.9
05	METALS	87.7	62.9	78.7	110.5	100.9	107.0	104.9	100.0	103.1	106.9	100.8	104.6	111.3	99.0	106.8	101.6	106.6	103.4
06	WOOD & PLASTICS	91.2	48.4	69.2	94.3	119.0	107.0	86.1	119.1	103.1	96.8	122.1	109.8	91.9	118.5	105.6	107.6	140.6	124.6
07	THERMAL & MOISTURE PROTECTION	98.7	47.0	74.2	120.9	117.3	119.2	105.5	109.0	107.1	101.7	111.2	106.2	111.8	116.6	114.1	111.4	131.7	121.0
08	DOORS & WINDOWS	97.0	45.7	84.7	103.2	115.1	106.0	102.0	113.3	104.7	102.0	114.9	105.1	98.3	114.8	102.3	106.2	130.0	112.0
09200	PLASTER & GYPSUM BOARD	92.9	47.7	63.7	93.6	119.7	110.4	93.6	119.7	110.4	92.6	122.7	112.1	85.2	119.7	107.5	98.4	141.6	126.3
095,098	CEILINGS & ACOUSTICAL TREATMENT	103.3	47.7	65.9	118.1	119.7	119.1	118.1	119.7	119.1	118.1	122.7	121.2	115.6	119.7	118.6	115.4	141.6	133.0
09600	FLOORING	109.7	49.2	94.7	124.3	107.0	120.0	119.7	71.0	107.7	120.5	142.9	126.0	116.8	107.0	114.3	112.4	121.9	114.8
097,099	WALL FINISHES, PAINTS & COATINGS	95.8	41.2	63.8	110.7	112.8	111.9	113.8	102.3	107.1	113.2	98.2	104.4	107.4	112.8	110.6	116.9	131.5	125.4
09	FINISHES	99.6	45.6	71.5	113.5	116.8	115.2	114.7	109.5	112.0	114.8	124.4	119.8	108.8	116.5	112.8	115.4	135.8	126.0
10 - 14	TOTAL DIV. 10000 - 14000	100.0	45.0	88.5	100.0	115.4	103.2	100.0	132.1	106.7	100.0	132.4	106.8	100.0	114.3	103.0	100.0	136.2	107.6
15	MECHANICAL	100.1	43.1	74.8	100.2	111.8	105.4	100.2	109.4	104.3	100.2	112.0	105.5	100.2	111.7	105.3	100.1	137.8	116.8
16	ELECTRICAL	97.7	45.7	61.6	95.0	109.2	104.9	95.4	97.3	96.7	94.1	99.5	97.9	109.8	116.5	114.5	113.6	133.9	127.7
01 - 16	WEIGHTED AVERAGE	95.3	50.2	73.5	104.2	112.6	108.2	103.7	108.6	106.1	104.8	111.4	108.0	103.4	113.4	108.3	110.9	127.9	119.1

DIVISION		CALIFORNIA																	
		OXNARD			REDDING			RIVERSIDE			SACRAMENTO			SAN DIEGO			SAN FRANCISCO		
		MAT.	INST.	TOTAL	MAT.	INST.	TOTAL	MAT.	INST.	TOTAL	MAT.	INST.	TOTAL	MAT.	INST.	TOTAL	MAT.	INST.	TOTAL
01590	EQUIPMENT RENTAL	.0	98.3	98.3	.0	99.3	99.3	.0	100.9	100.9	.0	101.2	101.2	.0	97.4	97.4	.0	106.8	106.8
02	SITE CONSTRUCTION	101.3	104.8	104.0	106.7	106.4	106.5	93.6	107.8	104.4	97.7	110.6	107.5	102.5	102.1	102.2	132.2	110.7	115.9
03100	CONCRETE FORMS & ACCESSORIES	102.2	120.3	117.9	101.6	122.8	120.0	104.9	120.2	118.1	108.6	123.1	121.2	103.8	112.3	111.2	110.5	139.5	135.7
03200	CONCRETE REINFORCEMENT	106.5	114.2	111.0	103.2	114.4	109.7	105.5	114.2	110.6	100.0	114.5	108.4	110.5	114.1	112.6	112.6	115.6	114.4
03300	CAST-IN-PLACE CONCRETE	105.4	119.4	111.0	119.4	114.3	117.3	103.0	120.3	110.0	110.9	114.5	112.4	107.8	104.2	106.4	129.5	119.8	125.6
03	CONCRETE	112.5	117.8	115.1	123.0	117.2	120.1	111.3	118.1	114.7	116.8	117.3	117.1	116.2	109.1	112.7	127.8	126.8	127.3
04	MASONRY	112.4	113.0	112.8	113.8	115.2	114.7	85.4	114.2	103.0	120.7	115.3	117.4	102.4	111.5	108.0	150.6	132.4	139.5
05	METALS	104.4	100.3	102.9	108.8	100.6	105.8	110.5	100.6	106.9	97.7	100.2	98.6	110.5	99.7	106.6	106.8	107.8	107.2
06	WOOD & PLASTICS	93.9	119.1	106.9	94.7	122.1	108.7	94.3	119.0	107.0	101.7	122.1	112.2	97.5	108.8	103.3	107.6	140.8	124.7
07	THERMAL & MOISTURE PROTECTION	110.8	114.4	112.5	111.5	114.3	112.8	119.9	115.0	117.6	122.1	114.0	118.3	113.9	106.9	110.6	111.4	133.7	122.0
08	DOORS & WINDOWS	100.8	115.1	104.3	103.6	117.1	106.9	103.2	115.1	106.0	117.8	117.2	117.7	104.2	108.9	105.3	110.7	130.1	115.3
09200	PLASTER & GYPSUM BOARD	93.6	119.7	110.4	93.1	122.7	112.2	93.3	119.7	110.3	96.6	122.7	113.5	89.1	109.0	102.0	100.1	141.6	126.9
095,098	CEILINGS & ACOUSTICAL TREATMENT	118.1	119.7	119.1	125.1	122.7	123.5	116.7	119.7	118.7	122.5	122.7	122.6	115.1	109.0	111.0	120.7	141.6	134.8
09600	FLOORING	119.7	107.0	116.6	119.0	112.8	117.5	124.2	107.0	120.0	115.2	112.8	114.6	118.5	116.2	118.0	112.4	121.9	114.8
097,099	WALL FINISHES, PAINTS & COATINGS	113.2	105.5	108.7	113.2	126.1	120.8	110.7	112.8	111.9	115.3	126.1	121.6	111.9	112.8	112.4	116.9	140.6	130.8
09	FINISHES	114.5	116.1	115.3	115.8	122.3	119.2	112.9	116.8	115.0	115.6	122.4	119.2	111.4	113.0	112.3	116.7	137.0	127.3
10 - 14	TOTAL DIV. 10000 - 14000	100.0	115.8	103.3	100.0	132.4	106.8	100.0	115.4	103.2	100.0	132.5	106.8	100.0	114.4	103.0	100.0	136.9	107.7
15	MECHANICAL	100.2	111.8	105.4	100.2	111.4	105.2	100.2	111.8	105.3	100.0	117.4	107.8	100.1	111.8	105.3	100.1	163.1	128.1
16	ELECTRICAL	99.2	110.7	107.2	104.4	100.9	102.0	95.1	103.6	101.0	103.9	100.9	101.8	92.8	94.0	93.6	109.3	151.8	138.8
01 - 16	WEIGHTED AVERAGE	104.7	111.7	108.1	107.8	111.5	109.6	103.8	111.0	107.3	107.5	113.1	110.2	105.3	106.3	105.8	112.3	137.3	124.4

DIVISION		CALIFORNIA												COLORADO					
		SAN JOSE			SANTA BARBARA			STOCKTON			VALLEJO			COLORADO SPRINGS			DENVER		
		MAT.	INST.	TOTAL	MAT.	INST.	TOTAL	MAT.	INST.	TOTAL	MAT.	INST.	TOTAL	MAT.	INST.	TOTAL	MAT.	INST.	TOTAL
01590	EQUIPMENT RENTAL	.0	100.0	100.0	.0	99.7	99.7	.0	99.3	99.3	.0	101.9	101.9	.0	96.5	96.5	.0	101.7	101.7
02	SITE CONSTRUCTION	137.9	100.6	109.6	101.7	107.1	105.8	99.7	106.6	104.9	94.6	110.8	106.9	106.6	98.4	100.4	105.0	107.8	107.1
03100	CONCRETE FORMS & ACCESSORIES	104.4	138.8	134.3	102.9	120.3	118.0	96.9	122.8	119.4	110.6	137.4	133.9	90.6	79.6	81.1	100.2	80.0	82.6
03200	CONCRETE REINFORCEMENT	103.5	115.3	110.4	106.5	114.3	111.0	107.0	114.5	111.4	98.6	115.4	108.4	99.0	83.3	89.8	98.4	84.1	90.1
03300	CAST-IN-PLACE CONCRETE	123.6	118.5	121.5	105.0	119.2	110.7	104.9	114.4	108.8	116.2	115.9	116.1	109.2	84.8	99.4	102.1	84.4	95.0
03	CONCRETE	121.4	126.1	123.8	112.3	117.8	115.0	111.9	117.3	114.6	119.4	124.3	121.8	103.8	82.4	93.2	108.4	82.6	95.5
04	MASONRY	149.1	128.9	136.7	107.8	113.9	111.5	112.0	117.3	115.2	86.1	130.6	113.3	106.1	69.1	83.5	105.6	84.2	92.5
05	METALS	111.1	108.0	110.0	104.2	100.4	102.8	106.5	101.2	104.6	100.9	102.3	101.4	101.5	87.1	96.3	105.1	88.3	98.9
06	WOOD & PLASTICS	102.0	140.5	121.8	93.9	119.1	106.9	89.9	122.1	106.4	99.8	140.3	120.6	90.9	80.6	85.6	99.0	80.5	89.5
07	THERMAL & MOISTURE PROTECTION	107.3	133.1	119.5	106.8	113.2	109.8	110.9	113.2	112.0	125.1	130.7	127.8	106.5	81.0	94.4	105.9	84.6	95.8
08	DOORS & WINDOWS	93.1	129.9	102.0	102.0	115.1	105.2	101.3	117.1	105.1	119.2	129.8	121.7	97.3	86.3	94.7	98.7	86.2	95.7
09200	PLASTER & GYPSUM BOARD	94.1	141.6	124.8	93.6	119.7	110.4	94.6	122.7	112.8	97.1	141.6	125.9	86.5	80.0	82.3	95.4	80.1	85.5
095,098	CEILINGS & ACOUSTICAL TREATMENT	109.6	141.6	131.1	118.1	119.7	119.1	118.1	122.7	121.2	122.5	141.6	135.3	99.8	80.0	86.5	98.4	80.1	86.1
09600	FLOORING	116.6	121.9	117.9	119.7	112.8	118.0	119.7	112.9	118.0	119.5	121.9	120.1	110.9	95.6	107.1	111.9	95.6	107.9
097,099	WALL FINISHES, PAINTS & COATINGS	114.9	129.7	123.6	113.2	105.5	108.7	113.2	100.7	105.9	113.3	131.5	124.0	110.6	60.2	81.1	110.6	76.8	90.8
09	FINISHES	113.7	135.6	125.1	114.7	117.1	116.0	114.6	119.4	117.1	114.2	135.2	125.1	100.2	79.5	89.4	101.3	82.3	91.4
10 - 14	TOTAL DIV. 10000 - 14000	100.0	136.1	107.5	100.0	115.8	103.3	100.0	132.4	106.8	100.0	134.6	107.2	100.0	85.8	97.0	100.0	85.3	96.9
15	MECHANICAL	100.2	146.6	120.8	100.2	111.8	105.4	100.2	120.0	105.5	100.0	128.7	112.8	100.2	81.3	91.8	100.1	84.5	93.2
16	ELECTRICAL	109.9	145.2	134.4	91.9	109.5	104.2	103.4	110.5	108.3	99.8	114.0	109.6	95.2	86.5	89.2	97.3	92.4	93.9
01 - 16	WEIGHTED AVERAGE	109.7	131.4	120.2	104.0	111.9	107.8	105.2	113.2	109.1	106.0	122.8	114.1	100.7	83.4	92.4	102.2	87.9	95.3

Figure 25.10

Historical Cost Indexes

The table below lists both the Means Historical Cost Index based on Jan. 1, 1993 = 100 as well as the computed value of an index based on Jan. 1, 2002 costs. Since the Jan. 1, 2002 figure is estimated, space is left to write in the actual index figures as they become available through either the quarterly "Means Construction Cost Indexes" or as printed in the "Engineering News-Record." To compute the actual index based on Jan. 1, 2002 = 100, divide the Historical Cost Index for a particular year by the actual Jan. 1, 2002 Construction Cost Index. Space has been left to advance the index figures as the year progresses.

Year	Historical Cost Index Jan. 1, 1993 = 100		Current Index Based on Jan. 1, 2002 = 100		Year	Historical Cost Index Jan. 1, 1993 = 100	Current Index Based on Jan. 1, 2002 = 100		Year	Historical Cost Index Jan. 1, 1993 = 100	Current Index Based on Jan. 1, 2002 = 100	
	Est.	Actual	Est.	Actual		Actual	Est.	Actual		Actual	Est.	Actual
Oct 2002					July 1987	87.7	69.5		July 1969	26.9	21.3	
July 2002					1986	84.2	66.8		1968	24.9	19.7	
April 2002					1985	82.6	65.5		1967	23.5	18.6	
Jan 2002	126.1		100.0	100.0	1984	82.0	65.0		1966	22.7	18.0	
July 2001		125.1	99.2		1983	80.2	63.6		1965	21.7	17.2	
2000		120.9	95.9		1982	76.1	60.4		1964	21.2	16.8	
1999		117.6	93.3		1981	70.0	55.5		1963	20.7	16.4	
1998		115.1	91.3		1980	62.9	49.9		1962	20.2	16.0	
1997		112.8	89.5		1979	57.8	45.8		1961	19.8	15.7	
1996		110.2	87.4		1978	53.5	42.4		1960	19.7	15.6	
1995		107.6	85.3		1977	49.5	39.3		1959	19.3	15.3	
1994		104.4	82.8		1976	46.9	37.2		1958	18.8	14.9	
1993		101.7	80.7		1975	44.8	35.5		1957	18.4	14.6	
1992		99.4	78.9		1974	41.4	32.8		1956	17.6	14.0	
1991		96.8	76.8		1973	37.7	29.9		1955	16.6	13.2	
1990		94.3	74.8		1972	34.8	27.6		1954	16.0	12.7	
1989		92.1	73.1		1971	32.1	25.5		1953	15.8	12.5	
1988		89.9	71.3		1970	28.7	22.8		1952	15.4	12.2	

Adjustments to Costs

The Historical Cost Index can be used to convert National Average building costs at a particular time to the approximate building costs for some other time.

Example:

Estimate and compare construction costs for different years in the same city.

To estimate the National Average construction cost of a building in 1970, knowing that it cost $900,000 in 2002:

INDEX in 1970 = 28.7

INDEX in 2002 = 126.1

Note: The City Cost Indexes for Canada can be used to convert U.S. National averages to local costs in Canadian dollars.

Time Adjustment using the Historical Cost Indexes:

$$\frac{\text{Index for Year A}}{\text{Index for Year B}} \times \text{Cost in Year B} = \text{Cost in Year A}$$

$$\frac{\text{INDEX 1970}}{\text{INDEX 2002}} \times \text{Cost 2002} = \text{Cost 1970}$$

$$\frac{28.7}{126.1} \times \$900,000 = .228 \times \$900,000 = \$205,200$$

The construction cost of the building in 1970 is $205,200.

Figure 25.11

Part III

Estimating Examples

Chapter 26

Unit Price Estimating Example

The Unit Price Estimate is the most detailed and most accurate type of estimate. The estimator should have working drawings and sufficient information to complete this estimate effectively. In commercial renovation, however, it is not enough to have the plans and specifications. The estimator must also perform a thorough evaluation of the site to understand how the existing conditions will affect the work.

This chapter presents an example of a Unit Price estimate for a hypothetical commercial project. The example begins with a description of the project and includes a discussion of the site evaluation as well as the existing conditions for each division. The individual items in the estimate may not represent every item that will be found in a renovation project. The example does provide a basis for understanding, evaluating, and estimating commercial renovation as a whole. Using this example as a guideline for actual projects, the reader will realize that consideration must be given to all Building, Fire, Health, and Safety Codes, as well as to regulations effective in a given locality.

The example assumes that working drawings and specifications have been provided, and refers to them.

Project Description

The sample renovation project involves the conversion of a turn-of-the-century mill building into retail and commercial office space. The building is located in the downtown area of a small city. The exterior walls are brick, and the floor systems are cast steel columns and wood beams with heavy wood decking. The roof structure is heavy timber trusses.

The building was originally used for manufacturing and has recently been utilized primarily for warehousing. The building has not been well maintained and is in a general state of disrepair. A retail tenant currently occupying the premises is to remain in operation throughout the renovation.

Figures 26.1 through 26.7 are plans, a section, and an elevation of the existing building. Figures 26.8 through 26.13 provide the same information for the proposed renovations. The owner has secured one office tenant for half of the third floor. Because the owner must know the costs for pricing future tenant renovations, the costs for the tenant improvements are estimated separately in each appropriate division. All costs in the sample are from *Means Repair & Remodeling Cost Data*.

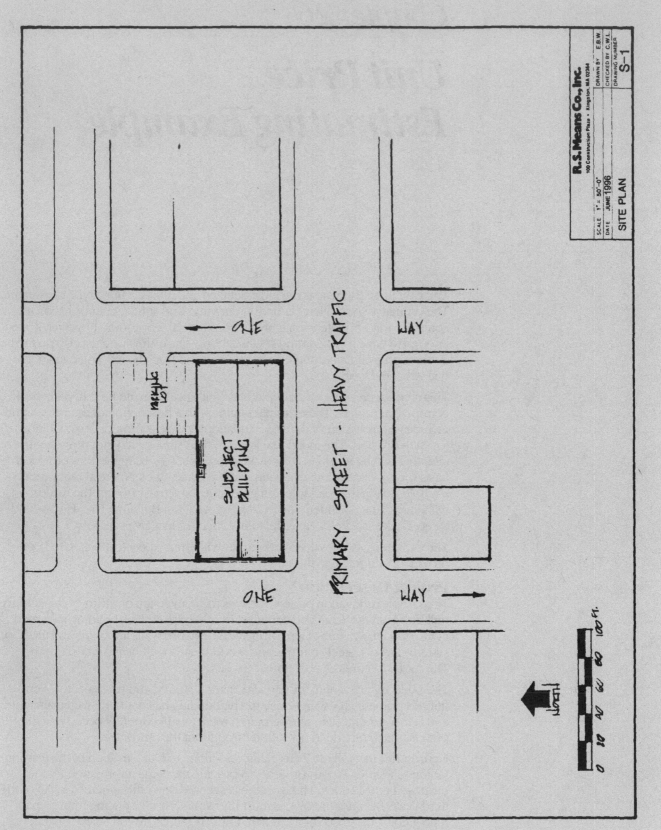

Figure 26.1

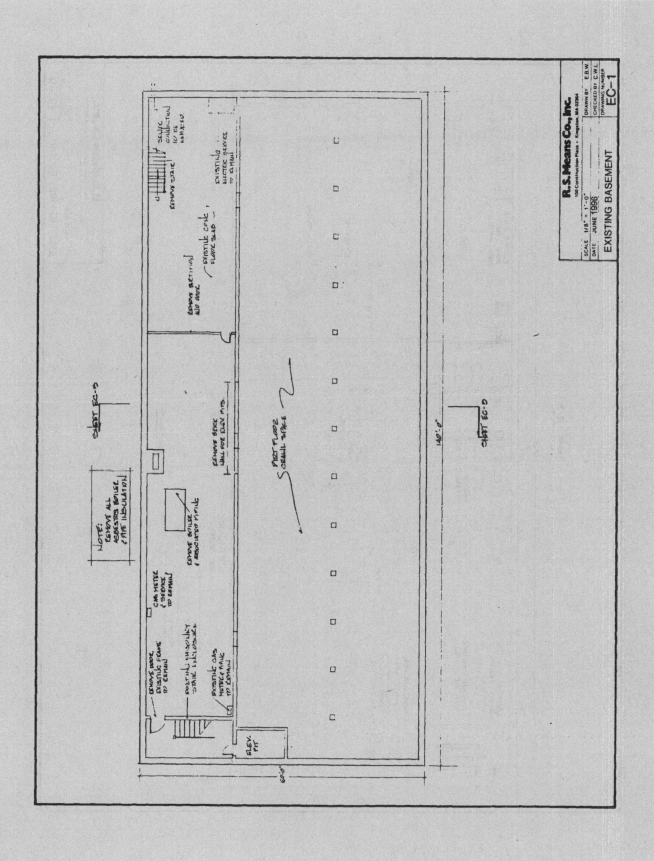

Figure 26.2

257

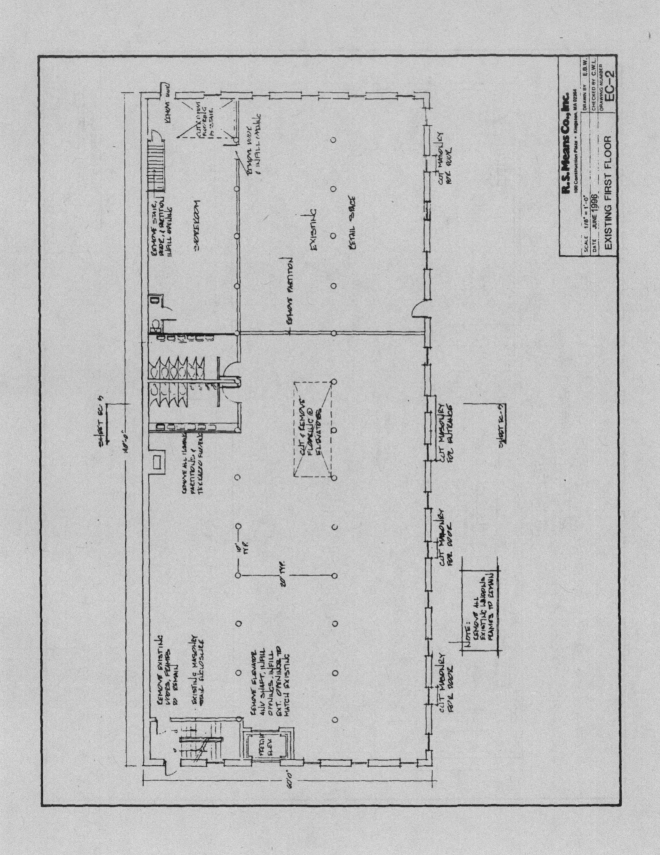

Figure 26.3

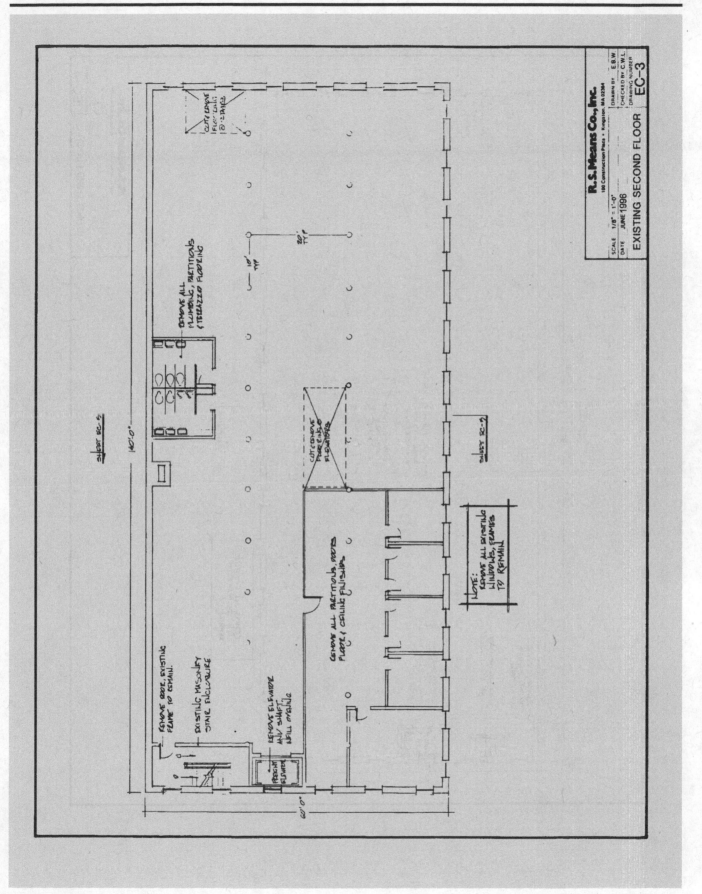

Figure 26.4

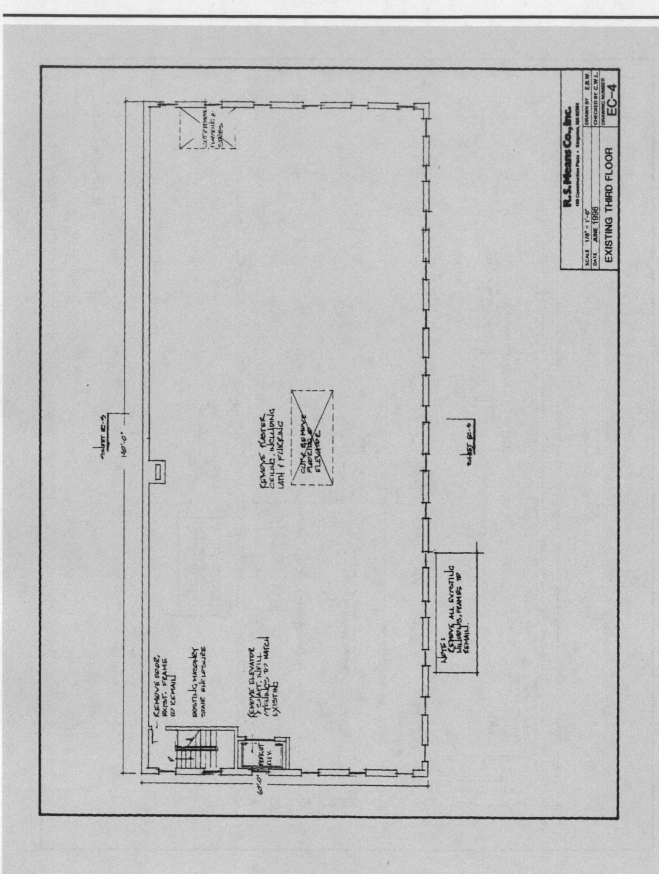

Figure 26.5

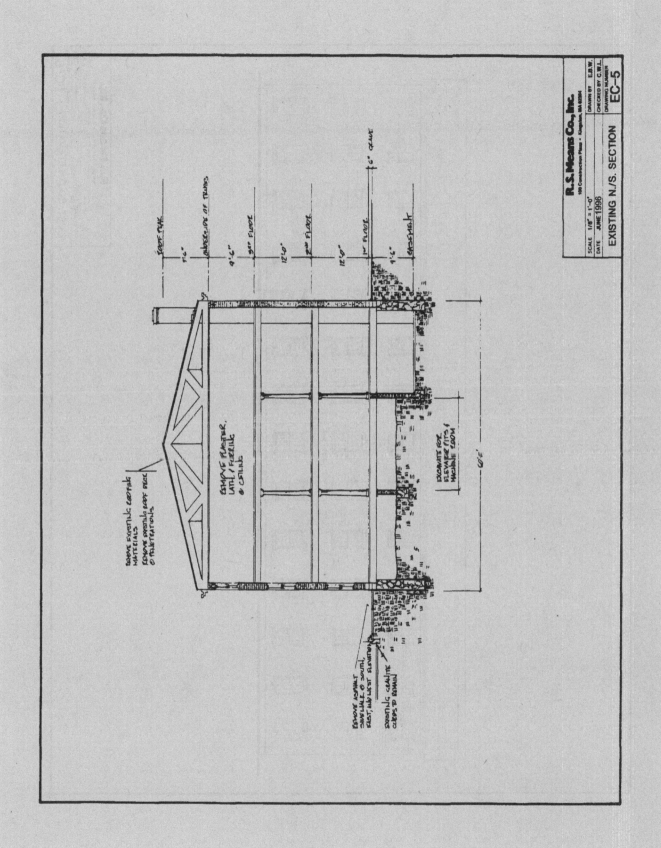

Figure 26.6

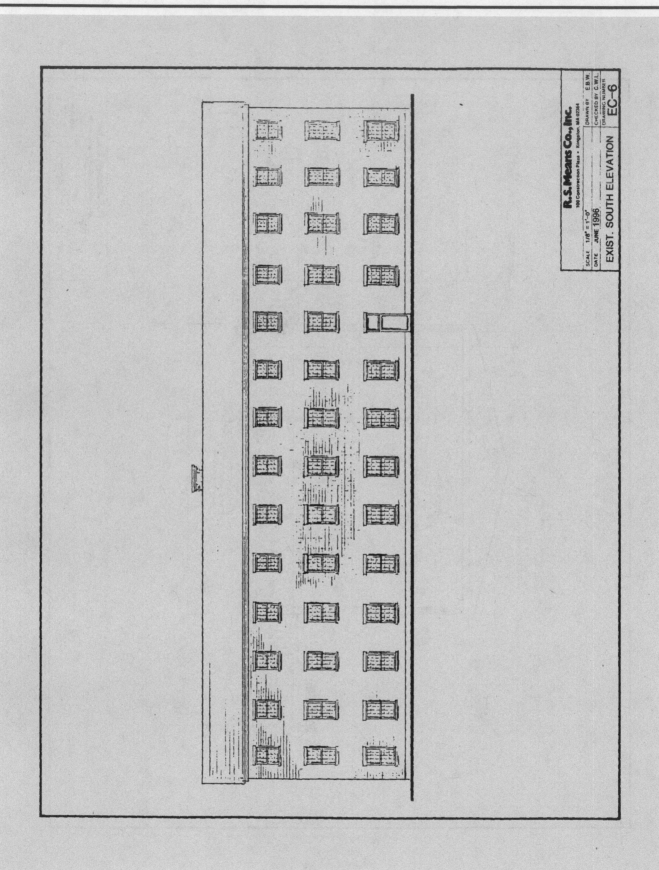

Figure 26.7

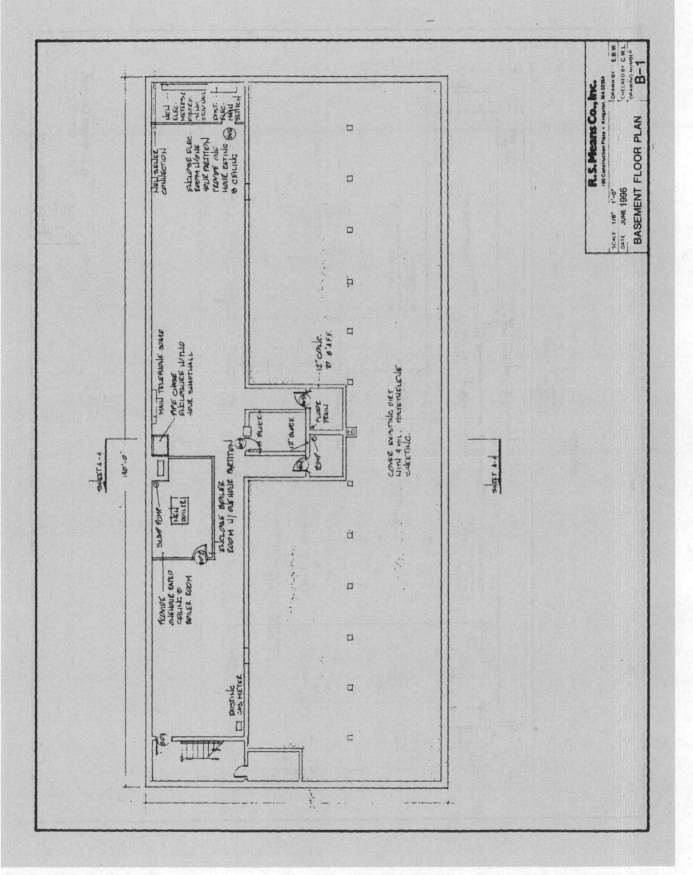

Figure 26.8

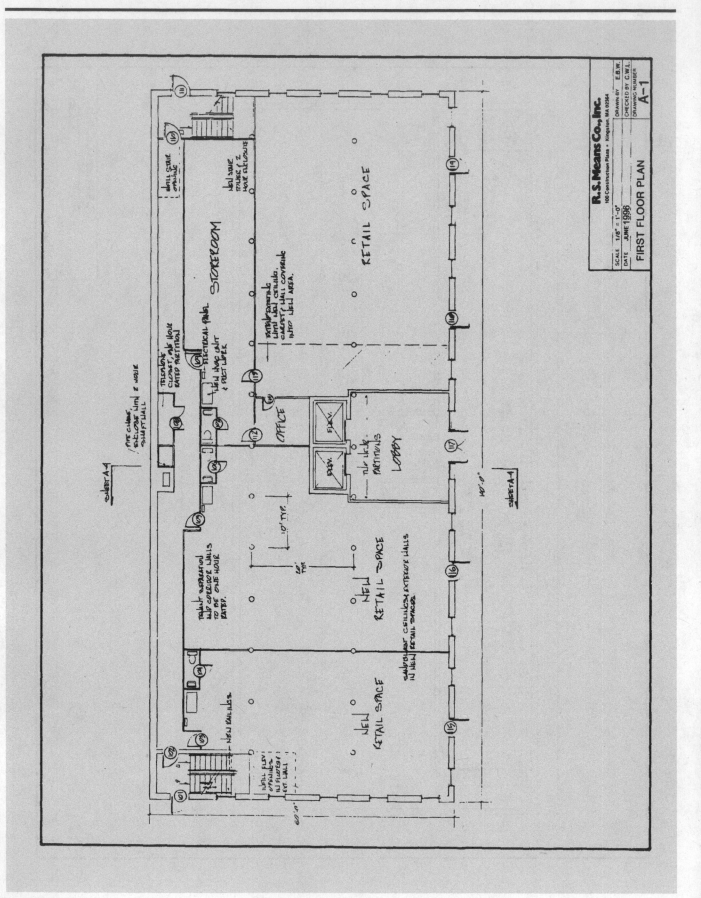

Figure 26.9

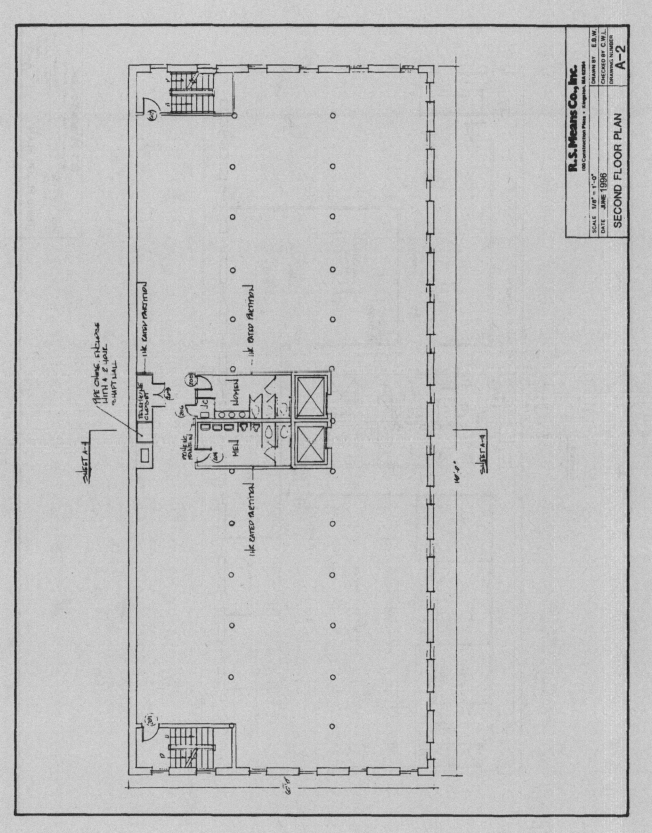

Figure 26.10

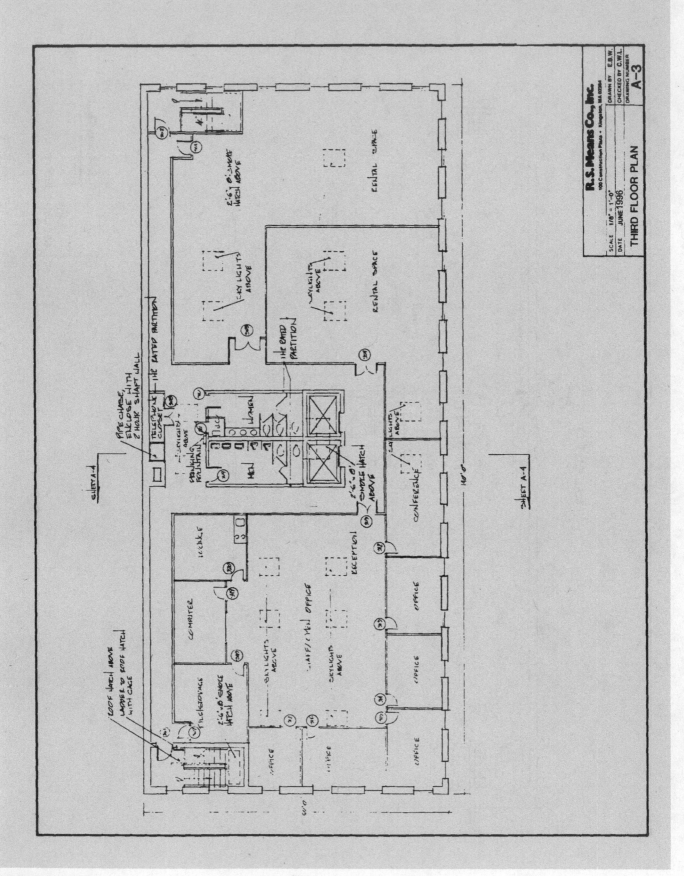

Figure 26.11

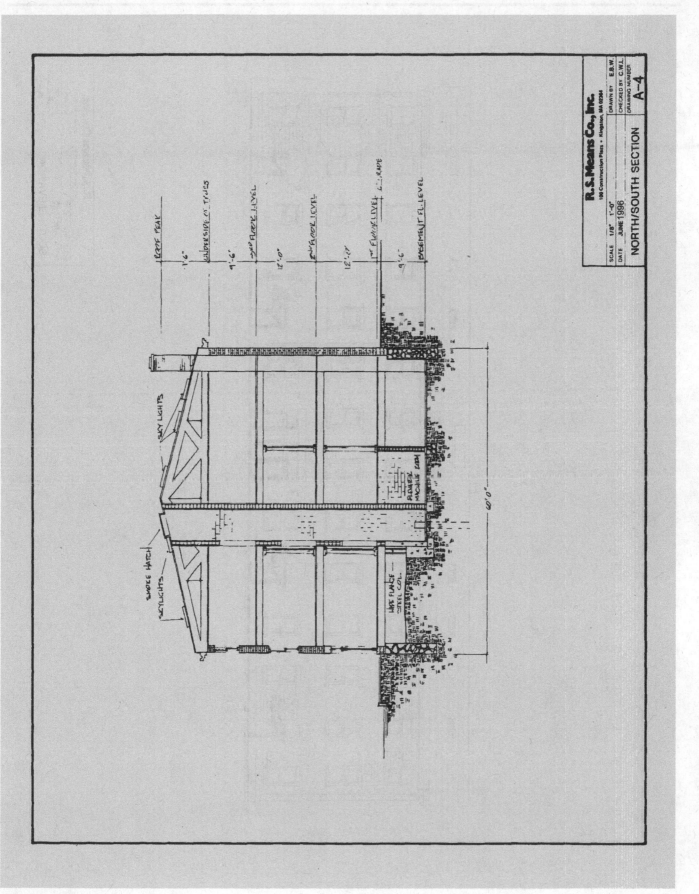

Figure 26.12

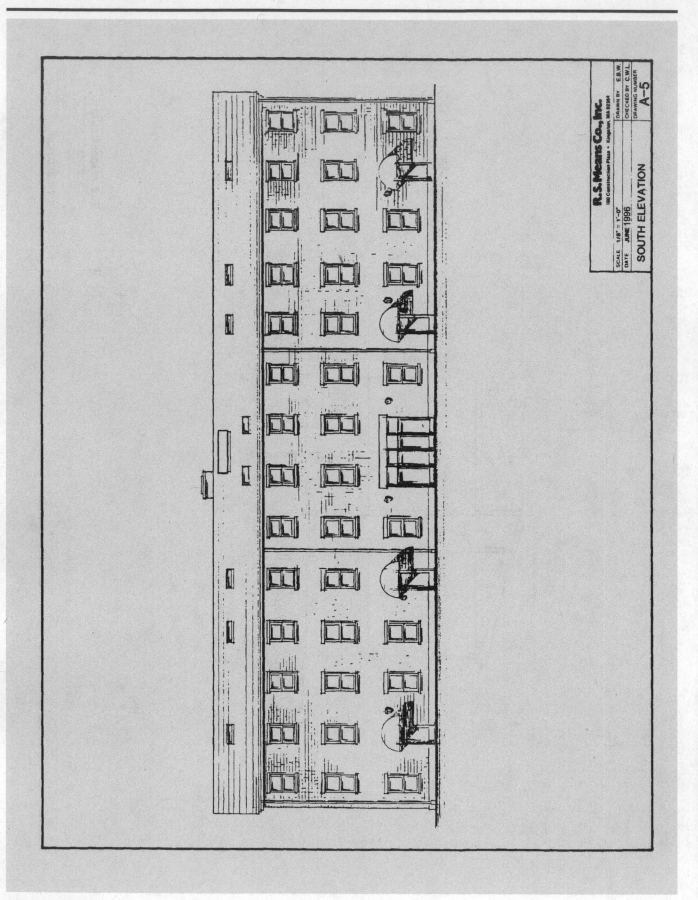

Figure 26.13

Division 1: General Requirements

Requirements of a renovation estimate should include the costs for all items that are not directly part of the physical construction (e.g., permits, insurance, and bonds) as well as those direct costs that cannot be allocated to a particular division (e.g., cleanup, temporary construction, and scaffolding) used by many trades.

Many of the items to be included in the General Requirements are dependent upon the total cost and/or the time duration of the project. Therefore, they cannot be quantified or priced until completion of the Estimate Summary.

Figures 26.14a and 26.14b are a Project Overhead Summary form, which lists more items to be included in General Requirements. This form will be used throughout the appropriate divisions of the sample estimate, wherever such items are to be included.

Please note that "Main Office Expense" and "Contingencies" are among the final items of Figure 26.14b. Some contractors include prorated costs for these items as indirect job costs within the estimate. Most contractors, however, include these costs as part of the overhead and profit percentages added at the Estimate Summary. The latter method is used in this sample estimate.

Included in one column are the costs for equipment, fees, rentals, and other items that are not labor or purchased material costs. All such costs receive the same markup in the Estimate Summary (usually 10% for supervision and handling). The material costs also receive a 10% markup, as well as added sales tax. These percentages may vary depending upon local practice.

"Factors" that affect the costs are included in each division and are discussed in Chapter 6 of this book. The application of these factors is crucial to effective estimating for remodeling and renovation. They help to determine added costs and restrictions caused by existing conditions. These factors are illustrated in Figure 26.15. The notations in the highlighted area are those that will be used throughout the sample estimate.

Division 2: Site Work and Demolition

After a quick glance at the floor plans of existing conditions, the estimator at first concludes that site work and demolition costs are minimal. But, upon investigation of the proposed plans and specifications, and upon evaluation of the site, it is determined that these costs are significant. Please refer to Figures 26.16 to 26.19 throughout the following discussion.

Asbestos removal has become a job only for licensed and experienced companies. A quotation is required and is shown in Figure 26.20. When the quotation is by telephone, the estimator must be sure to obtain all pertinent information. In this case, the estimator must determine what work required by the general contractor is not included in the quote. A factor for dust protection must be added to the costs. Figure 26.16 shows how the costs are included in the estimate. It is also important for the estimator to determine how long the work will take, because other workers will not be allowed in the area during the asbestos removal. This will affect the Project Schedule.

Even when specific items are not listed as individual line items in *Means Repair & Remodeling Cost Data*, the costs can be derived from the information that is available. In Figure 26.16, the costs for cutting the mill-type flooring for the elevator, stair, and roof openings have been calculated by estimating the time required and by determining the appropriate crew (F-2) to perform the work. Referring to the Crew Lists in *Means Repair & Remodeling Cost Data* provides the labor and equipment costs to be used. The estimator should make a note that the flooring that is removed should be saved to fill in the openings where the freight elevator is to be removed.

PROJECT OVERHEAD SUMMARY

PROJECT _____

LOCATION _____ ARCHITECT _____ DATE _____

QUANTITIES BY: _____ PRICES BY: _____ EXTENSIONS BY: _____ CHECKED BY: _____

DESCRIPTION	QUANTITY	UNIT	MATERIAL/EQUIPMENT		LABOR		TOTAL COST	
			UNIT	TOTAL	UNIT	TOTAL	UNIT	TOTAL
Job Organization: Superintendent								
Project Manager								
Timekeeper & Material Clerk								
Clerical								
Safety, Watchman & First Aid								
Travel Expense: Superintendent								
Project Manager								
Engineering: Layout								
Inspection/Quantities								
Drawings								
CPM Schedule								
Testing: Soil								
Materials								
Structural								
Equipment: Cranes								
Concrete Pump, Conveyor, Etc.								
Elevators, Hoists								
Freight & Hauling								
Loading, Unloading, Erecting, Etc.								
Maintenance								
Pumping								
Scaffolding								
Small Power Equipment/Tools								
Field Offices: Job Office								
Architect/Owner's Office								
Temporary Telephones								
Utilities								
Temporary Toilets								
Storage Areas & Sheds								
Temporary Utilities: Heat								
Light & Power								
Water								
PAGE TOTALS								

Page 1 of 2

Figure 26.14a

DESCRIPTION	QUANTITY	UNIT	MATERIAL/EQUIPMENT		LABOR		TOTAL COST	
			UNIT	TOTAL	UNIT	TOTAL	UNIT	TOTAL
Totals Brought Forward								
Winter Protection: Temp. Heat/Protection								
Snow Plowing								
Thawing Materials								
Temporary Roads								
Signs & Barricades: Site Sign								
Temporary Fences								
Temporary Stairs, Ladders & Floors								
Photographs								
Clean Up								
Dumpster								
Final Clean Up								
Punch List								
Permits: Building								
Misc.								
Insurance: Builders Risk								
Owner's Protective Liability								
Umbrella								
Unemployment Ins. & Social Security								
Taxes								
City Sales Tax								
State Sales Tax								
Bonds								
Performance								
Material & Equipment								
Main Office Expense								
Special Items								
TOTALS:								

Figure 26.14b

01200 | Price & Payment Procedures

01250 | Contract Modification Procedures

			CREW	DAILY OUTPUT	LABOR-HOURS	UNIT	2002 BARE COSTS MAT.	LABOR	EQUIP.	TOTAL	TOTAL INCL O&P	
200	0010	**CONTINGENCIES** for estimate at conceptual stage				Project					25%	**200**
	0050	Schematic stage									20%	
	0100	Preliminary working drawing stage (Design Dev.)									15%	
	0150	Final working drawing stage				↓					8%	
400	0010	**FACTORS** Cost adjustments	R01250 -010									**400**
	0100	Add to construction costs for particular job requirements										
	0500	Cut & patch to match existing construction, add, minimum >1				Costs	2%	3%				
	0550	Maximum				↓	5%	9%				
	0800	Dust protection, add, minimum >2				Costs	1%	2%				
	0850	Maximum					4%	11%				
	1100	Equipment usage curtailment, add, minimum >3					1%	1%				
	1150	Maximum					3%	10%				
	1400	Material handling & storage limitation, add, minimum >4					1%	1%				
	1450	Maximum					6%	7%				
	1700	Protection of existing work, add, minimum >5					2%	2%				
	1750	Maximum					5%	7%				
	2000	Shift work requirements, add, minimum >6						5%				
	2050	Maximum						30%				
	2300	Temporary shoring and bracing, add, minimum >7					2%	5%				
	2350	Maximum					5%	12%				
	2400	Work inside prisons and high security areas, add, minimum >8						30%				
	2450	Maximum				↓		50%				
500	0010	**JOB CONDITIONS** Modifications to total										**500**
	0020	project cost summaries										
	0100	Economic conditions, favorable, deduct				Project					2%	
	0200	Unfavorable, add									5%	
	0300	Hoisting conditions, favorable, deduct									2%	
	0400	Unfavorable, add									5%	
	0500	General Contractor management, experienced, deduct									2%	
	0600	Inexperienced, add									10%	
	0700	Labor availability, surplus, deduct									1%	
	0800	Shortage, add									10%	
	0900	Material storage area, available, deduct									1%	
	1000	Not available, add									2%	
	1100	Subcontractor availability, surplus, deduct									5%	
	1200	Shortage, add									12%	
	1300	Work space, available, deduct									2%	
	1400	Not available, add				↓					5%	

01290 | Payment Procedures

			CREW	DAILY OUTPUT	LABOR-HOURS	UNIT	2002 BARE COSTS MAT.	LABOR	EQUIP.	TOTAL	TOTAL INCL O&P	
800	0010	**TAXES** Sales tax, State, average	R01100 -090			%	4.70%					**800**
	0050	Maximum					7%					
	0200	Social Security, on first $80,400 of wages	R01100 -100					7.65%				
	0300	Unemployment, MA, combined Federal and State, minimum						2.10%				
	0350	Average						7%				
	0400	Maximum				↓		8%				

Figure 26.15

Repair and Remodeling Estimating Methods Example

Project No.: 001
Project Name: Commercial Renovation
Location: Estimating Methods Book

Title: ase Estimate
Architect: As Shown

Quantities: DEF
Take-off: ABC
Checked: GHI

Line Number	Description	Qty.	Unit	Material Unit Cost	Material Extension	Labor Unit Cost	Labor Extension	Equipment Unit Cost	Equipment Extension	Subcontract Unit Cost	Subcontract Extension
	Division 2 Site Construction										
132811200010	Bulk asbestos removal	1	WRITTEN QUOTE							4420.00	4420.00
012504000850	Factors, dust protection, add, maximum		Costs							11%	486.20
022256901220	Roofing & siding demo, dk, rf, wood, bds, tongue and groove, 2" x 10"	2	Day	2 Carpenters		480.00	960.00				
022208751200	Site demo, no hauling, masonry walls, blk or tile, brick, solid	180	C.F.			1.61	289.80	0.96	172.80		
022208751200	Site demo, no hauling, masonry walls, blk or tile, brick, solid	314	C.F.			1.61	505.54	0.96	301.44		
012504001750	Factors, protection of existing work, add, maximum		Costs	5%		7%	122.87				
012504002350	Factors, temporary shoring and bracing, add, maximum		Costs	5%		12%	210.84				
	Remove elevator and machinery	1	WRITTEN QUOTE							1040.00	1040.00
022208751200	Site demo, no hauling, masonry walls, blk or tile, brick, solid	1040	C.F.			1.61	1674.40	0.96	998.40		
012504000850	Factors, dust protection, add, maximum		Costs	4%		11%	184.18				
012504001750	Factors, protection of existing work, add, maximum		Costs	5%		7%	117.21				
022256903000	Roofing & siding dml, built-up 5 ply roof, no gravel	8660	S.F.			0.60	5196.00				

Figure 26.16

Repair and Remodeling Estimating Methods Example

Project No.: 001
Project Name: Commercial Renovation
Location: Estimating Methods Book

Title: see Estimate
Architect: As Shown

Quantities: DEF
Take-off: ABC
Checked: GHI

Line Number	Description	Qty.	Unit	Material Unit Cost	Material Extension	Labor Unit Cost	Labor Extension	Equipment Unit Cost	Equipment Extension	Subcontract Unit Cost	Subcontract Extension
022253101580	Ceiling demo, tile, wd fiber, 12" x 12", on suspension sys, incl sys	2440	S.F.			0.49	1195.60				
022253101000	Ceiling dml, plaster, lime & horse hair, on wood lath, incl lath	8400	S.F.			0.54	4536.00				
022253400200	Door demo, doors, exterior, 1-3/4" thick, single, 3' x 7' high	3	Ea.			11.75	35.25				
022253400500	Door demo, doors, interior, 1-3/8" thick, single, 3' x 7' high	15	Ea.			9.40	141.00				
022253400200	Door demo, doors, exterior, 1-3/4" thick, single, 3' x 7' high	4	Ea.			11.75	47.00				
022253401000	Walls & partitions demo, drywall, nailed	400	S.F.			0.19	76.00				
022253906600	Framing demo, wood framing, studs, 2" x 4"	680	L.F.			0.19	129.20				
012504000800	Factors, dust protection, add, minimum		Costs	1%		2%	4.10				
012504001700	Factors, protection of existing work, add, minimum		Costs	2%		2%	4.10				
012504002050	Factors, shift work requirements, add, maximum		Costs			30%	61.56				
022253401000	Walls & partitions demo, drywall, nailed	1920	S.F.			0.19	364.80				
022253403000	Walls & partitions demo, plaster, lime and horsehair, on wood lath	1560	S.F.			0.47	733.20				
022253906600	Framing demo, wood framing, studs, 2" x 4"	3306	L.F.			0.19	628.14				
022253800400	Flooring demo, carpet, bonded, including surface scraping	1800	S.F.			0.19	342.00				
022253802640	Flooring demo, tile, terrazzo, cast in place	640	S.F.			1.25	800.00				
022253906200	Framing demo, wood framing, stairs and stringers, minimum	15	Riser			9.40	141.00				

Figure 26.17

Repair and Remodeling Estimating Methods Example

Project No.: 001
Project Name: Commercial Renovation
Location: Estimating Methods Book

Title: ase Estimate
Architect: As Shown

Quantities: DEF
Take-off: ABC
Checked: GHI

Line Number	Description	Qty.	Unit	Material Unit Cost	Material Extension	Labor Unit Cost	Labor Extension	Equipment Unit Cost	Equipment Extension	Subcontract Unit Cost	Subcontract Extension
02225850202020	Window demolition, wood, including trim, to 25 S.F.	75	Ea.			10.40	780.00				
150556001400	Plumbing dml, fixtures w/10' pipe, water closet, floor mounted	16	Ea.			36.00	576.00				
150556001200	Plumbing dml, fixtures w/10' pipe, lavatory, wall hung	19	Ea.			29.00	551.00				
150556001520	Plumbing dml, fixtures w/10' pipe, urinal, wall mounted	6	Ea.			41.00	246.00				
02225840380800	Walls & partitions demo, toilet partitions, slate or marble	19	Ea.			37.50	712.50				
02225730044000	Rubbish handling, chute, circular, prefabricated steel, 30" diameter	30	L.F.	35.00	1050.00	19.30	579.00				
02225730204000	Rubbish handling, load, haul to chute & dumping into chute, 100' haul	180	C.Y.			22.50	4050.00				
02220875400000	Site dml, sidewalk removal, bituminous, 2.5" thick	249	S.Y.			1.89	470.61	0.55	136.95		
150553000010	HVAC dml										
150556000010	Plumbing dml										
160553000010	Electrical demolition										

Removed by salvage company at no cost.

Figure 26.18

Repair and Remodeling Estimating Methods Example

Project No.: 001
Project Name: Commercial Renovation
Location: Estimating Methods Book

Title: see Estimate
Architect: As Shown

Quantities: DEF
Take-off: ABC
Checked: GHI

Line Number	Description	Qty.	Unit	Material Unit Cost	Material Extension	Labor Unit Cost	Labor Extension	Equipment Unit Cost	Equipment Extension	Subcontract Unit Cost	Subcontract Extension
02775275031 0	Sidewalks, conc, 3000 PSI, CIP w/ 6x6 mesh, broom fin, no base, 4" T	2241	S.F.	1.15	2577.15	1.10	2465.10				
02315900150 0	Excavate trench, by hand with pick & shovel 2' to 6' deep, heavy soil	4	C.Y.			47.00	188.00				
02315100010 0	Backfill, by hand, no compaction, heavy soil	4	C.Y.			17.05	68.20				
02315100080 0	Backfill, by hand, compaction in 12" layers, hand tamp, add to above	4	C.Y.			5.50	22.00				
02530780208 0	Piping, drainage & sewage, PVC, 10' lengths, s.d.r. 35, 8" diam	10	L.F.	4.40	44.00	1.91	19.10				
01250400205 0	Factors, shift work requirements, add, maximum		Costs			30%	89.19				
02315440001 0	Excavating, structural, hand, pits to 6' deep, sandy soil	60	C.Y.			23.50	1410.00				
01250400235 0	Factors, temporary shoring and bracing, add, maximum		Costs	5%		12%	169.20				
	Division 2 Total				3671.15		30695.49		1609.59		5946.20

Figure 26.19

TELEPHONE QUOTATION

PROJECT	Office Renovation	DATE	
FIRM QUOTING	Asbestos Removal Co.	TIME	
ADDRESS		PHONE	
ITEM QUOTED		BY	
		RECEIVED BY	ABC

WORK INCLUDED	AMOUNT OF QUOTATION
Removal of asbestos pipe	
& boiler insulation in basement	
(includes required permits)	
Lump Sum	3,400
Job requires 3 Workers / 2 Days	

DELIVERY TIME				TOTAL BID	3,400

DOES QUOTATION INCLUDE THE FOLLOWING:			If NO is checked, determine the following:	
STATE & LOCAL SALES TAXES	☐ YES	☐ NO	N/A MATERIAL VALUE	
DELIVERY TO THE JOB SITE	☐ YES	☐ NO	N/A WEIGHT	
COMPLETE INSTALLATION	☐ YES	☐ NO	N/A QUANTITY	
COMPLETE SECTION AS PER PLANS & SPECIFICATIONS	☐ YES	☒ NO	DESCRIBE BELOW	

EXCLUSIONS AND QUALIFICATIONS
Does not include dust protection
of existing Tenant
Add factor to quote: 2+   Add 11% on Labor

ADDENDA ACKNOWLEDGMENT	TOTAL ADJUSTMENTS	
	ADJUSTED TOTAL BID	

ALTERNATES	
ALTERNATE NO.	
ALTERNATE NO.	
ALTERNATE NO.	
ALTERNATE NO.	
ALTERNATE NO.	
ALTERNATE NO.	
ALTERNATE NO.	

Figure 26.20

The estimator must use experience and good judgment when applying the "Factors" to the estimate. The project includes openings to be cut in the exterior masonry walls for the lobby entrance and retail doorways, as shown in Figure 26.3. Figure 26.21 contains the appropriate line item. The estimator knows that this masonry removal will have to be performed carefully because the edges of the openings will have to be rebuilt as finished surfaces. The estimator chooses the appropriate factors, as shown in Figure 26.22, and records them in Figure 26.16.

The existing freight elevator is to be used for hauling materials and is to remain in operation for as long as possible. This forces a delay of the demolition of the shaft and elevator, as well as of the construction of the third-floor office. Since the new windows and other work will be in place when the shaft is demolished, factors are applied for the protection of existing work and for dust protection. The estimator should also note, for scheduling purposes, that workers should place all materials on the upper floors before removal of the elevator.

The removal of the wall at the existing retail space will also entail extra labor expense to protect existing work, and the work must be performed after business hours. Note that on Figure 26.17 the total Factors (43% for labor) are added only to the labor cost. The 30% added for overtime affects the cost.

Other work, from other divisions, is required when the masonry is removed at the openings. For example, needling will be necessary at the wall above the entrance opening; and all opening jambs require toothing and rebuilding, using existing brick that has been removed. When items such as these are encountered, the estimator should make notes on separate sheets, as in Figure 26.23, so that items will not be omitted.

Salvaged materials may help to reduce the costs of demolition. In this case, a wrecking subcontractor is to remove all piping, wiring, radiators, and the boiler, at no cost. Ingenuity and legwork by the estimator can result in lower costs for the client and lower prices for competitive bids.

When all items in Division 2 have been entered, the estimator should review the division to note and include all related items to be the entered on the Project Overhead Summary, as shown in Figures 26.24a and 26.24b. The costs for a dumpster and the scaffolding, for example, are included in Division 1 in this estimate because they are to be used by different trades throughout the job, not just for demolition. Note that only quantities have been listed for these and other items. The total costs are dependent on the time extent of the project and will be entered when the Project Schedule has been completed.

Division 3: Concrete

The cast-in-place for the project is limited to the footings, walls, and slab in the basement for the new elevator shafts. These items can be estimated in two ways using *Means Repair & Remodeling Cost Data*.

The individual components, forms, reinforcing, labor, and concrete can be priced separately. The costs for the floor slab are shown in Figure 26.25. No forms are required because the new floor slab is to be placed flush with the existing basement floor. The "out" for the sump pump is too small to be deducted. At this point, the sump pumps and associated floor drain should be listed on a sheet for Division 15: Mechanical (Figure 26.26). For small pours, the estimator should be aware of minimum concrete costs.

The second method for pricing the cast-in-place concrete is shown in Figure 26.27. *Means Repair & Remodeling Cost Data* provides costs for complete concrete installations, including forms, reinforcing, concrete and placement. This method is used for the footings and walls in Figure 26.25.

02200 | Site Preparation

02220 | Site Demolition

		CREW	DAILY OUTPUT	LABOR-HOURS	UNIT	2002 BARE COSTS MAT.	LABOR	EQUIP.	TOTAL	TOTAL INCL O&P	
875	0010	**SITE DEMOLITION** No hauling, abandon catch basin or manhole	B-6	7	3.429	Ea.		87.50	25.50	113	171	**875**
	0020	Remove existing catch basin or manhole, masonry		4	6			153	44.50	197.50	299	
	0030	Catch basin or manhole frames and covers, stored		13	1.846			47	13.75	60.75	92	
	0040	Remove and reset		7	3.429			87.50	25.50	113	171	
	0100	Roadside delineators, remove only	B-80	175	.183			4.71	3.20	7.91	11.15	
	0110	Remove and reset	"	100	.320			8.25	5.60	13.85	19.55	
	0400	Minimum labor/equipment charge	B-6	4	6	Job		153	44.50	197.50	299	
	0600	Fencing, barbed wire, 3 strand	2 Clab	430	.037	L.F.		.87		.87	1.45	
	0650	5 strand	"	280	.057			1.34		1.34	2.22	
	0700	Chain link, posts & fabric, remove only, 8' to 10' high	B-6	445	.054			1.38	.40	1.78	2.68	
	0800	Guiderail, corrugated steel, remove only	B-80A	100	.240			5.65	1.82	7.47	11.30	
	0850	Remove and reset	"	40	.600			14.05	4.56	18.61	28.50	
	0860	Guide posts, remove only	B-80B	120	.267	Ea.		6.70	1.71	8.41	12.80	
	0870	Remove and reset	B-55	50	.480	"		11.40	14.70	26.10	35	
	0890	Minimum labor/equipment charge	2 Clab	4	4	Job		94		94	155	
	0900	Hydrants, fire, remove only	B-21A	5	8	Ea.		234	94.50	328.50	480	
	0950	Remove and reset	"	2	20	"		585	237	822	1,200	
	0990	Minimum labor/equipment charge	2 Plum	2	8	Job		288		288	450	
	1000	Masonry walls, block or tile, solid, remove	B-5	1,800	.031	C.F.		.81	.48	1.29	1.85	
	1100	Cavity wall		2,200	.025			.66	.39	1.05	1.51	
	1200	Brick, solid		900	.062			1.61	.96	2.57	3.69	
	1300	With block back-up		1,130	.050			1.29	.77	2.06	2.95	
	1400	Stone, with mortar		900	.062			1.61	.96	2.57	3.69	
	1500	Dry set		1,500	.037			.97	.58	1.55	2.22	
	1600	Median barrier, precast concrete, remove and store	B-3	430	.112	L.F.		2.86	3.80	6.66	8.85	
	1610	Remove and reset	"	390	.123	"		3.15	4.18	7.33	9.75	
	1650	Minimum labor/equipment charge	A-1	4	2	Job		47	15.15	62.15	94	
	1710	Pavement removal, bituminous roads, 3" thick	B-38	690	.058	S.Y.		1.55	1.26	2.81	3.90	
	1750	4" to 6" thick		420	.095			2.54	2.07	4.61	6.40	
	1800	Bituminous driveways		640	.063			1.67	1.36	3.03	4.20	
	1900	Concrete to 6" thick, hydraulic hammer, mesh reinforced		255	.157			4.18	3.41	7.59	10.55	
	2000	Rod reinforced		200	.200			5.35	4.35	9.70	13.45	
	2100	Concrete, 7" to 24" thick, plain		33	1.212	C.Y.		32.50	26.50	59	81.50	
	2200	Reinforced		24	1.667	"		44.50	36.50	81	112	
	2250	Minimum labor/equipment charge		6	6.667	Job		178	145	323	450	
	2300	With hand held air equipment, bituminous, to 6" thick	B-39	1,900	.025	S.F.		.63	.09	.72	1.13	
	2320	Concrete to 6" thick, no reinforcing		1,600	.030			.75	.11	.86	1.34	
	2340	Mesh reinforced		1,400	.034			.85	.13	.98	1.54	
	2360	Rod reinforced		765	.063			1.56	.23	1.79	2.82	
	2390	Minimum labor/equipment charge	B-38	6	6.667	Job		178	145	323	450	
	2400	Curbs, concrete, plain	B-6	360	.067	L.F.		1.70	.50	2.20	3.32	
	2500	Reinforced		275	.087			2.23	.65	2.88	4.34	
	2600	Granite		360	.067			1.70	.50	2.20	3.32	
	2700	Bituminous		528	.045			1.16	.34	1.50	2.26	
	2790	Minimum labor/equipment charge		6	4	Job		102	30	132	199	
	2900	Pipe removal, sewer/water, no excavation, 12" diameter		175	.137	L.F.		3.51	1.02	4.53	6.80	
	2930	15" diameter		150	.160			4.09	1.19	5.28	7.95	
	2960	24" diameter		120	.200			5.10	1.49	6.59	9.95	
	3000	36" diameter		90	.267			6.80	1.98	8.78	13.30	
	3200	Steel, welded connections, 4" diameter		160	.150			3.84	1.12	4.96	7.50	
	3300	10" diameter		80	.300			7.65	2.23	9.88	14.95	
	3390	Minimum labor/equipment charge		3	8	Job		205	59.50	264.50	400	
	3500	Railroad track removal, ties and track	B-13	330	.170	L.F.		4.32	2.20	6.52	9.45	
	3600	Ballast	B-14	500	.096	C.Y.		2.38	.36	2.74	4.30	
	3700	Remove and re-install, ties & track using new bolts & spikes		50	.960	L.F.		24	3.57	27.57	43	
	3800	Turnouts using new bolts and spikes		1	48	Ea.		1,200	179	1,379	2,150	

Figure 26.21

279

01200 | Price & Payment Procedures

01250 | Contract Modification Procedures

			CREW	DAILY OUTPUT	LABOR-HOURS	UNIT	2002 BARE COSTS				TOTAL INCL O&P	
							MAT.	LABOR	EQUIP.	TOTAL		
200	0010	**CONTINGENCIES** for estimate at conceptual stage				Project					25%	200
	0050	Schematic stage									20%	
	0100	Preliminary working drawing stage (Design Dev.)									15%	
	0150	Final working drawing stage				↓					8%	
400	0010	**FACTORS** Cost adjustments R01250-010										400
	0100	Add to construction costs for particular job requirements										
	0500	Cut & patch to match existing construction, add, minimum >1				Costs	2%	3%				
	0550	Maximum				↓	5%	9%				
	0800	Dust protection, add, minimum >2				Costs	1%	2%				
	0850	Maximum					4%	11%				
	1100	Equipment usage curtailment, add, minimum >3					1%	1%				
	1150	Maximum					3%	10%				
	1400	Material handling & storage limitation, add, minimum >4					1%	1%				
	1450	Maximum					6%	7%				
	1700	Protection of existing work, add, minimum >5					2%	2%				
	1750	Maximum					5%	7%				
	2000	Shift work requirements, add, minimum >6						5%				
	2050	Maximum						30%				
	2300	Temporary shoring and bracing, add, minimum >7					2%	5%				
	2350	Maximum					5%	12%				
	2400	Work inside prisons and high security areas, add, minimum >8						30%				
	2450	Maximum				↓		50%				
500	0010	**JOB CONDITIONS** Modifications to total										500
	0020	project cost summaries										
	0100	Economic conditions, favorable, deduct				Project					2%	
	0200	Unfavorable, add									5%	
	0300	Hoisting conditions, favorable, deduct									2%	
	0400	Unfavorable, add									5%	
	0500	General Contractor management, experienced, deduct									2%	
	0600	Inexperienced, add									10%	
	0700	Labor availability, surplus, deduct									1%	
	0800	Shortage, add									10%	
	0900	Material storage area, available, deduct									1%	
	1000	Not available, add									2%	
	1100	Subcontractor availability, surplus, deduct									5%	
	1200	Shortage, add									12%	
	1300	Work space, available, deduct									2%	
	1400	Not available, add				↓					5%	

01290 | Payment Procedures

			CREW	DAILY OUTPUT	LABOR-HOURS	UNIT	MAT.	LABOR	EQUIP.	TOTAL	TOTAL INCL O&P	
800	0010	**TAXES** Sales tax, State, average R01100-090				%	4.70%					800
	0050	Maximum					7%					
	0200	Social Security, on first $80,400 of wages R01100-100						7.65%				
	0300	Unemployment, MA, combined Federal and State, minimum						2.10%				
	0350	Average						7%				
	0400	Maximum				↓		8%				

Figure 26.22

Repair and Remodeling Estimating Methods Example

Project No.: 001		Title: ase Estimate				Quantities: DEF							
Project Name: Commercial Renovation		Architect: As Shown				Take-off: ABC							
Location: Estimatting Methods Book						Checked: GHI							
				Material		Labor		Equipment		Subcontract			
Line Number	Description	Qty.	Unit	Unit Cost	Extension	Unit Cost	Extension	Unit Cost	Extension	Unit Cost	Extension		
	Division 4 Masonry												
04106001080	Needle beam masonry, incl shoring 10' x 10' opening, brick, solid, 16" thk												
04106002000	Needle beam masonry, add for additional floors of shoring												
04108000520	Toothing masonry, brickwork, hard mortar												

Figure 26.23

PROJECT   Commercial Renovation   ESTIMATE NO. 001

LOCATION   ARCHITECT   DATE 2002

QUANTITIES BY: ABC   PRICES BY RSM   EXTENSIONS BY   DEF   CHECKED BY   GHI

DESCRIPTION			Material		Labor		Equip, Fees, Rental	
Job Organization: Superintendent								
Project Manager								
Timekeeper & Material Clerk								
Clerical								
Safety, Watchman, First Aid								
Travel Expense: Superintendent								
Project Manager								
Engineering: Layout								
Inspection / Quantities								
Drawings								
CPM Schedule								
Testing: Soil								
Materials								
Structural								
Equipment: Cranes								
Concrete Pump, Conveyor, Etc.								
Elevators, Hoists								
Freight & Hauling								
Loading, Unloading, Erecting, Etc.								
Maintenance								
Pumping								
Scaffolding 01540-750-0090	119	C.S.F.						
Small Power Equipment / Tools								
Field Facilities: Job Office								
Architects / Owners Office								
Temporary Telephones								
Utilities (Power) 01510-800-0450	280	C.S.F.						
Temporary Toilets								
Storage Areas & Sheds								
Temporary Utilities: Heat								
Light & Power 01510-800-0350	280	C.S.F.						
Water								
PAGE TOTALS								

Figure 26.24a

Division 1

DESCRIPTION				Material			Labor			Equip, Fees, Rental		
Winter Protection: Temp. Heat / Protection												
Snow Plowing												
Thawing Materials												
Temporary Roads												
Signs & Barricades: Site Sign												
Temporary Fences 01560-100-1000	270	L.F.										
Temporary Stairs, Ladders & Floors												
Photographs												
Clean Up												
Dumpster 02225-730-0800		Week										
Final Clean Up												
Punch List												
Permits: Building												
Misc.												
Insurance: Builders Risk												
Owner's Protective Liability												
Umbrella												
Unemployment Ins. & Social Security												
Taxes												
City Sales Tax												
State Sales Tax												
Bonds												
Performance												
Material & Equipment												
Main Office Expense												
Special Items												
TOTALS:												

Figure 26.24b

Repair and Remodeling Estimating Methods Example

Project No.: 001
Project Name: Commercial Renovation
Location: Estimating Methods Book

Title: ase Estimate
Architect: As Shown

Quantities: DEF
Take-off: ABC
Checked: GHI

Line Number	Description	Qty.	Unit	Material Unit Cost	Material Extension	Labor Unit Cost	Labor Extension	Equipment Unit Cost	Equipment Extension	Subcontract Unit Cost	Subcontract Extension
	Division 3 Concrete										
033102403850	Concrete in place, footings, spread, over 5 C.Y.	12	C.Y.	92.00	1104.00	40.00	480.00	0.44	5.28		
033102404260	Concrete in place, grade walls, 12" thick, 8' high	18	C.Y.	111.00	1998.00	93.00	1674.00	11.15	200.70		
012504001150	Factors, equipment usage curtailment, add, maximum		Costs	3%		10%					
032202000100	Welded wire fabric, sheets, 6 x 6 - W1.4 x W1.4 (10 x 10), 21 lb/CSF	4	C.S.F.	7.00	28.00	15.65	62.60				
033107004300	Placing conc, incl vib, slab on grade, 4" thick, direct chute	5	C.Y.			10.75	53.75	0.65	3.25		
033102002000150	Concrete, ready mix, regular weight, 3000 psi	5	C.Y.	69.00	345.00						
034101001200	Beams, precast, rectangular, 20' span, 12" x 20"	1	Ea.	955.00	955.00	75.00	75.00	53.00	53.00		
012504001150	Factors, equipment usage curtailment, add, maximum		Costs	3%	28.65	10%	7.50				
	Division 3 Total				4458.65		2352.85		262.23		

Figure 26.25

Repair and Remodeling Estimating Methods Example

Project No.: 001
Project Name: Commercial Renovation
Location: Estimating Methods Book

Title: ase Estimate
Architect: As Shown

Quantities: DEF
Take-off: ABC
Checked: GHI

Line Number	Description	Qty.	Unit	Material		Labor		Equipment		Subcontract	
				Unit Cost	Extension	Unit Cost	Extension	Unit Cost	Extension	Unit Cost	Extension
	Division 15 Mechanical										
1544094407100	Pumps, submersible, sump pump, auto, plastic, 1-1/4" discharge, 1/4 HP	2	Ea.							191.00	382.00
1515530002040	Drains, flr, medium dty, C.I., D flange, 7" dia top, 2" and 3" pipe size	1	Ea.							150.00	150.00

Figure 26.26

03300 | Cast-In-Place Concrete

03310 | Structural Concrete

		CREW	DAILY OUTPUT	LABOR-HOURS	UNIT	2002 BARE COSTS				TOTAL INCL O&P
						MAT.	LABOR	EQUIP.	TOTAL	
2300	Waffle const., 30" domes, 125 psf Sup. Load, 20' span R03310 -010	C-14B	37.07	5.611	C.Y.	197	169	19.35	385.35	520
2350	30' span		44.07	4.720		180	142	16.30	338.30	455
2500	One way joists, 30" pans, 125 psf Sup. Load, 15' span R03310 -100		27.38	7.597		231	229	26	486	665
2550	25' span		31.15	6.677		216	201	23	440	600
2700	One way beam & slab, 125 psf Sup. Load, 15' span R04210 -055		20.59	10.102		186	305	35	526	755
2750	25' span		28.36	7.334		169	221	25.50	415.50	585
2900	Two way beam & slab, 125 psf Sup. Load, 15' span		24.04	8.652		174	261	30	465	660
2950	25' span		35.87	5.799		148	175	20	343	475
3100	Elevated slabs including finish, not									
3110	including forms or reinforcing									
3150	Regular concrete, 4" slab	C-8	2,613	.021	S.F.	.94	.56	.26	1.76	2.23
3200	6" slab		2,585	.022		1.45	.57	.26	2.28	2.81
3250	2-1/2" thick floor fill		2,685	.021		.64	.55	.25	1.44	1.87
3300	Lightweight, 110# per C.F., 2-1/2" thick floor fill		2,585	.022		.74	.57	.26	1.57	2.02
3400	Cellular concrete, 1-5/8" fill, under 5000 S.F.		2,000	.028		.49	.74	.34	1.57	2.11
3450	Over 10,000 S.F.		2,200	.025		.39	.67	.31	1.37	1.86
3500	Add per floor for 3 to 6 stories high		31,800	.002			.05	.02	.07	.10
3520	For 7 to 20 stories high		21,200	.003			.07	.03	.10	.15
3800	Footings, spread under 1 C.Y.	C-14C	38.07	2.942	C.Y.	100	84.50	.94	185.44	252
3850	Over 5 C.Y.		81.04	1.382		92	40	.44	132.44	168
3900	Footings, strip, 18" x 9", plain		41.04	2.729		91	78.50	.87	170.37	232
3950	36" x 12", reinforced		61.55	1.820		92.50	52.50	.58	145.58	190
4000	Foundation mat, under 10 C.Y.		38.67	2.896		124	83.50	.93	208.43	276
4050	Over 20 C.Y.		56.40	1.986		110	57	.64	167.64	217
4200	Grade walls, 8" thick, 8' high	C-14D	45.83	4.364		126	130	15.65	271.65	370
4250	14' high		27.26	7.337		149	219	26.50	394.50	560
4260	12" thick, 8' high		64.32	3.109		111	93	11.15	215.15	288
4270	14' high		40.01	4.999		119	149	17.95	285.95	400
4300	15" thick, 8' high		80.02	2.499		104	74.50	8.95	187.45	248
4350	12' high		51.26	3.902		106	117	14	237	325
4500	18' high		48.85	4.094		116	122	14.70	252.70	345
4520	Handicap access ramp, railing both sides, 3' wide	C-14H	14.58	3.292	L.F.	139	98	2.44	239.44	320
4525	5' wide		12.22	3.928		146	117	2.91	265.91	360
4530	With 6" curb and rails both sides, 3' wide		8.55	5.614		146	167	4.15	317.15	445
4535	5' wide		7.31	6.566		149	195	4.86	348.86	495
4650	Slab on grade, not including finish, 4" thick	C-14E	60.75	1.449	C.Y.	84.50	43	.58	128.08	167
4700	6" thick	"	92	.957	"	81	28.50	.38	109.88	138
4751	Slab on grade, incl. troweled finish, not incl. forms									
4760	or reinforcing, over 10,000 S.F., 4" thick slab	C-14F	3,425	.021	S.F.	.92	.57	.01	1.50	1.95
4820	6" thick slab		3,350	.021		1.35	.58	.01	1.94	2.43
4840	8" thick slab		3,184	.023		1.85	.61	.01	2.47	3.02
4900	12" thick slab		2,734	.026		2.77	.72	.01	3.50	4.21
4950	15" thick slab		2,505	.029		3.48	.78	.01	4.27	5.10
5000	Slab on grade, incl. textured finish, not incl. forms									
5001	or reinforcing, 4" thick slab	C-14G	2,873	.019	S.F.	.90	.52	.01	1.43	1.84
5010	6" thick		2,590	.022		1.40	.58	.01	1.99	2.49
5020	8" thick		2,320	.024		1.83	.65	.02	2.50	3.07
5200	Lift slab in place above the foundation, incl. forms,									
5210	reinforcing, concrete and columns, minimum	C-14B	2,113	.098	S.F.	4.96	2.97	.34	8.27	10.75
5250	Average		1,650	.126		5.60	3.80	.43	9.83	13
5300	Maximum		1,500	.139		6.35	4.18	.48	11.01	14.50
5500	Lightweight, ready mix, including screed finish only,									
5510	not including forms or reinforcing									
5550	1:4 for structural roof decks	C-14B	260	.800	C.Y.	102	24	2.76	128.76	155
5600	1:6 for ground slab with radiant heat	C-14F	92	.783		96.50	21.50	.38	118.38	140
5650	1:3:2 with sand aggregate, roof deck	C-14B	260	.800		101	24	2.76	127.76	154

Figure 26.27

The precast beam that will serve as the lintel at the lobby entrance cannot be lowered into place with a crane. The beam must be slid into place by hand and winch. A factor for the added labor is included in the estimate. The specifications call for the beam to receive a special stucco finish to match the existing limestone window lintels. This is included in Division 9: Finishes.

During the estimate, the owner asks for an alternate price to lower the dirt floor to create an area for tenant storage. While the estimator knows from experience that such a project would not be cost effective, the owner needs to know the costs. In a short period of time, the estimator can calculate fairly accurate costs, without plans and specifications, by visualizing the work to be performed. The costs for the alternate are shown in Figure 26.28. This price does not include the costs for continuing the new stairway to the basement (required by code for basement occupancy) or the resulting costs for relocating the electrical service. After a phone call to the owner, the estimator does not have to spend any more time.

Division 4: Masonry

In commercial renovation, new masonry work is usually limited, compared to the extent of repair and restoration of existing masonry, which takes extensive time and labor. Normal methods of work must be altered to accommodate existing conditions. Figure 26.16 in Division 2 lists costs for removing the brick at the new exterior openings. The demolition of the brick for the lobby entrance cannot occur until the structure above is supported by needles, as shown in Figure 26.29. Such an operation should always be planned and supervised by a structural engineer. The costs are included in Figure 26.31.

Included in the costs for removing the exterior brick is a factor for protection of existing work (Figure 26.15). This factor is applied for protection of the remaining adjacent surfaces and because the existing brick must be removed carefully to be reused to construct jambs at the new exterior openings. Brick in older buildings usually cannot be matched with new brick. Since the old brick is to be used, there are no material costs for the specified brick masonry. There is no specific line item for rebuilding jambs to match existing conditions at new openings, using existing materials. The estimator must use good judgment to determine the most accurate costs. The work is similar to constructing brick columns, so line 04810-170-0300 in Figure 26.30 is chosen. Since the work requires care to match existing conditions, a Factor is included, as shown in Figure 26.31. Estimate sheets for Division 4: Masonry are illustrated in Figures 26.31 to 26.33.

Unless the contractor has a great deal of experience in masonry restoration, a knowledgeable architect or an historic preservationist should be consulted regarding masonry cleaning and repair in older buildings. It is very easy to damage existing masonry if the wrong methods and materials are used. In this example, the work is very clearly specified. High-pressure water with no chemical cleaners is required (see Figure 26.32). The cleaning is to be performed during normal working hours. Pedestrian protection and tarpaulins are included (although not directly specified) to prevent the high volumes of water and mist from causing complaints or damage. Also, the type of mortar should be defined in detail. Modern mortars have strengths and expansion/contraction properties that are different from former types. These modern properties may cause damage to old brick.

Sandblasting is often rejected as a method of cleaning exterior masonry because it destroys the weatherproof integrity of the masonry surface. For interior surfaces, however, it is widely used. In the sample project, the interior faces of the perimeter walls and wood ceiling of the new retail spaces, the entire second-floor ceiling, and the interior faces of the perimeter walls of the third

Repair and Remodeling Estimating Methods Example

Project No.: 001
Project Name: Underpinning Alternate
Location: Estimating Methods Book

Title: ase Estimate
Architect: As Shown

Quantities: DEF
Take-off: ABC
Checked: GHI

Line Number	Description	Qty.	Unit	Material Unit Cost	Material Extension	Labor Unit Cost	Labor Extension	Equipment Unit Cost	Equipment Extension	Subcontract Unit Cost	Subcontract Extension
022501000100	Underpinning foundations, 5' to 16' below grade, 100 to 500CY	13	C.Y.	193.00	2509.00	675.00	8775.00	150.00	1950.00		
022501000900	Underpinning foundations, for under 50 CY, add		C.Y.	10%	250.90	40%	3510.00				
012504001150	Factors, equipment usage curtailment, add, maximum		Costs	3%	75.27	10%	877.50				
012504002350	Factors, temporary shoring and bracing, add, maximum		Costs	5%	125.45	12%	1053.00				
023154400010	Excavating, structural, hand, pits to 6' deep, sandy soil	700	C.Y.			23.50	16450.00				
012504001150	Factors, equipment usage curtailment, add, maximum		Costs	3%		10%	1645.00				
012504002350	Factors, temporary shoring and bracing, add, maximum		Costs	5%		12%	1974.00				
015901000900	Rent conveyer concrete portable, gas 16"wide 46'long	10	Day+					471.20	4712.00		
032202000100	Welded wire fabric, sheets, 6 x 6 - W1.4 x W1.4 (10 x 10), 21 lb/CSF	56	C.S.F.	14.00	784.00	31.30	1752.80				
033107004300	Placing conc, incl vib, slab on grade, 4" thick, direct chute	69	C.Y.			21.50	1483.50	1.30	89.70		
033107005610	Placing concrete, wheel dumping, walking cart, 50' haul, add	69	C.Y.			13.30	917.70	2.90	200.10		
012504000500	Factors, cut & patch to match existing construction, add, minimum		Costs	2%		3%	72.04				
012504001100	Factors, equipment usage curtailment, add, minimum		Costs	1%		1%	24.01				
033102200150	Concrete, ready mix, regular weight, 3000 psi	69	C.Y.	138.00	9522.00						
	Subtotal				13266.62		38534.55		6951.80		
	Overhead and Profit			10%	1326.66	64.3%	24777.72	10%	695.18		
	Subtotal				14593.28		63312.27		7646.98		
	General Contractors Overhead and Profit			10%	1459.33	10%	6331.23	10%	764.7		
	Total				16052.61		69643.50		8411.68		
	Grand Total										94107.79

Figure 26.28

floor tenant space are to be sandblasted. A subcontractor will perform the work; but, because the estimator has not yet received quotations, the costs must be calculated. Workers must take precautionary measures and perform the job after business hours, because the building is occupied by the existing retail tenant and is located in an active urban area. The appropriate Factors are added to the costs. The Factors (43%) are applied to labor costs only (see Figures 26.15 and 26.33). Because the work is to be subcontracted, the increase must be figured to include overhead and profit for the subcontractor. Note that a Factor of only 13% is added to the labor-hours. This is because the overtime Factor (30%) does not entail more time expended, only greater expense.

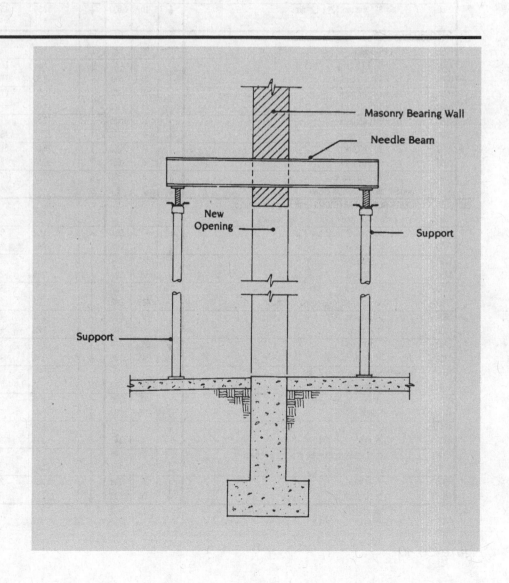

Figure 26.29

04700 | Simulated Masonry

04710 | Simulated Brick

			CREW	DAILY OUTPUT	LABOR-HOURS	UNIT	2002 BARE COSTS				TOTAL INCL O&P	
							MAT.	LABOR	EQUIP.	TOTAL		
600	0010	**SIMULATED BRICK** Aluminum, baked on colors	1 Carp	200	.040	S.F.	2.35	1.20		3.55	4.58	600
	0050	Fiberglass panels		200	.040		2.45	1.20		3.65	4.69	
	0100	Urethane pieces cemented in mastic		150	.053		4.60	1.60		6.20	7.70	
	0150	Vinyl siding panels		200	.040		1.85	1.20		3.05	4.03	

04800 | Masonry Assemblies

04810 | Unit Masonry Assemblies

			CREW	DAILY OUTPUT	LABOR-HOURS	UNIT	2002 BARE COSTS				TOTAL INCL O&P	
							MAT.	LABOR	EQUIP.	TOTAL		
160	0010	**CHIMNEY** See Div. 03310 for foundation, add to prices below										160
	0100	Brick, 16" x 16", 8" flue, scaff. not incl.	D-1	18.20	.879	V.L.F.	14.20	23.50		37.70	54.50	
	0150	16" x 20" with one 8" x 12" flue		16	1		22.50	27		49.50	68.50	
	0200	16" x 24" with two 8" x 8" flues		14	1.143		33	31		64	87	
	0250	20" x 20" with one 12" x 12" flue		13.70	1.168		30	31.50		61.50	84.50	
	0300	20" x 24" with two 8" x 12" flues		12	1.333		37	36		73	99.50	
	0350	20" x 32" with two 12" x 12" flues		10	1.600		52.50	43		95.50	129	
170	0010	**COLUMNS** Brick, scaffolding not included										170
	0050	8" x 8", 9 brick	D-1	56	.286	V.L.F.	3.51	7.70		11.21	16.45	
	0100	12" x 8", 13.5 brick		.37	.432		5.25	11.70		16.95	25	
	0200	12" x 12", 20 brick		25	.640		7.80	17.30		25.10	36.50	
	0300	16" x 12", 27 brick		19	.842		10.50	22.50		33	48.50	
	0400	16" x 16", 36 brick		14	1.143		14.05	31		45.05	66	
	0500	20" x 16", 45 brick		11	1.455		17.55	39.50		57.05	83.50	
	0600	20" x 20", 56 brick		9	1.778		22	48		70	103	
	0700	24" x 20", 68 brick		7	2.286		26.50	61.50		88	130	
	0800	24" x 24", 81 brick		6	2.667		31.50	72		103.50	153	
	1000	36" x 36", 182 brick		3	5.333		71	144		215	315	
	9000	Minimum labor/equipment charge		2	8	Job		216		216	355	
210	0010	**CONCRETE BLOCK, PARTITIONS** Scaffolding not included										210
	1000	Lightweight block, tooled joints, 2 sides, hollow										
	1100	Not reinforced, 8" x 16" x 4" thick	D-8	440	.091	S.F.	1.07	2.52		3.59	5.30	
	1150	6" thick		410	.098		1.46	2.70		4.16	6	
	1200	8" thick		385	.104		1.79	2.88		4.67	6.65	
	1250	10" thick		370	.108		2.36	2.99		5.35	7.50	
	1300	12" thick	D-9	350	.137		2.79	3.70		6.49	9.10	
	1500	Reinforced alternate courses, 4" thick	D-8	435	.092		1.22	2.55		3.77	5.50	
	1600	6" thick		405	.099		1.51	2.74		4.25	6.15	
	1650	8" thick		380	.105		1.85	2.92		4.77	6.80	
	1700	10" thick		365	.110		2.43	3.04		5.47	7.65	
	1750	12" thick	D-9	345	.139		2.86	3.76		6.62	9.30	
	4000	Regular block, tooled joints, 2 sides, hollow										
	4100	Not reinforced, 8" x 16" x 4" thick	D-8	430	.093	S.F.	.90	2.58		3.48	5.20	
	4150	6" thick		400	.100		1.33	2.77		4.10	6	
	4200	8" thick		375	.107		1.44	2.95		4.39	6.40	
	4250	10" thick		360	.111		2.06	3.08		5.14	7.30	
	4300	12" thick	D-9	340	.141		2.11	3.81		5.92	8.55	
	4500	Reinforced alternate courses, 8" x 16" x 4" thick	D-8	425	.094		.97	2.61		3.58	5.35	
	4550	6" thick		395	.101		1.41	2.81		4.22	6.15	
	4600	8" thick		370	.108		1.52	2.99		4.51	6.55	
	4650	10" thick		355	.113		2.14	3.12		5.26	7.45	

Figure 26.30

Repair and Remodeling Estimating Methods Example

Project No.: 001
Project Name: Commercial Renovation
Location: Estimating Methods Book

Title: ase Estimate
Architect: As Shown

Quantities: DEF
Take-off: ABC
Checked: GHI

Line Number	Description	Qty.	Unit	Material		Labor		Equipment		Subcontract	
				Unit Cost	Extension	Unit Cost	Extension	Unit Cost	Extension	Unit Cost	Extension
	Division 4 Masonry										
049106001080	Needle beam masonry, incl shoring 10' x 10' opening, brick, solid, 16" thk	4	Ea.	72.50	290.00	212.00	848.00	39.50	158.00		
049106002000	Needle beam masonry, add for additional floors of shoring	8	Ea.	38.00	304.00	96.50	772.00				
049108000520	Toothing masonry, brickwork, hard mortar	80	V.L.F.			6.25	500.00				
044144002100	Limestone, sills, lintels, jambs, trim, stops, sugarcube finish, avg	5	C.F.	50.50	252.50	57.00	285.00	23.50	117.50		
048101700300	Column, brick, scaff. not incl, 16" x 12", 27 brick/ VLF	80	V.L.F.			22.50	1800.00				
012504000550	Factors, cut & patch to match existing construction, add, maximum	5%	Costs			9%					
048108501050	Walls, brick, common, 4" x 2-2/3" x 8", 16" thk wall, 27/SF	42	S.F.			14.75	619.50				
012504000550	Factors, cut & patch to match existing construction, add, maximum	5%	Costs			9%	55.76				

Figure 26.31

Repair and Remodeling Estimating Methods Example

Project No.: 001
Project Name: Commercial Renovation
Location: Estimating Methods Book

Title: ase Estimate
Architect: As Shown

Quantities: DEF
Take-off: ABC
Checked: GHI

Line Number	Description	Qty.	Unit	Material Unit Cost	Material Extension	Labor Unit Cost	Labor Extension	Equipment Unit Cost	Equipment Extension	Subcontract Unit Cost	Subcontract Extension
049302200440	Masonry cleaning, high pressure water only, maximum	11900	S.F.			0.95	11305.00	0.18	2142.00		
049302204200	Masonry cleaning, add for pedestrian protection		Job			10%	1130.50	10%	214.20		
015408000300	Tarpaulins, reinforced polyethylene, 4 mils thick, white, clr or black	1000	S.F.	0.14	140.00						
	Move tarpaulins, 6 moves (2 Clab)	6	Hrs.			46.90	281.40				
049107200320	Pointing masonry, cut and repoint brick, hard mortar, common bond	6000	S.F.	0.25	1500.00	3.17	19020.00				
049107200700	Pointing masonry, cut and repoint brick, stonework, hard mortar	370	L.F.	0.33	122.10	1.74	643.80				
049102001100	Caulking masonry, no scaf, 1/2" x 1/2" jt, cut out/re-caulk, polyurethane	780	L.F.	0.34	265.20	1.95	1521.00				
049107200320	Pointing masonry, cut and repoint brick, hard mortar, common bond	140	S.F.	0.25	35.00	3.17	443.80				

Figure 26.32

Repair and Remodeling Estimating Methods Example

Project No.: 001	Title: see Estimate	Quantities: DEF
Project Name: Commercial Renovation	Architect: As Shown	Take-off: ABC
Location: Estimating Methods Book		Checked: GHI

Line Number	Description	Qty.	Unit	Material Unit Cost	Material Extension	Labor Unit Cost	Labor Extension	Equipment Unit Cost	Equipment Extension	Subcontract Unit Cost	Subcontract Extension
048102104200	CMU partition, no scaf/reinf, rglr, hollow, 8" x 16" x 8" thk	288	S.F.	1.44	414.72	2.95	849.60				
051204802100	Lintels, steel angles, 3-1/2" x 3", 1/4" thick, 4'-6" long	8	Ea.	11.90	95.20	9.40	75.20				
048102104300	CMU partition, no scaf/reinf, rglr, hollow, 8" x 16" x 12" thk	2696	S.F.	2.11	5688.56	3.81	10271.76				
051204802600	Lintels, steel angles, 4" x 3-1/2", 1/4" thick, 5'-0" long	12	Ea.	15.20	182.40	11.60	139.20				
049302201420	Masonry cleaning, sandblast, dry system, average	2100	S.F.							1.25	2625.00
012504400850	Factors, dust protection, add, maximum		Costs						Average of material and labor factors.	24.5%	643.13
012504401700	Factors, protection of existing work, add, minimum		Costs								
012504402050	Factors, shift work requirements, add, maximum		Costs								
049302201420	Masonry cleaning, sandblast, dry system, average	11800	S.F.							1.25	14750.00
012504400850	Factors, dust protection, add, maximum		Costs						Average of material and labor factors.	24.5%	3613.75
012504401700	Factors, protection of existing work, add, minimum		Costs								
012504402050	Factors, shift work requirements, add, maximum		Costs								
	Division 4 Total				9269.68		50723.52		2631.70		21631.88

Figure 26.33

Division 5: Metals

The specified wide flange steel column at the elevator pit is 4' high with special bearing plates. This is a unique item and requires a steel fabricator for the material cost. The quotation would most likely come from the subcontractor who will be installing the new stairs. The installation must be visualized in order to estimate it properly, because this is a single, odd item. Figures 26.34 and 26.35 are the estimate sheets for Division 5.

Figure 26.26 lists costs for structural steel. Note on line 05120-720-9000 that the minimum labor and equipment charges are $930 and $710, respectively. If the estimator does not use common sense, the total installation costs for this single, small column could be $1640. When using *Means Repair & Remodeling Cost Data* or any other data source, the estimator must be aware of the crew and equipment necessary for installation. The crew listed for the minimum labor and equipment charges in Figure 26.36 is E-2, shown in Figure 26.37. The minimum charge is for a half day. Obviously, this kind of crew and equipment is not required for the column installation in the sample estimate. The specifications call for expansion bolts into the concrete footing and lag bolts into the wood beam above. These are easily priced, as shown in Figure 26.34. Note that drilling the holes is a separate item. The crew used for these prices is one carpenter. There is a minimum charge of $60 for the expansion anchors, substantially less than $1640. A little extra time and sound judgment may help to prevent expensive mistakes. The temporary shoring and bracing is included in Division 2, Figure 26.19.

It should be noted that there is relatively little structural metal work in this sample project or in any efficiently designed commercial renovation. This keeps the costs down and makes renovation a favorable alternative, in terms of cost, to new construction. Whether the work is planned by an architect or by the contractor, the designer should try to work with and around the existing structural elements.

The new stairway, ladder, and railings are to be subcontracted to a steel fabricator. The estimator has received quotations, but feels that they are not quite right (either too high or too low), based upon past experience. The costs are included in the estimate as a cross-check (see Figure 26.34). These costs are subtotaled so that the quotation may be substituted easily in the Estimate Summary.

The specifications (and building codes) require that the contractor place fire extinguishers at every level when welding. Since the extinguishers will be in place throughout the project for all trades, they are included in Division 1. Figures 26.38a and 26.38b show the Project Overhead Summary as items are added throughout the estimate.

Division 6: Wood and Plastics

The use of wood in commercial renovation is often dependent upon two factors: the type of existing construction and the local building code requirements for fire resistance. When the use of wood (whether structural, rough, or finish) is permitted in buildings of fire-resistant construction, fire-retardant treatment is usually specified. Because the interior structure of the sample project is wood, such treatment is not required. Light-gauge metal studs are becoming more common in commercial renovation because they are lightweight and easy to install. It is faster and less costly for a carpenter, with a pair of tin snips and a screw gun, to cut and install metal (rather than wood) studs to conform to the variety of existing conditions in older buildings. Also, with the wide use of materials such as metal door frames and vinyl base,

Repair and Remodeling Estimating Methods Example

Project No.: 001
Project Name: Commercial Renovation
Location: Estimating Methods Book

Title: see Estimate
Architect: As Shown

Quantities: DEF
Take-off: ABC
Checked: GHI

Line Number	Description	Qty.	Unit	Material Unit Cost	Material Extension	Labor Unit Cost	Labor Extension	Equipment Unit Cost	Equipment Extension	Subcontract Unit Cost	Subcontract Extension
	Division 5 Metals										
	COLUMNS										
051202600010	Material only quote	1	Ea.	325.00	325.00						
050903801100	Expansion shield, zinc, 3/4" diameter, 3-15/16" long, double	4	Ea.	7.35	29.40	3.69	14.76				
050904600300	Lag screws, steel, 5/8" diameter, 3" long	4	Ea.	0.62	2.48	2.00	8.00				
050903400500	Drilling, layout, no anchor, 4" deep, 3/4" dia, conc/brick walls/flr	4	Ea.	0.14	0.56	5.35	21.40				
060908000800	Timber connectors, drilling bolt holes in timber, 1/2" diameter	48	Inch			0.53	25.44				
055145000020	Ladder, steel, 20" wide, bolted to concrete, with cage	12	V.L.F.							104.00	1248.00
055207000945	Railing, wall rail, steel pipe, 1-1/2" diameter, galvanized	78	L.F.							19.80	1544.40
055207000945	Railing, wall rail, steel pipe, 1-1/2" diameter, galvanized	52	L.F.							19.80	1029.60
055177000300	Stair, cement fill metal pan, picket rail, 4'-0" wide	36	Riser							269.00	9684.00
055177001500	Stair, landing, steel pan, conventional	72	S.F.							36.50	2628.00

Figure 26.34

Repair and Remodeling Estimating Methods Example

Project No.: 001			**Title:** see Estimate				**Quantities:** DEF			
Project Name: Commercial Renovation			**Architect:** As Shown				**Take-off:** ABC			
Location: Estimating Methods Book							**Checked:** GHI			

Line Number	Description	Qty.	Unit	Material		Labor		Equipment		Subcontract	
				Unit Cost	Extension	Unit Cost	Extension	Unit Cost	Extension	Unit Cost	Extension
060908001800	Timber connectors, joist & beam hgr, 16 ga galv, 4" x 6" - 4" x 10"	76	Ea.	2.03	154.28	1.55	117.80				
050905800300	Studs, powder charges for above, low velocity	3	C	16.65	49.95						
050905800600	Studs, drive pins & studs, 1/4" & 3/8" diam., to 3" long, minimum	3	C	21.50	64.50						
	Division 5 Total				626.17		187.40				16134.00

Figure 26.35

296

05100 | Structural Metal Framing

05120 | Structural Steel

		CREW	DAILY OUTPUT	LABOR-HOURS	UNIT	2002 BARE COSTS MAT.	LABOR	EQUIP.	TOTAL	TOTAL INCL O&P	
480	0300	2,000 to 4,000 lb.	1 Bric	640	.013	Lb.	.43	.38		.81	1.10	480
	0500	For built-up angles and plates, add to above					.15			.15	.17	
	0700	For engineering, add to above					.06			.06	.07	
	0900	For galvanizing, add to above, under 500 lb.					.26			.26	.29	
	0950	500 to 2,000 lb.					.24			.24	.26	
	1000	Over 2,000 lb.					.21			.21	.23	
	2000	Steel angles, 3-1/2" x 3", 1/4" thick, 2'-6" long	1 Bric	47	.170	Ea.	6.60	5.20		11.80	15.80	
	2100	4'-6" long		26	.308		11.90	9.40		21.30	28.50	
	2500	3-1/2" x 3-1/2" x 5/16", 5'-0" long		18	.444		17.65	13.55		31.20	41.50	
	2600	4" x 3-1/2", 1/4" thick, 5'-0" long		21	.381		15.20	11.60		26.80	35.50	
	2700	9'-0" long		12	.667		27.50	20.50		48	63	
	2800	4" x 3-1/2" x 5/16", 7'-0" long		12	.667		26.50	20.50		47	62	
	2900	5" x 3-1/2" x 5/16", 10'-0" long		8	1		42.50	30.50		73	97	
	3500	For precast concrete lintels, see div. 03480-400										
	9000	Minimum labor/equipment charge	1 Bric	4	2	Job		61		61	99.50	
520	0010	PIPE SUPPORT FRAMING										520
	0020	Under 10#/L.F.	E-4	3,900	.008	Lb.	.82	.29	.02	1.13	1.46	
	0200	10.1 to 15#/L.F.		4,300	.007		.81	.26	.02	1.09	1.39	
	0400	15.1 to 20#/L.F.		4,800	.007		.80	.23	.02	1.05	1.33	
	0600	Over 20#/L.F.		5,400	.006		.78	.21	.02	1.01	1.27	
680	0010	STRUCTURAL STEEL PROJECTS Bolted, unless noted otherwise										680
	0700	Offices, hospitals, etc., steel bearing, 1 to 2 stories	E-5	10.30	7.767	Ton	1,225	262	147	1,634	1,975	
	1300	Industrial bldgs., 1 story, beams & girders, steel bearing [R05120-210]		12.90	6.202		1,225	209	117	1,551	1,850	
	1400	Masonry bearing		10	8		1,225	270	151	1,646	2,000	
	1600	1 story with roof trusses, steel bearing		10.60	7.547		1,450	254	142	1,846	2,225	
	1700	Masonry bearing		8.30	9.639		1,450	325	182	1,957	2,400	
720	0010	STRUCTURAL STEEL Bolted, incl. fabrication [R05120-210]										720
	0050	Beams, W 6 x 9	E-2	720	.078	L.F.	6.60	2.58	1.97	11.15	14.10	
	0100	W 8 x 10 [R05120-215]		720	.078		7.35	2.58	1.97	11.90	14.90	
	0150	W 10 x 15		720	.078		11.05	2.58	1.97	15.60	18.95	
	0200	Columns, W 6 x 15		540	.104		11.95	3.44	2.63	18.02	22	
	0250	W 8 x 31		540	.104		24.50	3.44	2.63	30.57	36	
	0500	Girders, W 12 x 22		900	.062		16.15	2.06	1.58	19.79	23.50	
	0550	W 14 x 26		900	.062		19.10	2.06	1.58	22.74	26.50	
	0600	W 16 x 31		900	.062		23	2.06	1.58	26.64	30.50	
	0700	Joists (bar joists, H or K series), span to 30'	E-7	30,000	.003	Lb.	.51	.09	.05	.65	.78	
	0750	Span to 50'	"	20,000	.004	"	.50	.13	.08	.71	.89	
	9000	Minimum labor/equipment charge	E-2	2	28	Job		930	710	1,640	2,450	

05200 | Metal Joists

05210 | Steel Joists

		CREW	DAILY OUTPUT	LABOR-HOURS	UNIT	2002 BARE COSTS MAT.	LABOR	EQUIP.	TOTAL	TOTAL INCL O&P	
600	0010	OPEN WEB JOISTS, Truckload lots										600
	0020	K series, horizontal bridging, spans up to 30', minimum	E-7	15	5.333	Ton	820	180	107	1,107	1,350	
	0050	Average		12	6.667		920	225	133	1,278	1,550	
	0080	Maximum		9	8.889		1,100	300	178	1,578	1,975	
	0410	Span 30' to 50', minimum		17	4.706		805	159	94	1,058	1,275	
	0440	Average		17	4.706		905	159	94	1,158	1,375	

Figure 26.36

Crew C-24

Crew No.	Bare Costs Hr.	Daily	Incl. Subs O & P Hr.	Daily	Cost Per Labor-Hour Bare Costs	Incl. O&P
2 Skilled Worker Foremen	$32.95	$527.20	$54.15	$866.40	$31.06	$50.67
6 Skilled Workers	30.95	1485.60	50.85	2440.80		
1 Equip. Oper. (crane)	32.35	258.80	51.15	409.20		
1 Equip. Oper. Oiler	26.65	213.20	42.15	337.20		
1 Truck Crane, 150 Ton		1690.00		1859.00	21.13	23.24
80 L.H., Daily Totals		$4174.80		$5912.60	$52.19	$73.91

Crew C-25

Crew No.	Bare Costs Hr.	Daily	Incl. Subs O & P Hr.	Daily	Cost Per Labor-Hour Bare Costs	Incl. O&P
2 Rodmen (reinf.)	$34.25	$548.00	$60.20	$963.20	$27.02	$47.93
2 Rodmen Helpers	19.80	316.80	35.65	570.40	$27.02	$47.93
32 L.H., Daily Totals		$864.80		$1533.60	$27.02	$47.93

Crew D-1

Crew No.	Bare Costs Hr.	Daily	Incl. Subs O & P Hr.	Daily	Cost Per Labor-Hour Bare Costs	Incl. O&P
1 Bricklayer	$30.50	$244.00	$49.85	$398.80	$27.00	$44.13
1 Bricklayer Helper	23.50	188.00	38.40	307.20		
16 L.H., Daily Totals		$432.00		$706.00	$27.00	$44.13

Crew D-2

Crew No.	Bare Costs Hr.	Daily	Incl. Subs O & P Hr.	Daily	Cost Per Labor-Hour Bare Costs	Incl. O&P
3 Bricklayers	$30.50	$732.00	$49.85	$1196.40	$27.91	$45.67
2 Bricklayer Helpers	23.50	376.00	38.40	614.40		
.5 Carpenter	30.00	120.00	49.70	198.80		
44 L.H., Daily Totals		$1228.00		$2009.60	$27.91	$45.67

Crew D-3

Crew No.	Bare Costs Hr.	Daily	Incl. Subs O & P Hr.	Daily	Cost Per Labor-Hour Bare Costs	Incl. O&P
3 Bricklayers	$30.50	$732.00	$49.85	$1196.40	$27.81	$45.48
2 Bricklayer Helpers	23.50	376.00	38.40	614.40		
.25 Carpenter	30.00	60.00	49.70	99.40		
42 L.H., Daily Totals		$1168.00		$1910.20	$27.81	$45.48

Crew D-4

Crew No.	Bare Costs Hr.	Daily	Incl. Subs O & P Hr.	Daily	Cost Per Labor-Hour Bare Costs	Incl. O&P
1 Bricklayer	$30.50	$244.00	$49.85	$398.80	$26.83	$43.44
2 Bricklayer Helpers	23.50	376.00	38.40	614.40		
1 Equip. Oper. (light)	29.80	238.40	47.10	376.80		
1 Grout Pump, 50 C.F./hr		104.95		115.45		
1 Hoses & Hopper		26.80		29.50		
1 Accessories		11.20		12.30	4.47	4.91
32 L.H., Daily Totals		$1001.35		$1547.25	$31.30	$48.35

Crew D-5

Crew No.	Bare Costs Hr.	Daily	Incl. Subs O & P Hr.	Daily	Cost Per Labor-Hour Bare Costs	Incl. O&P
1 Bricklayer	$30.50	$244.00	$49.85	$398.80	$30.50	$49.85
8 L.H., Daily Totals		$244.00		$398.80	$30.50	$49.85

Crew D-6

Crew No.	Bare Costs Hr.	Daily	Incl. Subs O & P Hr.	Daily	Cost Per Labor-Hour Bare Costs	Incl. O&P
3 Bricklayers	$30.50	$732.00	$49.85	$1196.40	$27.12	$44.35
3 Bricklayer Helpers	23.50	564.00	38.40	921.60		
.25 Carpenter	30.00	60.00	49.70	99.40		
50 L.H., Daily Totals		$1356.00		$2217.40	$27.12	$44.35

Crew D-7

Crew No.	Bare Costs Hr.	Daily	Incl. Subs O & P Hr.	Daily	Cost Per Labor-Hour Bare Costs	Incl. O&P
1 Tile Layer	$29.15	$233.20	$45.85	$366.80	$26.25	$41.30
1 Tile Layer Helper	23.35	186.80	36.75	294.00		
16 L.H., Daily Totals		$420.00		$660.80	$26.25	$41.30

Crew D-8

Crew No.	Bare Costs Hr.	Daily	Incl. Subs O & P Hr.	Daily	Cost Per Labor-Hour Bare Costs	Incl. O&P
3 Bricklayers	$30.50	$732.00	$49.85	$1196.40	$27.70	$45.27
2 Bricklayer Helpers	23.50	376.00	38.40	614.40		
40 L.H., Daily Totals		$1108.00		$1810.80	$27.70	$45.27

Crew D-9

Crew No.	Bare Costs Hr.	Daily	Incl. Subs O & P Hr.	Daily	Cost Per Labor-Hour Bare Costs	Incl. O&P
3 Bricklayers	$30.50	$732.00	$49.85	$1196.40	$27.00	$44.13
3 Bricklayer Helpers	23.50	564.00	38.40	921.60		
48 L.H., Daily Totals		$1296.00		$2118.00	$27.00	$44.13

Crew D-10

Crew No.	Bare Costs Hr.	Daily	Incl. Subs O & P Hr.	Daily	Cost Per Labor-Hour Bare Costs	Incl. O&P
1 Bricklayer Foreman	$32.50	$260.00	$53.15	$425.20	$28.47	$46.19
1 Bricklayer	30.50	244.00	49.85	398.80		
2 Bricklayer Helpers	23.50	376.00	38.40	614.40		
1 Equip. Oper. (crane)	32.35	258.80	51.15	409.20		
1 Truck Crane, 12.5 Ton		473.20		520.50	11.83	13.01
40 L.H., Daily Totals		$1612.00		$2368.10	$40.30	$59.20

Crew D-11

Crew No.	Bare Costs Hr.	Daily	Incl. Subs O & P Hr.	Daily	Cost Per Labor-Hour Bare Costs	Incl. O&P
1 Bricklayer Foreman	$32.50	$260.00	$53.15	$425.20	$28.83	$47.13
1 Bricklayer	30.50	244.00	49.85	398.80		
1 Bricklayer Helper	23.50	188.00	38.40	307.20		
24 L.H., Daily Totals		$692.00		$1131.20	$28.83	$47.13

Crew D-12

Crew No.	Bare Costs Hr.	Daily	Incl. Subs O & P Hr.	Daily	Cost Per Labor-Hour Bare Costs	Incl. O&P
1 Bricklayer Foreman	$32.50	$260.00	$53.15	$425.20	$27.50	$44.95
1 Bricklayer	30.50	244.00	49.85	398.80		
2 Bricklayer Helpers	23.50	376.00	38.40	614.40		
32 L.H., Daily Totals		$880.00		$1438.40	$27.50	$44.95

Crew D-13

Crew No.	Bare Costs Hr.	Daily	Incl. Subs O & P Hr.	Daily	Cost Per Labor-Hour Bare Costs	Incl. O&P
1 Bricklayer Foreman	$32.50	$260.00	$53.15	$425.20	$28.73	$46.77
1 Bricklayer	30.50	244.00	49.85	398.80		
2 Bricklayer Helpers	23.50	376.00	38.40	614.40		
1 Carpenter	30.00	240.00	49.70	397.60		
1 Equip. Oper. (crane)	32.35	258.80	51.15	409.20		
1 Truck Crane, 12.5 Ton		473.20		520.50	9.86	10.84
48 L.H., Daily Totals		$1852.00		$2765.70	$38.59	$57.61

Crew E-1

Crew No.	Bare Costs Hr.	Daily	Incl. Subs O & P Hr.	Daily	Cost Per Labor-Hour Bare Costs	Incl. O&P
1 Welder Foreman	$36.25	$290.00	$67.90	$543.20	$33.43	$59.72
1 Welder	34.25	274.00	64.15	513.20		
1 Equip. Oper. (light)	29.80	238.40	47.10	376.80		
1 Gas Welding Machine		88.60		97.45	3.69	4.06
24 L.H., Daily Totals		$891.00		$1530.65	$37.12	$63.78

Crew E-2

Crew No.	Bare Costs Hr.	Daily	Incl. Subs O & P Hr.	Daily	Cost Per Labor-Hour Bare Costs	Incl. O&P
1 Struc. Steel Foreman	$36.25	$290.00	$67.90	$543.20	$33.18	$59.69
4 Struc. Steel Workers	34.25	1096.00	64.15	2052.80		
1 Equip. Oper. (crane)	32.35	258.80	51.15	409.20		
1 Equip. Oper. Oiler	26.65	213.20	42.15	337.20		
1 Crane, 90 Ton		1421.00		1563.10	25.38	27.91
56 L.H., Daily Totals		$3279.00		$4905.50	$58.56	$87.60

Crew E-3

Crew No.	Bare Costs Hr.	Daily	Incl. Subs O & P Hr.	Daily	Cost Per Labor-Hour Bare Costs	Incl. O&P
1 Struc. Steel Foreman	$36.25	$290.00	$67.90	$543.20	$34.92	$65.40
1 Struc. Steel Worker	34.25	274.00	64.15	513.20		
1 Welder	34.25	274.00	64.15	513.20		
1 Gas Welding Machine		88.60		97.45	3.69	4.06
24 L.H., Daily Totals		$926.60		$1667.05	$38.61	$69.46

Crew E-4

Crew No.	Bare Costs Hr.	Daily	Incl. Subs O & P Hr.	Daily	Cost Per Labor-Hour Bare Costs	Incl. O&P
1 Struc. Steel Foreman	$36.25	$290.00	$67.90	$543.20	$34.75	$65.09
3 Struc. Steel Workers	34.25	822.00	64.15	1539.60		
1 Gas Welding Machine		88.60		97.45	2.77	3.05
32 L.H., Daily Totals		$1200.60		$2180.25	$37.52	$68.14

Figure 26.37

PROJECT Commercial Renovation

ESTIMATE NO. 001

LOCATION   ARCHITECT   DATE 2002

QUANTITIES BY: ABC   PRICES BY RSM   EXTENSIONS BY   DEF   CHECKED BY GHI

DESCRIPTION			Material			Labor			Equip, Fees, Rental		
Job Organization: Superintendent 01310-700-0260											
Project Manager											
Timekeeper & Material Clerk											
Clerical											
Safety, Watchman, First Aid											
Travel Expense: Superintendent											
Project Manager											
Engineering: Layout											
Inspection / Quantities											
Drawings											
CPM Schedule											
Testing: Soil											
Materials											
Structural											
Equipment: Cranes											
Concrete Pump, Conveyor, Etc.											
Elevators, Hoists											
Freight & Hauling											
Loading, Unloading, Erecting, Etc.											
Maintenance											
Pumping											
Scaffolding 01540-750-0090	119	C.S.F.									
Small Power Equipment / Tools											
Field Facilities: Job Office											
Architects / Owners Office											
Temporary Telephones											
Utilities (Power) 01510-800-0450	280	C.S.F.									
Temporary Toilets											
Storage Areas & Sheds											
Temporary Utilities: Heat											
Light & Power 01510-800-0350	280	C.S.F.									
Water											
PAGE TOTALS											

Figure 26.38a

Division 1

DESCRIPTION			Material		Labor		Equip, Fees, Rental	
Winter Protection: Temp. Heat / Protection								
Snow Plowing								
Thawing Materials								
Temporary Roads								
Signs & Barricades: Site Sign								
Temporary Fences	01560-100-1000	270	L.F.					
Temporary Stairs, Ladders & Floors								
Photographs								
Clean Up								
Dumpster	02225-730-0800		Week					
Final Clean Up	01740-500-0100	28	M.S.F.					
Punch List								
Permits: Building	01310-150-0010	1	Job					
Misc.								
Insurance: Builders Risk								
Owner's Protective Liability								
Umbrella								
Unemployment Ins. & Social Security								
Taxes								
City Sales Tax								
State Sales Tax								
Bonds								
Performance								
Material & Equipment								
Main Office Expense								
Special Items								
Fire Extinguishers	10525-300-2080	4	Ea.					
TOTALS:								

Figure 26.38b

wood is becoming less prevalent as a primary material in commercial renovation as well as in new construction. On some projects, a carpenter might not work with wood at all.

In this example, metal studs are specified for all interior, nonload-bearing partitions. Wood is specified only to match existing work, and for blocking and framing for fixtures and equipment. Estimate sheets for Division 6 are illustrated in Figures 26.39, 26.40 and 26.41.

To frame the floor openings at the freight elevator, contractors use planking that has been removed from other areas, so no material costs are included. The planks must be cut to match existing conditions. This work will occur in the late stages of the renovation, so a Factor for protection of existing work must be applied.

The roof openings for hatches and skylights are also framed using existing materials. In Division 2, the costs for cutting the openings were calculated by determining the crew size and the time required. The same method and crew are used to price the framing, because the two will be done together.

The exterior walls of older masonry buildings are usually very thick. Oversized windowsills are required. The plans specify solid oak for the sills, as shown in Figure 26.42. A material price is obtained from a local supplier. Note that 1 x 10 (nominal) boards are needed, cut to 8″ wide and surfaced on three sides. Installation is similar to that of normal window stools, so line item 06220-600-5100 in Figure 26.43, is used for labor only. A Factor is added for cutting to conform to the variations in wall thickness and window installation.

In Figure 26.41, the quantities for the sills and grounds are 116 L.F. and 606 L.F., respectively. It is important to record how these, and all quantities, are derived for cross-checking and for future reference. A Quantity Sheet, as shown in Figure 26.44, is useful to keep track of this information. Sketches are helpful to show how dimensions are obtained.

The first-floor windows receive no sills or grounds because the windows at the existing retail space are already finished and trimmed, and the interior faces of the walls in the new retail spaces will be sandblasted. The existing stone sills will remain exposed.

All windows on the second floor, except in the stairways, will receive grounds and sills. On the third floor, the walls in the tenant space will be sandblasted, so only the remaining eleven windows are included. Walls in the stairways will also be sandblasted.

In order to price rough hardware for the project, the material costs for rough hardware are shown in Figure 26.45 as a minimum and a maximum. The maximum percentage is used because material costs are slightly low, due to the use of existing materials for some of the work.

Division 7: Thermal and Moisture Protection

Four types of insulation are specified in the project. The urethane roof insulation is included in the roofing subcontract. The polystyrene insulation is installed on the interior of the foundation walls at the crawl space. The takeoff for these types is straightforward. The exterior wall fiberglass (for thermal protection) and the interior wall fiberglass (for reduced sound transmission) are an integral part of the drywall partition systems. As the quantities are calculated, the estimator should note and record the wall types and dimensions. This information will save time when estimating Division 9: Finishes. Estimate sheets for Division 7 are illustrated in Figures 26.46 to 26.48.

The polyethylene vapor barrier in the basement is an item that might easily be overlooked. Many such items in a complex package of construction documents

Repair and Remodeling Estimating Methods Example

Project No.: 001
Project Name: Commercial Renovation
Location: Estimating Methods Book

Title: ase Estimate
Architect: As Shown

Quantities: DEF
Take-off: ABC
Checked: GHI

Line Number	Description	Qty.	Unit	Material		Labor		Equipment		Subcontract	
				Unit Cost	Extension	Unit Cost	Extension	Unit Cost	Extension	Unit Cost	Extension
	Division 6 Wood and Plastics										
06110100 2660	Blocking, miscellaneous, to wood construction, 2" x 8"	0.125	M.B.F.	650.00	81.25	890.00	111.25				
06110100 2620	Blocking, miscellaneous, to wood construction, 2" x 4"	0.314	M.B.F.	535.00	167.99	1400.00	439.60				
06110550 0602	Framing 2x4 studs 12' high 16" OC	10	L.F.	4.70	47.00	6.00	60.00				
06160800 0702	Plywd shtng on wal int std 5/8 thick	196	S.F.	0.57	111.72	0.46	90.16				
06110520 1102	Framg flor plnks 3"thk s4s T&G 3x6	639	B.F.			0.46	293.94				
01250400 0500	Factors, cut & patch to match existing construction, add, minimum		Costs	2%		3%	8.82				
01250400 1750	Factors, protection of existing work, add, maximum		Costs	5%		7%	20.58				
06110550 0010	Framing, roofs [2 Carpenters]	1.5	Day			480.00	720.00				
01250400 0550	Factors, cut & patch to match existing construction, add, maximum		Costs	5%		9%	64.80				
06160900 0102	Undrlymnt pywd und grade 1/2" thick	390	SF Flr.	0.64	249.60	0.33	128.70				
06160900 0102	Undrlymnt pywd und grade 1/2" thick	2080	SF Flr.	0.64	1331.20	0.33	686.40				

Figure 26.39

Repair and Remodeling Estimating Methods Example

Project No.: 001				**Title: ase Estimate**			**Quantities: DEF**		
Project Name: Commercial Renovation				**Architect: As Shown**			**Take-off: ABC**		
Location: Estimating Methods Book							**Checked: GHI**		

Line Number	Description	Qty.	Unit	Material		Labor		Equipment		Subcontract	
				Unit Cost	Extension	Unit Cost	Extension	Unit Cost	Extension	Unit Cost	Extension
06110555360	Framing, roofs, hip and valley rafters, 2" x 8", ordinary	220	L.F.	0.87	191.40	0.67	147.40				
06110555770	Framing, roofs, for dormers or complex roofs, add					50%	73.70				
012504000550	Factors, cut & patch to match existing construction, add, maximum		Costs	5%		9%	19.90				
012504001700	Factors, protection of existing work, add, minimum		Costs	2%	191.40	2%	221.10				
06160800302	Plywd shtng on roof cdx 3/4" thick	150	S.F.	0.66	99.00	0.40	60.00				
012504001450	Factors, material handling & storage limitation, add, maximum		Costs	6%	5.94	7%	4.20				
060908250200	Rough hardware, average % of carpentry material, maximum			1.50%	37.29						

Figure 26.40

Repair and Remodeling Estimating Methods Example

Project No.: 001	Title: ase Estimate		Quantities: DEF
Project Name: Commercial Renovation	Architect: As Shown		Take-off: ABC
Location: Estimating Methods Book			Checked: GHI

Line Number	Description	Qty.	Unit	Material		Labor		Equipment		Subcontract	
				Unit Cost	Extension	Unit Cost	Extension	Unit Cost	Extension	Unit Cost	Extension
062702000600	Shelving, plywood, 3/4" thick w/lumber edge, 12" wide	30	L.F.	1.24	37.20	3.20	96.00				
06410408100	Vanities, vanity bases, 2 doors, 30" high, 21" deep, 36" wide	4	Ea.	247.00	988.00	36.00	144.00				
060557401000	Counter top, square edge, plastic face, 7/8" thick, with splash	12	L.F.	26.00	312.00	8.00	96.00				
060557401900	Counter top, stock, for cut outs, std, add, minimum	6	Ea.	2.63	15.78	7.50	45.00				
062208000010	Moldings, window and door [Material quote.]	116		2.63	305.08						
062208005100	Moldings, stool caps, stock pine, 1-1/16" x 3-1/4"	116	L.F.			1.60	185.60				
012504000500	Factors, cut & patch to match existing construction, add, minimum		Costs	2%	6.10	3%	5.57				
061107000102	Grounds for cswk 1"x2" on masonry	606	L.F.	0.21	127.26	0.84	509.04				
	Division 6 Totals				4314.78		4231.76				

Figure 26.41

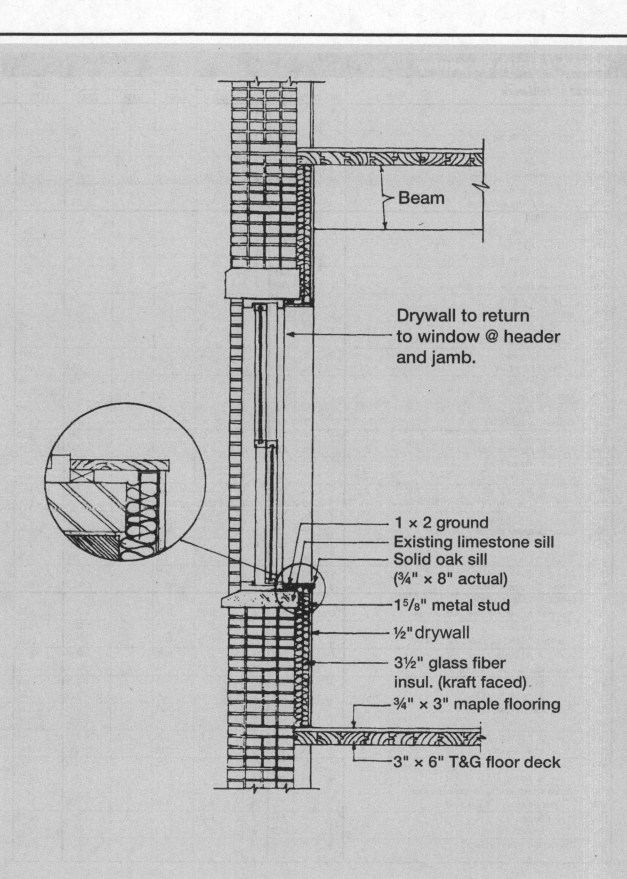

Beam

Drywall to return
to window @ header
and jamb.

1 × 2 ground
Existing limestone sill
Solid oak sill
(¾" × 8" actual)

1⅝" metal stud

½" drywall

3½" glass fiber
insul. (kraft faced)

¾" × 3" maple flooring

3" × 6" T&G floor deck

Figure 26.42

			CREW	DAILY OUTPUT	LABOR-HOURS	UNIT	2002 BARE COSTS				TOTAL INCL O&P
06220	**Millwork**						MAT.	LABOR	EQUIP.	TOTAL	
	3800	Screen	▼	395	.020	▼	.32	.61		.93	1.36
	3850										
	4100	Verge board, sterling pine, 1" x 4"	1 Carp	200	.040	L.F.	.47	1.20		1.67	2.51
	4200	1" x 6"		200	.040		.72	1.20		1.92	2.78
	4300	2" x 6"		165	.048		1.40	1.45		2.85	3.95
	4400	2" x 8"	▼	165	.048	▼	1.86	1.45		3.31	4.46
	4700	For redwood trim, add					200%				
	9000	Minimum labor/equipment charge	1 Carp	4	2	Job		60		60	99.50
700	0010	**MOLDINGS, TRIM**									**700**
	0200	Astragal, stock pine, 11/16" x 1-3/4"	1 Carp	255	.031	L.F.	.79	.94		1.73	2.43
	0250	1-5/16" x 2-3/16"		240	.033		2.74	1		3.74	4.67
	0800	Chair rail, stock pine, 5/8" x 2-1/2"		270	.030		.85	.89		1.74	2.41
	0900	5/8" x 3-1/2"		240	.033		1.27	1		2.27	3.06
	1000	Closet pole, stock pine, 1-1/8" diameter		200	.040		.80	1.20		2	2.87
	1100	Fir, 1-5/8" diameter		200	.040		1.20	1.20		2.40	3.31
	3300	Half round, stock pine, 1/4" x 1/2"		270	.030		.20	.89		1.09	1.69
	3350	1/2" x 1"	▼	255	.031	▼	.35	.94		1.29	1.95
	3400	Handrail, fir, single piece, stock, hardware not included									
	3450	1-1/2" x 1-3/4"	1 Carp	80	.100	L.F.	1.22	3		4.22	6.30
	3470	Pine, 1-1/2" x 1-3/4"		80	.100		1.07	3		4.07	6.15
	3500	1-1/2" x 2-1/2"		76	.105		1.44	3.16		4.60	6.85
	3600	Lattice, stock pine, 1/4" x 1-1/8"		270	.030		.33	.89		1.22	1.83
	3700	1/4" x 1-3/4"		250	.032		.35	.96		1.31	1.98
	3800	Miscellaneous, custom, pine, 1" x 1"		270	.030		.30	.89		1.19	1.80
	3900	1" x 3"		240	.033		.61	1		1.61	2.33
	4100	Birch or oak, nominal 1" x 1"		240	.033		.47	1		1.47	2.18
	4200	Nominal 1" x 3"		215	.037		1.60	1.12		2.72	3.61
	4400	Walnut, nominal 1" x 1"		215	.037		.77	1.12		1.89	2.70
	4500	Nominal 1" x 3"		200	.040		2.30	1.20		3.50	4.52
	4700	Teak, nominal 1" x 1"		215	.037		1.09	1.12		2.21	3.05
	4800	Nominal 1" x 3"		200	.040		3.10	1.20		4.30	5.40
	4900	Quarter round, stock pine, 1/4" x 1/4"		275	.029		.22	.87		1.09	1.69
	4950	3/4" x 3/4"		255	.031	▼	.42	.94		1.36	2.02
	5600	Wainscot moldings, 1-1/8" x 9/16", 2' high, minimum		76	.105	S.F.	5.90	3.16		9.06	11.75
	5700	Maximum		65	.123	"	13.50	3.69		17.19	21
	9000	Minimum labor/equipment charge	▼	4	2	Job		60		60	99.50
800	0010	**MOLDINGS, WINDOW AND DOOR**									**800**
	2800	Door moldings, stock, decorative, 1-1/8" wide, plain	1 Carp	17	.471	Set	28.50	14.10		42.60	54.50
	2900	Detailed		17	.471	"	68	14.10		82.10	98.50
	3150	Door trim set, 1 head and 2 sides, pine, 2-1/2 wide	▼	5.90	1.356	Opng.	12.70	40.50		53.20	81.50
	3170	4-1/2" wide	1 Carp	5.30	1.509	Opng.	24	45.50		69.50	101
	3200	Glass beads, stock pine, 1/4" x 11/16"		285	.028	L.F.	.30	.84		1.14	1.73
	3250	3/8" x 1/2"		275	.029		.36	.87		1.23	1.85
	3270	3/8" x 7/8"		270	.030		.40	.89		1.29	1.91
	4850	Parting bead, stock pine, 3/8" x 3/4"		275	.029		.29	.87		1.16	1.77
	4870	1/2" x 3/4"		255	.031		.36	.94		1.30	1.96
	5000	Stool caps, stock pine, 11/16" x 3-1/2"		200	.040		1.31	1.20		2.51	3.43
	5100	1-1/16" x 3-1/4"		150	.053	▼	1.97	1.60		3.57	4.82
	5300	Threshold, oak, 3' long, inside, 5/8" x 3-5/8"		32	.250	Ea.	6.20	7.50		13.70	19.30
	5400	Outside, 1-1/2" x 7-5/8"	▼	16	.500	"	25	15		40	52.50
	5900	Window trim sets, including casings, header, stops,									
	5910	stool and apron, 2-1/2" wide, minimum	1 Carp	13	.615	Opng.	14.85	18.45		33.30	47
	5950	Average		10	.800		25.50	24		49.50	68.50
	6000	Maximum		6	1.333	▼	38	40		78	108
	9000	Minimum labor/equipment charge	▼	4	2	Job		60		60	99.50

Figure 26.43

QUANTITY SHEET

Division 6

		SHEET NO.	1 of 1
PROJECT **Commercial Renovation**		ESTIMATE NO.	**02-1**
LOCATION	ARCHITECT	DATE	**2002**
TAKE OFF BY **ABC**	EXTENSIONS BY	CHECKED BY	

DESCRIPTION	NO.	L (DIM)	W (DIM)	H (DIM)	Grounds Sill	UNIT	Grounds Jamb	UNIT	Grounds Head	UNIT	Oak Sills	UNIT
Window Sills & Grounds												
Basement		NONE										
First Floor		NONE										
Second Floor (22 Windows)												
Grounds												
Sill	22	3.5'			77	L.F.						
6'-0" Jamb	44	6.0'					264	L.F.				
Head	22	3.5'							77	L.F.		
Oak Sills	22	3.5'									77	L.F.
3'-6"												
Third Floor (11 Windows)												
Grounds												
Sill	11	3.5'			39	L.F.						
5'-0" Jamb	22	5.0'					110	L.F.				
Head	11	3.5'							39	L.F.		
3'-6" Oak Sills	11	3.5'									39	L.F.
					116	L.F.	374	L.F.	116	L.F.	116	L.F.
Total Grounds							606	L.F.				
Total Sills											116	L.F.

Figure 26.44

307

06050 | Basic Wood / Plastic Materials / Methods

06090 | Wood & Plastic Fastenings

			CREW	DAILY OUTPUT	LABOR-HOURS	UNIT	MAT.	LABOR	EQUIP.	TOTAL	TOTAL INCL O&P	
825	0010	ROUGH HARDWARE Average % of carpentry material, minimum					.50%					825
	0200	Maximum					1.50%					
850	0010	BRACING										850
	0300	Let-in, "T" shaped, 22 ga. galv. steel, studs at 16" O.C.	1 Carp	5.80	1.379	C.L.F.	40	41.50		81.50	113	
	0400	Studs at 24" O.C.		6	1.333		40	40		80	111	
	0500	16 ga. galv. steel straps, studs at 16" O.C.		6	1.333		60	40		100	133	
	0600	Studs at 24" O.C.	▼	6.20	1.290	▼	60	38.50		98.50	130	

06100 | Rough Carpentry

06110 | Wood Framing

			CREW	DAILY OUTPUT	LABOR-HOURS	UNIT	MAT.	LABOR	EQUIP.	TOTAL	TOTAL INCL O&P	
100	0010	BLOCKING										100
	2600	Miscellaneous, to wood construction										
	2620	2" x 4"	1 Carp	.17	47.059	M.B.F.	535	1,400		1,935	2,925	
	2625	Pneumatic nailed		.21	38.095		535	1,150		1,685	2,475	
	2660	2" x 8"		.27	29.630		650	890		1,540	2,200	
	2665	Pneumatic nailed	▼	.33	24.242	▼	650	725		1,375	1,925	
	2720	To steel construction										
	2740	2" x 4"	1 Carp	.14	57.143	M.B.F.	535	1,725		2,260	3,425	
	2780	2" x 8"		.21	38.095	"	650	1,150		1,800	2,625	
	9000	Minimum labor/equipment charge	▼	4	2	Job		60		60	99.50	
150	0012	BRACING Let-in, with 1" x 6" boards, studs @ 16" O.C.	1 Carp	150	.053	L.F.	.51	1.60		2.11	3.21	150
	0202	Studs @ 24" O.C.	"	230	.035	"	.51	1.04		1.55	2.29	
200	0012	BRIDGING Wood, for joists 16" O.C., 1" x 3"	1 Carp	130	.062	Pr.	.29	1.85		2.14	3.37	200
	0017	Pneumatic nailed		170	.047		.29	1.41		1.70	2.65	
	0102	2" x 3" bridging		130	.062		.37	1.85		2.22	3.47	
	0107	Pneumatic nailed		170	.047		.37	1.41		1.78	2.75	
	0302	Steel, galvanized, 18 ga., for 2" x 10" joists at 12" O.C.		130	.062		.59	1.85		2.44	3.70	
	0402	24" O.C.		140	.057		2.20	1.71		3.91	5.25	
	0902	Compression type, 16" O.C., 2" x 8" joists		200	.040		1.26	1.20		2.46	3.38	
	1002	2" x 12" joists	▼	200	.040	▼	1.41	1.20		2.61	3.54	
505	0010	FRAMING, BEAMS & GIRDERS	[R06100 -010]									505
	1002	Single, 2" x 6"	2 Carp	700	.023	L.F.	.57	.69		1.26	1.76	
	1007	Pneumatic nailed	[R06110 -030]	812	.020		.57	.59		1.16	1.60	
	1022	2" x 8"		650	.025		.87	.74		1.61	2.17	
	1027	Pneumatic nailed		754	.021		.87	.64		1.51	2	
	1042	2" x 10"		600	.027		1.24	.80		2.04	2.70	
	1047	Pneumatic nailed		696	.023		1.24	.69		1.93	2.51	
	1062	2" x 12"		550	.029		1.70	.87		2.57	3.32	
	1067	Pneumatic nailed		638	.025		1.70	.75		2.45	3.12	
	1082	2" x 14"		500	.032		1.91	.96		2.87	3.70	
	1087	Pneumatic nailed		580	.028		1.91	.83		2.74	3.48	
	1102	3" x 8"		550	.029		2.25	.87		3.12	3.92	
	1122	3" x 10"		500	.032		2.81	.96		3.77	4.68	
	1142	3" x 12"		450	.036		3.37	1.07		4.44	5.50	
	1162	3" x 14"	▼	400	.040		3.93	1.20		5.13	6.30	
	1170	4" x 6"	F-3	1,100	.036		2.56	1.11	.55	4.22	5.25	
	1182	4" x 8"		1,000	.040		3.41	1.22	.61	5.24	6.40	
	1202	4" x 10"	▼	950	.042	▼	4.27	1.28	.64	6.19	7.50	

Figure 26.45

Repair and Remodeling Estimating Methods Example

Project No.: 001
Project Name: Commercial Renovation
Location: Estimating Methods Book
Title: ase Estimate
Architect: As Shown
Quantities: DEF
Take-off: ABC
Checked: GHI

Line Number	Description	Qty.	Unit	Material Unit Cost	Material Extension	Labor Unit Cost	Labor Extension	Equipment Unit Cost	Equipment Extension	Subcontract Unit Cost	Subcontract Extension
	Division 7 Thermal and Moisture Protection										
072109500080	Wall/ceiling insul, fbgls, kraft faced, batt/blkt,3.5" thk,R11,15" W	4799	S.F.	0.23	1103.77	0.15	719.85				
072109500820	Wall/ceiling insul, fbgls, unfaced, batt/blkt, 3.5" thk, R11, 15" W	10131	S.F.	0.21	2127.51	0.18	1823.58				
072106000700	Perimeter insulation, polystyrene, expanded, 2" thick, R8	880	S.F.	0.37	325.60	0.36	316.80				
012504000500	Factors, cut & patch to match existing construction, add, minimum		Costs	2%	6.51	3%	9.50				
072601000700	Building paper, polyethylene vapor barrier, std, .004" thick	52	Sq.	2.19	113.88	6.50	338.00				
073101000305	Asphalt shingles, laminated, class A,240-260 lbs/sq, pneumatic nailed	87	Sq.							112.00	9744.00
073101000800	Asphalt shingles, #15 felt underlayment	83	Sq.							8.35	693.05
073101000850	Asphalt shingles, polyethylene and rubberized asph. underlayment	4	Sq.							58.50	234.00
073101000900	Asphalt shingles, ridge shingles	140	L.F.							1.95	273.00
077104500200	Drip edge, aluminum, 8" wide, mill finish	140	L.F.							1.32	184.80
077104002500	Downspouts, copper, stock, 3" x 4"	272	L.F.	6.30	1713.60	1.94	527.68				
077104002800	Downspouts, copper, wire strainers, rectangular, 3" x 4"	8	Ea.	4.44	35.52	1.94	15.52				
077106501300	Gutters, copper, K type, 16 oz, stock, 5" wide	280	L.F.	3.13	876.40	2.34	655.20				

Figure 26.46

Repair and Remodeling Estimating Methods Example

Project No.: 001
Project Name: Commercial Renovation
Location: Estimating Methods Book

Title: ase Estimate
Architect: As Shown

Quantities: DEF
Take-off: ABC
Checked: GHI

Line Number	Description	Qty.	Unit	Material		Labor		Equipment		Subcontract	
				Unit Cost	Extension	Unit Cost	Extension	Unit Cost	Extension	Unit Cost	Extension
07720700000500	Roof hatches, 2'-6" x 3', w/curb, 1" fbgls insul, al curb & cover	1	Ea.	410.00	410.00	93.50	93.50				
07720700001400	Roof hatches, insul, galv st curb & al cover, 2'-6" x 8'-0"	1	Ea.	935.00	935.00	142.00	142.00				
07720850000300	Smoke hatches, for 8'-0" long, add to roof hatches from div 07720-700		Ea.	10%	93.50	5%	7.10				
	Division 7 Total				7741.29		4648.73				11128.85

Figure 26.47

Repair and Remodeling Estimating Methods Example

Project No.: 001
Project Name: Tenant Estimate
Location: Estimating Methods Book

Title: Base Estimate
Architect: As Shown

Quantities: DEF
Take-off: ABC
Checked: GHI

Line Number	Description	Qty.	Unit	Material Unit Cost	Material Extension	Labor Unit Cost	Labor Extension	Equipment Unit Cost	Equipment Extension	Subcontract Unit Cost	Subcontract Extension
072109500820	Wall/ceiling insul, fbgls, unfaced, batt/blkt, 3.5" thk, R11, 15"W	3015	S.F.	0.21	633.15	0.18	542.70				
086208000600	Skylight, plstc domes, flush/curb mtd, 10+ units, no curb, 10 to 20 SF, dbl	14	S.F.	15.45	216.30	2.97	41.58				
086208001800	Skylight, plstc domes, for integral insulated 9" curbs, double, add			30%	64.89						
	Division 7 Total				914.34		584.28				

Figure 26.48

may be shown only once on the plans and not even mentioned directly in the specifications. The estimator must be careful and thorough in order to avoid omissions.

The roofing and roof insulation are estimated for the complete roof surface, with no deductions for the hatches or skylights. These accessories will require extra labor for cutting and flashing. When bids are submitted, the estimator must be sure that the quotes are for the roofing and materials, as they are exactly specified. There are many materials and methods of installation, especially for single-ply roofs.

Division 7 is the first instance where work for the third-floor tenant is included in the estimate. In Figure 26.48, the items are priced separately so that the total tenant costs can be determined easily at the end of the estimate. The tenant has requested that an additional skylight be installed. Note that curbs must be added to the cost of the skylights.

During the estimating process, the estimator should begin to think about scheduling and the progression of the work. Figure 26.49 is a preliminary schedule for Division 2 through Division 7. The schedule, at this point, is used only as a reference and will undoubtedly be changed as the estimate is completed. The estimator should note such items as the demolition of the freight elevator and shaft, which will occur out of a normal sequence. The sidewalk is scheduled near the end of the project, because the scaffolding must be removed and all exterior work completed before installation. (See Chapter 8, "Pre-Bid Scheduling.") Proper preparation of an accurate Project Schedule is very important when determining the allocation of workers for a project. It is preferable to keep the workforce as constant as possible throughout the project. The schedule also provides a basis for determining costs of time-related items in Division 1.

Division 8: Doors, Windows and Glass

A proper set of plans and specifications will have door, window, and hardware schedules. These schedules should provide all information that is necessary for the quantity takeoff. Figure 26.50 shows portions of the door schedule for the sample project. There are two common types of door schedules. The first, illustrated in Figure 26.50, lists each door separately. The second lists each type of door (usually accompanied by elevations) that have common characteristics. Hardware is usually only designated by "set" on the schedules. These "sets" are described elsewhere on the plans or in the specifications. The estimator should cross-check quantities obtained from the schedules by counting and marking each item on the plans.

Specifications often state that all doors, frames, and hardware be "as specified" or an "approved equal." This means that if the estimator can find similar quality products at lower cost, the architect may accept these products as alternatives. Specifications usually require that the contractor submit shop drawings, schedules (as prepared by the contractor or supplier), or "cuts" (product literature). Most door, frame, and hardware suppliers provide these services to the contractor. It is recommended that the estimator obtain material prices from a supplier, because prices vary and fluctuate widely.

Installation costs, also, may vary due to restrictive existing conditions in renovation. During the site evaluation, the estimator should note any areas where such restrictions require the addition of Factors. In the sample project, the metal frames with transoms at the stairway must be installed in existing masonry openings. These openings most likely will not conform exactly to the sizes of new metal door frames. A Factor is added to match existing conditions. The existing interior door frames at the west stairway are to remain and to be reused. Again, a Factor must be added to the appropriate door

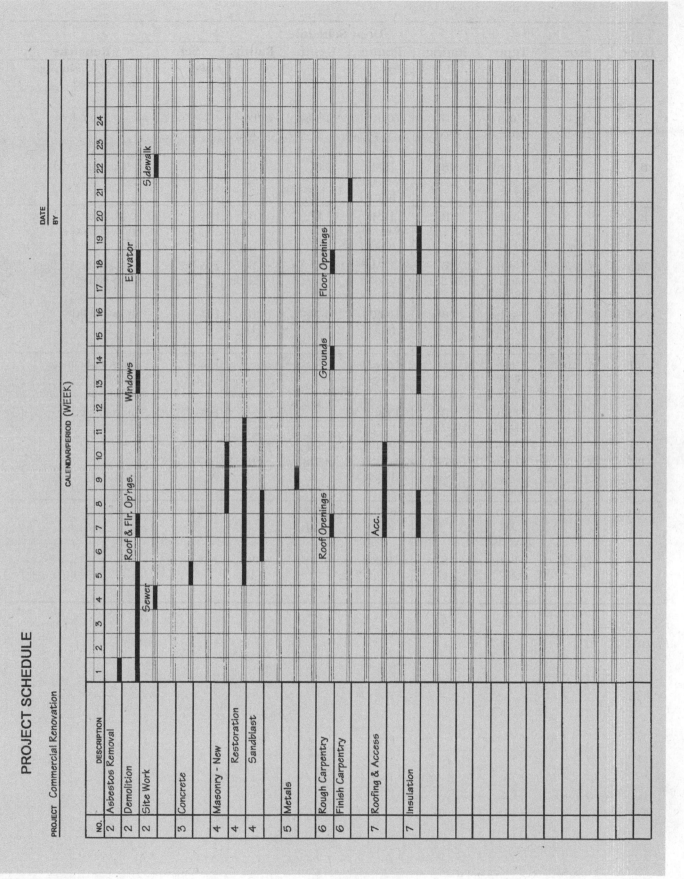

Figure 26.49

Door Schedule								
Door	Size	Type	Rating	Frame	Depth	Rating	Set	Remarks
B01	3^0 x 6^8	Flush Steel 18 ga.	"B" 1-1/2 Hr.	Exist'g.	–	–	HW-1	10" x 10" Vision Lite Shop-Primed
B02	3^0 x 6^8	Flush Steel 18 ga.	"B" 1 Hr.	HMKD 16 ga.	4-7/8"	"B" 1 Hr.	HW-2	Shop-Primed
B03	3^0 x 6^8	Flush Steel 18 ga.	"B" 1 Hr.	H.M. Welded 16 ga.	8"	"B" 1 Hr.	HW-3	w/Masonry Anchors Shop-Primed
B04	3^0 x 6^8	Flush Steel 18 ga.	"B" 1-1/2 Hr.	H.M. Welded 16 ga.	8"	"B" 1-1/2 Hr.	HW-3	w/Masonry Anchors Shop-Primed
B05	3^0 x 6^8	Flush Steel 18 ga.	"B" 1-1/2 Hr.	H.M. Welded 16 ga.	8"	"B" 1-1/2 Hr.	HW-3	w/Masonry Anchors Shop-Primed
B06	3^0 x 6^8	Flush Steel 18 ga.	"B" 1 Hr.	HMKD 16 ga.	4-7/8"	"B" 1 Hr.	HW-2	Shop-Primed
101	3^0 x 6^8	Flush Steel 18 ga.	–	HMKD 16 ga.	4-7/8"	–	HW-4	Transom Frame Above w/Masonry Anchors
102	3^0 x 6^8	Flush Steel 18 ga.	"B" 1-1/2 Hr.	Exist'g.	–	–	HW-1	Shop-Primed
103	3^0 x 6^8	Flush Oak Face	"B" 1 Hr.	HMKD 16 ga.	4-7/8"	"B" 1 Hr.	HW-5	
104	2^6 x 6^8	Flush Oak Face SC	–	HMKD 16 ga.	4-5/8"	–	HW-6	
105	3^0 x 6^8	Flush Oak Face	"B" 1 Hr.	HMKD 16 ga.	4-7/8"	"B" 1 Hr.	HW-5	

Figure 26.50

installations to match existing conditions. (See Figure 26.52.) Estimate sheets for Division 8 are illustrated in Figures 26.51 to 26.55.

Note that the tenant entrance door costs, in Figure 26.52, have a deduction for doors with a height of 6′ 8″. This information is obtained from Figure 26.56.

The new windows for the project are not standard sizes and require a special black finish. A quotation from a subcontractor is obtained and is included in Figure 26.53. There are no factors added separately because any special installation or material costs should be included in the subcontract price. It is important, therefore, that the subcontractor examine the existing conditions thoroughly before submitting the quote. The estimator must be sure that no "surprises" will be encountered during installation.

As with the windows, the retail entrance doors and lobby entrance require a black finish. These items will be furnished and installed by a subcontractor. The added costs for the black finish are provided as a percentage of the bare material cost, as shown in Figure 26.57. Also, the Factors for matching existing conditions and protecting existing work (applied to bare costs) must be added to the total, including the installing subcontractor's overhead and profit. The calculations are shown in Figure 26.58.

The factors do not apply to the costs for the concealed panic devices. These items are installed in the door and frame at the supplier's shop and affect only the material cost of the door units. However, the installation of the concealed closers is affected by the existing conditions, so the Factors do apply. The black finish cost is also included before the Factors are calculated, because any required shims or spacers to conform to the existing opening will have the same black finish. Similar calculations are required for the lobby entrance storefront.

Storefront systems can be estimated in two ways, using *Means Repair & Remodeling Cost Data*. The first method separates the system into components (tube framing, glass, doors, and hardware), and prices each item individually. The second method uses the data provided for complete systems, as shown in Figure 26.59. This example estimating project uses the latter method. When pricing is used for a system such as this, the estimator must be sure that all specified items are included in the system, or added if necessary.

The estimate sheets (Figures 26.51 to 26.55) have been organized so that groups of similar items are priced together. The estimator can use this information to determine the percentage total job costs, or the costs per square foot of the different components of commercial renovation. These percentages and square foot costs may be helpful when estimating future jobs.

The tenant work for Division 8 is priced separately, as in Division 7. The estimator must remember, for scheduling purposes, that all materials for the second and third floors should be in place before the freight elevator is removed.

Division 9: Finishes

In commercial renovation, Division 9 usually represents a major portion of the work. The seven estimate sheets for the finishes are shown in Figures 26.60 to 26.66.

The estimator must be careful when taking off quantities for metal stud and drywall partitions. It would be simple if each type and size of stud had a different drywall application. But, different size studs will have the same drywall treatment, and vice versa. Thoroughly completed plans and specifications will often have a schedule of the wall types as well as a Room Finish Schedule. In either case, the different wall types should be marked on the plans, with different colors, during the quantity takeoff. The estimator should determine

Repair and Remodeling Estimating Methods Example

Project No.: 001	Title: ase Estimate
Project Name: Commercial Renovation	Architect: As Shown
Location: Estimating Methods Book	
	Quantities: DEF
	Take-off: ABC
	Checked: GHI

Line Number	Description	Qty.	Unit	Material Unit Cost	Material Extension	Labor Unit Cost	Labor Extension	Equipment Unit Cost	Equipment Extension	Subcontract Unit Cost	Subcontract Extension
	Division 8 Doors and Windows										
081108203600	Steel frs, knock dn, 14 Ga., up to 5-3/4" D, 7'-0" hi, 4'-0" W, sgl	6	Ea.	63.50	381.00	32.00	192.00				
081108205400	Steel frames knock down, 14 ga, "B" label, 5.75" D, 7' H, 4' W, sgl	13	Ea.	87.00	1131.00	32.00	416.00				
081108204900	Steel frames, knock down, for welded frames, add	3	Ea.	26.00	78.00						
081108205440	Steel frames knock down, 14 ga, "B" label, 5.75" D, 7' H, 8' W, dbl	6	Ea.	108.00	636.00	40.00	240.00				
081108205800	Steel frames, 14 Ga. "B" label, 6-3/4" D, 7'-0" high, 4'-0" wide, single	3	Ea.	96.00	288.00	32.00	96.00				
081108205400	Steel frames knock down, 14 ga, "B" label, 5.75" D, 7' H, 4' W, sgl	2	Ea.	87.00	174.00	32.00	64.00				
081108207900	Steel frames, transom lite frames, fixed, add	9	S.F.	25.00	225.00	3.10	27.90				
081108206200	Steel frames knock down, 14 ga, "B" label, 8.75" D, 7' H, 4' W, sgl	3	Ea.	108.00	324.00	32.00	96.00				

Figure 26.51

Repair and Remodeling Estimating Methods Example

Project No.: 001	Title: ase Estimate	Quantities: DEF
Project Name: Commercial Renovation	Architect: As Shown	Take-off: ABC
Location: Estimating Methods Book		Checked: GHI

Line Number	Description	Qty.	Unit	Material		Labor		Equipment		Subcontract	
				Unit Cost	Extension	Unit Cost	Extension	Unit Cost	Extension	Unit Cost	Extension
081102001020	Coml st doors, hol core, 1-3/4" thk, full pnl, 18 Ga., 3'-0" x 6'-8"	2	Ea.	182.00	364.00	30.00	60.00				
081103000140	Fire door, st, fl, "B" lbl, 90 min, full pnl, 18 Ga., 3'-0" x 6'-8"	4	Ea.	192.00	768.00	30.00	120.00				
012504000550	Factors, cut & patch to match existing construction, add, maximum		Costs	5%	38.40	9%	10.80				
081103000140	Fire door, st, fl, "B" lbl, 90 min, full pnl, 18 Ga., 3'-0" x 6'-8"	8	Ea.	192.00	1536.00	30.00	240.00				
081103000240	Fire door, steel, flush, for vision lite, add	7	Ea.	18.00	126.00						
082109500140	Wood FR dr,part core,7 ply stile, "b" lbl,1 hr,oak face,2'-6"x6'-8"	6	Ea.	219.00	1314.00	34.50	207.00				
082109500190	Wood FR dr,part core,7 ply stile, "b" lbl,1 hr,oak face,3'-0"x7'-0"	17	Ea.	237.00	4029.00	40.00	680.00				
082109502460	Wd dr,cust arch "b" lbl,fl,1-3/4"thk,sol core,6'-8"h,deduct from 7'dr	17	Ea.	-13.25	-225.25						
082109002240	Wood door, arch fl, int, 5 ply particle core, oak face, 3'-0" x 6'-8"	3	Ea.	99.00	297.00	34.50	103.50				
082109002220	Wood door, arch fl, int, 5 ply particle core, oak face, 2'-6" x 6'-8"	3	Ea.	90.00	270.00	32.00	96.00				

Figure 26.52

Repair and Remodeling Estimating Methods Example

Project No.: 001
Project Name: Commercial Renovation
Location: Estimating Methods Book

Title: ase Estimate
Architect: As Shown

Quantities: DEF
Take-off: ABC
Checked: GHI

Line Number	Description	Qty.	Unit	Material Unit Cost	Material Extension	Labor Unit Cost	Labor Extension	Equipment Unit Cost	Equipment Extension	Subcontract Unit Cost	Subcontract Extension
085201200010	Aluminum windows, w/frame & glazing, grade C	70	WRITTEN QUOTE							687.00	48090.00
06208000600	Skylight, plstc domes,flush/curb mtd,10+ units,no curb,10 to 20 SF,dbl	196	S.F.	15.45	3028.20	2.97	582.12				
06208001800	Skylight, plstc domes, for integral insulated 9" curbs, double, add			30%	908.46						
087103002400	Door closer, backcheck & adj pwr, top jamb mount, all sizes, reg arm	28	Ea.	139.00	3892.00	40.00	1120.00				
087103401600	Doorstops, door bumper, floor type, aluminum	21	Ea.	3.87	81.27	7.50	157.50				
087105201430	Hinges, full mortise, high frequency, brs base, 4-1/2" x 4-1/2", us10	23	Pr.	41.50	954.50						
087105500500	Kick plate, 6" high, for 3' door, bronze	4	Ea.	20.00	80.00	16.00	64.00				
087106500100	Lockset, std dty, cylindrical, with sectional trim, non-keyed, privacy	3	Ea.	45.00	135.00	20.00	60.00				
087106500400	Lockset, std dty, cylindrical, w/sectional trim, keyed, sgl cyl function	13	Ea.	65.50	851.50	24.00	312.00				
087107002130	Mortise lockset, cast knobs & full esc trim, keyed, sgl cyl, typ, max	8	Ea.	375.00	3000.00	34.50	276.00				
087107500210	Panic device, for rim locks, sgl dr, bar&vert rod, outside key&pull	2	Ea.	580.00	1160.00	60.00	120.00				
087107500020	Panic device, for rim locks, sgl door, exit only, outside key & pull	7	Ea.	380.00	2660.00	48.00	336.00				
087107800500	Push-pull, push plate, .050" thick, 4" x 16", bronze	4	Ea.	11.40	45.60	20.00	80.00				
087208000100	Threshold, Aluminum, 8" wide, 1/2" thick	2	Ea.	29.00	58.00	20.00	40.00				

Figure 26.53

Repair and Remodeling Estimating Methods Example

Project No.: 001
Project Name: Commercial Renovation
Location: Estimating Methods Book

Title: see Estimate
Architect: As Shown

Quantities: DEF
Take-off: ABC
Checked: GHI

Line Number	Description	Qty.	Unit	Material Unit Cost	Material Extension	Labor Unit Cost	Labor Extension	Equipment Unit Cost	Equipment Extension	Subcontract Unit Cost	Subcontract Extension
08411140200	Aluminum doors & frs, hdwe & closer, 3'-6" x 10' opng, 3' hi transom	4	Ea.							1350.00	5400.00
08411140500	Alum dr/fr,w/hdw/closer,no glass,for black finish,add	4	Leaf							36%	1944.00
08411140900	Aluminum doors & frs, add to abv for W stile doors, cncl closer, add	4	Leaf							522.50	2090.00
01250400550	Factors, cut & patch to match existing construction, add, maximum		Costs							9%	849.06
01250400700	Factors, protection of existing work, add, minimum		Costs			Average % for material and labor					
08411140600	Alum door & frs, add to abv for W stile door, cncl panic device, add	4	Leaf							1023.00	4092.00
08810280800	Float glass, 1/4" thick, tempered, clear	84	S.F.							12.55	1054.20
08810280600	Float glass, 1/4" thick, plain, clear	36	S.F.							11.65	419.40
08810280800	Float glass, 1/4" thick, tempered, clear	18	S.F.							12.55	225.90
08411650100	Storeft, alum FR,3/8" pl gl,6'x7' dr W/hdwe,institutional GR	135	S.F.							25.50	3442.50
08411650600	Storefront systems, alum FR, clr 3/8" pl gl, for black anod fin, add	135	S.F.							30%	1032.75
01250400550	Factors, cut & patch to match existing construction, add, maximum		Costs			Average % for material and labor				9.5%	425.15
01250400200	Factors, shift work requirements, add, minimum		Costs								
	Division 8 Total				29608.66		5796.82				69064.96

Figure 26.54

319

Repair and Remodeling Estimating Methods Example

Project No.: 001
Project Name: Tenant Estimate
Location: Estimating Methods Book

Title: Base Estimate
Architect: As Shown

Quantities: DEF
Take-off: ABC
Checked: GHI

Line Number	Description	Qty.	Unit	Material Unit Cost	Material Extension	Labor Unit Cost	Labor Extension	Equipment Unit Cost	Equipment Extension	Subcontract Unit Cost	Subcontract Extension
082109002240	Wood door, arch fl, int, 5 ply particle core, oak face, 3'-0" x 6'-8"	9	Ea.	99	891.00	34.50	310.50				
081108203600	Steel frs, knock dn, 14 Ga., up to 5-3/4" D, 7'-0" hi, 4'-0" W, sgl	9	Ea.	63.5	571.50	32.00	288.00				
087105201430	Hinges, full mortise, high frequency, brs base, 4-1/2" x 4-1/2", us10	12	Pr.	41.5	498.00						
087103401600	Doorstops, door bumper, floor type, aluminum	9	Ea.	3.87	34.83	7.50	67.50				
087106501000	Lockset, heavy duty with sectional trim, non-keyed, passages	7	Ea.	105	735.00	20.00	140.00				
087106501400	Lockset, heavy duty with sectional trim, keyed, single cyl function	2	Ea.	156	312.00	24.00	48.00				
089117000200	Tube FR, for wdw wl&str frts,alum,stk,plain tube,mill fin,1-3/4"x4-1/2"	20	L.F.	8.85	177.00	5.05	101.00				
089117000450	Tube framing, for window walls and store fronts, glass bead	20	L.F.	1.68	33.60	2.00	40.00				
089117008020	Tube framing, al, for black finish, add			27%	47.79						
088102600800	Float glass, 1/4" thick, tempered, clear	24	S.F.	5.55	133.20	4.00	96.00				
	Division 8 Total				3433.92		1091.00				

Figure 26.55

08210 | Wood Doors

		CREW	DAILY OUTPUT	LABOR-HOURS	UNIT	2002 BARE COSTS				TOTAL INCL O&P	
						MAT.	LABOR	EQUIP.	TOTAL		
950	0190	3'-0" x 7'-0"	2 Carp	12	1.333	Ea.	237	40		277	325	**950**
	0200	4'-0" x 7'-0"		12	1.333		310	40		350	405	
	0240	Walnut face, 2'-6" x 6'-8"		14	1.143		288	34.50		322.50	370	
	0280	3'-0" x 6'-8"		13	1.231		294	37		331	385	
	0290	3'-0" x 7'-0"		12	1.333		305	40		345	405	
	0300	4'-0" x 7'-0"		12	1.333		415	40		455	520	
	0440	M.D. overlay on hardboard, 2'-6" x 6'-8"		15	1.067		192	32		224	264	
	0480	3'-0" x 6'-8"		14	1.143		200	34.50		234.50	277	
	0490	3'-0" x 7'-0"		13	1.231		211	37		248	293	
	0500	4'-0" x 7'-0"		12	1.333		258	40		298	350	
	0540	H.P. plastic laminate, 2'-6" x 6'-8"		13	1.231		257	37		294	345	
	0590	3'-0" x 7'-0"		11	1.455		273	43.50		316.50	375	
	0600	4'-0" x 7'-0"		10	1.600		350	48		398	460	
	0740	90 minutes, birch face, 1-3/4" x 2'-6" x 6'-8"		14	1.143		217	34.50		251.50	296	
	0780	3'-0" x 6'-8"		13	1.231		222	37		259	305	
	0790	3'-0" x 7'-0"		12	1.333		233	40		273	325	
	0800	4'-0" x 7'-0"		12	1.333		315	40		355	415	
	0840	Oak face, 2'-6" x 6'-8"		14	1.143		205	34.50		239.50	283	
	0880	3'-0" x 6'-8"		13	1.231		213	37		250	296	
	0890	3'-0" x 7'-0"		12	1.333		224	40		264	315	
	0900	4'-0" x 7'-0"		12	1.333		310	40		350	405	
	0940	Walnut face, 2'-6" x 6'-8"		14	1.143		282	34.50		316.50	365	
	0980	3'-0" x 6'-8"		13	1.231		289	37		326	380	
	0990	3'-0" x 7'-0"		12	1.333		300	40		340	400	
	1000	4'-0" x 7'-0"		12	1.333		435	40		475	545	
	1140	M.D. overlay on hardboard, 2'-6" x 6'-8"		15	1.067		223	32		255	299	
	1180	3'-0" x 6'-8"		14	1.143		229	34.50		263.50	310	
	1190	3'-0" x 7'-0"		13	1.231		238	37		275	325	
	1200	4'-0" x 7'-0"		12	1.333		325	40		365	420	
	1240	For 8'-0" height, add					41			41	45	
	1260	For 8'-0" height walnut, add					54			54	59.50	
	1340	H.P. plastic laminate, 2'-6" x 6'-8"	2 Carp	13	1.231		278	37		315	365	
	1380	3'-0" x 6'-8"		12	1.333		292	40		332	385	
	1390	3'-0" x 7'-0"		11	1.455		295	43.50		338.50	400	
	1400	4'-0" x 7'-0"		10	1.600		385	48		433	500	
	2200	Custom architectural "B" label, flush, 1-3/4" thick, birch,										
	2210	Solid core										
	2220	2'-6" x 7'-0"	2 Carp	15	1.067	Ea.	330	32		362	420	
	2260	3'-0" x 7'-0"		14	1.143		335	34.50		369.50	425	
	2300	4'-0" x 7'-0"		13	1.231		460	37		497	565	
	2420	4'-0" x 8'-0"		11	1.455		535	43.50		578.50	665	
	2460	For 6'-8" high door, deduct from 7'-0" door					13.25			13.25	14.60	
	2480	For oak veneer, add					50%					
	2500	For walnut veneer, add					75%					
	9000	Minimum labor/equipment charge	1 Carp	4	2	Job		60		60	99.50	
960	0010	**WOOD FRAMES**										**960**
	0400	Exterior frame, incl. ext. trim, pine, 5/4 x 4-9/16" deep	2 Carp	375	.043	L.F.	3.92	1.28		5.20	6.45	
	0420	5-3/16" deep		375	.043		6.75	1.28		8.03	9.55	
	0440	6-9/16" deep		375	.043		8.75	1.28		10.03	11.75	
	0600	Oak, 5/4 x 4-9/16" deep		350	.046		8.40	1.37		9.77	11.50	
	0620	5-3/16" deep		350	.046		9.50	1.37		10.87	12.70	
	0640	6-9/16" deep		350	.046		10.55	1.37		11.92	13.85	
	0800	Walnut, 5/4 x 4-9/16" deep		350	.046		9.90	1.37		11.27	13.15	
	0820	5-3/16" deep		350	.046		14.35	1.37		15.72	18.05	
	0840	6-9/16" deep		350	.046		16.95	1.37		18.32	21	

Figure 26.56

08411 | Aluminum Framed Storefront

		CREW	DAILY OUTPUT	LABOR-HOURS	UNIT	MAT.	LABOR	EQUIP.	TOTAL	TOTAL INCL O&P	
140	**0010**	**ALUMINUM DOORS & FRAMES** Entrance, narrow stile, including										**140**
	0015	Standard hardware, clear finish, not incl. glass, 2'-6" x 7'-0" opng.	2 Sswk	2	8	Ea.	435	274		709	990	
	0020	3'-0" x 7'-0" opening		2	8		435	274		709	995	
	0030	3'-6" x 7'-0" opening		2	8		450	274		724	1,000	
	0100	3'-0" x 10'-0" opening, 3' high transom		1.80	8.889		710	305		1,015	1,350	
	0200	3'-6" x 10'-0" opening, 3' high transom.		1.80	8.889		700	305		1,005	1,350	
	0280	5'-0" x 7'-0" opening		2	8		745	274		1,019	1,325	
	0300	6'-0" x 7'-0" opening		1.30	12.308	Pr.	725	420		1,145	1,575	
	0400	6'-0" x 10'-0" opening, 3' high transom		1.10	14.545		650	500		1,150	1,650	
	0420	7'-0" x 7'-0" opening		1	16		785	550		1,335	1,900	
	0500	Wide stile, 2'-6" x 7'-0" opening	2 Sswk	2	8	Ea.	665	274		939	1,250	
	0520	3'-0" x 7'-0" opening		2	8		655	274		929	1,225	
	0540	3'-6" x 7'-0" opening		2	8		685	274		959	1,275	
	0560	5'-0" x 7'-0" opening		2	8		1,050	274		1,324	1,675	
	0580	6'-0" x 7'-0" opening		1.30	12.308	Pr.	1,000	420		1,420	1,900	
	0600	7'-0" x 7'-0" opening		1	16	"	1,150	550		1,700	2,275	
	1100	For full vision doors, with 1/2" glass, add				Leaf	55%					
	1200	For non-standard size, add					67%					
	1300	Light bronze finish, add					36%					
	1400	Dark bronze finish, add					18%					
	1500	For black finish, add					36%					
	1600	Concealed panic device, add					930			930	1,025	
	1700	Electric striker release, add				Opng.	239			239	263	
	1800	Floor check, add				Leaf	710			710	780	
	1900	Concealed closer, add				"	475			475	520	
	2000	Flush 3' x 7' insulated, 12"x 12" lite, clear finish	2 Sswk	2	8	Ea.	900	274		1,174	1,500	
	9000	Minimum labor/equipment charge	2 Carp	4	4	Job		120		120	199	
600	**0010**	**STAINLESS STEEL AND GLASS** Entrance unit, narrow stiles										**600**
	0020	3' x 7' opening, including hardware, minimum	2 Sswk	1.60	10	Opng.	4,600	345		4,945	5,700	
	0050	Average		1.40	11.429		4,975	390		5,365	6,200	
	0100	Maximum		1.20	13.333		5,325	455		5,780	6,725	
	1000	For solid bronze entrance units, statuary finish, add					60%					
	1100	Without statuary finish, add					45%					
	2000	Balanced doors, 3' x 7', economy	2 Sswk	.90	17.778	Ea.	6,225	610		6,835	8,000	
	2100	Premium		.70	22.857	"	10,700	785		11,485	13,300	
	9000	Minimum labor/equipment charge		2	8	Job		274		274	515	
650	**0010**	**STOREFRONT SYSTEMS** Aluminum frame, clear 3/8" plate glass,										**650**
	0020	incl. 3' x 7' door with hardware (400 sq. ft. max. wall)										
	0500	Wall height to 12' high, commercial grade	2 Glaz	150	.107	S.F.	12	3.20		15.20	18.35	
	0600	Institutional grade		130	.123		16	3.69		19.69	23.50	
	0700	Monumental grade		115	.139		23	4.17		27.17	32	
	1000	6' x 7' door with hardware, commercial grade		135	.119		12.25	3.56		15.81	19.25	
	1100	Institutional grade		115	.139		16.80	4.17		20.97	25.50	
	1200	Monumental grade		100	.160		31	4.80		35.80	42	
	1500	For bronze anodized finish, add					15%					
	1600	For black anodized finish, add					30%					
	1700	For stainless steel framing, add to monumental					75%					
	9000	Minimum labor/equipment charge	2 Glaz	1	16	Job		480		480	775	

08460 | Automatic Entrance Doors

		CREW	DAILY OUTPUT	LABOR-HOURS	UNIT	MAT.	LABOR	EQUIP.	TOTAL	TOTAL INCL O&P	
600	**0010**	**SLIDING ENTRANCE** 12' x 7'-6" opng., 5' x 7' door, 2 way traf.,										**600**
	0020	mat activated, panic pushout, incl. operator & hardware,										
	0030	not including glass or glazing	2 Glaz	.70	22.857	Opng.	5,900	685		6,585	7,600	
	9000	Minimum labor/equipment charge	"	.70	22.857	Job		685		685	1,100	

Figure 26.57

quantities for these "assembled" wall types before attempting to figure quantities of individual components. During the takeoff, the estimator should also make notes on existing conditions that will require added factors. In the sample project, the wood beams at the ceilings occur every 10'. The studs and drywall will have to be cut to fit around the beams. No deduction for material is made for this type of condition, but extra labor expense is included.

When the quantities of the different wall types have been determined, the estimator should list each individual component and the appropriate quantities. Then the total quantities of each component for all wall types are added and entered on the estimate sheet. As a cross-check, the resulting totals for studs and drywall can be compared. When comparing, remember that the drywall quantities will be twice those for some of the studs, because of application on two sides. This method will help to ensure that all items have been included. The comparison for the sample project is illustrated in Figure 26.68. Refer to Figures 26.60 and 26.61 for a cost analysis of metal studs and drywall, respectively.

In Figure 26.61, the 5/8" fire-resistant drywall is listed in two ways, taped and with no finish. The drywall that requires no finish is the first layer of a two-layer application at the new stairway enclosure. The estimator should visualize the work in order to include the proper costs.

The drywall returns at the window jambs and headers could easily be overlooked (Figure 26.42). For quantities, refer back to Figure 26.44, the Quantity Sheet, for the sills and grounds. The measurements have already been calculated and can be used for the drywall takeoff. The drywall installation at the window is almost all corners, and the appropriate finish work is included in the estimate (Figure 26.61).

Aluminum Entrance Doors							
		Material		**Labor**		**Total**	**Total w/ O&P**
Entrance Door (08411-140-0200)			$700		$305	$1,005	
Black Finish (08411-140-1500)	(36%)	252				252	
Concealed Closer (08411-140-1900)		475				475	
Subtotals			$1,427		$305	$1,732	
Factors: 1 max., 5 min.	(7%)	100		(11%)	34		
Subtotals			$1,527		$339		
O&P (10% Material, 93% Labor)	(10%)	153		(61.3%)	208		
Totals Including Overhead & Profit			$1,680		$547		$2,227

Figure 26.58

08400 | Entrances & Storefronts

08411 | Aluminum Framed Storefront

		CREW	DAILY OUTPUT	LABOR-HOURS	UNIT	2002 BARE COSTS MAT.	LABOR	EQUIP.	TOTAL	TOTAL INCL O&P	
140	0500	Wide stile, 2'-6" x 7'-0" opening	2 Sswk	2	8	Ea.	665	274		939	1,250	**140**
	0520	3'-0" x 7'-0" opening		2	8		655	274		929	1,225	
	0540	3'-6" x 7'-0" opening		2	8		685	274		959	1,275	
	0560	5'-0" x 7'-0" opening		2	8		1,050	274		1,324	1,675	
	0580	6'-0" x 7'-0" opening		1.30	12.308	Pr.	1,000	420		1,420	1,900	
	0600	7'-0" x 7'-0" opening		1	16	"	1,150	550		1,700	2,275	
	1100	For full vision doors, with 1/2" glass, add				Leaf	55%					
	1200	For non-standard size, add					67%					
	1300	Light bronze finish, add					36%					
	1400	Dark bronze finish, add					18%					
	1500	For black finish, add					36%					
	1600	Concealed panic device, add					930			930	1,025	
	1700	Electric striker release, add				Opng.	239			239	263	
	1800	Floor check, add				Leaf	710			710	780	
	1900	Concealed closer, add				"	475			475	520	
	2000	Flush 3' x 7' Insulated, 12"x 12" lite, clear finish	2 Sswk	2	8	Ea.	900	274		1,174	1,500	
	9000	Minimum labor/equipment charge	2 Carp	4	4	Job		120		120	199	
600	0010	**STAINLESS STEEL AND GLASS** Entrance unit, narrow stiles										**600**
	0020	3' x 7' opening, including hardware, minimum	2 Sswk	1.60	10	Opng.	4,600	345		4,945	5,700	
	0050	Average		1.40	11.429		4,975	390		5,365	6,200	
	0100	Maximum		1.20	13.333		5,325	455		5,780	6,725	
	1000	For solid bronze entrance units, statuary finish, add					60%					
	1100	Without statuary finish, add					45%					
	2000	Balanced doors, 3' x 7', economy	2 Sswk	.90	17.778	Ea.	6,225	610		6,835	8,000	
	2100	Premium		.70	22.857	"	10,700	785		11,485	13,300	
	9000	Minimum labor/equipment charge		2	8	Job		274		274	515	
650	0010	**STOREFRONT SYSTEMS** Aluminum frame, clear 3/8" plate glass,										**650**
	0020	incl. 3' x 7' door with hardware (400 sq. ft. max. wall)										
	0500	Wall height to 12' high, commercial grade	2 Glaz	150	.107	S.F.	12	3.20		15.20	18.35	
	0600	Institutional grade		130	.123		16	3.69		19.69	23.50	
	0700	Monumental grade		115	.139		23	4.17		27.17	32	
	1000	6' x 7' door with hardware, commercial grade		135	.119		12.25	3.56		15.81	19.25	
	1100	Institutional grade		115	.139		16.80	4.17		20.97	25.50	
	1200	Monumental grade		100	.160		31	4.80		35.80	42	
	1500	For bronze anodized finish, add					15%					
	1600	For black anodized finish, add					30%					
	1700	For stainless steel framing, add to monumental					75%					
	9000	Minimum labor/equipment charge	2 Glaz	1	16	Job		480		480	775	

08460 | Automatic Entrance Doors

		CREW	DAILY OUTPUT	LABOR-HOURS	UNIT	MAT.	LABOR	EQUIP.	TOTAL	TOTAL INCL O&P	
600	0010	**SLIDING ENTRANCE** 12' x 7'-6" opng., 5' x 7' door, 2 way traf.,										**600**
	0020	mat activated, panic pushout, incl. operator & hardware,										
	0030	not including glass or glazing	2 Glaz	.70	22.857	Opng.	5,900	685		6,585	7,600	
	9000	Minimum labor/equipment charge	"	.70	22.857	Job		685		685	1,100	
650	0010	**SLIDING PANELS**										**650**
	0020	Mall fronts, aluminum & glass, 15' x 9' high	2 Glaz	1.30	12.308	Opng.	2,325	370		2,695	3,150	
	0100	24' x 9' high		.70	22.857		3,350	685		4,035	4,800	
	0200	48' x 9' high, with fixed panels		.90	17.778		6,250	535		6,785	7,725	
	0500	For bronze finish, add					17%					
	9000	Minimum labor/equipment charge	2 Glaz	1	16	Job		480		480	775	

08480 | Balanced Entrance Doors

		CREW	DAILY OUTPUT	LABOR-HOURS	UNIT	MAT.	LABOR	EQUIP.	TOTAL	TOTAL INCL O&P	
150	0010	**BALANCED DOORS**										**150**
	0020	Hardware & frame, alum. & glass, 3' x 7', econ.	2 Sswk	.90	17.778	Ea.	4,725	610		5,335	6,350	

Figure 26.59

Repair and Remodeling Estimating Methods Example

Project No.: 001
Project Name: Commercial Renovation
Location: Estimating Methods Book
Title: ase Estimate
Architect: As Shown
Quantities: DEF
Take-off: ABC
Checked: GHI

Line Number	Description	Qty.	Unit	Material		Labor		Equipment		Subcontract	
				Unit Cost	Extension	Unit Cost	Extension	Unit Cost	Extension	Unit Cost	Extension
	Division 9 Finishes										
091101002000	Metal studs ptn, 10' H, N.L.B., galv, 25 ga 1-5/8" W, 16" OC	7144	S.F.	0.16	1143.04	0.48	3429.12				
091101002300	Metal studs ptn, 10' H, N.L.B., galv, 25 ga 3-5/8" W, 16" OC	11775	S.F.	0.19	2237.25	0.50	5887.50				
091101002500	Metal studs ptn, 10' H, N.L.B., galv, 25 ga, 6" W,16" OC	560	S.F.	0.31	173.60	0.51	285.60				
012504000500	Factors, cut & patch to match existing construction, add, minimum		Costs	2%	71.06	3%	288.07				
092701000900	Accessories, furring channel, galv. steel, 7/8" deep, standard	26.38	C.L.F.	7.35	457.69	92.50	2440.15				
092608000060	Shaft wl,25g jtrk&c-h,24"oc,1"corebd sft lnr,2hr,2yr,5/8"f rtd gyp	552	S.F.	0.89	491.28	2.18	1203.36				
092608000900	Shaft wall, for taping & finishing, add per side	552	S.F.	0.04	22.08	0.23	126.96				

Figure 26.60

Repair and Remodeling Estimating Methods Example

Project No.: 001
Project Name: Commercial Renovation
Location: Estimating Methods Book

Title: ase Estimate
Architect: As Shown

Quantities: DEF
Take-off: ABC
Checked: GHI

Line Number	Description	Qty.	Unit	Material Unit Cost	Material Extension	Labor Unit Cost	Labor Extension	Equipment Unit Cost	Equipment Extension	Subcontract Unit Cost	Subcontract Extension
092507000350	Gypsum plasterboard, nailed/scrd to std, 1/2" thk, on wl, std, taped&fin (level)	13292	S.F.	0.25	3323.00	0.50	6646.00				
092507002100	Drywall, gyp plasbd, nailed/scrd to std, 5/8" thk, on wl, fire res, no fin	3288	S.F.	0.22	723.36	0.24	789.12				
092507002150	Drywall, gyp plasbd, nailed/scrd to std, 5/8" thk, on wl, fire res, taped&fin	22014	S.F.	0.26	5723.64	0.50	11007.00				
092507003150	Drywall, gyp plasbd, nailed/scrd, 5/8" thk, on clgs, fire res, taped&fin	490	S.F.	0.26	127.40	0.63	308.70				
092507005200	Drywall, for high ceilings, over 8' high, add	39084	S.F.			0.16	6253.44				
012504000500	Factors, cut & patch to match existing construction, add, minimum		Costs	2%	197.95	3%	750.13				
012504001400	Factors, material handling & storage limitation, add, minimum		Costs	1%	98.97	1%	250.04				
092701000400	Accessories, corner bead, galvanized steel, 1-1/4" x 1-1/4"	14.4	C.L.F.	9.25	133.20	68.50	986.40				
092507005350	Drywall, for finishing corners, inside or outside, add	2588	L.F.	0.06	155.28	0.44	1138.72				
092507000300	Gypsum plasterboard, 1/2" thk, on walls, std, no finish incl	404	S.F.	0.21	84.84	0.24	96.96				
012504000500	Factors, cut & patch to match existing construction, add, minimum		Costs	2%	1.70	3%	2.91				
012504001400	Factors, material handling & storage limitation, add, minimum		Costs	1%	0.85	1%	0.97				
092701000400	Accessories, corner bead, galvanized steel, 1-1/4" x 1-1/4"	6.06	C.L.F.	9.25	56.06	68.50	415.11				
092507005350	Drywall, for finishing corners, inside or outside, add	1344	L.F.	0.06	80.64	0.44	591.36				

Figure 26.61

Repair and Remodeling Estimating Methods Example

Project No.: 001
Project Name: Commercial Renovation
Location: Estimating Methods Book
Title: see Estimate
Architect: As Shown

Quantities: DEF
Take-off: ABC
Checked: GHI

Line Number	Description	Qty.	Unit	Material Unit Cost	Material Extension	Labor Unit Cost	Labor Extension	Equipment Unit Cost	Equipment Extension	Subcontract Unit Cost	Subcontract Extension
09310101001300	Ceramic tile, sanitary cove base, 6" x 4-1/4" high, thin set	252	L.F.							9.25	2331.00
09310103003400	Ceramic tile, flr,porcelain type,1 color,group 2,2"x2" or 2"x1",thin set	684	S.F.							7.85	5369.40
09310105005400	Ceramic tile, walls, interior, thin set, 4-1/4" x 4-1/4" tile	1008	S.F.							5.80	5846.40
09310101002500	Ceramic tile, bullnose trim, 4-1/4" x 4-1/4", thin set	252	L.F.							7.95	2003.40
09680080003200	Carpet, com grs, direct cem, nylon, plush, 46 oz., med to hwy traffic	129	S.Y.							33.50	4321.50
09658100007350	Resilient, vinyl composition tile, 1/8" thick, marbleized	153	S.F.							1.81	276.93
09658100001150	Resilient, base, cove, rbr or vinyl, .080" thick, std colors, 4" hi	993	L.F.							1.68	1668.24
09658100001630	Resilient, base, corners, 4" high	60	Ea.							2.37	142.20

Figure 26.62

Repair and Remodeling Estimating Methods Example

Project No.: 001
Project Name: Commercial Renovation
Location: Estimating Methods Book
Title: ase Estimate
Architect: As Shown
Quantities: DEF
Take-off: ABC
Checked: GHI

Line Number	Description	Qty.	Unit	Material Unit Cost	Material Extension	Labor Unit Cost	Labor Extension	Equipment Unit Cost	Equipment Extension	Subcontract Unit Cost	Subcontract Extension
09130100000300	Suspension systems for boards & tile, class A, 15/16" T bar, 2 x 2 grid	2837	S.F.	0.74	2099.38	0.37	1049.69				
09510700003740	Suspd acst clg bd,NO spnsn,mnrl fbr,24"x24"(48"),reveal edge,ptd,3/4"t	2837	S.F.	1.58	4482.46	0.42	1191.54				
	Lobby finishes	1	WRITTEN QUOTE							22340.00	22340.00
09651100001000	Stair treads & risers, rubber, grip strip safety tread, colors, 3/16"	192	L.F.	7.60	1459.20	1.94	372.48				
09651100001900	Stair treads & risers, rubber, risers, 7" high, 1/8" thk, coved	192	L.F.	2.35	451.20	0.93	178.56				
09651100001300	Stair treads and RS, rubber, ldgs, smooth sheet rubber, 3/16" thick	106	S.F.	4.50	486.00	1.94	209.52				
01250400000500	Factors, cut & patch to match existing construction, add, minimum		Costs	2%	47.93	3%	22.82				
09648100007600	Wood, refinish floor, sand, 2 cts poly, wax, hard wood, max	3190	S.F.	0.98	3126.20	1.44	4593.60				
09220200000100	Stucco, 3 coats 1" thick, float finish, on masonry constr, no mesh incl.	4	S.Y.	2.03	8.12						
09220200001550	Stucco, minimum labor&equipment charge	1	Job		225.00	225.00	225.00				
09220200000700	Stucco, 3 coats 1" thk, for 3/4" thk, for coloring&spcl fin, add, max	4	S.Y.	1.26	5.04						

Figure 26.63

Repair and Remodeling Estimating Methods Example

Project No.: 001
Project Name: Commercial Renovation
Location: Estimating Methods Book

Title: aee Estimate
Architect: As Shown

Quantities: DEF
Take-off: ABC
Checked: GHI

Line Number	Description	Qty.	Unit	Material Unit Cost	Material Extension	Labor Unit Cost	Labor Extension	Equipment Unit Cost	Equipment Extension	Subcontract Unit Cost	Subcontract Extension
099103201800	Flush door&frame,3'x7',varnish,3 coats,brshwk,sand after 1st coat	58	Ea.	3.47	201.26	24.00	1392.00				
099103201000	Flush door & frame, 3' x 7', oil paint, 1 coat	28	Ea.	1.97	55.16	21.50	602.00				
099109200240	Walls&clgs,conc/dry w/plas,oil base,prmr/sealer CT,sm fin,roller	6680	S.F.	0.04	267.20	0.11	734.80				
099109200840	Walls&clgs,conc,dry w/ or plas,oil base,pnt 2 coats,sm fin,roller	6680	S.F.	0.09	601.20	0.27	1803.60				
099301000010	VARNISH 1 coat + sealer, on wood trim, no sanding included	77	S.F.	0.06	4.62	0.54	41.58				
097201003300	Wall covering, vinyl, fabric-backed, mdm weight, type 2,(20-24oz/SY)	7950	S.F.	0.68	5406.00	0.45	3577.50				
099109204100	Walls&clgs,mas or conc blk, silicone,water repellent,2 coats,spray	20460	S.F.	0.24	4910.40	0.11	2250.60				
099109204100	Walls&clgs,mas or conc blk, silicone,water repellent,2 coats,spray	2100	S.F.	0.24	504.00	0.11	231.00				
097201003300	Wall covering, vinyl, fabric-backed, mdm weight, type 2,(20-24oz/SY)	580	S.F.	0.68	394.40	0.45	261.00				
095107003740	Suspd acst clg bd,NO spnsn,mnrl fbr,24"x24"(48"),reveal edge,ptd,3/4"t	486	S.F.	1.58	767.88	0.42	204.12				
091301000300	Suspension systems for boards & tile, class A, 15/16" T bar, 2' x 2' grid	486	S.F.	0.74	359.64	0.37	179.82				
096808001100	Carpet, com grs, dir cem, nylon, lvl loop, 40 oz., med to hvy traffic	54	S.Y.							37.50	2025.00
012504000550	Factors, cut & patch to match existing construction, add, maximum	5%	Costs	5%	76.10	9%	58.04				
012504000850	Factors, dust protection, add, maximum	4%	Costs	4%	60.88	11%	70.94				
012504001750	Factors, protection of existing work, add, maximum	5%	Costs	5%	76.10	7%	45.15				
012504002050	Factors, shift work requirements, add, maximum	30%	Costs	30%		30%	193.48				
	Division 9 Total				41153.28		62386.46				46324.07

Figure 26.64

Repair and Remodeling Estimating Methods Example

Project No.: 001	Title: Base Estimate	Quantities: DEF
Project Name: Tenant Estimate	Architect: As Shown	Take-off: ABC
Location: Estimating Methods Book		Checked: GHI

Line Number	Description	Qty.	Unit	Material Unit Cost	Material Extension	Labor Unit Cost	Labor Extension	Equipment Unit Cost	Equipment Extension	Subcontract Unit Cost	Subcontract Extension
09110100 2300	Metal studs ptn, 10' H, N.L.B., galv, 25 ga 3-5/8" W, 16" OC	3015	S.F.	0.19	572.85	0.50	1507.50				
09250700 0350	Gypsum plasterboard, nailed/scrd to std,1/2" thk, on wl,std,taped&fin/level	6030	S.F.	0.25	1507.50	0.50	3015.00				
09250700 5200	Drywall, for high ceilings, over 8' high, add	6030	S.F.			0.16	964.80				
01250400 0500	Factors, cut & patch to match existing construction, add, minimum		Costs	2%	30.15	3%	90.45				
01250400 2000	Factors, shift work requirements, add, minimum		Costs	5%		5%	150.75				
09270100 0400	Accessories, corner bead, galvanized steel, 1-1/4" x 1-1/4"	0.3	C.L.F.	9.25	2.78	68.5	20.55				
09250700 5350	Drywall, for finishing corners, inside or outside, add	615	L.F.	0.06	36.90	0.44	270.60				
09910320 1800	Flush door&frame,3'x7',varnish,3 coats,brshwk,sand after 1st coat	16	Ea.	3.47	55.52	24	384.00				
09910920 0240	Walls&clgs,concl/dry wl/plas,oil base,prmr/sealer CT,sm fin,roller	2610	S.F.	0.04	104.40	0.11	287.10				
09910920 0840	Walls&clgs,conc,dry wl or plas,oil base,pnt 2 coats,sm fin,roller	2610	S.F.	0.09	234.90	0.27	704.70				
09720100 3000	Wall covering, vinyl, fabric-backed, lightweight, (12-15 oz/SY)	4320	S.F.	0.55	2376.00	0.34	1468.80				
09510700 3740	Suspd acst clg bd,NO spnsn,mnrl fbr,24"x24"(48"),reveal edge,ptd,3/4"t	607	S.F.	1.58	959.06	0.42	254.94				
09130100 0300	Suspension systems for boards & tile, class A, 15/16" T bar, 2' x 2' grid	607	S.F.	0.74	449.18	0.37	224.59				
09658100 7350	Resilient, vinyl composition tile, 1/8" thick, marbleized	607	S.F.							1.81	1098.67
09658100 1150	Resilient, base, cove, rbr or vinyl, .080" thick, std colors, 4" hi	482	L.F.							1.68	809.76
09680800 4500	Carpet, com grs, dir cem, wool, 50 oz., med to hvy traffic, M loop	222	S.Y.							76.00	16872
09648100 7600	Wood, refinish floor, sand, 2 cts poly, wax, hard wood, max	418	S.F.	0.98	409.64	1.44	601.92				
	Division 9 Total				6738.88		9945.70				18780.43

Figure 26.65

It is important, when estimating drywall, to be aware of units. Studs, drywall, and taping are priced by the square foot, furring and corner bead per one hundred linear feet, and finishing corners by the linear foot. Confusion of units may result in expensive mistakes.

The estimator must be constantly aware of how the existing conditions and Factors will affect the costs. For example, the wallcovering, ceiling, and carpet in the existing retail space must be installed after business hours, and precautions are necessary. The appropriate Factors and the costs are included in Figure 26.64. The work, without Factors, would cost $4,192. With the Factors, however, the cost is $4,773, an increase of 14%.

Division 10: Specialties

Specialties are items of construction that do not fall within other divisions and are permanently attached or built into the work. The materials are usually prefinished items installed at or near the end of the project. The estimate sheet for Division 10, Figure 26.67, does not include any factors. Previous work has "accounted for" or "corrected" discrepancies caused by existing conditions, and the items are installed at areas of new work. This is not to say that factors will never be applicable. For example, bathroom accessories may be specified at existing walls. If there is no backing, the wall must be cut and patched for proper installation.

The estimator must be very careful when examining the plans and specifications to be sure to include all required items that may be shown or described only once. Sometimes these items are included in a Room Finish Schedule. If there are no plans and specifications, the estimator should refer to past estimates or to a checklist.

Figure 26.66

Quantity Comparison of Studs vs. Drywall			
Metal Studs		Drywall	
1-5/8"	7,144 S.F.	1/2"	13,292 S.F.
3-5/8" (x2)	23,550	5/8"	22,014
6" (x2)	1,120	(disregard one layer on two lower walls and ceiling)	
Furring	3,492		
(convert to S.F.)			
	35,306 S.F.		35,306 S.F.

Repair and Remodeling Estimating Methods Example

Project No.: 001
Project Name: Commercial Renovation
Location: Estimating Methods Book

Title: ase Estimate
Architect: As Shown

Quantities: DEF
Take-off: ABC
Checked: GHI

Line Number	Description	Qty.	Unit	Material Unit Cost	Material Extension	Labor Unit Cost	Labor Extension	Equipment Unit Cost	Equipment Extension	Subcontract Unit Cost	Subcontract Extension
	Division 10 Specialties										
101551002500	Partitions, toilet, cubs, flr mounted, headrail braced, painted metal	6	Ea.	365.00	2190.00	80.00	480.00				
101551002900	P'n, toilet,cubs,flr mtd,for handicap units,incl 52" grab bars,add	4	Ea.	260.00	1040.00						
101551005300	Partitions toilet, urinal screen, fl mounted, painted metal	4	Ea.	164.00	656.00	60.00	240.00				
108201003800	Bathroom accessories, mir with 5" SS shelf, 3/4" sq. FR, 72" x 24"	4	Ea.	271.00	1084.00	40.00	160.00				
108201004200	Bathroom accessories, napkin/tampon dispenser, recessed	2	Ea.	390.00	780.00	16.00	32.00				
108201004600	Bathroom accessories, soap disp, chrome, surf mounted, liquid	11	Ea.	42.50	467.50	12.00	132.00				
108201006100	Bathroom accessories, toilet tissue disp, surf mounted,sst,single roll	9	Ea.	11.75	105.75	8.00	72.00				
108201006200	Bathroom accessories, toilet tissue disp, surf mounted,sst,double roll	4	Ea.	16.45	65.80	10.00	40.00				
108201006700	Bathroom accessories, towel dispenser, sst, surf mounted	7	Ea.	38.00	266.00	15.00	105.00				
108201008000	Bathroom accessories, waste recpts, stainless steel, with top, 13 gal	4	Ea.	169.00	676.00	24.00	96.00				
108204000010	MEDICINE CABINETS With mirror, st. st. frame, 16" x 22", unlighted	3	Ea.	68.50	205.50	17.15	51.45				
105351001260	Canopie door typ alum, .032", 48" projection 8' wide residential	4	Ea.							990.00	3960.00
	Division 10 Total				7536.55		1408.45				3960.00

Figure 26.67

Division 11: Architectural Equipment

This division includes prefabricated items or items that may be built or installed by specialty subcontractors. Often, the architect arranges to have the owner purchase architectural equipment. The estimator must include installation costs, if necessary. Figure 26.68 is the estimate sheet for Divisions 11, 12, and 13. The only work in Division 11 is the installation of the kitchen unit in the third-floor lounge.

Division 12: Furnishings

Furnishings are usually purchased by the owner. In the sample project, the blinds are included in the project specifications because all windows receive the same treatment. A construction contract will most likely not include furniture, but the estimator may be asked to provide budget prices.

Division 13: Special Construction

The estimator must carefully examine and analyze work in Division 13 on an item-by-item basis, obtaining prices from appropriate specialty subcontractors. Special construction often requires preparation work, excavation, and unloading, which may not be included in the subcontract. All such requirements must be included in the estimate.

Division 14: Conveying Systems

Installation of elevators in commercial renovation is often difficult at best. The estimator and the installing subcontractor must be thoroughly aware of the existing conditions, and they must be familiar with the complete installation process. Hydraulic elevators are used most often in commercial renovation, unless the buildings are tall.

In the sample project, the price included in Figure 26.69 is a budget price only. A firm subcontract price must be obtained. Such a bid from the elevator installer will itemize what work is included and what is excluded. For the exclusions, the estimator must either determine and estimate the appropriate costs or define the exclusions in the final project bid. For example, ledge or rock encountered in drilling for a hydraulic elevator piston is very expensive. The potential extra costs should be included as a possible addendum in the prime contract.

The costs for the elevator budget price are derived from Figure 26.70. Note that there is an addition for more than two stops. Factors are applied to the budget price because access for drilling is very restricted, and the handling of the drilling waste requires special consideration. However ridiculous it may seem, forgetting to double the elevator cost to include the two elevators is a common error.

Division 15: Mechanical

When mechanical plans and specifications are provided and prepared by an engineer, takeoff and pricing may be relatively simple. If no plans are available, the work should be estimated by an experienced subcontractor who will be sure to include all requirements.

In the sample project, the plumbing work is itemized and estimated. A subcontractor will perform the work. Costs, including overhead and profit, are used. The estimator must visualize the whole system and follow the path of the piping to be sure that existing conditions will not restrict installation. If no plans are available, a quick-riser diagram will be helpful.

Estimate sheets for Division 15 are illustrated in Figures 26.72 to 26.75. In order to include all components, the estimator should make a complete list of fixtures and equipment, including required faucets, fittings, and hanging hardware. All required backing should be included in Division 6.

Repair and Remodeling Estimating Methods Example

Project No.: 001   Title: see Estimate   Quantities: DEF
Project Name: Commercial Renovation   Architect: As Shown   Take-off: ABC
Location: Estimating Methods Book   Checked: GHI

Line Number	Description	Qty.	Unit	Material Unit Cost	Material Extension	Labor Unit Cost	Labor Extension	Equipment Unit Cost	Equipment Extension	Subcontract Unit Cost	Subcontract Extension
	Division 12 Furnishings										
124921000020	Blinds, interior, horizontal, 1" al. slats, solid color, stock	924	S.F.	2.76	2550.24	0.41	378.84				
124921000020	Blinds, interior, horizontal, 1" al. slats, solid color, stock	455	S.F.	2.76	1255.80	0.41	186.55				
	Division 12 Total				3806.04		565.39				

Repair and Remodeling Estimating Methods Example

Project No.: 001   Title: see Estimate   Quantities: DEF
Project Name: Commercial Renovation   Architect: As Shown   Take-off: ABC
Location: Estimating Methods Book   Checked: GHI

Line Number	Description	Qty.	Unit	Material Unit Cost	Material Extension	Labor Unit Cost	Labor Extension	Equipment Unit Cost	Equipment Extension	Subcontract Unit Cost	Subcontract Extension
	Division 11 Equipment										
1145450001660	Appliances, comb range, refrigerator, sk, microwave, oven & ice maker	1	Ea.			715.00	715.00				
	Division 11 Total						715.00				

Figure 26.68

Repair and Remodeling Estimating Methods Example

Project No.: 001
Project Name: Commercial Renovation
Location: Estimating Methods Book

Title: ase Estimate
Architect: As Shown

Quantities: DEF
Take-off: ABC
Checked: GHI

Line Number	Description	Qty.	Unit	Material		Labor		Equipment		Subcontract	
				Unit Cost	Extension	Unit Cost	Extension	Unit Cost	Extension	Unit Cost	Extension
	Division 14 Conveying Systems										
142102002050	Elevators, hyd pass, base unit, 1500 lb, 100 fpm, 2 stop, std finishes	2	Ea.							39700.00	79400.00
142102002375	Elevators, hydraulic passenger, for number of stops over 2, add	2	Stop							3925.00	7850.00
142102002350	Elevators, hydraulic passenger, for travel over 12 V.L.F., add	19	V.L.F.							470.00	8930.00
012504001150	Factors, equipment usage curtailment, add, maximum		Costs					Average % for material and labor		13%	12503.40
012504001450	Factors, material handling & storage limitation, add, maximum		Costs								
	Division 14 Total										108683.40

Figure 26.69

			CREW	DAILY OUTPUT	LABOR-HOURS	UNIT	2002 BARE COSTS				TOTAL INCL O&P
	14210	**Electric Traction Elevators**					MAT.	LABOR	EQUIP.	TOTAL	
200 0925	For class "C-2" loading, add	R14200 -100				Ea.	4,850			4,850	5,350	**200**
0950	For class "C-3" loading, add					↓	6,675			6,675	7,325
0975	For travel over 40 V.L.F., add	R14200 -200	2 Elev	7.25	2.207	V.L.F.	97	82		179	234
1000	For number of stops over 4, add		↓	.27	59.259	Stop	1,725	2,200		3,925	5,325
1025	Hydraulic freight, base unit, 2000 lb, 50 fpm, 2 stop, std. fin.	R14200 -300	↓	.10	160	Ea.	33,400	5,925		39,325	45,900
1050	For 2500 lb capacity, add						2,425			2,425	2,675
1075	For 3000 lb capacity, add	R14200 -400					3,775			3,775	4,150
1100	For 3500 lb capacity, add						5,700			5,700	6,275
1125	For 4000 lb capacity, add						6,100			6,100	6,700
1150	For 4500 lb capacity, add						7,125			7,125	7,850
1175	For 5000 lb capacity, add						9,725			9,725	10,700
1200	For 6000 lb capacity, add						10,300			10,300	11,300
1225	For 7000 lb capacity, add						15,700			15,700	17,200
1250	For 8000 lb capacity, add						17,600			17,600	19,300
1275	For 10000 lb capacity, add						18,700			18,700	20,500
1300	For 12000 lb capacity, add						22,600			22,600	24,800
1325	For 16000 lb capacity, add						29,400			29,400	32,300
1350	For 20000 lb capacity, add						32,600			32,600	35,900
1375	For increased speed, 100 fpm, add						700			700	770
1400	125 fpm, add						1,300			1,300	1,450
1425	150 fpm, add						2,475			2,475	2,725
1450	175 fpm, add						3,850			3,850	4,225
1475	For class "B" loading, add						1,600			1,600	1,775
1500	For class "C-1" loading, add						4,025			4,025	4,425
1525	For class "C-2" loading, add						4,825			4,825	5,300
1550	For class "C-3" loading, add					↓	6,625			6,625	7,275
1575	For travel over 20 V.L.F., add		2 Elev	7.25	2.207	V.L.F.	315	82		397	470
1600	For number of stops over 2, add			.27	59.259	Stop	445	2,200		2,645	3,925
1625	Electric pass., base unit, 2000 lb, 200 fpm, 4 stop, std. fin.		↓	.05	320	Ea.	63,500	11,900		75,400	88,500
1650	For 2500 lb capacity, add						2,600			2,600	2,850
1675	For 3000 lb capacity, add						3,950			3,950	4,350
1700	For 3500 lb capacity, add						5,625			5,625	6,200
1725	For 4000 lb capacity, add						5,775			5,775	6,350
1750	For 4500 lb capacity, add						7,575			7,575	8,325
1775	For 5000 lb capacity, add						9,650			9,650	10,600
1800	For increased speed, 250 fpm, geared electric, add						2,125			2,125	2,350
1825	300 fpm, geared electric, add						4,375			4,375	4,825
1850	350 fpm, geared electric, add						5,200			5,200	5,725
1875	400 fpm, geared electric, add						7,400			7,400	8,125
1900	500 fpm, gearless electric, add						34,700			34,700	38,200
1925	600 fpm, gearless electric, add						36,700			36,700	40,300
1950	700 fpm, gearless electric, add						40,100			40,100	44,100
1975	800 fpm, gearless electric, add					↓	44,200			44,200	48,600
2000	For travel over 40 V.L.F., add		2 Elev	7.25	2.207	V.L.F.	97	82		179	234
2025	For number of stops over 4, add			.27	59.259	Stop	2,175	2,200		4,375	5,800
2050	Hyd. pass., base unit, 1500 lb, 100 fpm, 2 stop, std. fin.		↓	.10	160	Ea.	27,800	5,925		33,725	39,700
2075	For 2000 lb capacity, add						500			500	550
2100	For 2500 lb capacity, add						1,100			1,100	1,200
2125	For 3000 lb capacity, add						2,775			2,775	3,050
2150	For 3500 lb capacity, add						4,700			4,700	5,150
2175	For 4000 lb capacity, add						5,425			5,425	5,950
2200	For 4500 lb capacity, add						6,550			6,550	7,200
2225	For 5000 lb capacity, add						9,125			9,125	10,100
2250	For increased speed, 125 fpm, add						750			750	825
2275	150 fpm, add						1,625			1,625	1,775
2300	175 fpm, add		↓			↓	2,850			2,850	3,125

Figure 26.70

Costs for water supply risers and waste vent and stack only are estimated, because rough-in piping is included with the fixtures. A Factor for working around existing conditions is applied to the entire piping cost.

Because of the scope and size of the sample project, a subcontractor bids the cost of heating, ventilation, and air conditioning. The estimator must be sure that all necessary work is included in the bids. For example, since the third-floor ceiling is open to the trusses and roof deck, ductwork will be exposed. The corridors must be heated and cooled, and the bathrooms require ventilation. The subcontractor should supply shop drawings of all work for clarification. At the owner's request, tenant costs are priced separately.

Division 16: Electrical

As with the heating, ventilation, and air conditioning, the electrical work is bid by a subcontractor (Figure 26.76). For purposes of this example, the itemized estimate for the tenant electrical work is Figure 26.77.

When estimating electrical work for commercial renovation, the estimator should visualize the installation and follow the proposed paths of the wiring during the site visit. Wrapping conduit around beams and columns may become very expensive. Such restrictive conditions exist in the tenant space. The ceiling is open to the roof deck. Light fixture wiring is exposed and carried in EMT. The conduit must be bent and neatly installed over, along, and around the exposed trusses. The exterior walls are sandblasted brick. Wiremold is specified. Factors must be applied to this work for conforming to existing conditions and for the protection of existing work, as shown in Figure 26.76 and 26.77.

The wiring for switches and receptacles in the perimeter offices must be concealed, because there is no suspended ceiling. Much more wire is needed because straight runs are not possible. The estimator must check to be sure

Calculations for Elevator Budget Price					
	Material		**Labor**	**Equipment**	**Total w/ O&P**
Elevator, includes 2 stops (142-014-2050)		$27,200		$ 4,950	
Add one stop (142-014-2375)		520		1,825	
Subtotals		27,720		6,775	
Factors: 9% Material, 17% Labor	(9%)	2,495	(17%)	1,152	
Subtotals		30,215		7,927	
O&P (10% Material, 56.9% Labor)	(10%)	3,022	(56.9%)	4,510	
Totals Including Overhead & Profit		$33,236		$12,437		$45,674
$45,674 x 2 Elevators = $91,348					

Figure 26.71

Repair and Remodeling Estimating Methods Example

Project No: 001
Project Name: Commercial Renovation
Location: Estimating Methods Book
Title: see Estimate
Architect: As Shown
Quantities: DEF
Take-off: ABC
Checked: GHI

Line Number	Description	Qty.	Unit	Material		Labor		Equipment		Subcontract	
				Unit Cost	Extension	Unit Cost	Extension	Unit Cost	Extension	Unit Cost	Extension
	Division 15 Mechanical										
151401005120	Backflow pvntr, OS&Y v,&4 tst cocks, rdcd press, flg, brz, 6" pipe	1	Ea.							11700.00	11700.00
154103002100	Faucets/fittings, lavatory faucet, centerset, without drain	15	Ea.							84.50	1267.50
154103002200	Faucets/fittings, lavatory faucet, centerset, for pop-up drain, add	15	Ea.							44.50	667.50
154103003000	Faucets/fittings, svce sink faucet, cast spout, pail hook, hose end	2	Ea.							109.00	218.00
154122002820	Drinking founts, wl mtd, non-rec, dual lvl for handicapped type	2	Ea.							1250.00	2500.00
154122003980	Drinking founts, wl mtd, non-recessed, for rgh-in, sply and W, add	2	Ea.							260.00	520.00
154184500600	Lavatory, w/trim, vanity top, porc enam on CI, 20" x 18"	6	Ea.							310.00	1860.00
154184504180	Lavatory, w/trim, wall hung, porc enam on CI, 20" x 18", single bowl	9	Ea.							320.00	2880.00
154184506960	Lavatories, w/ hung,hosp type,rgh-in,sply,w&vent for abv lavatories	15	Ea.							725.00	10875.00
154184000100	Laundry sinks, W/trim, porc enam on CI, Bl FR, 24" x 23", sgl compt	2	Ea.							425.00	850.00
154184009600	Laundry sinks, with trim, rgh-in, sply, W and vent, for all laundry sks	2	Ea.							470.00	940.00

Figure 26.72

Repair and Remodeling Estimating Methods Example

Project No.: 001
Project Name: Commercial Renovation
Location: Estimating Methods Book
Title: ase Estimate
Architect: As Shown
Quantities: DEF
Take-off: ABC
Checked: GHI

Line Number	Description	Qty.	Unit	Material Unit Cost	Material Extension	Labor Unit Cost	Labor Extension	Equipment Unit Cost	Equipment Extension	Subcontract Unit Cost	Subcontract Extension
154117003000	Urinals, wall hung, vitreous china, hanger & self-closing valve	4	Ea.							585.00	2340.00
154117003300	Urinals, wall hung, vitreous china, rough-in, supply, waste & vent	4	Ea.							390.00	1560.00
154189003100	Water clo, bowl only, w/flush valve, seat, wall hung	10	Ea.							515.00	5150.00
154189003200	Water clos, bowl only, seat, for rgh-in, sply, W & vent, sgl WC	10	Ea.							615.00	6150.00
154189001100	Water clo, tank, vit china, seat, fl mtd, supply w/stop, 2 piece	3	Ea.							305.00	915.00
154189001980	Water clos, tk type, vit china, incl seat, flr mtd, for rgh-in, sply, W	3	Ea.							405.00	1215.00
154802006040	Water htrs, com, gas flr, std cont, NO vent, 75 MBH input, 73 GPH	1	Ea.							1775.00	1775.00
154409407100	Pumps, submersible, sump pump, auto, plastic, 1-1/4" discharge, 1/4 HP	2	Ea.							191.00	382.00
151553002040	Drains, flr, medium dty, C.I., D flange, 7" dia top, 2" and 3" pipe size	1	Ea.							150.00	150.00

Figure 26.73

Repair and Remodeling Estimating Methods Example

Project No.: 001
Project Name: Commercial Renovation
Location: Estimating Methods Book

Title: ase Estimate
Architect: As Shown

Quantities: DEF
Take-off: ABC
Checked: GHI

Line Number	Description	Qty.	Unit	Material Unit Cost	Material Extension	Labor Unit Cost	Labor Extension	Equipment Unit Cost	Equipment Extension	Subcontract Unit Cost	Subcontract Extension
15108520490	Pipe, plstc, DWV, cplgs 10' o.c., hgrs 3 per 10', PVC, 6" dia	132	L.F.							30.00	3960.00
15108560500	DWV, PVC, sched 40, socket joints, 1/4 bend, 6"	16	Ea.							123.00	1968.00
15108560259	DWV, PVC, sched 40, socket joints, tee, sanitary, 6"	12	Ea.							190.00	2280.00
15107420280	Pipe cu solder jts, type L tubing, cplgs & hgrs 10' o.c., 2-1/2" dia	112	L.F.							23.50	2632.00
15107420200	Pipe cu solder joints, type L tubing, cplgs & hgrs 10' o.c., 1" dia	86	L.F.							9.55	821.30
15107460350	Pipe, cu ftg, wrought, sldr jt, cu x cu, 45< elb, 2.5"	6	Ea.							82.00	492.00
15107460310	Pipe, cu ftg, wrought, sldr jt, cu x cu, 45< elb, 1"	12	Ea.							31.50	378.00
15107460550	Pipe, cu fitng, wrt, sldr jts, cu x cu, T, 2-1/2"	4	Ea.							135.00	540.00
15107460510	Pipe, cu ftg, wrought, sldr jt, cu x cu, T, 1"	7	Ea.							50.00	350.00
15107460721	Pipe, cu fntg, wrt, sldr jts, cu x cu, cplg, 2-1/2"	8	Ea.							64.00	512.00
15107460710	Pipe, cu ftg, wrought, sldr jt, cu x cu, cplg, 1"	12	Ea.							28.00	312.00
01250400550	Factors, cut & patch to match existing construction, add, maximum		Costs					Average % for material and labor		7%	997.17

Figure 26.74

Repair and Remodeling Estimating Methods Example

Project No.: 001
Project Name: Tenant Estimate
Location: Estimating Methods Book

Title: Base Estimate
Architect: As Shown

Quantities: DEF
Take-off: ABC
Checked: GHI

Line Number	Description	Qty.	Unit	Material		Labor		Equipment		Subcontract	
				Unit Cost	Extension	Unit Cost	Extension	Unit Cost	Extension	Unit Cost	Extension
	Tenant HVAC (WRITTEN QUOTE)	1								20,970.00	20970
	Division 15 Total										20970.00

Repair and Remodeling Estimating Methods Example

Project No.: 001
Project Name: Commercial Renovation
Location: Estimating Methods Book

Title: see Estimate
Architect: As Shown

Quantities: DEF
Take-off: ABC
Checked: GHI

Line Number	Description	Qty.	Unit	Material		Labor		Equipment		Subcontract	
				Unit Cost	Extension	Unit Cost	Extension	Unit Cost	Extension	Unit Cost	Extension
157000000000	Heating/Ventilating/Air Conditioning Equipment	1	WRITTEN QUOTE							270450.00	270450.00
	Division 15 Total										339607.47

Figure 26.75

Repair and Remodeling Estimating Methods Example

Project No.: 001
Project Name: Commercial Renovation
Location: Estimating Methods Book

Title: ase Estimate
Architect: As Shown

Quantities: DEF
Take-off: ABC
Checked: GHI

Line Number	Description	Qty.	Unit	Material			Labor			Equipment			Subcontract	
				Unit Cost	Extension		Unit Cost	Extension		Unit Cost	Extension		Unit Cost	Extension
	Division 16 Electrical													
160000000000	Electrical	1	WRITTEN QUOTE										183386.00	183386.00
	Division 16 Total													183386.00

Figure 26.76

Repair and Remodeling Estimating Methods Example

Project No.: 001	Title: Base Estimate	Quantities: DEF
Project Name: Tenant Estimate	Architect: As Shown	Take-off: ABC
Location: Estimating Methods Book		Checked: GHI

Line Number	Description	Qty.	Unit	Material Unit Cost	Material Extension	Labor Unit Cost	Labor Extension	Equipment Unit Cost	Equipment Extension	Subcontract Unit Cost	Subcontract Extension
161322055020	Conduit, to 15' H, incl 2 termn, 2 elb & 11 bm clp per 100', (EMT), 3/4" dia	285	L.F.							3.98	1134.3
161322055220	Conduit, field bends, 45< to 90<, 3/4" diameter	92	Ea.							5.45	501.4
161338000100	Surface raceway, metal, straight section, #500	112	L.F.							5.10	571.2
161338003000	Surface raceway, fittings, switch box, No. 500	16	Ea.							36.00	576
012504000550	Factors, cut & patch to match existing construction, add, maximum		Costs							13%	361.78
012504001750	Factors, protection of existing work, add, maximum		Costs					Average % for material and labor			
164407200850	Panelboards, NQOD, 4 wire, 120/208 volts, 225 amp main lugs, 32 circs	1	Ea.							1,900.00	1900
161201209050	Armored cable, 600 volt, copper (MC) steel clad, #12, 3 wire	4.6	C.L.F.							300.00	1380
161366000152	Steel outlet box 4" square	57	Ea.							27.00	1539
161366000300	Outlet boxes, pressed steel, square, 4", plaster rings	57	Ea.							8.10	461.7
161409100500	Wiring devices, toggle switch, quiet type, single pole, 20 amp	14	Ea.							23.50	329
161409102460	Wiring device, receptacle, duplex, 120 volt, grounded, 15 amp	59	Ea.							12.20	719.8
165104400600	Interior ltg fxtr, flour, rec mtd, acryl lens, 2'w x 4'l, four 40 watt	10	Ea.							155.00	1550
165508200100	Track lighting, track, 1 circuit, 8' section	24	Ea.							148.00	3552
165104400010	Interior lighting fixtures (WRITTEN QUOTE)	48	Ea.							125	6000
	Division 16 Total										20576.18

Figure 26.77

that the subcontractor bids the work to conform to all code requirements. Even if the plans are prepared by an electrical engineer, approval by local authorities is usually necessary, especially for fire alarms.

Estimate Summary

When the work for all divisions is priced, the estimator should complete the Project Schedule so that time-related costs in the Project Overhead Summary can be determined. When preparing the schedule, the estimator must visualize the entire construction process so that the correct sequence of work is determined. Certain tasks must be completed before others are begun. Different trades will work simultaneously. Material deliveries will affect scheduling. All such variables must be incorporated into the Project Schedule, as shown in Figure 26.78. The labor-hour figures, which have been calculated for each division, are used to assist with scheduling. The estimator must be careful not to use the labor-hours for each division independently. Each division must be coordinated with related work.

The schedule shows that the project will last approximately six months. Time-dependent items, such as equipment rental and superintendent costs, may be included in the Project Overhead Summary. Some items are dependent on total job costs, such as permits and insurance. The total direct costs for the project must be determined as shown in the Condensed Estimate Summary, Figure 26.79. All costs can now be included and totalled on the Project Overhead Summary, Figure 26.80.

The estimator is now able to complete the Estimate Summary, as shown in Figure 26.81. Appropriate contingency, sales tax, and overhead and profit costs must be added to direct costs of the project. Ten percent is added to material, equipment, and subcontractor costs for handling and supervision. The overhead and profit percentage of 66.9% for labor is obtained from the information in Figure 26.82, as the average markup for skilled workers. Contractors should determine appropriate markups for their own companies, as discussed in Chapter 6, "Indirect Costs."

As part of the requirements for the estimate, the tenant improvement costs have been separated within each appropriate division, and they are included and totalled on a separate estimate sheet (Figure 26.83). Cost per square foot is shown at the top of the Estimate Summary. The owner may use this figure to budget costs for future tenant improvements in the building. This square foot cost figure should be used with discretion. Even in the same building, different areas will require special considerations because of existing conditions.

Note that totals in Figures 26.81 and 26.83 are recorded horizontally for each division, and vertically for the different cost categories. This method of recording the numbers provides a way of cross-checking the final calculations. This is an important feature, because a mistake at this final stage of the estimate might be very costly.

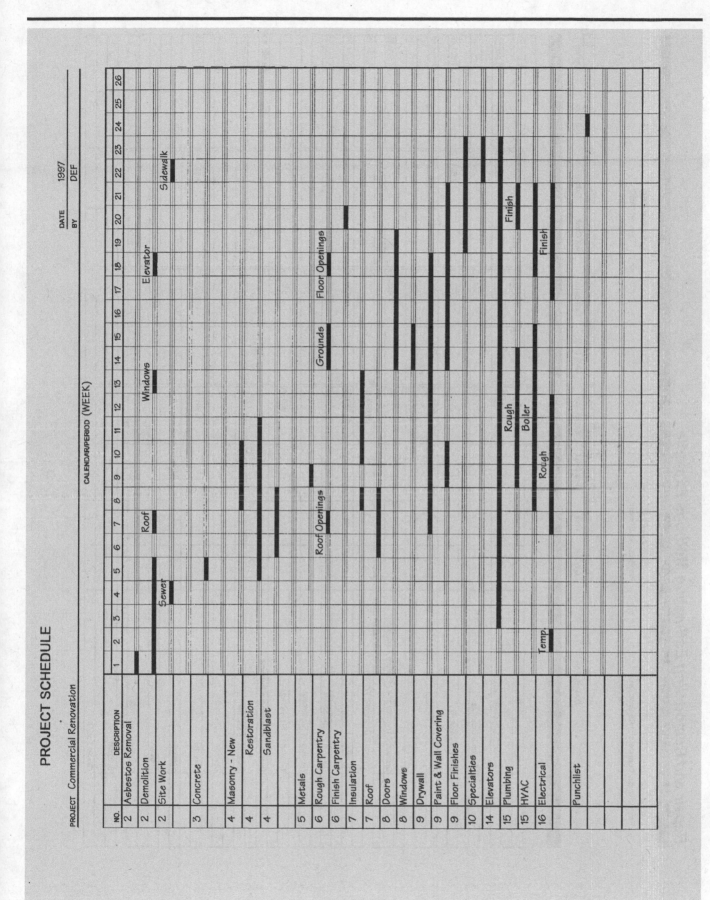

Figure 26.78

345

Repair and Remodeling Estimating Methods Example

Project No.: 001
Project Name: Commercial Renovation
Location: Estimating Methods Book

Title: Base Estimate
Architect: As Shown

Quantities: DEF.
Take-off: ABC
Checked: GHI

Line Number	Description	Qty.	Unit	Material Unit Cost	Material Extension	Labor Unit Cost	Labor Extension	Equipment Unit Cost	Equipment Extension	Subcontract Unit Cost	Subcontract Extension
	Division 1 General Requirements				4,085.10		39,992.00		94.64		7,679.20
	Division 2 Site Construction				3,671.15		30,895.49		1,609.59		5,946.20
	Division 3 Concrete				4,458.65		2,352.85		262.23		
	Division 4 Masonry				9,289.68		50,723.52		2,631.70		21,631.88
	Division 5 Metals				626.17		187.40				16,134.00
	Division 6 Wood and Plastics				4,314.78		4,231.76				
	Division 7 Thermal and Moisture Protection				7,741.29		4,648.73				11,128.85
	Division 8 Doors and Windows				28,608.68		5,796.82				69,064.96
	Division 9 Finishes				41,153.28		62,386.46				46,324.07
	Division 10 Specialties				7,536.55		1,408.45				3,960.00
	Division 11 Equipment						715.00				
	Division 12 Furnishings				3,806.04		565.39				
	Division 14 Conveying Systems										108,683.40
	Division 15 Mechanical										339,607.47
	Division 16 Electrical										183,386.00
	Subtotal				115,291.37		203,903.87		4,598.16		813,546.03
	Sales Tax			0.05							
	Subtotal										
	Overhead and Profit			0.10		0.64		0.10		0.10	
	Subtotal										
	Grand Total										

Figure 26.79

Repair and Remodeling Estimating Methods Example

Project No.: 001
Project Name: Commercial Renovation
Location: Estimating Methods Book

Title: ase Estimate
Architect: As Shown

Quantities: DEF
Take-off: ABC
Checked: GHI

Line Number	Description	Qty.	Unit	Material Unit Cost	Material Extension	Labor Unit Cost	Labor Extension	Equipment Unit Cost	Equipment Extension	Subcontract Unit Cost	Subcontract Extension
	Division 1 General Requirements										
01310700000260	Field personnel, superintendent, average	24	Week			1350.00	32400.00				
01540750000090	Scaffolding,steel tubular,bldg ext wall face,1 use per mo, 1 to 5 stories	119	C.S.F.	24.50	2915.50	30.00	3570.00				
01510800000450	Temp lighting power only, per month, max/month 23.6 KWH	280	CSF Flr							3.14	879.20
01510800000350	Temporary lighting, incl service lamps, wiring & outlets, minimum	280	CSF Flr	2.11	590.80	8.35	2338.00				
01560100001000	Barricades, guardrail, wooden, 3' high, 1" x 6", on 2" x 4" posts	270	L.F.	0.98	264.60	2.40	648.00				
02225730000800	Rubbish handling, dumpster, 30 c.y cap.(10 Tons), rent/wk, 1 dump/wk	22	Week								
01740500000100	Cleanup, floor area, final, by GC	28	M.S.F.	2.65	74.20	37.00	1036.00	3.38	94.64		
01310150000010	Permits rule of thumb, most cities, minimum		Job							.50%	6800.00
10525300002080	Fire extinguishers, dry chem, abc all purpose type, port, 9-1/2 lb.	4	Ea.	60.00	240.00						
	Division 1 Total				4085.10		39992.00		94.64		7679.20

Figure 26.80

Repair and Remodeling Estimating Methods Example

Project No.: 001
Project Name: Commercial Renovation
Location: Estimating Methods Book

Title: Base Estimate
Architect: As Shown

Quantities: DEF
Take-off: ABC
Checked: GHI

Line Number / Description	Qty.	Unit	Material Unit Cost	Material Extension	Labor Unit Cost	Labor Extension	Equipment Unit Cost	Equipment Extension	Subcontract Unit Cost	Subcontract Extension
Division 1 General Requirements										
Division 2 Site Construction				3,671.15		30,895.49		1,609.59		5,946.20
Division 3 Concrete				4,458.65		2,352.85		262.23		
Division 4 Masonry				9,289.68		50,723.52		2,631.70		21,631.88
Division 5 Metals				626.17		187.40				16,134.00
Division 6 Wood and Plastics				4,314.78		4,231.76				
Division 7 Thermal and Moisture Protection				7,741.29		4,648.73				11,128.85
Division 8 Doors and Windows				28,608.68		5,796.82				69,064.96
Division 9 Finishes				41,153.28		62,386.46				46,324.07
Division 10 Specialties				7,536.55		1,408.45				3,960.00
Division 11 Equipment						715.00				
Division 12 Furnishings				3,806.04		565.39				
Division 14 Conveying Systems										108,683.40
Division 15 Mechanical										339,607.47
Division 16 Electrical										183,386.00
Subtotal				111,206.27		163,911.87		4,503.52		805,866.83
Sales Tax			0.05							
Subtotal										
Overhead and Profit			0.10		0.64		0.10		0.10	
Subtotal										
Grand Total										

Figure 26.81

elow are the **average** installing contractor's percentage mark-ups
pplied to base labor rates to arrive at typical billing rates.

Column A: Labor rates are based on union wages averaged for 30 major
S. cities. Base rates including fringe benefits are listed hourly and
ily. These figures are the sum of the wage rate and employer-paid fringe
enefits such as vacation pay, employer-paid health and welfare costs,
ension costs, plus appropriate training and industry advancement
nds costs.

Column B: Workers' Compensation rates are the national average of
te rates established for each trade.

Column C: Column C lists average fixed overhead figures for all trades.
cluded are Federal and State Unemployment costs set at 7.0%; Social
curity Taxes (FICA) set at 7.65%; Builder's Risk Insurance costs set at
34%; and Public Liability costs set at 1.55%. All the percentages
cept those for Social Security Taxes vary from state to state as well as
om company to company.

Columns D and E: Percentages in Columns D and E are based on the
presumption that the installing contractor has annual billing of
$1,500,000 and up. Overhead percentages may increase with smaller
annual billing. The overhead percentages for any given contractor may
vary greatly and depend on a number of factors, such as the
contractor's annual volume, engineering and logistical support costs, and
staff requirements. The figures for overhead and profit will also vary
depending on the type of job, the job location, and the prevailing
economic conditions. All factors should be examined very carefully for
each job.

Column F: Column F lists the total of Columns B, C, D, and E.

Column G: Column G is Column A (hourly base labor rate) multiplied
by the percentage in Column F (O&P percentage).

Column H: Column H is the total of Column A (hourly base labor rate)
plus Column G (Total O&P).

Column I: Column I is Column H multiplied by eight hours.

		A		B	C	D	E	F		G	H		I
		Base Rate Incl. Fringes		Work- ers' Comp. Ins.	Average Fixed Over- head	Over- head	Profit	Total Overhead & Profit			Rate with O & P		
Abbr.	Trade	Hourly	Daily					%	Amount		Hourly	Daily	
kwk	Skilled Workers Average (35 trades)	$30.95	$247.60	16.8%	16.5%	16.0%	15%	64.3%	$19.90		$50.85	$406.80	
	Helpers Average (5 trades)	22.75	182.00	18.5				66.0	15.00		37.75	302.00	
	Foreman Average, Inside ($.50 over trade)	31.45	251.60	16.8				64.3	20.20		51.65	413.20	
	Foreman Average, Outside ($2.00 over trade)	32.95	263.60	16.8				64.3	21.20		54.15	433.20	
lab	Common Building Laborers	23.45	187.60	18.1				65.6	15.40		38.85	310.80	
sbe	Asbestos/Insulation Workers/Pipe Coverers	33.45	267.60	16.2				63.7	21.30		54.75	438.00	
oil	Boilermakers	36.25	290.00	14.7				62.2	22.55		58.80	470.40	
ric	Bricklayers	30.50	244.00	16.0				63.5	19.35		49.85	398.80	
rhe	Bricklayer Helpers	23.50	188.00	16.0				63.5	14.90		38.40	307.20	
arp	Carpenters	30.00	240.00	18.1				65.6	19.70		49.70	397.60	
efi	Cement Finishers	28.70	229.60	10.6				58.1	16.65		45.35	362.80	
lec	Electricians	35.45	283.60	6.7				54.2	19.20		54.65	437.20	
lev	Elevator Constructors	37.10	296.80	7.7				55.2	20.50		57.60	460.80	
ghv	Equipment Operators, Crane or Shovel	32.35	258.80	10.6				58.1	18.80		51.15	409.20	
qmd	Equipment Operators, Medium Equipment	31.20	249.60	10.6				58.1	18.15		49.35	394.80	
qlt	Equipment Operators, Light Equipment	29.80	238.40	10.6				58.1	17.30		47.10	376.80	
qol	Equipment Operators, Oilers	26.65	213.20	10.6				58.1	15.50		42.15	337.20	
qmm	Equipment Operators, Master Mechanics	32.80	262.40	10.6				58.1	19.05		51.85	414.80	
laz	Glaziers	30.00	240.00	13.8				61.3	18.40		48.40	387.20	
ath	Lathers	28.75	230.00	11.1				58.6	16.85		45.60	364.80	
arb	Marble Setters	30.10	240.80	16.0				63.5	19.10		49.20	393.60	
ill	Millwrights	31.75	254.00	10.6				58.1	18.45		50.20	401.60	
lstz	Mosaic & Terrazzo Workers	29.25	234.00	9.8				57.3	16.75		46.00	368.00	
ord	Painters, Ordinary	27.15	217.20	13.8				61.3	16.65		43.80	350.40	
sst	Painters, Structural Steel	27.90	223.20	48.4				95.9	26.75		54.65	437.20	
ape	Paper Hangers	27.10	216.80	13.8				61.3	16.60		43.70	349.60	
ile	Pile Drivers	29.80	238.40	24.9				72.4	21.60		51.40	411.20	
las	Plasterers	28.10	224.80	15.8				63.3	17.80		45.90	367.20	
lah	Plasterer Helpers	23.70	189.60	15.8				63.3	15.00		38.70	309.60	
lum	Plumbers	35.95	287.60	8.3				55.8	20.05		56.00	448.00	
odm	Rodmen (Reinforcing)	34.25	274.00	28.3				75.8	25.95		60.20	481.60	
ofc	Roofers, Composition	26.60	212.80	32.6				80.1	21.30		47.90	383.20	
ots	Roofers, Tile & Slate	26.75	214.00	32.6				80.1	21.45		48.20	385.60	
ohe	Roofers, Helpers (Composition)	19.80	158.40	32.6				80.1	15.85		35.65	285.20	
hee	Sheet Metal Workers	35.10	280.80	11.7				59.2	20.80		55.90	447.20	
pri	Sprinkler Installers	36.20	289.60	8.7				56.2	20.35		56.55	452.40	
tpi	Steamfitters or Pipefitters	36.20	289.60	8.3				55.8	20.20		56.40	451.20	
ton	Stone Masons	30.65	245.20	16.0				63.5	19.45		50.10	400.80	
swk	Structural Steel Workers	34.25	274.00	39.8				87.3	29.90		64.15	513.20	
lf	Tile Layers	29.15	233.20	9.8				57.3	16.70		45.85	366.80	
lh	Tile Layers Helpers	23.35	186.80	9.8				57.3	13.40		36.75	294.00	
rlt	Truck Drivers, Light	24.30	194.40	14.9				62.4	15.15		39.45	315.60	
rhv	Truck Drivers, Heavy	25.00	200.00	14.9				62.4	15.60		40.60	324.80	
swl	Welders, Structural Steel	34.25	274.00	39.8				87.3	29.90		64.15	513.20	
rck	*Wrecking	23.45	187.60	41.2				88.7	20.80		44.25	354.00	

ot included in averages

Figure 26.82

Repair and Remodeling Estimating Methods Example

Project No.: 001
Project Name: **Tenant Estimate**
Location: **Estimating Methods Book**

Title: **Base Estimate**
Architect: **As Shown**

Quantities: **DEF**
Take-off: **ABC**
Checked: **GHI**

Line Number / Description	Qty.	Unit	Material Unit Cost	Material Extension	Labor Unit Cost	Labor Extension	Equipment Unit Cost	Equipment Extension	Subcontract Unit Cost	Subcontract Extension
Division 1 General Requirements										
Division 2 Site Construction										
Division 3 Concrete										
Division 4 Masonry										
Division 5 Metals										
Division 6 Wood and Plastics										
Division 7 Thermal and Moisture Protection				914.34		584.28				
Division 8 Doors and Windows				3,433.92		1,091.00				
Division 9 Finishes				6,738.88		9,945.70				18,780.43
Division 10 Specialties										
Division 11 Equipment										
Division 12 Furnishings										
Division 14 Conveying Systems										20,970.00
Division 15 Mechanical										20,576.18
Division 16 Electrical										60,326.61
Subtotal				11,087.14		11,620.98				
Sales Tax			5%	554.36						
Subtotal				11,641.50		11,620.98				60,326.61
Overhead and Profit			10%	1,164.15	64.3%	7,472.29	10%		10%	6,032.66
Subtotal				12,805.65		19,093.27				66,359.27
Grand Total										98,258.19

Figure 26.83

350

Chapter 27

Assemblies Estimating Example

The Assemblies, or Systems, estimate is useful during the design development stage of a project. The estimator needs only certain parameters, and perhaps a preliminary floor plan, to complete the estimate effectively. The advantage of using an Assemblies estimate is the ability to develop costs quickly and to establish a budget before preparation of working drawings and specifications. The estimator can easily substitute one assembly for another to determine the most cost effective approach. The Assemblies estimate can be completed in much less time than the Unit Price estimate. Some accuracy is sacrificed, however, and the Assemblies estimate should be used only for budgetary purposes.

In remodeling and renovation, costs vary greatly from project to project because of different requirements and the restrictions caused by existing conditions. Budgets and cost control are becoming increasingly important before the project enters the final design process, when owners take on the expense of preparing working drawings and specifications. It is crucial that the estimator combine a thorough evaluation of the existing conditions with the design parameters in order to properly complete the Assemblies estimate. The estimator must rely on experience to be sure to include all requirements, because little information is provided. Applicable building and fire codes and local regulations must also be considered.

Prices used in the following example are from *Means Repair & Remodeling Cost Data*. This cost data book, published annually, eliminates the guesswork when it is necessary to price unknowns, by providing quick, reasonable, average prices for remodeling, renovation, and repair work. The Assemblies section contains over 2,000 costs for related assemblies, or systems. Components of the assemblies are fully detailed and accompanied by illustrations. Included also are factors used to adjust costs for restrictive existing conditions. The assemblies cost data is organized according to the 7 UNIFORMAT II divisions. Chapter 25 of *Means Repair and Remodeling Estimating* describes in detail how the cost book is organized and how the costs are derived.

The individual items in the sample estimate may not represent every item that will be found in a renovation project. However, the example does provide a basis for understanding, evaluating, and estimating commercial renovation as a whole. Using this example as a guideline for actual projects, the reader will realize that consideration must be given to all building, fire, health, and safety codes, as well as to regulations effective in a given locality.

Project Description

The sample renovation project involves the conversion of a twenty-five-year-old, two-story suburban office building into eight apartments. The owner feels that the building is not profitable as office space and wants to know how much it will cost to convert to apartments. An architect has prepared a preliminary floor plan for the ground level floor only. The estimator has the following information available: the findings of the site visit, items passed on in a few discussions, and the floor plan (shown in Figure 27.1).

The exterior of the building is concrete block, with single-pane, steel-frame windows. The roof is wood trusses with old, curling asphalt shingles. The ground floor is partially below grade. The ground floor structure is a concrete slab, and the second floor is wood joist. The requirements for the project and existing conditions will be discussed throughout the appropriate divisions.

Division A SUBSTRUCTURE

Preliminary floor plans often do not designate what is new work and what is existing. The estimator must determine the scope of work during the site visit. Footings are required for the retaining walls at the patios. Costs are determined from Figure 27.2. The footing system, as shown, is appropriate. Access for trucks, however, is restricted, so the concrete must be placed by hand. The appropriate factor for equipment usage curtailment is added. The quantities and costs are entered on the cost analysis in Figure 27.3.

Costs for the retaining wall are determined from Figure 27.4. The system, as shown, must be modified to meet the requirements of the project. The waterproofing, insulation, and anchor bolts are not needed. The costs for these items must be deducted from the complete system. Also, the wall is 6' high and must be attached to the existing building foundation. The calculations, including the appropriate factors, are shown in Figure 27.5. The costs are entered on the cost analysis (Figure 27.3).

Sections of the ground floor slab must be cut and removed to install plumbing pipes. The estimator must be sure to include the demolition costs in Division G of the Assemblies portion of *Means Repair & Remodeling Cost Data*. The system used to replace the concrete is shown in Figure 27.6. The forms are not required and are deducted, and the appropriate factors are added, as shown in Figure 27.7. The total costs are entered in Figure 27.3.

Division B SHELL

The requirements for the exterior of the building are determined through discussions with the owner.

1. Stucco over the existing concrete block
2. Insulation
3. New casement windows
4. New entrance door
5. New corridor exit doors
6. Patio doors at ground floor

These items can be easily priced, but the existing conditions will have a great effect on the work. The quantity of the stucco finish must be determined at the site. New door and window openings must be deducted. The costs for stucco are found in Figure 27.18 and are included in the estimate in Figure 27.19.

While the systems as described in *Means Repair & Remodeling Cost Data* may not conform exactly to job requirements, portions of the systems may be used as needed. The owner has requested that the building be well insulated, but he has not specified the type of insulation. That choice is left to the estimator. Figures 27.10 and 27.11 illustrate two types of exterior wall systems

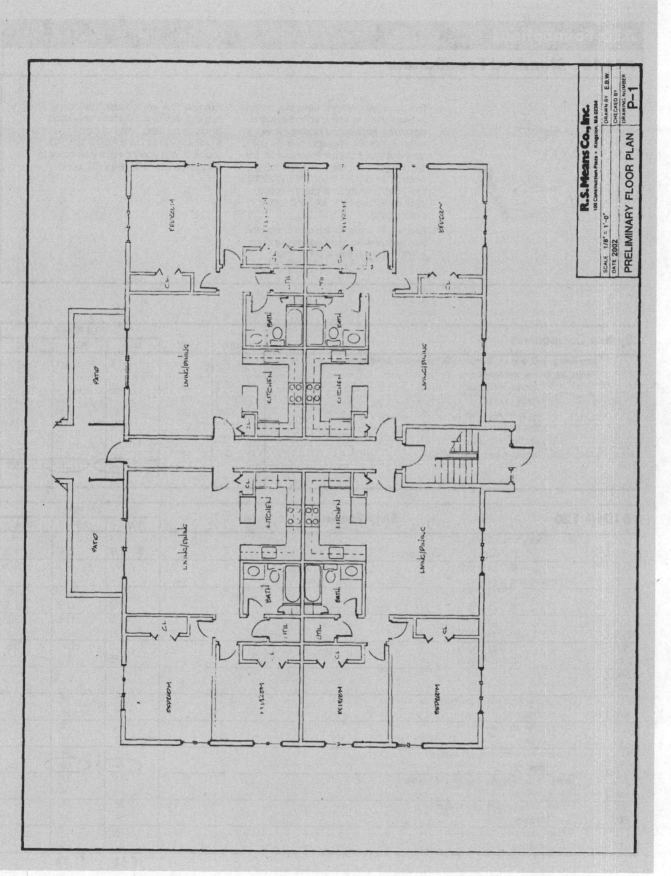

Figure 27.1

A1010   Standard Foundations

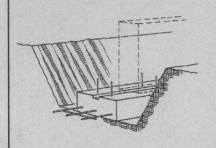

This page illustrates and describes a strip footing system including concrete, forms, reinforcing, keyway and dowels. Lines within System Components give the unit price and total price per linear foot for this system. Prices for alternate strip footing systems are on Line Items A1010 120 1500 thru 2500. Both material quantities and labor costs have been adjusted for the system listed.

Factors: To adjust for job conditions other than normal working situations use Lines A1010 120 2900 thru 4000.

Example: You are to install this footing, and due to a lack of accessibility, only hand tools can be used. Material handling is also a problem. Go to Lines A1010 120 3400 and 3600 and apply these percentages to the appropriate MAT. and INST. costs.

System Components	QUANTITY	UNIT	COST PER L.F. MAT.	COST PER L.F. INST.	COST PER L.F. TOTAL
Strip footing, 2'-0" wide x 1'-0" thick, 2000 psi concrete including forms Reinforcing, keyway, and dowels.					
Concrete, 2000 psi	.074	C.Y.	5.29		5.29
Placing concrete	.074	C.Y.		1.25	1.25
Forms, footing, 4 uses	2.000	S.F.	1.62	6.20	7.82
Reinforcing	3.170	Lb.	1.01	1.46	2.47
Keyway, 2" x 4", 4 uses	1.000	L.F.	.21	.75	.96
Dowels, #4 bars, 2' long, 24" O.C.	.500	Ea.	.23	1.01	1.24
TOTAL		L.F.	8.36	10.67	19.03

A1010 120	Strip Footing	MAT.	INST.	TOTAL
1400	Above system with the following:			
1500	2'-0" wide x 1' thick, 3000 psi concrete	8.70	10.65	19.35
1600	4000 psi concrete	9.10	10.65	19.75
1800	For alternate footing systems:			
1900	2'-6" wide x 1' thick, 2000 psi concrete	10	11.35	21.35
2000	3000 psi concrete	10.40	11.35	21.75
2100	4000 psi concrete	10.90	11.35	22.25
2300	3'-0" wide x 1' thick, 2000 psi concrete	11.45	12	23.45
2400	3000 psi concrete	11.95	12	23.95
2500	4000 psi concrete	12.55	12	24.55
2700				
2800				
2900	Cut & patch to match existing construction, add, minimum	2%	3%	
3000	Maximum	5%	9%	
3100	Dust protection, add, minimum	1%	2%	
3200	Maximum	4%	11%	
3300	Equipment usage curtailment, add, minimum	1%	1%	
3400	Maximum	3%	10%	
3500	Material handling & storage limitation, add, minimum	1%	1%	
3600	Maximum	6%	7%	
3700	Protection of existing work, add, minimum	2%	2%	
3800	Maximum	5%	7%	
3900	Shift work requirements, add, minimum		5%	
4000	Maximum		30%	
		8.61	11.74	

Figure 27.2

COST ANALYSIS

PROJECT	Apartment Renovation		ESTIMATE NO.	
ARCHITECT			DATE	2002
TAKE OFF BY: ABC	QUANTITIES BY: ABC	PRICES BY: RSM	EXTENSIONS BY: DEF	CHECKED BY GHI

DESCRIPTION	SOURCE/DIMENSIONS			QUANTITY	UNIT	MATERIAL		LABOR		EQ./TOTAL	
						UNIT COST	TOTAL	UNIT COST	TOTAL	UNIT COST	TOTAL
Division A: Substructure											
Footings	A1010	120	-	64	LF	8.61	551	11.74	751		
Retaining Wall	A2020	120	1600	64	LF	24.67	1,579	81.29	5,203		
Interior Slab	A1030	110	-	160	SF	1.43	229	1.71	274		
Division A: Totals							2,359		6,228		

Figure 27.3

A2020   Basement Walls

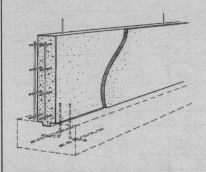

This page illustrates and describes a concrete wall system including concrete, placing concrete, forms, reinforcing, insulation, waterproofing and anchor bolts. Lines within System Components give the unit price and total price per linear foot for this system. Prices for alternate concrete wall systems are on Line Items A2020 120 1500 thru 2600. Both material quantities and labor costs have been adjusted for the system listed.

Factors: To adjust for job conditions other than normal working situations use Lines A2020 120 2900 thru 4000.

Example: You are to install this wall system where delivery of material is difficult. Go to Line A2020 120 3600 and apply these percentages to the appropriate MAT. and INST. costs.

System Components			COST PER L.F.		
	QUANTITY	UNIT	MAT.	INST.	TOTAL
Cast in place concrete foundation wall, 8" thick, 3' high, 2500 psi					
Concrete including forms, reinforcing, waterproofing, and anchor bolts.					
Concrete, 2500 psi, 8" thick, 3' high	.070	C.Y.	4.48		4.48
Forms, wall, 4 uses	6.000	S.F.	4.74	32.10	36.84
Reinforcing	6.000	Lb.	1.92	1.92	3.84
Placing concrete	.070	C.Y.		2.44	2.44
Waterproofing	3.000	S.F.	.33	2.31	2.64
Rigid insulaton, 1" polystyrene Deduct {	3.000	S.F.	.63	1.74	2.37
Anchor bolts, 1/2" diameter, 4' O.C.	.250	Ea.	.32	.53	.85
TOTAL		L.F.	12.42	41.04	53.46

A2020 120	Concrete Wall	COST PER L.F.		
		MAT.	INST.	TOTAL
1400	For alternate wall systems:			
1500	8" thick, 2500 psi concrete, 4' high	16.90	55	71.90
1600	6' high	25	82	107
1700	8' high	33.50	109	142.50
1800	3500 psi concrete, 4' high	18.30	55	73.30
1900	6' high	27.50	82	109.50
2000	8' high	36	109	145
2100	12" thick, 2500 psi concrete, 4' high	21.50	58	79.50
2200	6' high	31	86	117
2300	8' high	41	113	154
2400	3500 psi concrete, 4' high	23.50	58	81.50
2500	8' high	45.50	113	158.50
2600	10' high	56	142	198
2700				
2900	Cut & patch to match existing construction, add, minimum	2%	3%	
3000	Maximum	5%	9%	
3100	Dust protection, add, minimum	1%	2%	
3200	Maximum	4%	11%	
3300	Equipment usage curtailment, add, minimum	1%	1%	
3400	Maximum	3%	10%	
3500	Material handling & storage limitation, add, minimum	1%	1%	
3600	Maximum	6%	7%	
3700	Protection of existing work, add, minimum	2%	2%	
3800	Maximum	5%	7%	
3900	Shift work requirements, add, minimum		5%	
4000	Maximum		30%	
4100				

Figure 27.4

that have different interior insulation and finish treatments. By comparing the costs of the two treatments (and also the "R" values) the estimator is able to determine which is better for the particular application. This cost comparison is demonstrated in Figure 27.12. Note that the drywall in the second calculation is 1/2" substituted for 5/8", as shown, in order to compare the systems equally. The wood stud/fiberglass insulation system is chosen because of lower cost and higher "R" value.

By thinking ahead, the estimator should realize that the stud system will allow for easier installation of electrical receptacles and switches in the exterior walls. The furring system would require that the concrete block be chipped away at every box. This is a good example of the advantage of being able to visualize the whole job.

The doors and windows, with appropriate factors added, are taken from Figures 27.13 to 27.15. The entrance door and corridor exit door systems include panic hardware. The estimator must be sure that the hardware is as required. Figure 27.16, Hardware Selective Price Sheet, shows that the panic device is rim-type, exit only. A mortise lock for exit and entrance is required at the entrance and is substituted, as shown in Figure 27.17.

The existing roof-truss system is to remain, with some modifications. New shingles and roof trim are specified, and the owner wants 9" of fiberglass insulation installed. The costs for these items are obtained from Figures 27.18 and 27.19. Remember that the drywall ceiling will be included in Division C. The costs for new aluminum gutters and downspouts are found in the Roof Accessory Selective Price Sheet (see Figures 27.20 and 27.21). The calculations for adding the appropriate factors are shown in Figure 27.22, and the total costs for Division B of the Assemblies Section of *Means Repair & Remodeling Cost Data* are entered on Figure 27.14.

Already, the advantage of speed in Systems estimating can be seen. In a relatively short period of time, a large portion of the renovation has been estimated.

Calculations for Retaining Wall			
	Material	**Install.**	**Total**
Concrete Wall (A2020 120 1600)	$25.00	$82.00	$107.00
Deducts:			
Waterproofing	(0.33)	(2.31)	(2.64)
Insulation	(0.63)	(1.74)	(2.37)
Anchor Bolts	(0.32)	(0.53)	(0.85)
Subtotal	23.72	77.42	101.14
Factors: 4% Material, 5% Installation	0.95	3.87	4.82
Total Costs per L.F.	$24.67	$81.29	$105.96

Figure 27.5

A1030 Slab on Grade

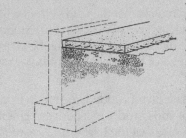

This page illustrates and describes a slab on grade system including slab, bank run gravel, bulkhead forms, placing concrete, welded wire fabric, vapor barrier, steel trowel finish and curing paper. Lines within System Components give the unit price and total price per square foot for this system. Prices for alternate slab on grade systems are on Line Items A1030 110 1500 thru 2600. Both material quantities and labor costs have been adjusted for the system listed.

Factors: To adjust for job conditions other than normal working situations use Lines A1030 110 2900 thru 4000.

Example: You are to install the system at a site where protection of the existing building is required. Go to Line A1030 110 3800 and apply these percentages to the appropriate MAT. and INST. costs.

System Components	QUANTITY	UNIT	COST PER S.F.		
			MAT.	INST.	TOTAL
Ground slab, 4″ thick, 3000 psi concrete, 4″ granular base, vapor barrier **Welded wire fabric, screed and steel trowel finish.**					
Concrete, 4″ thick, 3000 psi concrete	.012	C.Y.	.91		.91
Bank run gravel, 4″ deep	.074	C.Y.	.19	.06	.25
Polyethylene vapor barrier, 10 mil.	.011	C.S.F.	.07	.12	.19
Bulkhead forms, expansion material Deduct	.100	L.F.	.05	.25	.30
Welded wire fabric, 6 x 6 - #10/10	.011	C.S.F.	.08	.30	.38
Place concrete	.012	C.Y.		.22	.22
Screed & steel trowel finish	1.000	S.F.		.66	.66
TOTAL			1.30	1.61	2.91

A1030 110	Interior Slab on Grade	COST PER S.F.		
		MAT.	INST.	TOTAL
1400	Above system with the following:			
1500	4″ thick slab, 3000 psi concrete, 6″ deep bank run gravel	2.05	1.61	3.66
1600	12″ deep bank run gravel	2.97	1.63	4.60
1700				
1800				
1900				
2000	For alternate slab systems:			
2100	5″ thick slab, 3000 psi concrete, 6″ deep bank run gravel	2.28	1.66	3.94
2200	12″ deep bank run gravel	3.20	1.68	4.88
2300				
2400				
2500	6″ thick slab, 3000 psi concrete, 6″ deep bank run gravel	2.58	1.74	4.32
2600	12″ deep bank run gravel	3.50	1.76	5.26
2700				
2900	Cut & patch to match existing construction, add, minimum	2%	3%	
3000	Maximum	5%	9%	
3100	Dust protection, add, minimum	1%	2%	
3200	Maximum	4%	11%	
3300	Equipment usage curtailment, add, minimum	1%	1%	
3400	Maximum	3%	10%	
3500	Material handling & storage limitation, add, minimum	1%	1%	
3600	Maximum	6%	7%	
3700	Protection of existing work, add, minimum	2%	2%	
3800	Maximum	5%	7%	
3900	Shift work requirements, add, minimum		5%	
4000	Maximum		30%	

14%   26%

Figure 27.6

Division C INTERIORS

While no complete floor systems are specified for the project, note that the superstructure systems in *Means Repair & Remodeling Cost Data* (Figure 27.23) include floor and ceiling finishes. The local building code requires that apartment ceilings have a one-hour fire rating. The Ceiling Selective Price Sheet in Figure 27.24 is used to determine the costs for the rated ceiling. Painting costs are taken from Figure 27.23. The floor finishes will be included in Division C Interiors, of the Assemblies Section of *Means Repair & Remodeling Cost Data*.

During the site visit, the estimator must also determine the requirements for an exterior exit stairway from the second-floor corridor. There is no indication of this item on the plan, so it is left to the estimator to itemize the requirements and calculate costs. Figure 27.25 is used to determine the costs for the stairway. The costs are entered on the cost analysis for Division C Interiors in Figure 27.26.

In Systems estimating, the estimator must often make choices of methods and materials. Experience and a thorough evaluation of the existing conditions are important for making the correct choices.

The most important thing to remember when performing an Assemblies estimate is to be sure to include all items. Since design data is usually limited, the estimator may easily overlook items that are assumed and must be anticipated. Each of the previous five divisions have included relatively few items. In commercial renovation, interior construction encompasses a great deal of work and must be carefully planned and estimated.

The major portion of the interior construction is wood stud partitions. The following three types of partitions are required for the renovation: interior partitions within the apartments, one-hour firewalls between units and at the corridors, and furring and drywall at the existing stair enclosure. Figure 27.27 illustrates the appropriate system for the interior partitions. The insulation is included for soundproofing. The costs are entered on the estimate sheet for Division C, in Figure 27.28. The firewalls are essentially the same system as the interior partitions. The drywall, however, must be 5/8" fire resistant. Costs

Substructure			
	Material	**Install.**	**Total**
Interior Slab (A1030 110)	$ 1.30	$ 1.61	$ 2.91
Deduct Forms	(0.05)	(0.25)	(0.30)
Subtotal	1.25	1.36	2.61
Factors: 14% Materials, 26% Installation	0.18	0.35	0.54
Total Costs per S.F.	$ 1.43	$ 1.71	$ 3.15

Figure 27.7

B2020 126	Selective Price Sheet	COST PER S.F.		
		MAT.	INST.	TOTAL
0100	Exterior surface, masonry, concrete block, standard 4" thick	.99	4.21	5.20
0200	6" thick	1.46	4.53	5.99
0300	8" thick	1.59	4.83	6.42
0400	12" thick	2.32	6.25	8.57
0500	Split rib, 4" thick	2.11	5.25	7.36
0600	8" thick	3.22	6.05	9.27
0700	Brick running bond, standard size, 6.75/S.F.	2.96	8.25	11.21
0800	Buff, 6.75/S.F.	3.14	8.25	11.39
0900	Stucco, on frame	.45	3.80	4.25
1000	On masonry	.25	2.96	3.21
1100	Metal, aluminum, horizontal, plain	1.35	1.54	2.89
1200	Insulated	1.53	1.54	3.07
1300	Vertical, plain	1.19	1.54	2.73
1400	Insulated	1.24	1.61	2.85
1500	Wood, beveled siding, "A" grade cedar, 1/2" x 6"	1.82	1.59	3.41
1600	1/2" x 8"	2.09	1.45	3.54
1700	Shingles, 16" #1 red, 7-1/2" exposure	1.11	1.94	3.05
1800	18" perfections, 7-1/2" exposure	1.61	1.62	3.23
1900	Handsplit, 10" exposure	1.52	1.59	3.11
2000	White cedar, 7-1/2" exposure	.85	1.99	2.84
2100	Vertical, board & batten, redwood	3.43	1.99	5.42
2200	White pine	.74	1.45	2.19
2300	T. & G. boards, redwood, 1" x 4"	2.93	2.65	5.58
2400	1' x 8"	2.52	2.12	4.64
2500				
2600	Interior surface, drywall, taped & finished, standard, 1/2"	.28	.82	1.10
2700	5/8" thick	.29	.82	1.11
2800	Fire resistant, 1/2" thick	.29	.82	1.11
2900	5/8" thick	.29	.82	1.11
3000	Moisture resistant, 1/2" thick	.29	.82	1.11
3100	5/8" thick	.34	.82	1.16
3200	Core board, 1" thick	.50	1.66	2.16
3300	Plaster, gypsum, 2 coats	.38	1.90	2.28
3400	3 coats	.54	2.30	2.84
3500	Perlite or vermiculite, 2 coats	.40	2.17	2.57
3600	3 coats	.66	2.73	3.39
3700	Gypsum lath, standard, 3/8" thick	.43	.48	.91
3800	1/2" thick	.43	.51	.94
3900	Fire resistant, 3/8" thick	.43	.58	1.01
4000	1/2" thick	.48	.62	1.10
4100	Metal lath, diamond, 2.5 lb.	.18	.48	.66
4200	Rib, 3.4 lb.	.40	.58	.98
4300	Framing metal studs including top and bottom			
4400	Runners, walls 10' high			
4500	24" O.C., non load bearing 20 gauge, 2-1/2" wide	.22	.66	.88
4600	3-5/8" wide	.25	.67	.92
4700	4" wide	.27	.67	.94
4800	6" wide	.33	.69	1.02
4900	Load bearing 18 gauge, 2-1/2" wide	.36	.75	1.11
5000	3-5/8" wide	.43	.76	1.19
5100	4" wide	3.89	7.70	11.59
5200	6" wide	.58	.79	1.37
5300	16" O.C., non load bearing 20 gauge, 2-1/2" wide	.30	1.02	1.32
5400	3-5/8" wide	.34	1.03	1.37
5500	4" wide	.37	1.05	1.42
5600	6" wide	.45	1.06	1.51
5700	Load bearing 18 gauge, 2-1/2" wide	.49	1.04	1.53
5800	3-5/8" wide	.59	1.05	1.64

Figure 27.8

COST ANALYSIS

PROJECT	**Apartment Renovation**		SHEET NO.	
ARCHITECT			ESTIMATE NO.	
			DATE	**2002**

TAKE OFF BY: **ABC**	QUANTITIES BY: **ABC**	PRICES BY: **RSM**	EXTENSIONS BY: **DEF**	CHECKED BY **GHI**

DESCRIPTION	SOURCE/DIMENSIONS			QUANTITY	UNIT	MATERIAL		LABOR		EQ./TOTAL	
						UNIT COST	TOTAL	UNIT COST	TOTAL	UNIT COST	TOTAL
Division B: Shell											
Stucco	B2020	126	1000	3,120	SF	0.25	780	2.96	9,235		
Interior Treatment	B2010	150	-	4,420	SF	1.43	6,321	2.70	11,934		
Patio Doors	B2030	510	1100	2	EA	950	1,900	360	720		
Windows - Type I	B2020	114	1600	16	EA	653	10,448	149	2,384		
Type II	B2020	114	1700	14	EA	1,122	15,708	239	3,346		
Entrance Door	B2030	125	-	1	EA	1,493	1,493	966	966		
Corridor Doors	B2030	125	1600	2	EA	887	1,774	355	710		
Shingles & Trim	B3010	150	-	4,216	SF	0.63	2,656	1.11	4,680		
Insulation	B3010	160	2900	4,000	SF	0.66	2,640	0.35	1,400		
Downspouts	A5.9	500	0100	56	LF	1.23	69	2.48	139		
Gutters	A5.9	500	6900	164	LF	1.33	218	3.73	612		
Division B: Totals							44,008		36,127		

Figure 27.9

B2010 Exterior Walls

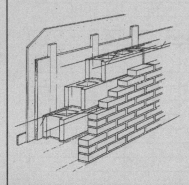

This page illustrates and describes a masonry wall, brick-stone system including brick, concrete block, durawall, insulation, plasterboard, taped and finished, furring, baseboard and painting interior. Lines within System Components give the unit price and total price per square foot for this system. Prices for alternate masonry wall, brick-stone systems are on Line Item B2010 129 1500 thru 2500. Both material quantities and labor costs have been adjusted for the system listed.

Factors: To adjust for job conditions other than normal working situations use Lines B2010 129 3100 thru 4200.

Example: You are to install the system without damaging the existing work. Go to Line B2010 129 3900 and apply these percentages to the appropriate MAT. and INST. costs.

System Components	QUANTITY	UNIT	COST PER S.F.		
			MAT.	INST.	TOTAL
Face brick, 4"thick, concrete block back-up, reinforce every second course, 3/4"insulation, furring, 1/2"drywall, taped, finish, and painted, baseboard					
Face brick, 4" brick	1.000	S.F.	2.96	8.25	11.21
Concrete back-up block, reinforced 8" thick	1.000	S.F.	1.71	4.58	6.29
3/4" rigid polystyrene insulation	1.000	S.F.	.37	.50	.87
Furring, 1" x 3", wood, 16" O.C.	1.000	L.F.	.21	.80	1.01
Drywall, 1/2" thick	1.000	S.F.	.23	.40	.63
Taping & finishing	1.000	S.F.	.04	.40	.44
Painting, 2 coats	1.000	S.F.	.15	.46	.61
Baseboard, wood, 9/16" x 2-5/8"	.100	L.F.	.13	.17	.30
Paint baseboard, primer + 1 coat enamel	.100	L.F.	.01	.04	.05
TOTAL		S.F.	5.81	15.60	21.41

B2010 129	Masonry Wall, Brick - Stone	COST PER S.F.		
		MAT.	INST.	TOTAL
1400	For alternate exterior wall systems:			
1500	Face brick, Norman, 4" x 2-2/3" x 12" (4.5 per S.F.)	6.70	13	19.70
1600	Roman, 4" x 2" x 12" (6.0 per S.F.)	7.80	14.60	22.40
1700	Engineer, 4" x 3-1/5" x 8" (5.63 per S.F.)	5.10	14.30	19.40
1800	S.C.R., 6" x 2-2/3" x 12" (4.5 per S.F.)	7.40	13.20	20.60
1900	Jumbo, 6" x 4" x 12" (3.0 per S.F.)	6.75	11.50	18.25
2000	Norwegian, 6" x 3-1/5" x 12" (3.75 per S.F.)	5.15	12.20	17.35
2100				
2200				
2300	Stone, veneer, fieldstone, 6" thick	8.40	13.25	21.65
2400	Marble, 2" thick	39.50	23.50	63
2500	Limestone, 2" thick	16.20	15.95	32.15
2600				
2700				
3100	Cut & patch to match existing construction, add, minimum	2%	3%	
3200	Maximum	5%	9%	
3300	Dust protection, add, minimum	1%	2%	
3400	Maximum	4%	11%	
3500	Equipment usage curtailment, add, minimum	1%	1%	
3600	Maximum	3%	10%	
3700	Material handling & storage limitation, add, minimum	1%	1%	
3800	Maximum	6%	7%	
3900	Protection of existing work, add, minimum	2%	2%	
4000	Maximum	5%	7%	
4100	Shift work requirements, add, minimum		5%	
4200	Maximum		30%	

Figure 27.10

B20 Exterior Enclosure

B2010 Exterior Walls

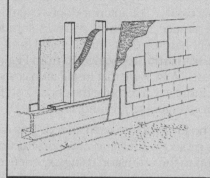

This page illustrates and describes a wood frame exterior wall system including wood studs, sheathing, felt, insulation, plasterboard, taped and finished, baseboard and painted interior. Lines within System Components give the unit price and total price per square foot for this system. Prices for alternate wood frame exterior wall systems are on Line Items B2010 150 1700 thru 2700. Both material quantities and labor costs have been adjusted for the system listed.

Factors: To adjust for job conditions other than normal working situations use Lines B2010 150 3100 thru 4200.

Example: You are to install the system with need for complete temporary bracing. Go to Line B2010 150 4200 and apply these percentages to the appropriate MAT. and INST. costs.

System Components	QUANTITY	UNIT	COST PER S.F.		
			MAT.	INST.	TOTAL
Wood stud wall, cedar shingle siding, building paper, plywood sheathing, Insulation, 5/8" drywall, taped, finished and painted, baseboard.					
2" x 4" wood studs, 16" O.C.	.100	L.F.	.45	.80	1.25
1/2" CDX sheathing	1.000	S.F.	.54	.57	1.11
18" No. 1 red cedar shingles, 7-1/2" exposure	.008	C.S.F.	1.03	1.42	2.45
15# felt paper	.010	C.S.F.	.02	.11	.13
3-1/2" fiberglass insulation	1.000	S.F.	.37	.25	.62
5/8" drywall	1.000	S.F.	.24	.40	.64
For taping and finishing joints, add	1.000	S.F.	.04	.40	.44
Baseboard trim, stock pine, 9/16" x 3-1/2", painted	.100	L.F.	.13	.17	.30
Paint baseboard, primer + 1 coat enamel	.100	L.F.	.01	.04	.05
Paint, 2 coats, interior	1.000	S.F.	.15	.46	.61
TOTAL		S.F.	2.98	4.62	7.60

B2010 150	Wood Frame Exterior Wall	COST PER S.F.		
		MAT.	INST.	TOTAL
1600	For alternate exterior wall systems:			
1700	Aluminum siding, horizontal clapboard	3.30	4.74	8.04
1800	Cedar bevel siding, 1/2" x 6", vertical , painted	3.77	4.79	8.56
1900	Redwood siding 1" x 4" to 1" x 6" vertical, T & G	5.40	5.20	10.60
2000	Board and batten	4.17	4.73	8.90
2100	Ship lap siding	4.11	4.79	8.90
2200	Plywood, grooved (T1-11) fir	2.90	4.38	7.28
2300	Redwood	4.30	4.38	8.68
2400	Southern yellow pine	2.84	4.38	7.22
2500	Masonry on stud wall, stucco, wire and plaster	2.39	7	9.39
2600	Stone veneer	7.50	9.10	16.60
2700	Brick veneer, brick $275 per M	4.91	11.45	16.36
2800				
2900				
3100	Cut & patch to match existing construction, add, minimum	2%	3%	
3200	Maximum	5%	9%	
3300	Dust protection, add, minimum	1%	2%	
3400	Maximum	4%	11%	
3500	Material handling & storage limitation, add, minimum	1%	1%	
3600	Maximum	6%	7%	
3700	Protection of existing work, add, minimum	2%	2%	
3800	Maximum	5%	7%	
3900	Shift work requirements, add, minimum		5%	
4000	Maximum		30%	
4100	Temporary shoring and bracing, add, minimum	2%	5%	
4200	Maximum	5%	12%	

Figure 27.11

for the drywall are found in Figure 27.29 and incorporated into the system in Figure 27.30. Note that the costs in Figure 27.29 are per single square foot and must be doubled when substituted into the system price.

There is no door schedule for the project. The estimator must be sure to establish the different door types, hardware sets, and quantities of each. When only one, or a "typical," floor plan is provided, it is easy to forget to multiply for repetitive floors. Cross-checking is necessary. The interior door system is found in Figure 27.31. The apartment entrance doors and stairway corridor doors must both be fire-rated, but different hardware sets are required from those, as shown in Figure 27.32. The correct hardware is selected from Figure 27.33. The calculations are shown in Figure 27.34. Note that installation costs for the hinges are already included in the installation costs for the door. Costs for the remaining door types are obtained from Figures 27.35 and 27.36.

Throughout the estimate for Division C, no factors have been added. This is because the interior construction for this project is similar to new construction, unimpeded by existing conditions. It is up to the estimator, using experience and discretion, to determine if and when factors are required for a particular project.

The estimator must be constantly aware of the project as a whole to be sure to include all requirements. The heating and cooling system is forced air. The

Comparison of Assemblies from Figures 27.10 and 27.11	Material	Install.	Total
Polystyrene Insulation (3/4″)	$0.37	$0.50	$0.87
Furring	0.21	0.80	1.01
Drywall (1/2″)	0.23	0.40	0.63
Taping	0.04	0.40	0.44
Paint	0.15	0.46	0.61
Baseboard	0.13	0.17	0.30
Paint Baseboard	0.01	0.04	0.05
Subtotal	1.14	2.77	3.91
Factors: 2% Material, 3% Installation	0.02	0.08	0.10
Total Costs per S.F. (R-Value 4)	$1.16	$2.85	$4.01
Wood Studs	$0.45	$0.80	$1.25
Fiberglass Insulation (3-1/2″)	0.37	0.25	0.62
Drywall (Price for 1/2″)	0.24	0.40	0.64
Taping	0.04	0.40	0.44
Paint	0.15	0.46	0.61
Baseboard	0.13	0.17	0.30
Paint Baseboard	0.01	0.04	0.05
Subtotal	1.39	2.52	3.91
Factors: 2% Materials, 3% Installation	0.03	0.07	
Total Costs per S.F. (R-Value 11)	$1.43	$2.70	$4.13

Figure 27.12

B20 Exterior Enclosure

B2030 Exterior Doors

This page illustrates and describes sliding door systems including a sliding door, frame, interior and exterior trim with exterior staining. Lines within System Components give the unit price and total price on a cost each basis for this system. Prices for alternate sliding door systems are on Line Items B2030 510 1100 thru 2400. Both material quantities and labor costs have been adjusted for the system listed.

Factors: To adjust for job conditions other than normal working situations use Lines B2030 510 2700 thru 4000.

Example: You are to install the system with temporary shoring and bracing. Go to Line B2030 510 3900 and apply these percentages to the appropriate MAT. and INST. costs.

System Components	QUANTITY	UNIT	COST EACH MAT.	COST EACH INST.	COST EACH TOTAL
Sliding wood door, 6'-0" x 6'-8", with wood frame, interior and exterior Trim and exterior staining.					
Sliding wood door, standard, 6'-0" x 6'-8", insulated glass	1.000	Ea.	760	199	959
Interior & exterior trim	1.000	Set	15	33.20	48.20
Stain door	1.000	Ea.	1.82	20.50	22.32
Stain trim	1.000	Ea.	2	8.80	10.80
TOTAL		Ea.	778.82	261.50	1,040.32

B2030 510	Doors, Sliding - Patio	COST EACH MAT.	COST EACH INST.	COST EACH TOTAL
1000	For alternate sliding door systems:			
1100	Wood, standard, 8'-0" x 6'-8", insulated glass	905	330	1,235
1200	12'-0" x 6'-8"	1,300	405	1,705
1300	Vinyl coated, 6'-0" x 6'-8"	1,075	262	1,337
1400	8'-0" x 6'-8"	1,250	330	1,580
1500	12'-0" x 6'-8"	1,975	405	2,380
1700	Aluminum, standard, 6'-0" x 6'-8", insulated glass	605	281	886
1800	8'-0" x 6'-8"	1,225	355	1,580
1900	12'-0" x 6'-8"	1,475	425	1,900
2000	Anodized, 6'-0" x 6'-8"	1,425	281	1,706
2100	8'-0" x 6'-8"	1,425	355	1,780
2200	12'-0" x 6'-8"	2,350	425	2,775
2300				
2400	Deduct for single glazing	50		50
2500				
2700	Cut & patch to match existing construction, add, minimum	2%	3%	
2800	Maximum	5%	9%	
2900	Dust protection, add, minimum	1%	2%	
3000	Maximum	4%	11%	
3100	Equipment usage curtailment, add, minimum	1%	1%	
3200	Maximum	3%	10%	
3300	Material handling & storage limitation, add, minimum	1%	1%	
3400	Maximum	6%	7%	
3500	Protection of existing, work, add, minimum	2%	2%	
3600	Maximum	5%	7%	
3700	Shift work requirements, add, minimum		5%	
3800	Maximum		30%	
3900	Temporary shoring and bracing, add, minimum	2%	5%	
4000	Maximum	5%	12%	
		950.25	359.70	

Figure 27.13

B2020 Exterior Windows

This page illustrates and describes a wood window system including double hung wood window, exterior and interior trim, hardware and insulating glass. Lines within System Components give the unit price and total price on a cost each basis for this system. Prices for alternate wood window systems are on Line Items B2020 114 1300 thru 2300. Both material quantities and labor costs have been adjusted for the system listed.

Factors: To adjust for job conditions other than normal working situations use Lines B2020 114 3100 thru 4000.

Example: You are to install the system during evening hours only. Go to Line B2020 114 4000 and apply this percentage to the appropriate INST. cost.

System Components			COST EACH		
	QUANTITY	UNIT	MAT.	INST.	TOTAL
Double hung wood window 2'-0" x 3'-0", exterior and interior trim, Hardware, glazed with insulating glass.					
2'-0" x 3'-0" double hung wood window, with insulating glass	1.000	Ea.	243	40	283
Exterior and interior trim	1.000	Set	16.30	30.50	46.80
Hardware	1.000	Set	1.97	16.55	18.52
TOTAL		Ea.	261.27	87.05	348.32

B2020 114	Windows - Wood		COST EACH		
			MAT.	INST.	TOTAL
1200	For alternate window systems:				
1300	Double hung, 3'-0" x 4'-0"		325	101	426
1400	4'-0" x 4'-6"		415	133	548
1500	Casement 2'-0" x 3'-0"		242	87	329
1600	2 leaf, 4'-0" x 4'-0"	Type I	640	145	785
1700	3 leaf, 6'-0" x 6'-0"	Type II	1,100	232	1,332
1800	Awning, 2'-10" x 1'-10"		256	87	343
1900	3'-6" x 2'-4"		330	101	431
2000	4'-0" x 3'-0"		500	133	633
2100	Horizontal sliding 3'-0" x 2'-0"		205	87	292
2200	4'-0" x 3'-6"		250	101	351
2300	6'-0" x 5'-0"		430	133	563
2400					
2500					
2600					
2700					
2800					
2900					
3100	Cut & patch to match existing construction, add, minimum		2%	3%	
3200	Maximum		5%	9%	
3300	Dust protection, add, minimum		1%	2%	
3400	Maximum		4%	11%	
3500	Material handling & storage limitation, add, minimum		1%	1%	
3600	Maximum		6%	7%	
3700	Protection of existing work, add, minimum		2%	2%	
3800	Maximum		5%	7%	
3900	Shift work requirements, add, minimum			5%	
4000	Maximum			30%	

Figure 27.14

B20 Exterior Enclosure

B2030 Exterior Doors

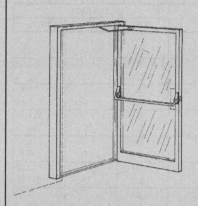

This page illustrates and describes a commercial metal door system, including a single aluminum and glass door, narrow stiles, jamb, hardware weatherstripping, panic hardware and closer. Lines within System Components give the unit price and total price on a cost each basis for this system. Prices for alternate commercial metal door systems are on Line Items B2030 125 1300 thru 2500. Both material quantities and labor costs have been adjusted for the system listed.

Factors: To adjust for job conditions other than normal working situations, use Lines B2030 125 3100 thru 4000.

Example: You are to install the system and cut and patch to match existing construction. Go to Line B2030 125 3200 and apply these percentages to the appropriate MAT. and INST. costs.

System Components	QUANTITY	UNIT	COST EACH		
			MAT.	INST.	TOTAL
Single aluminum and glass door, 3'-0"x7'-0", with narrow stiles, ext. jamb. Weatherstripping, 1/2" tempered insul. glass, panic hardware, and closer.					
Aluminum door, 3'-0" x 7'-0" x 1-3/4", narrow stiles	1.000	Ea.	480	515	995
Tempered insulating glass, 1/2" thick	20.000	S.F.	369	282	651
Panic hardware	1.000	Set	370	66.50	436.50
Automatic closer	1.000	Ea.	80	61	141
Entrance TOTAL		Ea.	1,299	924.50	2,223.50

B2030 125	Doors, Metal - Commercial		COST EACH		
			MAT.	INST.	TOTAL
1200	For alternate door systems:				
1300	Single aluminum and glass with transom, 3'-0" x 10'-0"		1,750	1,100	2,850
1400	Anodized aluminum and glass, 3'-0" x 7'-0"		1,475	1,100	2,575
1500	With transom, 3'-0" x 10'-0"		2,025	1,300	3,325
1600	Steel, deluxe, hollow metal 3'-0" x 7'-0" Corridor		870	345	1,215
1700	With transom 3'-0" x 10'-0"		1,125	390	1,515
1800	Fire door, "A" label, 3'-0" x 7'-0"		895	345	1,240
1900	Double, aluminum and glass, 6'-0" x 7'-0"		2,175	1,575	3,750
2000	With transom, 6'-0" x 10'-0"		2,400	1,950	4,350
2100	Anodized aluminum and glass 6'-0" x 7'-0"		2,450	1,850	4,300
2200	With transom, 6'-0" x 10'-0"		2,650	2,275	4,925
2300	Steel, deluxe, hollow metal, 6'-0" x 7'-0"		1,425	605	2,030
2400	With transom, 6'-0" x 10'-0"		1,925	695	2,620
2500	Fire door, "A" label, 6'-0" x 7'-0"		1,450	600	2,050
2800					
2900					
3100	Cut & patch to match existing construction, add, minimum		2%	3%	
3200	Maximum		5%	9%	
3300	Dust protection, add, minimum		1%	2%	
3400	Maximum		4%	11%	
3500	Material handling & storage limitation, add, minimum		1%	1%	
3600	Maximum		6%	7%	
3700	Protection of existing work, add, minimum		2%	2%	
3800	Maximum		5%	7%	
3900	Shift work requirements, add, minimum			5%	
4000	Maximum			30%	

Figure 27.15

B2020 128	Selective Price Sheet	COST EACH		
		MAT.	INST.	TOTAL
0100	Door closer, rack and pinion	123	61	184
0200	Backcheck and adjustable power	127	66.50	193.50
0300	Regular, hinge face mount, all sizes, regular arm	153	61	214
0400	Hold open arm	165	61	226
0500	Top jamb mount, all sizes, regular arm	153	66.50	219.50
0600	Hold open arm	165	66.50	231.50
0700	Stop face mount, all sizes, regular arm	153	61	214
0800	Hold open arm	164	61	225
0900	Fusible link, hinge face mount, all sizes, regular arm	164	61	225
1000	Hold open arm	176	61	237
1100	Top jamb mount, all sizes, regular arm	164	66.50	230.50
1200	Hold open arm	176	66.50	242.50
1300	Stop face mount, all sizes, regular arm	164	61	225
1400	Hold open arm	175	61	236
1500				
1600				
1700	Door stops			
1800				
1900	Holder & bumper, floor or wall	30	12.45	42.45
2000	Wall bumper	7.40	12.45	19.85
2100	Floor bumper	4.26	12.45	16.71
2200	Plunger type, door mounted	26	12.45	38.45
2300	Hinges, full mortise, material only, per pair			
2400	Low frequency, 4-1/2" x 4-1/2", steel base, USP	14.80		14.80
2500	Brass base, US10	34		34
2600	Stainless steel base, US32	56		56
2700	Average frequency, 4-1/2" x 4-1/2", steel base, USP	19.70		19.70
2800	Brass base, US10	40.50		40.50
2900	Stainless steel base, US32	67.50		67.50
3000	High frequency, 4-1/2" x 4-1/2", steel base, USP	41.50		41.50
3100	Brass base, US10	46		46
3200	Stainless steel base, US32	104		104
3300	Kick plate			
3400				
3500	6" high, for 3'-0" door, aluminum	15.80	26.50	42.30
3600	Bronze	22	26.50	48.50
3700	Panic device			
3800				
3900	For rim locks, single door, exit	370	66.50	436.50
4000	Outside key and pull	420	79.50	499.50
4100	Bar and vertical rod, exit only	535	79.50	614.50
4200	Outside key and pull	640	99.50	739.50
4300	Lockset			
4400				
4500	Heavy duty, cylindrical, passage doors	115	33	148
4600	Classroom	256	49.50	305.50
4700	Bedroom, bathroom, and inner office doors	146	33	179
4800	Apartment, office, and corridor doors	203	40	243
4900	Standard duty, cylindrical, exit doors	72	40	112
5100	Passage doors	50	33	83
5200	Public restroom, classroom, & office doors	99	49.50	148.50
5300	Deadlock, mortise, heavy duty	126	44	170
5400	Double cylinder	139	44	183
5500	Entrance lock, cylinder, deadlocking latch	115	44	159
5600	Deadbolt	140	49.50	189.50
5700	Commercial, mortise, wrought knob, keyed, minimum	161	49.50	210.50
5800	Maximum	297	57	354
5900	Cast knob, keyed, minimum	217	44	261

Figure 27.16

ductwork will have to be concealed by a suspended ceiling. At this point, or in Division D of the Assemblies section of *Means Repair & Remodeling Cost Data*, a rough duct layout should be made to identify areas that will require a suspended ceiling. The system to be used is shown in Figure 27.37. A factor is added because the installation is in small spaces, in the bathrooms and closets. This will entail a high waste factor for materials, as well as added labor expense.

Painting is included in the systems for all the required interior construction except for the furring and drywall system, which was "assembled." The ceilings, in Division C of the Assemblies section of *Means Repair & Remodeling Cost Data*, also included painting. The costs for painting, and for other interior finishes that may be substituted for those included in the systems, are found in Figure 27.38.

Without specifications, the estimator must also choose the type of carpeting to be used. This decision should be based on experience, or made with the advice of a floorcovering subcontractor. The costs are found in Figure 27.39.

Division D SERVICES

While no conveying systems are specified for the sample project, such work is often included in commercial renovation. Accessibility for handicapped persons has become a consideration in most Building Codes, and wheelchair lifts are becoming more common. The Assemblies pages of *Means Repair & Remodeling Cost Data* contain costs for hydraulic elevators. When other conveying systems are required, refer to the Unit Price pages of *Means Repair & Remodeling Cost Data*, as in Figure 27.40.

The mechanical systems for the sample project include the bathrooms and the HVAC systems.

A system similar to the bathrooms of the sample project is found in Figure 27.41. The water meter and controls are already in place and are not added to the costs. The calculations for the bathrooms are in Figure 27.42. The rough-in for the kitchen sink is not included in the systems price for kitchens (Division E Equipment & Furnishings) and is obtained from the Unit Price section (Division

Hardware Substitution			
	Material	**Install.**	**Total**
Exterior Door (B2030-125)	$1,299.00	$924.50	$2,223.50
Deduct Panic (B2030-125)	(370.00)	(66.50)	(436.50)
Add Panic (B2020-128-4100)	535.00	79.50	614.50
Subtotal	1,464.00	937.50	2,401.50
Factors: 2% Materials, 3% Installation	29.28	28.13	57.41
Total	$1,493.28	$965.63	$2,458.91

Figure 27.17

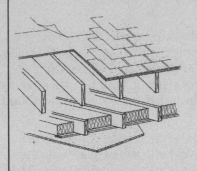

This page illustrates and describes a wood frame roof system including rafters, ceiling joists, sheathing, building paper, asphalt shingles, roof trim, furring, insulation, plaster and paint. Lines within System Components give the unit price and total price per square foot for this system. Prices for alternate wood frame roof systems are on Line Items B3010 150 1900 thru 2700. Both material quantities and labor costs have been adjusted for the system listed.

Factors: To adjust for job conditions other than normal working situations use Lines B3010 150 3300 thru 4000.

Example: You are to install the system while protecting existing work. Go to Line B3010 150 3800 and apply these percentages to the appropriate MAT. and INST. costs.

System Components	QUANTITY	UNIT	MAT.	INST.	TOTAL
Wood frame roof system, 4 in 12 pitch, including rafters, sheathing, Shingles, insulation, drywall, thin coat plaster, and painting.					
Rafters, 2" x 6", 16" O.C., 4 in 12 pitch	1.080	L.F.	.67	.86	1.53
Ceiling joists, 2" x 6", 16 O.C.	1.000	L.F.	.62	.64	1.26
Sheathing, 1/2" CDX	1.080	S.F.	.58	.62	1.20
Building paper, 15# felt	.011	C.S.F.	.03	.12	.15
Asphalt shingles, 240#	.011	C.S.F.	.45	.84	1.29
Roof trim	.100	L.F.	.15	.18	.33
Furring, 1" x 3", 16" O.C.	1.000	L.F.	.21	1.14	1.35
Fiberglass insulation, 6" batts	1.000	S.F.	.45	.29	.74
Gypsum board, 1/2" thick	1.000	S.F.	.23	.44	.67
Thin coat plaster	1.000	S.F.	.08	.50	.58
Paint, roller, 2 coats	1.000	S.F.	.15	.46	.61
TOTAL		S.F.	3.62	6.09	9.71

B3010 150	Wood Frame Roof & Ceiling	MAT.	INST.	TOTAL
1800	For alternate roof systems:			
1900	Rafters 16" O.C., 2" x 8"	3.98	6.10	10.08
2000	2" x 10"	4.43	6.55	10.98
2100	2" x 12"	4.97	6.70	11.67
2200	Rafters 24" O.C., 2" x 6"	3.30	5.70	9
2300	2" x 8"	3.56	5.70	9.26
2400	2" x 10"	3.90	6.05	9.95
2500	2" x 12"	4.30	6.15	10.45
2600	Roof pitch, 6 in 12, add	3%	10%	
2700	8 in 12, add	5%	12%	
2800				
2900				
3000				
3100				
3300	Cut & patch to match existing construction, add, minimum	2%	3%	
3400	Maximum	5%	9%	
3500	Material handling & storage limitation, add, minimum	1%	1%	
3600	Maximum	6%	7%	
3700	Protection of existing work, add, minimum	2%	2%	
3800	Maximum	5%	7%	
3900	Shift work requirements, add, minimum		5%	
4000	Maximum		30%	

Figure 27.18

B30 Roofing

B3010 Roof Coverings

B3010 160	Selective Price Sheet	COST PER S.F.		
		MAT.	INST.	TOTAL
0100	Roofing, built-up, asphalt roll roof, 3 ply organic/mineral surface	.39	1.05	1.44
0200	3 plies glass fiber felt type iv, 1 ply mineral surface	.59	1.14	1.73
0300	Cold applied, 3 ply		.39	.39
0400	Coal tar pitch, 4 ply tarred felt	1.12	1.36	2.48
0500	Mopped, 3 ply glass fiber	.93	1.50	2.43
0600	4 ply organic felt	1.12	1.36	2.48
0700	Elastomeric, hypalon, neoprene unreinforced	2.26	2.51	4.77
0800	Polyester reinforced	2.28	2.97	5.25
0900	Neoprene, 5 coats 60 mils	5	8.80	13.80
1000	Over 10,000 S.F.	4.68	4.55	9.23
1100	PVC, traffic deck sprayed	1.46	4.55	6.01
1200	With neoprene	1.55	1.84	3.39
1300	Shingles, fiber cement, strip, 14" x 30", 325#/sq.	2.68	1.74	4.42
1500	Shake, 9.35" x 16" 500#/sq.	2.43	1.74	4.17
1600				
1700	Asphalt, strip, 210-235#/sq.	.31	.70	1.01
1800	235-240#/sq.	.41	.77	1.18
1900	Class A laminated	.44	.85	1.29
2000	Class C laminated	.54	.96	1.50
2100	Slate, buckingham, 3/16" thick	5.95	2.20	8.15
2200	Black, 1/4" thick	7.90	2.20	10.10
2300	Wood, shingles, 16" no. 1, 5" exp.	1.66	1.59	3.25
2400	Red cedar, 18" perfections	1.76	1.45	3.21
2500	Shakes, 24", 10" exposure	1.52	1.59	3.11
2600	18", 8-1/2" exposure	1.07	1.99	3.06
2700	Insulation, ceiling batts, fiberglass, 3-1/2" thick, R11	.25	.25	.50
2800	6" thick, R19	.36	.29	.65
2900	9" thick, R30	.66	.35	1.01
3000	12" thick, R38	.84	.40	1.24
3100	Mineral, 3-1/2" thick, R13	.29	.25	.54
3200	Fiber, 6" thick, R19	.43	.25	.68
3300	Roof deck, fiberboard, 1" thick, R2.78	.37	.48	.85
3400	Mineral, 2" thick, R5.26	.75	.48	1.23
3500	Perlite boards, 3/4" thick, R2.08	.36	.48	.84
3600	2" thick, R5.26	.63	.55	1.18
3700	Polystyrene extruded, R5.26, 1" thick,	.25	.26	.51
3800	2" thick R10	.40	.31	.71
3900	40 PSI compressive strength, 1" thick R5	.39	.26	.65
4000	Tapered for drainage	.55	.27	.82
4100	Foamglass, 1 1/2" thick R4.55	1.61	.48	2.09
4200	3" thick R9.00	3.43	.55	3.98
4300	Ceiling, plaster, gypsum, 2 coats	.38	2.17	2.55
4400	3 coats	.54	2.55	3.09
4500	Perlite or vermiculite, 2 coats	.40	2.55	2.95
4600	3 coats	.66	3.19	3.85
4700	Gypsum lath, plain 3/8" thick	.43	.48	.91
4800	1/2" thick	.43	.51	.94
4900	Firestop, 3/8" thick	.43	.58	1.01
5000	1/2" thick	.48	.62	1.10
5100	Metal lath, rib, 2.75 lb.	.28	.54	.82
5200	3.40 lb.	.40	.58	.98
5300	Diamond, 2.50 lb.	.18	.54	.72
5400	3.40 lb.	.28	.68	.96
5500	Drywall, taped and finished standard, 1/2" thick	.28	1.04	1.32
5600	5/8" thick	.29	1.04	1.33
5700	Fire resistant, 1/2" thick	.29	1.04	1.33
5800	5/8" thick	.29	1.04	1.33
5900	Water resist., 1/2" thick	.29	1.04	1.33

Figure 27.19

B3020 240	Selective Price Sheet	COST PER L.F.		
		MAT.	INST.	TOTAL
0100	Downspouts per L.F., aluminum, enameled .024" thick, 2" x 3"	1.23	2.48	3.71
0200	3" x 4"	1.75	3.19	4.94
0300	Round .025" thick, 3" diam.	.87	2.35	3.22
0400	4" diam.	1.66	3.19	4.85
0500	Copper, round 16 oz. stock, 2" diam.	4.81	2.35	7.16
0600	3" diam.	4.47	2.35	6.82
0700	4" diam.	5.45	3.08	8.53
0800	5" diam.	7.65	3.44	11.09
0900	Rectangular, 2" x 3"	5.25	2.35	7.60
1000	3" x 4"	6.90	3.08	9.98
1100	Lead coated copper, round, 2" diam.	6.45	2.35	8.80
1200	3" diam.	5.60	2.35	7.95
1300	4" diam.	7.40	3.08	10.48
1400	5" diam.	9.25	3.44	12.69
1500	Rectangular, 2" x 3"	8.10	2.35	10.45
1600	3" x 4"	8.75	3.08	11.83
1700	Steel galvanized, round 28 gauge, 3" diam.	.80	2.35	3.15
1800	4" diam.	1.18	3.08	4.26
1900	5" diam.	1.54	3.44	4.98
2000	6" diam.	2.13	4.26	6.39
2100	Rectangular, 2" x 3"	.69	2.35	3.04
2200	3" x 4"	1.35	3.08	4.43
2300	Elbows, aluminum, round, 3" diam.	2.18	4.47	6.65
2400	4" diam.	3.29	4.47	7.76
2500	Rectangular, 2" x 3"	1	4.47	5.47
2600	3" x 4"	3.52	4.47	7.99
2700	Copper, round 16 oz., 2" diam.	12.30	4.47	16.77
2800	3" diam.	6.95	4.47	11.42
2900	4" diam.	11.75	4.47	16.22
3100	Rectangular, 2" x 3"	5.65	4.47	10.12
3200	3" x 4"	10.25	4.47	14.72
3300	Drip edge per L.F., aluminum, 5" wide	.22	.99	1.21
3400	8" wide	.33	.99	1.32
3500	28" wide	3.76	3.98	7.74
3600				
3700	Steel galvanized, 5" wide	.24	.99	1.23
3800	8" wide	.36	.99	1.35
3900				
4000				
4100				
4200				
4300	Flashing 12" wide per S.F., aluminum, mill finish, .013" thick	.39	2.64	3.03
4400	.019" thick	.85	2.64	3.49
4500	.040" thick	1.60	2.64	4.24
4600	.050" thick	2.02	2.64	4.66
4700	Copper, mill finish, 16 oz.	2.53	3.33	5.86
4800	20 oz.	4.29	3.48	7.77
4900	24 oz.	5.15	3.65	8.80
5000	32 oz.	6.85	3.83	10.68
5100	Lead, 2.5 lb./S.F., 12" wide	3.15	2.84	5.99
5200	Over 12" wide	3.58	2.84	6.42
5300	Lead-coated copper, fabric backed, 2 oz.	1.55	1.16	2.71
5400	5 oz.	1.78	1.16	2.94
5500	Mastic backed, 2 oz.	1.21	1.16	2.37
5600	5 oz.	1.50	1.16	2.66
5700	Paper backed, 2 oz.	1.05	1.16	2.21
5800	3 oz.	1.23	1.16	2.39
5900	Polyvinyl chloride, black, .010" thick	.15	1.34	1.49

Figure 27.20

B30 Roofing

B3020 Roof Openings

B3020 240	Selective Price Sheet	COST PER L.F.		
		MAT.	INST.	TOTAL
6000	.020" thick	.24	1.34	1.58
6100	.030" thick	.32	1.34	1.66
6200	.056" thick	.76	1.34	2.10
6300	Steel, galvanized, 20 gauge	.80	2.95	3.75
6400	30 gauge	.34	2.40	2.74
6500	Stainless, 32 gauge, .010" thick	2.38	2.47	4.85
6600	28 gauge, .015" thick	2.81	2.47	5.28
6700	26 gauge, .018" thick	3.58	2.47	6.05
6800	24 gauge, .025" thick	4.64	2.47	7.11
6900	Gutters per L.F., aluminum, 5" box, .027" thick	1.33	3.73	5.06
7000	.032" thick	1.28	3.73	5.01
7100	Copper, half round, 4" wide	3.80	3.73	7.53
7200	6" wide	5.55	3.89	9.44
7300	Steel, 26 gauge galvanized, 5" wide	1.05	3.73	4.78
7400	6" wide	1.55	3.73	5.28
7500	Wood, treated hem-fir, 3" x 4"	6.90	3.98	10.88
7600	4" x 5"	8	3.98	11.98
7700	Reglet per L.F., aluminum, .025" thick	1.03	1.77	2.80
7800	Copper, 10 oz.	1.78	1.77	3.55
7900	Steel, galvanized, 24 gauge	.78	1.77	2.55
8000	Stainless, .020" thick	1.76	1.77	3.53
8100	Counter flashing 12" wide per L.F., aluminum, .032" thick	1.33	2.98	4.31
8200	Copper, 10 oz.	3.73	2.98	6.71
8300	Steel, galvanized, 24 gauge	.69	2.98	3.67
8400	Stainless, .020" thick	3.10	2.98	6.08

Figure 27.21

15 of *Means Repair & Remodeling Cost Data*. Overhead and profit should be added to the bare material and labor costs.

The owner has requested that the estimate include cooling, so a ducted forced air system is the only choice. The estimator has learned from the site visit that the gas service is of adequate capacity. The system chosen is shown in Figure 27.43. Note that the flue, in Figure 27.43, is connected to an existing chimney. There are no chimneys in the existing building, so metal flues must be included in the estimate. The costs for the flues are obtained from Division 15 of the Unit Price section of *Means Repair & Remodeling Cost Data*. Overhead and profit must be added to the costs in Figure 27.44.

The mechanical and electrical work would, most likely, be performed by a subcontractor. The material and installation costs in the Assemblies pages of *Means Repair & Remodeling Cost Data* include the overhead and profit of the installing contractor. When certain work is to be done by a subcontractor, the estimator should add a percentage for supervision by the general contractor. This percentage is usually 10% and may be added within the appropriate division costs rather than at the Estimate Summary.

Since there are no electrical plans, the estimator must use experience and sound judgment to determine the requirements and quantities of the electrical work.

The existing electrical service is old and inadequate and will be removed. The project calls for each apartment to be metered separately with a 100-Amp. service.

The appropriate system, as shown in Figure 27.45, must be modified. Obviously, the building will not have eight individual service entrances and grounding systems. The weathercap, service entrance cable, and grounding system are deleted. Each apartment will have a meter socket, disconnect, and panelboard. To replace the deleted items, the estimator must determine costs for one large service entrance, meter trough, and the distribution feeders to each remote panelboard. Since this is a unique installation, the estimator can

Roofing Calculations			
	Material	**Install.**	**Total**
Shingles (B3010 150)	$0.45	$0.84	$1.29
Roof Trim (B3010 150)	0.15	0.18	0.33
Subtotal	0.60	1.02	1.62
Factors: 5% Materials, 9% Installation	0.03	0.09	0.11
Total per S.F.	$0.63	$1.11	$1.73
Insulation (B3010 160 2900)	$0.66	$0.35	$1.01
Factors: 8% Materials, 10% Installation	0.05	0.04	0.09
Total per S.F.	$0.71	$0.39	$1.10

Figure 27.22

B1010 Floor Construction

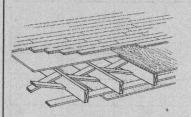

This page illustrates and describes a wood joist floor system including wood joist, oak floor, sub-floor, bridging, sand and finish floor, furring, plasterboard, taped, finished and painted ceiling. Lines within System Components give the unit price and total price per square foot for this system. Prices for alternate wood joist floor systems are on Line Items B1010 263 1700 thru 2300. Both material quantities and labor costs have been adjusted for the system listed.

Factors: To adjust for job conditions other than normal working situations use Lines B1010 263 2700 thru 4000.

Example: You are to install the system during off peak hours, 6 P.M. to 2 A.M. Go to Line B1010 263 3800 and apply this percentage to the appropriate INST. costs.

System Components	QUANTITY	UNIT	COST PER S.F. MAT.	COST PER S.F. INST.	COST PER S.F. TOTAL
Wood joists, 2" x 8", 16" O.C.,oak floor (sanded & finished),1/2" sub Floor, 1"x 3" bridging, furring, 5/8" drywall, taped, finished and painted.					
Wood joists, 2" x 8", 16" O.C.	1.000	L.F.	.95	.72	1.67
C.D.X. plywood sub floor, 1/2" thick	1.000	S.F.	.54	.53	1.07
Bridging 1" x 3"	.150	Pr.	.05	.46	.51
Oak flooring, No. 1 common	1.000	S.F.	3.85	2.34	6.19
Sand and finish floor	1.000	S.F.	.72	1.05	1.77
Wood furring, 1" x 3", 16" O.C.	1.000	L.F.	.21	1.14	1.35
Gypsum drywall, 5/8" thick	1.000	S.F.	.24	.44	.68
Tape and finishing	1.000	S.F.	.04	.40	.44
Paint ceiling	1.000	S.F.	.15	.46	.61
TOTAL		S.F.	6.75	7.54	14.29

B1010 263	Floor-Ceiling, Wood Joists	COST PER S.F. MAT.	COST PER S.F. INST.	COST PER S.F. TOTAL
1600	For alternate wood joist systems:			
1700	16" on center, 2" x 6" joists	6.40	7.45	13.85
1800	2" x 10"	7.15	7.70	14.85
1900	2" x 12"	7.65	7.75	15.40
2000	2" x 14"	7.90	7.85	15.75
2100	12" on center, 2" x 10" joists	7.50	7.90	15.40
2200	2" x 12"	8.15	7.95	16.10
2300	2" x 14"	8.45	8.10	16.55
2400				
2500				
2700	Cut & patch to match existing construction, add, minimum	2%	3%	
2800	Maximum	5%	9%	
2900	Dust protection, add, minimum	1%	2%	
3000	Maximum	4%	11%	
3100	Equipment usage curtailment, add, minimum	1%	1%	
3200	Maximum	3%	10%	
3300	Material handling & storage limitation, add, minimum	1%	1%	
3400	Maximum	6%	7%	
3500	Protection of existing work, add, minimum	2%	2%	
3600	Maximum	5%	7%	
3700	Shift work requirements, add, minimum		5%	
3800	Maximum		30%	
3900	Temporary shoring and bracing, add, minimum	2%	5%	
4000	Maximum	5%	12%	

Figure 27.23

C30 Interior Finishes

C3030 Ceiling Finishes

C3030 145	Selective Price Sheet	COST PER S.F.		
		MAT.	INST.	TOTAL
0100	Ceiling, plaster, gypsum, 2 coats	.38	2.17	2.55
0200	3 coats	.54	2.55	3.09
0300	Perlite or vermiculite, 2 coats	.40	2.55	2.95
0400	3 coats	.66	3.19	3.85
0500	Gypsum lath, plain, 3/8" thick	.43	.48	.91
0600	1/2" thick	.43	.51	.94
0700	Firestop, 3/8" thick	.43	.58	1.01
0800	1/2" thick	.48	.62	1.10
0900	Metal lath, rib, 2.75 lb.	.28	.54	.82
1000	3.40 lb.	.40	.58	.98
1100	Diamond, 2.50 lb.	.18	.54	.72
1200	3.40 lb.	.28	.68	.96
1300				
1400				
1500	Drywall, standard, 1/2" thick, no finish included	.23	.44	.67
1600	Taped and finished	.28	1.04	1.32
1700	5/8" thick, no finish included	.24	.44	.68
1800	Taped and finished	.29	1.04	1.33
1900	Fire resistant, 1/2" thick, no finish included	.24	.44	.68
2000	Taped and finished	.29	1.04	1.33
2100	5/8" thick, no finish included	.24	.44	.68
2200	Taped and finished	.29	1.04	1.33
2300	Water resistant, 1/2" thick, no finish included	.24	.44	.68
2400	Taped and finished	.29	1.04	1.33
2500	5/8" thick, no finish included	.30	.44	.74
2600	Taped and finished	.34	1.04	1.38
2700	Finish, instead of taping			
2800	For thin coat plaster, add	.08	.50	.58
2900	Finish, textured spray, add	.06	.46	.52
3000				
3100				
3200				
3300	Tile, stapled or glued, mineral fiber plastic coated, 5/8" thick	.87	.40	1.27
3400	3/4" thick	1.35	.40	1.75
3500	Wood fiber, 1/2" thick	.83	.99	1.82
3600	3/4" thick	1.12	.99	2.11
3700	Suspended, fiberglass, film faced, 5/8" thick	.57	.64	1.21
3800	3" thick	1.38	.88	2.26
3900	Mineral fiber, 5/8" thick, standard	.67	.59	1.26
4000	Aluminum	1.53	.66	2.19
4100	Wood fiber, reveal edge, 1" thick			
4200	3" thick			
4300	Framing, metal furring, 3/4" channels, 12" O.C.	.18	1.74	1.92
4400	16" O.C.	.16	1.26	1.42
4500	24" O.C.	.11	.87	.98
4600				
4700	1-1/2" channels, 12" O.C.	.27	1.92	2.19
4800	16" O.C.	.24	1.40	1.64
4900	24" O.C.	.16	.94	1.10
5000				
5100				
5200				
5300	Ceiling suspension systems, for tile,			
5400	Concealed "Z" bar, 12" module	.40	.76	1.16
5500	Class A, "T" bar 2'-0" x 4'-0" grid	.71	.50	1.21
5600	2'-0" x 2'-0" grid	.82	.61	1.43
5700	Carrier channels for lighting fixtures, add	.37	.86	1.23
5800				

Figure 27.24

C2010 Stair Construction

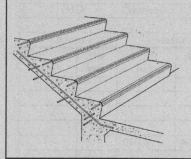

This page illustrates and describes a stair system based on a cost per flight price. Prices for various stair systems are on Line Items C2010 130 0700 thru 3200. Both material quantities and labor costs have been adjusted for the system listed.

Factors: To adjust for job conditions other than normal working situations use Lines C2010 130 3500 thru 4200.

Example: You are to install the system during evenings only. Go to Line C2010 130 4200 and apply this percentage to the appropriate MAT. and INST. costs.

System Components	QUANTITY	UNIT	COST PER FLIGHT		
			MAT.	INST.	TOTAL
Below are various stair systems based on cost per flight of stairs, no side Walls. Stairs are 4'-0" wide, railings are included unless otherwise noted					

C2010 130	Stairs	COST PER FLIGHT		
		MAT.	INST.	TOTAL
0700	Concrete, cast in place, no nosings, no railings, 12 risers	340	1,400	1,740
0800	24 risers	675	2,775	3,450
0900	Add for 1 intermediate landing	87	385	472
1000	Concrete, cast in place, with nosings, no railings, 12 risers	725	1,625	2,350
1100	24 risers	1,450	3,275	4,725
1200	Add for 1 intermediate landing	119	405	524
1300	Steel, grating tread, safety nosing, 12 risers	1,525	820	2,345
1400	24 risers	3,050	1,650	4,700
1500	Add for intermediate landing	370	218	588
1600	Steel, cement fill pan tread, 12 risers	2,275	820	3,095
1700	24 risers	4,575	1,650	6,225
1800	Add for intermediate landing	370	218	588
1900	Spiral, industrial, 4' - 6" diameter, 12 risers	2,300	585	2,885
2000	24 risers	4,575	1,175	5,750
2100	Wood, box stairs, oak treads, 12 risers	2,450	610	3,060
2200	24 risers	4,900	1,225	6,125
2300	Add for 1 intermediate landing	132	102	234
2400	Wood, basement stairs, no risers, 12 steps	695	219	914
2500	24 steps	1,375	440	1,815
2600	Add for 1 intermediate landing	24.50	19.45	43.95
2700	Wood, open, rough sawn cedar, 12 steps	995	365	1,360
2800	24 steps	2,000	730	2,730
2900	Add for 1 intermediate landing	21	19.90	40.90
3000	Wood, residential, oak treads, 12 risers	1,600	1,975	3,575
3100	24 risers	3,200	3,950	7,150
3200	Add for 1 intermediate landing	120	92	212
3500	Dust protection, add, minimum	1%	2%	
3600	Maximum	4%	11%	
3700	Material handling & storage limitation, add, minimum	1%	1%	
3800	Maximum	6%	7%	
3900	Protection of existing work, add, minimum	2%	2%	
4000	Maximum	5%	7%	
4100	Shift work requirements, add, minimum		5%	
4200	Maximum		30%	

1016   384.90

Figure 27.25

COST ANALYSIS

PROJECT	**Apartment Renovation**						SHEET NO.				
							ESTIMATE NO.				
ARCHITECT							DATE			**2002**	
TAKE OFF BY: **ABC**	QUANTITIES BY: **ABC**		PRICES BY: **RSM**	EXTENSIONS BY: **DEF**			CHECKED BY	**GHI**			

DESCRIPTION	SOURCE/DIMENSIONS			QUANTITY	UNIT	MATERIAL		LABOR		EQ./TOTAL	
						UNIT COST	TOTAL	UNIT COST	TOTAL	UNIT COST	TOTAL
Division C: Interiors											
Ceiling Drywall	C3030	145	2200	5,808	SF	0.29	1,684	1.04	6,040		
Paint	B1010	263	-	5,808	SF	0.15	871	0.46	2,672		
Stairs	C2010	130	2700	1	Flght	1,016	1,016	385	385		
Division C: Totals							3,572		9,097		

Figure 27.26

378

C1010 Partitions

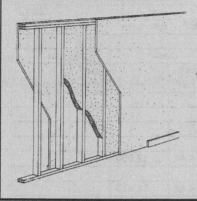

This page illustrates and describes a wood stud partition system including wood studs with plates, gypsum plasterboard – taped and finished, insulation, baseboard and painting. Lines within System Components give the unit price and total price per square foot for this system. Prices for alternate wood stud partition systems are on Line Items C1010 132 1300 thru 2700. Both material quantities and labor costs have been adjusted for the system listed.

Factors: To adjust for job conditions other than normal working situations use Lines C1010 132 2900 thru 4000.

Example: You are to install the system where material handling and storage present a serious problem. Go to Line C1010 132 3400 and apply these percentages to the appropriate MAT. and INST. costs.

System Components	QUANTITY	UNIT	COST PER S.F. MAT.	COST PER S.F. INST.	COST PER S.F. TOTAL
Wood stud wall, 2"x4", 16" O.C., dbl. top plate, sngl bot. plate, 5/8" dwl.					
Taped, finished and painted on 2 faces, insulation, baseboard, wall 8' high					
Wood studs, 2" x 4", 16" O.C., 8' high	1.000	S.F.	.49	.99	1.48
Gypsum drywall, 5/8" thick	2.000	S.F.	.48	.80	1.28
Taping and finishing	2.000	S.F.	.08	.80	.88
Insulation, 3-1/2" fiberglass batts	1.000	S.F.	.37	.25	.62
Baseboard	.200	L.F.	.31	.42	.73
Paint baseboard, primer + 2 coats	.200	L.F.	.04	.26	.30
Painting, roller, 2 coats	2.000	S.F.	.30	.92	1.22
TOTAL		S.F.	2.07	4.44	6.51

C1010 132	Partitions, Wood Stud	COST PER S.F. MAT.	COST PER S.F. INST.	COST PER S.F. TOTAL
1200	For alternate wood stud systems:			
1300	2" x 3" studs, 8' high, 16" O.C.	2.04	4.39	6.43
1400	24" O.C.	1.94	4.20	6.14
1500	10' high, 16" O.C.	2.01	4.21	6.22
1600	24" O.C.	1.91	4.05	5.96
1700	2" x 4" studs, 8' high, 24" O.C.	1.96	4.24	6.20
1800	10' high, 16" O.C.	2.03	4.25	6.28
1900	24" O.C.	1.93	4.09	6.02
2000	12' high, 16" O.C.	1.99	4.25	6.24
2100	24" O.C.	1.86	4.09	5.95
2200	2" x 6" studs, 8' high, 16" O.C.	2.35	4.56	6.91
2300	24" O.C.	2.18	4.31	6.49
2400	10' high, 16" O.C.	2.30	4.34	6.64
2500	24" O.C.	2.13	4.14	6.27
2600	12' high, 16" O.C.	2.24	4.36	6.60
2700	24" O.C.	2.07	4.16	6.23
2900	Cut & patch to match existing construction, add, minimum	2%	3%	
3000	Maximum	5%	9%	
3100	Dust protection, add, minimum	1%	2%	
3200	Maximum	4%	11%	
3300	Material handling & storage limitation, add, minimum	1%	1%	
3400	Maximum	6%	7%	
3500	Protection of existing work, add, minimum	2%	2%	
3600	Maximum	5%	7%	
3700	Shift work requirements, add, minimum		5%	
3800	Maximum		30%	
3900	Temporary shoring and bracing, add, minimum	2%	5%	
4000	Maximum	5%	12%	

Figure 27.27

COST ANALYSIS

PROJECT **Apartment Renovation**

ARCHITECT

TAKE OFF BY: **ABC** QUANTITIES BY: **ABC** PRICES BY: **RSM** EXTENSIONS BY: **DEF** CHECKED BY **GHI**

SHEET NO.

ESTIMATE NO.

DATE **2002**

DESCRIPTION	SOURCE/DIMENSIONS			QUANTITY	UNIT	MATERIAL		LABOR		EQ./TOTAL	
						UNIT COST	TOTAL	UNIT COST	TOTAL	UNIT COST	TOTAL
Division C Interiors											
Interior Partitions	C1010	132	-	7,208	SF	2.07	14,921	4.44	32,004		
Firewalls	C1010	132	-	2,584	SF	2.07	5,349	4.44	11,473		
	C3010	120	3400								
Furred Stair Walls	B2010	129	-	527	SF	0.92	485	2.88	1,518		
	C3010	210	400								
Interior Doors	C1020	106	1700	24	EA	198	4,752	181	4,344		
Apt. & Stair Doors	C1020	110	1600	10	EA	714	7,140	238	2,380		
	C3010	130	-								
Bath Closet Doors	C1020	108	1600	8	EA	247	1,976	175	1,400		
Utility Room Doors	C1020	108	1800	8	EA	289	2,312	181	1,448		
Coat Closet Doors	C1020	112	0700	8	EA	146	1,168	158	1,264		
Bdrm. Closet Door	C1020	112	0800	16	EA	233	3,728	212	3,392		
Suspended Ceiling	C3030	125	-	1,216	SF	1.40	1,702	1.31	1,593		
Painting	C3010	210	0400	527	SF	0.15	79	0.61	321		
Corridor Carpet	C3020	430	0200	304	SF	3.61	1,097	0.54	164		
Apartment Carpet	C3020	430	0500	5,824	SF	3.11	18,113	0.58	3,378		
Bath & Kit. Vinyl	C3020	430	2700	1,280	SF	2.20	2,816	1.59	2,035		
Division C Totals							65,638		66,714		

Figure 27.28

C30 Interior Finishes

C3010 Wall Finishes

C3010 120	Selective Price Sheet	COST PER S.F.		
		MAT.	INST.	TOTAL
0100	Lath, gypsum perforated			
0200				
0300	Regular, 3/8" thick	.43	.48	.91
0400	1/2" thick	.44	.58	1.02
0500	Fire resistant, 3/8" thick	.43	.58	1.01
0600	1/2" thick	.48	.62	1.10
0700	Foil back, 3/8" thick	.44	.54	.98
0800	1/2" thick	.48	.58	1.06
0900				
1000	Metal lath			
1100	Diamond painted, 2.5 lb.	.18	.48	.66
1200	3.4 lb.	.28	.54	.82
1300	Rib painted, 2.75 lb	.28	.54	.82
1400	3.40 lb	.40	.58	.98
1500				
1600				
1700	Plaster, gypsum, 2 coats	.38	1.90	2.28
1800	3 coats	.54	2.30	2.84
1900	Perlite/vermiculite, 2 coats	.40	2.17	2.57
2000	3 coats	.66	2.73	3.39
2100	Bondcrete, 1 coat	.40	.99	1.39
2200				
2500				
2600				
2700	Drywall, standard, 3/8" thick, no finish included	.21	.40	.61
2800	1/2" thick, no finish included	.23	.40	.63
2900	Taped and finished	.28	.82	1.10
3000	5/8" thick, no finish included	.24	.40	.64
3100	Taped and finished	.29	.82	1.11
3200	Fire resistant, 1/2" thick, no finish included	.24	.40	.64
3300	Taped and finished	.29	.82	1.11
3400	5/8" thick, no finish included	.24	.40	.64
3500	Taped and finished	.29	.82	1.11
3600	Water resistant, 1/2" thick, no finish included	.24	.40	.64
3700	Taped and finished	.29	.82	1.11
3800	5/8" thick, no finish included	.30	.40	.70
3900	Taped and finished	.34	.82	1.16
4000	Finish, instead of taping			
4100	For thin coat plaster, add	.08	.50	.58
4200	Finish, textured spray, add	.06	.46	.52

Figure 27.29

C1010 Partitions

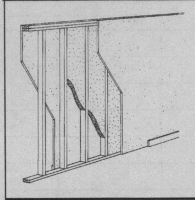

This page illustrates and describes a wood stud partition system including wood studs with plates, gypsum plasterboard – taped and finished, insulation, baseboard and painting. Lines within System Components give the unit price and total price per square foot for this system. Prices for alternate wood stud partition systems are on Line Items C1010 132 1300 thru 2700. Both material quantities and labor costs have been adjusted for the system listed.

Factors: To adjust for job conditions other than normal working situations use Lines C1010 132 2900 thru 4000.

Example: You are to install the system where material handling and storage present a serious problem. Go to Line C1010 132 3400 and apply these percentages to the appropriate MAT. and INST. costs.

System Components	QUANTITY	UNIT	COST PER S.F. MAT.	COST PER S.F. INST.	COST PER S.F. TOTAL
Wood stud wall, 2"x4", 16" O.C., dbl. top plate, sngl bot. plate, 5/8" dwl. Taped, finished and painted on 2 faces, insulation, baseboard, wall 8' high					
Wood studs, 2" x 4", 16" O.C., 8' high	1.000	S.F.	.49	.99	1.48
Gypsum drywall, 5/8" thick Fire Resistant	2.000	S.F.	.48	.80	1.28
Taping and finishing	2.000	S.F.	.08	.80	.88
Insulation, 3-1/2" fiberglass batts	1.000	S.F.	.37	.25	.62
Baseboard	.200	L.F.	.31	.42	.73
Paint baseboard, primer + 2 coats	.200	L.F.	.04	.26	.30
Painting, roller, 2 coats	2.000	S.F.	.30	.92	1.22
TOTAL		S.F.	2.07	4.44	6.51

C1010 132	Partitions, Wood Stud	COST PER S.F. MAT.	COST PER S.F. INST.	COST PER S.F. TOTAL
1200	For alternate wood stud systems:			
1300	2" x 3" studs, 8' high, 16" O.C.	2.04	4.39	6.43
1400	24" O.C.	1.94	4.20	6.14
1500	10' high, 16" O.C.	2.01	4.21	6.22
1600	24" O.C.	1.91	4.05	5.96
1700	2" x 4" studs, 8' high, 24" O.C.	1.96	4.24	6.20
1800	10' high, 16" O.C.	2.03	4.25	6.28
1900	24" O.C.	1.93	4.09	6.02
2000	12' high, 16" O.C.	1.99	4.25	6.24
2100	24" O.C.	1.86	4.09	5.95
2200	2" x 6" studs, 8' high, 16" O.C.	2.35	4.56	6.91
2300	24" O.C.	2.18	4.31	6.49
2400	10' high, 16" O.C.	2.30	4.34	6.64
2500	24" O.C.	2.13	4.14	6.27
2600	12' high, 16" O.C.	2.24	4.36	6.60
2700	24" O.C.	2.07	4.16	6.23
2900	Cut & patch to match existing construction, add, minimum	2%	3%	
3000	Maximum	5%	9%	
3100	Dust protection, add, minimum	1%	2%	
3200	Maximum	4%	11%	
3300	Material handling & storage limitation, add, minimum	1%	1%	
3400	Maximum	6%	7%	
3500	Protection of existing work, add, minimum	2%	2%	
3600	Maximum	5%	7%	
3700	Shift work requirements, add, minimum		5%	
3800	Maximum		30%	
3900	Temporary shoring and bracing, add, minimum	2%	5%	
4000	Maximum	5%	12%	

Figure 27.30

C1020 Interior Doors

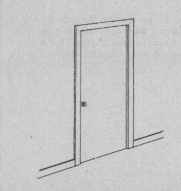

This page illustrates and describes flush interior door systems including hollow core door, jamb, header and trim with hardware. Lines within System Components give the unit price and total price on a cost each basis for this system. Prices for alternate flush interior door systems are on Line items C1020 106 1100 thru 2400. Both material quantities and labor costs have been adjusted for the system listed.

Factors: To adjust for job conditions other than normal working situations use Lines C1020 106 2700 thru 3800.

Example: You are to install the system in an area where dust protection is vital. Go to Line C1020 106 3000 and apply these percentages to the appropriate MAT. and INST. costs.

System Components	QUANTITY	UNIT	COST EACH		
			MAT.	INST.	TOTAL
Single hollow core door, include jamb, header, trim and hardware, painted.					
Hollow core Lauan, 1-3/8" thick, 2'-0" x 6'-8", painted	1.000	Ea.	28	44	72
Paint door and frame, 1 coat	1.000	Ea.	1.82	20.50	22.32
Wood jamb, 4-9/16" deep	1.000	Set	91.20	33.92	125.12
Trim, casing	1.000	Set	24	53.12	77.12
Hardware, hinges	1.000	Set	14.80		14.80
Hardware, lockset	1.000	Set	13.85	25	38.85
TOTAL		Ea.	173.67	176.54	350.21

C1020 106	Doors, Interior Flush, Wood	COST EACH		
		MAT.	INST.	TOTAL
1000	For alternate door systems:			
1100	Lauan (Mahogany) hollow core, 1-3/8" x 2'-6" x 6'-8"	181	179	360
1200	2'-8" x 6'-8"	184	181	365
1300	3'-0" x 6'-8"	189	188	377
1500	Birch, hollow core, 1-3/8" x 2'-0" x 6'-8"	186	177	363
1600	2'-6" x 6'-8"	196	179	375
1700	2'-8" x 6'-8"	198	181	379
1800	3'-0" x 6'-8"	207	188	395
1900	Solid core, pre-hung, 1-3/8" x 2'-6" x 6'-8"	325	175	500
2000	2'-8" x 6'-8"	330	177	507
2100	3'-0" x 6'-8"	340	183	523
2200				
2400	For metal frame instead of wood, add	50%	20%	
2600				
2700	Cut & patch to match existing construction, add, minimum	2%	3%	
2800	Maximum	5%	9%	
2900	Dust protection, add, minimum	1%	2%	
3000	Maximum	4%	11%	
3100	Equipment usage curtailment, add, minimum	1%	1%	
3200	Maximum	3%	10%	
3300	Material handling & storage limitation, add, minimum	1%	1%	
3400	Maximum	6%	7%	
3500	Protection of existing work, add, minimum	2%	2%	
3600	Maximum	5%	7%	
3700	Shift work requirements, add, minimum		5%	
3800	Maximum		30%	
3900				

Figure 27.31

C1020 Interior Doors

This page illustrates and describes interior metal door systems including a metal door, metal frame and hardware. Lines within System Components give the unit price and total price on a cost each for this system. Prices for alternate interior metal door systems are on Line Items C1020 110 1100 thru 2100. Both material quantities and labor costs have been adjusted for the system listed.

Factors: To adjust for job conditions other than normal working situations use Lines C1020 110 2900 thru 4000.

Example: You are to install the system while protecting existing construction. Go to Line C1020 110 3700 and apply these percentages to the appropriate MAT. and INST. costs.

System Components		QUANTITY	UNIT	COST EACH MAT.	COST EACH INST.	TOTAL
Single metal door, including frame and hardware.						
Hollow metal door, 1-3/8" thick, 2'-6" x 6'-8", painted		1.000	Ea.	210	53	263
Metal frame, 5-3/4" deep		1.000	Set	78	49.50	127.50
Paint door and frame, 1 coat		1.000	Ea.	1.82	20.50	22.32
Hardware, hinges	Deduct	1.000	Set	57.50		57.50
Hardware, passage lockset	Deduct	1.000	Set	13.85	25	38.85
	TOTAL		Ea.	361.17	148	509.17

C1020 110	Doors, Interior Flush, Metal	COST EACH MAT.	COST EACH INST.	TOTAL
1000	For alternate systems:			
1100	Hollow metal doors, 1-3/8" thick, 2'-8" x 6'-8"	365	148	513
1200	3'-0" x 7'-0"	355	156	511
1300				
1400	Interior fire door, 1-3/8" thick, 2'-6" x 6'-8"	410	148	558
1500	2'-8" x 6'-8"	410	148	558
1600	3'-0" x 7'-0"	410	156	566
1700				
1800	Add to fire doors:			
1900	Baked enamel finish	30%	15%	
2000	Galvanizing	15%		
2200				
2300				
2400				
2900	Cut & patch to match existing construction, add, minimum	2%	3%	
3000	Maximum	5%	9%	
3100	Dust protection, add, minimum	1%	2%	
3200	Maximum	4%	11%	
3300	Equipment usage curtailment, add, minimum	1%	1%	
3400	Maximum	3%	10%	
3500	Material handling & storage limitation, add, minimum	1%	1%	
3600	Maximum	6%	7%	
3700	Protection of existing work, add, minimum	2%	2%	
3800	Maximum	5%	7%	
3900	Shift work requirements, add, minimum		5%	
4000	Maximum		30%	

Figure 27.32

C30 Interior Finishes

C3010 Wall Finishes

C3010 130	Selective Price Sheet	COST EACH		
		MAT.	INST.	TOTAL
0100	Door closer, rack and pinion	123	61	184
0200	Backcheck and adjustable power	127	66.50	193.50
0300	Regular, hinge face mount, all sizes, regular arm	153	61	214
0400	Hold open arm	165	61	226
0500	Top jamb mount, all sizes, regular arm	153	66.50	219.50
0600	Hold open arm	165	66.50	231.50
0700	Stop face mount, all sizes, regular arm	153	61	214
0800	Hold open arm	164	61	225
0900	Fusible link, hinge face mount, all sizes, regular arm	164	61	225
1000	Hold open arm	176	61	237
1100	Top jamb mount, all sizes, regular arm	164	66.50	230.50
1200	Hold open arm	176	66.50	242.50
1300	Stop face mount, all sizes, regular arm	164	61	225
1400	Hold open arm	175	61	236
1500				
1600				
1700	Door stops			
1800				
1900	Holder & bumper, floor or wall	30	12.45	42.45
2000	Wall bumper	7.40	12.45	19.85
2100	Floor bumper	4.26	12.45	16.71
2200	Plunger type, door mounted	26	12.45	38.45
2300	Hinges, full mortise, material only, per pair			
2400	Low frequency, 4-1/2" x 4-1/2", steel base, USP	14.80		14.80
2500	Brass base, US10	34		34
2600	Stainless steel base, US32	56		56
2700	Average frequency, 4-1/2" x 4-1/2", steel base, USP	19.70		19.70
2800	Brass base, US10	40.50		40.50
2900	Stainless steel base, US32	67.50		67.50
3000	High frequency, 4-1/2" x 4-1/2", steel base, USP	41.50		41.50
3100	Brass base, US10	46		46
3200	Stainless steel base, US32	104		104
3300	Kick plate			
3400				
3500	6" high, for 3'-0" door, aluminum	15.80	26.50	42.30
3600	Bronze	22	26.50	48.50
3700	Panic device			
3800				
3900	For rim locks, single door, exit	370	66.50	436.50
4000	Outside key and pull	420	79.50	499.50
4100	Bar and vertical rod, exit only	535	79.50	614.50
4200	For touch bar	640	99.50	739.50
4300	Lockset			
4400				
4500	Heavy duty, cylindrical, passage doors	115	33	148
4600	Classroom	256	49.50	305.50
4700	Bedroom, bathroom, and inner office doors	146	33	179
4800	Apartment, office, and corridor doors	203	40	243
4900	Standard duty, cylindrical, exit doors	72	40	112
5000	Inner office doors			
5100	Passage doors	50	33	83
5200	Public restroom, classroom, & office doors	99	49.50	148.50
5300	Deadlock, mortise, heavy duty	126	44	170
5400	Double cylinder	139	44	183
5500	Entrance lock, cylinder, deadlocking latch	115	44	159
5600	Deadbolt	140	49.50	189.50
5700	Commercial, mortise, wrought knob, keyed, minimum	161	49.50	210.50
5800	Maximum	297	57	354

Figure 27.33

get a budget price from an electrical subcontractor, usually with only a telephone call. The costs for one grounding rod, clamp, and cable must also be included. The calculations are shown in Figure 27.46. The appropriate lighting fixtures are shown in Figures 27.47 and 27.48. The conduit is not required for this installation and is deleted from the cost.

Figure 27.49 gives wiring prices for various electrical devices using different types of wire. Note that wiring is included in costs for the lighting fixtures, so the costs for lighting wiring from Figure 27.49 are not included at this point. The cost analysis for Division D is shown in Figure 27.50. As with the mechanical costs, 10% is added for supervision of the electrical subcontractor.

Division E EQUIPMENT & FURNISHINGS

This division contains costs for various systems that are used in renovation but are not included in other divisions. The kitchen system, as shown in Figure 27.51, must be modified to meet the requirements of the sample project. A disposal unit, range, and refrigerator must be added, the compactor deleted, and the cabinets changed. Costs for these items are found in Figures 27.52a and 27.52b.

There are 19 L.F. of cabinets in each kitchen of the sample project. Since the types and sizes of the cabinets are not specified, the costs for 20 L.F. in Figure 27.51 are adjusted. The calculations for the kitchen are found in Figure 27.53. Mailboxes are to be installed. Costs for mailboxes are not found in the Systems pages, so Unit Prices from Figure 27.54 are used. The bare material and labor costs must be changed to include the installing contractor's overhead and profit. Figure 27.55 is the cost analysis for Division E.

Division G BUILDING SITEWORK

The amount of site work in renovation will vary greatly from project to project. In the sample project, the site work is relatively extensive. The scope of work is developed through discussions with the owner. Excavation is required for the retaining wall and footing in Division A of the Assemblies section of *Means Repair & Remodeling Cost Data*. The existing parking lot is to be

Calculations for Apartment and Stairway Doors			
	Material	**Install.**	**Total**
Apartment Entrance and Stair Doors (C1020 110 1600)	$410.00	$156.00	$566.00
Deduct Hardware (C1020 110)	(71.35)	(25.00)	(96.35)
Subtotal	338.65	131.00	469.65
Add:			
Closer (C3010 130 0500)	153.00	66.50	219.50
Hinges (C3010 130 2700)	19.70		19.70
Lockset (C3010 130 4800)	203.00	40.00	243.00
Total	$714.35	$237.50	$951.85

Figure 27.34

C1020 Interior Doors

This page illustrates and describes interior, solid and louvered door systems including a pine panel door, wood jambs, header, and trim with hardware. Lines within System Components give the unit price and total price on a cost each basis for this system. Prices for alternate interior, solid and louvered systems are on Line Items C1020 108 1200 thru 2400. Both material quantities and labor costs have been adjusted for the system listed.

Factors: To adjust for job conditions other than normal working situations use Lines C1020 108 2900 thru 4000.

Example: You are to install the system during night hours only. Go to Line C1020 108 4000 and apply these percentages to the appropriate INST. costs.

System Components	QUANTITY	UNIT	COST EACH		
			MAT.	INST.	TOTAL
Single interior door, including jamb, header, trim and hardware, painted.					
Solid pine panel door, 1-3/8" thick, 2'-0" x 6'-8"	1.000	Ea.	129	44	173
Paint door and frame, 1 coat	1.000	Ea.	1.82	20.50	22.32
Wooden jamb, 4-5/8" deep	1.000	Set	91.20	33.92	125.12
Trim, casing	1.000	Set	24	53.12	77.12
Hardware, hinges	1.000	Set	14.80		14.80
Hardware, lockset	1.000	Set	13.05	25	38.85
TOTAL		Ea.	274.67	176.54	451.21

C1020 108	Doors, Interior Solid & Louvered		COST EACH		
			MAT.	INST.	TOTAL
1000					
1100	For alternate door systems:				
1200	Solid pine, painted raised panel, 1-3/8" x 2'-6" x 6'-8"		294	179	473
1300	2'-8" x 6'-8"		310	181	491
1400	3'-0" x 6'-8"		320	188	508
1500					
1600	Louvered pine, painted 1'-6" x 6'-8" Bath Closet		247	175	422
1700	2'-0" x 6'-8"		276	179	455
1800	2'-6" x 6'-8" Utility Room		289	181	470
1900	3'-0" x 6'-8"		310	188	498
2200	For prehung door, deduct		5%	30%	
2300					
2400	For metal frame instead of wood, add		50%	20%	
2500					
2900	Cut & patch to match existing construction, add, minimum		2%	3%	
3000	Maximum		5%	9%	
3100	Dust protection, add, minimum		1%	2%	
3200	Maximum		4%	11%	
3300	Equipment usage curtailment, add, minimum		1%	1%	
3400	Maximum		3%	10%	
3500	Material handling & storage limitation, add, minimum		1%	1%	
3600	Maximum		6%	7%	
3700	Protection of existing work, add, minimum		2%	2%	
3800	Maximum		5%	7%	
3900	Shift work requirements, add, minimum			5%	
4000	Maximum			30%	

Figure 27.35

C1020 Interior Doors

This page illustrates and describes an interior closet door system including an interior closet door, painted, with trim and hardware. Prices for alternate interior closet door systems are on Line Items C1020 112 0500 thru 2200. Both material quantities and labor costs have been adjusted for the system listed.

Factors: To adjust for job conditions other than normal working situations use Lines C1020 112 2900 thru 4000.

Example: You are to install the system and match the existing construction. Go to Line C1020 112 2900 and apply these percentages to the appropriate MAT. and INST. costs.

C1020 112	Doors, Closet	COST PER SET		
		MAT.	INST.	TOTAL
0350	Interior closet door painted, including frame, trim and hardware, prehung.			
0400	Bi-fold doors			
0500	Pine paneled, 3'-0" x 6'-8"	245	146	391
0600	6'-0" x 6'-8"	415	196	611
0700	Birch, hollow core, 3'-0" x 6'-8" *Coat Closet*	146	158	304
0800	6'-0" x 6'-8" *Bedroom Closet*	233	212	445
0900	Lauan, hollow core, 3'-0" x 6'-8"	136	146	282
1000	6'-0" x 6'-8"	211	196	407
1100	Louvered pine, 3'-0" x 6'-8"	213	146	359
1200	6'-0" x 6'-8"	350	196	546
1300				
1400	Sliding, bi-passing closet doors			
1500	Pine paneled, 4'-0" x 6'-8"	465	155	620
1600	6'-0" x 6'-8"	570	196	766
1700	Birch, hollow core, 4'-0" x 6'-8"	288	155	443
1800	6'-0" x 6'-8"	340	196	536
1900	Lauan, hollow core, 4'-0" x 6'-8"	266	155	421
2000	6'-0" x 6'-8"	305	196	501
2100	Louvered pine, 4'-0" x 6'-8"	480	155	635
2200	6'-0" x 6'-8"	585	196	781
2300				
2400				
2500				
2600				
2700				
2800				
2900	Cut & patch to match existing construction, add, minimum	2%	3%	
3000	Maximum	5%	9%	
3100	Dust protection, add, minimum	1%	2%	
3200	Maximum	4%	11%	
3300	Equipment usage curtailment, add, minimum	1%	1%	
3400	Maximum	3%	10%	
3500	Material handling & storage limitation, add, minimum	1%	1%	
3600	Maximum	6%	7%	
3700	Protection of existing work, add, minimum	2%	2%	
3800	Maximum	5%	7%	
3900	Shift work requirements, add, minimum		5%	
4000	Maximum		30%	

Figure 27.36

C3030 Ceiling Finishes

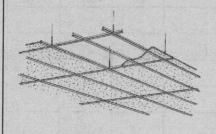

This page illustrates suspended acoustical board systems including acoustic ceiling board, hangers, and T bar suspension. Lines within System Components give the unit price and total price per square foot for this system. Prices for alternate suspended acoustical board systems are on Line Items C3030 125 1100 thru 2400. Both material quantities and labor costs have been adjusted for the system listed.

Factors: To adjust for job conditions other than normal working situations use Lines C3030 125 2900 thru 4000.

Example: You are to install the system and protect existing construction. Go to Line C3030 125 3800 and apply these percentages to the appropriate MAT. and INST. costs.

System Components	QUANTITY	UNIT	COST PER S.F.		
			MAT.	INST.	TOTAL
Suspended acoustical ceiling board installed on exposed grid system.					
Fiberglass boards, film faced, 2' x 4', 5/8" thick	1.000	S.F.	.57	.64	1.21
Hangers, #12 wire	1.000	S.F.	.05	.06	.11
T bar suspension system, 2' x 4' grid	1.000	S.F.	.71	.50	1.21
TOTAL		S.F.	1.33	1.20	2.53

C3030 125	Ceiling, Suspended Acoustical	COST PER S.F.		
		MAT.	INST.	TOTAL
1000	For alternate suspended ceiling systems:			
1100	2' x 4' grid, mineral fiber board, aluminum faced, 5/8" thick	1.76	1.17	2.93
1200	Standard faced	1.43	1.15	2.58
1300	Plastic faced	1.81	1.55	3.36
1400	Fiberglass, film faced, 3" thick, R11	2.14	1.44	3.58
1500	Grass cloth faced, 3/4" thick	2.61	1.36	3.97
1600	1" thick	2.66	1.38	4.04
1700	1-1/2" thick, nubby face	3.19	1.40	4.59
2200				
2300				
2400	Add for 2' x 2' grid system	.16	.12	.28
2500				
2600				
2700				
2900	Cut & patch to match existing construction, add, minimum	2%	3%	
3000	Maximum	5%	9%	
3100	Dust protection, add, minimum	1%	2%	
3200	Maximum	4%	11%	
3300	Equipment usage curtailment, add, minimum	1%	1%	
3400	Maximum	3%	10%	
3500	Material handling & storage limitation, add, minimum	1%	1%	
3600	Maximum	6%	7%	
3700	Protection of existing work, add, minimum	2%	2%	
3800	Maximum	5%	7%	
3900	Shift work requirements, add, minimum		5%	
4000	Maximum		30%	
		1.40	1.31	

Figure 27.37

C30 Interior Finishes

C3010 Wall Finishes

C3010 210	Selective Price Sheet	COST PER S.F.		
		MAT.	INST.	TOTAL
0100	Painting, on plaster or drywall, brushwork, primer and 1 ct.	.10	.52	.62
0200	Primer and 2 ct.	.15	.82	.97
0300	Rollerwork, primer and 1 ct.	.10	.44	.54
0400	Primer and 2 ct.　　Furred Stairwell Walls	.15	.61	.76
0500	Woodwork incl. puttying, brushwork, primer and 1 ct.	.10	.78	.88
0600	Primer and 2 ct.	.15	1.03	1.18
0700	Wood trim to 6" wide, enamel, primer and 1 ct.	.10	.44	.54
0800	Primer and 2 ct.	.15	.56	.71
0900	Cabinets and casework, enamel, primer and 1 ct.	.10	.88	.98
1000	Primer and 2 ct.	.15	1.08	1.23
1100	On masonry or concrete, latex, brushwork, primer and 1 ct.	.18	.73	.91
1200	Primer and 2 ct.	.24	1.05	1.29
1300	For block filler, add	.11	.90	1.01
1400				
1500	Varnish, wood trim, sealer 1 ct., sanding, puttying, quality work	.13	1.58	1.71
1600	Meduim work	.10	1.23	1.33
1700	Without sanding	.13	.19	.32
1800				
1900	Wall coverings, wall paper, at $9.70 per double roll, average workmanship	.31	.55	.86
2000	At $20.00 per double roll, average workmanship	.67	.65	1.32
2100	At $44.00 per double roll, quality workmanship	1.61	.80	2.41
2200				
2300	Grass cloths with lining paper, minimum	.67	.87	1.54
2400	Maximum	2.15	1	3.15
2500	Vinyl, fabric backed, light weight	.61	.55	1.16
2600	Medium weight	.75	.73	1.48
2700	Heavy weight	1.19	.80	1.99
2800				
2900	Cork tiles, 12" x 12", 3/16" thick	2.75	1.46	4.21
3000	5/16" thick	3.12	1.49	4.61
3100	Granular surface, 12" x 36", 1/2" thick	1.03	.91	1.94
3200	1" thick	1.34	.94	2.28
3300	Aluminum foil	.88	1.27	2.15
3400				
3500	Tile, ceramic, adhesive set, 4-1/4" x 4-1/4"	2.30	3.48	5.78
3600	6" x 6"	2.83	3.30	6.13
3700	Decorated, 4-1/4" x 4-1/4", minimum	3.85	2.45	6.30
3800	Maximum	42.50	3.67	46.17
3900	For epoxy grout, add	.36	.83	1.19
4000	Pregrouted sheets	4.57	2.75	7.32
4100	Glass mosaics, 3/4" tile on 12" sheets, minimum	16.80	9.05	25.85
4200	Color group 8	55	10.35	65.35
4300	Metal, tile pattern, 4' x 4' sheet, 24 ga., nailed			
4400	Stainless steel	24.50	1.55	26.05
4500	Aluminized steel	13.10	1.55	14.65
4600				
4700	Brick, interior veneer, 4" face brick, running bond, minimum	2.92	8.40	11.32
4800	Maximum	2.92	8.40	11.32
4900	Simulated, urethane pieces, set in mastic	5.05	2.65	7.70
5000	Fiberglass panels	2.70	1.99	4.69
5100	Wall coating, on drywall, thin coat, plain	.08	.50	.58
5200	Stipple	.08	.50	.58
5300	Textured spray	.06	.46	.52
5400				
5500	Paneling not incl. furring or trim, hardboard, tempered, 1/8" thick	.34	1.59	1.93
5600	1/4" thick	.43	1.59	2.02
5700	Plastic faced, 1/8" thick	.59	1.59	2.18
5800	1/4" thick	.79	1.59	2.38

Figure 27.38

C3020 Floor Finishes

C3020 430	Selective Price Sheet		COST PER S.F.		
			MAT.	INST.	TOTAL
0100	Flooring, carpet, acrylic, 26 oz. light traffic		1.70	.54	2.24
0200	35 oz. heavy traffic	Corridors	3.61	.54	4.15
0300	Nylon anti-static, 15 oz. light traffic		1.13	.72	1.85
0400	22 oz. medium traffic		2.22	.54	2.76
0500	26 oz. heavy traffic	Apartments	3.11	.58	3.69
0600	28 oz. heavy traffic		3.50	.58	4.08
0700	Tile, foamed back, needle punch		2.65	.64	3.29
0800	Tufted loop		1.12	.64	1.76
0900	Wool, 36 oz. medium traffic		7.85	.58	8.43
1000	42 oz. heavy traffic		7.95	.58	8.53
1100	Composition, epoxy, with colored chips, minimum		2.41	3	5.41
1200	Maximum		2.93	4.13	7.06
1300	Trowelled, minimum		3.11	3.61	6.72
1400	Maximum		4.53	4.21	8.74
1500	Terrazzo, 1/4" thick, chemical resistant, minimum		5.25	4.21	9.46
1600	Maximum		8.15	5.60	13.75
1700	Resilient, asphalt tile, 1/8" thick		1.06	.92	1.98
1800	Conductive flooring, rubber, 1/8" thick		2.84	1.16	4
1900	Cork tile 1/8" thick, standard finish		3.80	1.16	4.96
2000	Urethane finish		4.82	1.16	5.98
2100	PVC sheet goods for gyms, 1/4" thick		3.99	4.59	8.58
2200	3/8" thick		4.50	6.10	10.60
2300	Vinyl composition 12" x 12" tile, plain, 1/16" thick		.80	.73	1.53
2400	1/8" thick		2.05	.73	2.78
2500	Vinyl tile, 12" x 12" x 1/8" thick, minimum		2.20	.73	2.93
2600	Maximum		8.30	.73	9.03
2700	Vinyl sheet goods, backed, .093" thick	Bath + Kitchen	2.20	1.59	3.79
2900	Slate, random rectangular, 1/4" thick		11.55	4.71	16.26
3000	1/2" thick		10.75	6.70	17.45
3100	Natural cleft, irregular, 3/4" thick		4.96	7.65	12.61
3200	For sand rubbed finish, add		5.95		5.95
3300	Terrazzo, cast in place, bonded 1-3/4" thick, gray cement		2.68	6.50	9.18
3400	White cement		3.06	6.50	9.56
3500	Not bonded 3" thick, gray cement		3.36	7.30	10.66
3600	White cement		3.66	7.30	10.96
3700	Precast, 12" x 12" x 1" thick		16.35	24.50	40.85
3800	1-1/4" thick		18.80	24.50	43.30
3900	16" x 16" x 1-1/4" thick		20.50	30.50	51
4000	1-1/2" thick		18.65	33.50	52.15
4100	Marble travertine, standard, 12" x 12" x 3/4" thick		10.60	11	21.60
4200					
4300	Tile, ceramic, natural clay, thin set		3.98	3.61	7.59
4400	Porcelain, thin set		4.37	3.48	7.85
4500	Specialty, decorator finish		9.85	3.61	13.46
4600					
4700	Quarry, red, mud set, 4" x 4" x 1/2" thick		4.19	5.50	9.69
4800	6" x 6" x 1/2" thick		3.20	4.72	7.92
4900	Brown, imported, 6" x 6" x 3/4" thick		4.98	5.50	10.48
5000	8" x 8" x 1" thick		5.65	6	11.65
5100	Slate, vermont, thin set, 6" x 6" x 1/4" thick		4.55	3.67	8.22
5200					
5300	Wood, maple strip, 25/32" x 2-1/4", finished, select		4.32	2.53	6.85
5400	2nd and better		3.05	2.53	5.58
5500	Oak, 25/32" x 2-1/4" finished, clear		3.10	2.53	5.63
5600	No. 1 common		3.98	2.53	6.51
5700	Parquet, standard, 5/16", finished, minimum		3.09	2.68	5.77
5800	Maximum		5.80	4.17	9.97
5900	Custom, finished, minimum		14.75	3.98	18.73

Figure 27.39

14100 | Dumbwaiters

14110 | Manual Dumbwaiters

			CREW	DAILY OUTPUT	LABOR-HOURS	UNIT	2002 BARE COSTS				TOTAL INCL O&P	
							MAT.	LABOR	EQUIP.	TOTAL		
400	0010	**DUMBWAITERS** 2 stop, hand, minimum	2 Elev	.75	21.333	Ea.	2,275	790		3,065	3,725	400
	0100	Maximum		.50	32	"	5,200	1,175		6,375	7,550	
	0300	For each additional stop, add	↓	.75	21.333	Stop	825	790		1,615	2,125	

14120 | Electric Dumbwaiters

			CREW	DAILY OUTPUT	LABOR-HOURS	UNIT	2002 BARE COSTS				TOTAL INCL O&P	
							MAT.	LABOR	EQUIP.	TOTAL		
400	0010	**DUMBWAITERS** 2 stop, electric, minimum	2 Elev	.13	123	Ea.	5,775	4,575		10,350	13,500	400
	0100	Maximum		.11	145	"	17,300	5,400		22,700	27,400	
	0600	For each additional stop, add	↓	.54	29.630	Stop	2,550	1,100		3,650	4,500	

14200 | Elevators

14210 | Electric Traction Elevators

				CREW	DAILY OUTPUT	LABOR-HOURS	UNIT	2002 BARE COSTS				TOTAL INCL O&P	
								MAT.	LABOR	EQUIP.	TOTAL		
100	0012	**ELEVATORS OR LIFTS**											100
	7000	Residential, cab type, 1 floor, 2 stop, minimum		2 Elev	.20	80	Ea.	8,075	2,975		11,050	13,500	
	7100	Maximum			.10	160		13,700	5,925		19,625	24,200	
	7200	2 floor, 3 stop, minimum			.12	133		12,000	4,950		16,950	20,900	
	7300	Maximum			.06	266		19,600	9,900		29,500	36,900	
	7700	Stair climber (chair lift), single seat, minimum			1	16		3,925	595		4,520	5,225	
	7800	Maximum			.20	80		5,375	2,975		8,350	10,500	
	8000	Wheelchair lift, minimum			1	16		5,375	595		5,970	6,825	
	8500	Maximum			.50	32		12,700	1,175		13,875	15,900	
	8700	Stair lift, minimum			1	16		10,600	595		11,195	12,600	
	8900	Maximum		↓	.20	80	↓	16,800	2,975		19,775	23,100	
200	0010	**ELEVATORS**	R14200 -100										200
	0020	For multi-story buildings, housing project, minimum					% total					2.50%	
	0100	Maximum										4.50%	
	0300	Office building, minimum	R14200 -200									2.50%	
	0400	Maximum										10%	
	0425	Electric freight, base unit, 4000 lb, 200 fpm, 4 stop, std. fin.	R14200 -300	2 Elev	.05	320	Ea.	66,000	11,900		77,900	91,500	
	0450	For 5000 lb capacity, add						4,750			4,750	5,225	
	0500	For 6000 lb capacity, add	R14200 -400					8,350			8,350	9,200	
	0525	For 7000 lb capacity, add						11,200			11,200	12,300	
	0550	For 8000 lb capacity, add						15,800			15,800	17,300	
	0575	For 10000 lb capacity, add						18,800			18,800	20,700	
	0600	For 12000 lb capacity, add						22,900			22,900	25,200	
	0625	For 16000 lb capacity, add						27,500			27,500	30,300	
	0650	For 20000 lb capacity, add						30,400			30,400	33,400	
	0675	For increased speed, 250 fpm, add						9,075			9,075	9,975	
	0700	300 fpm, geared electric, add						11,300			11,300	12,500	
	0725	350 fpm, geared electric, add						13,400			13,400	14,700	
	0750	400 fpm, geared electric, add						15,100			15,100	16,600	
	0775	500 fpm, gearless electric, add						19,000			19,000	20,900	
	0800	600 fpm, gearless electric, add						21,000			21,000	23,100	
	0825	700 fpm, gearless electric, add						24,500			24,500	26,900	
	0850	800 fpm, gearless electric, add						27,200			27,200	29,900	
	0875	For class "B" loading, add						1,650			1,650	1,800	
	0900	For class "C-1" loading, add		↓			↓	4,075			4,075	4,475	

Figure 27.40

D2010 Plumbing Fixtures

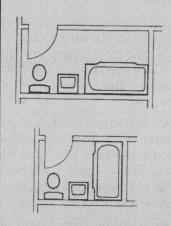

This page illustrates and describes a three fixture bathroom system including a water closet, tub, lavatory, accessories and service piping. Lines within System Components give the unit price and total price on a cost each basis for this system. Prices for an alternate three fixture bathroom system are on Line Item D2010 959 1700. Both material quantities and labor costs have been adjusted for the system listed.

Factors: To adjust for job conditions other than normal working situations use Lines D2010 959 2900 thru 4000.

Example: You are to install the system and protect all existing work. Go to Line D2010 959 3800 and apply these percentages to the appropriate MAT. and INST. costs.

System Components	QUANTITY	UNIT	COST EACH		
			MAT.	INST.	TOTAL
Three fixture bathroom incl. water closet, bathtub, lavatory, accessories, And necessary service piping to install this system in 2 walls.					
Water closet, floor mounted, 2 piece, close coupled	1.000	Ea.	154	152	306
Rough-in waste & vent for water closet	1.000	Set	129.27	245.52	374.79
Bathtub, P.E. cast iron 5' long with accessories	1.000	Ea.	410	183	593
Rough-in waste & vent for bathtub	1.000	Set	128.25	370.50	498.75
Lavatory, 20" x 18" P.E. cast iron with accessories	1.000	Ea.	218	101	319
Rough-in waste & vent for lavatory	1.000	Set	240	485	725
Accessories					
Toilet tissue dispenser, chrome, single roll	1.000	Ea.	12.95	13.25	26.20
18" long stainless steel towel bar	2.000	Ea.	70	34.60	104.60
Medicine cabinet with mirror, 20" x 16", unlighted	1.000	Ea.	75.50	28.50	104
TOTAL		System	1,437.97	1,613.37	3,051.34

D2010 959	Plumbing - Three Fixture Bathroom	COST EACH		
		MAT.	INST.	TOTAL
1600				
1700	Above system installed in one wall with all necessary service piping	1,425	1,575	3,000
2400	NOTE: PLUMBING APPROXIMATIONS			
2500	WATER CONTROL: water meter, backflow preventer,			
2600	Shock absorbers, vacuum breakers, mixer....10 to 15% of fixtures			
2700	PIPE AND FITTINGS: .... 30 to 60% of fixtures	30%	30%	
2800				
2900	Cut & patch to match existing construction, add, minimum	2%	3%	
3000	Maximum	5%	9%	
3100	Dust protection, add, minimum	1%	2%	
3200	Maximum	4%	11%	
3300	Equipment usage curtailment, add, minimum	1%	1%	
3400	Maximum	3%	10%	
3500	Material handling & storage limitation, add, minimum	1%	1%	
3600	Maximum	6%	7%	
3700	Protection of existing work, add, minimum	2%	2%	
3800	Maximum	5%	7%	
3900	Shift work requirements, add, minimum		5%	
4000	Maximum		30%	
		7%	11%	

Figure 27.41

resurfaced, and new loam and seeding is specified for the entire lawn. Demolition costs are also included in Division F of the Assemblies section of *Means Repair & Remodeling Cost Data*.

The estimator must be very careful when evaluating the site of an existing building for excavation work. Utility locations must be identified at the site. A waterline break may be a very expensive mistake.

After thorough examination, the estimator has determined that there should be no utilities at the area to be excavated. The costs are shown in Figure 27.56. Note that the costs as shown are for 360 C.Y. (a full day's cost) and per cubic yard. A half-day rental is usually a minimum for heavy equipment. Even though the amount to be excavated is 91 C.Y. (less than one-half of 360 C.Y.), the costs for half a day are included in the estimate.

The concrete slab at the patios was not included in Division A Substructure because the individual components required more closely resemble a sidewalk than an interior floor slab. The costs are derived from Figure 27.57. The edge forms are deducted because only a very small amount of forming will be necessary, compared to a normal sidewalk. The factors are added because access for the truck is impossible and the concrete will be placed by hand.

The parking lot is to receive only a new topping coat. The existing base and binder are adequate. Therefore, only those costs that apply are taken from Figure 27.58. The loam and seed costs are shown in Figure 27.59. In order to avoid any possible confusion, it is important that the estimator determine those areas that are to receive loam and seed. A quick sketch of the site will help to define the limits of the work.

Depending upon the precision necessary for the budget Assemblies Estimate, the estimator may elect to use a lump sum cost for demolition based on past experience and the thorough evaluation of the site. If greater accuracy is required, Figure 27.60 provides costs for selective demolition. In this example, a lump sum figure is used.

The costs for Division G Building Sitework are summarized on the estimate sheet in Figure 27.61.

Bathroom Calculations			
	Material	**Install.**	**Total**
Bathroom (D2010 959 1700)	$1,425.00	$1,575.00	$3,000.00
Pipe and Fittings (30%) (D2010 959 2700)	427.50	472.50	900.00
Subtotal	1,852.50	2,047.50	3,900.00
Factors: 7% Materials, 11% Installation	129.68	225.23	354.90
Total	$1,982.18	$2,272.73	$4,254.90

Figure 27.42

D3020 Heat Generating Systems

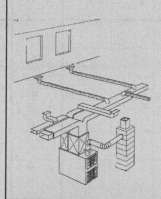

This page illustrates and describes a gas fired forced air system including a gas fired furnace, ductwork, registers and hookups. Lines within System Components give the unit price and total price per square foot for this system. Prices for alternate gas fired forced air systems are on Line Items D3020 124 1500 thru 2600. Both material quantities and labor costs have been adjusted for the system listed.

Factors: To adjust for job conditions other than normal working situations use Lines D3020 124 2900 thru 4000.

Example: You are to install the system with material handling and storage limitations. Go to Line D3020 124 3500 and apply these percentages to the appropriate MAT. and INST. costs.

System Components	QUANTITY	UNIT	COST PER S.F. MAT.	COST PER S.F. INST.	COST PER S.F. TOTAL
Gas fired hot air heating system including furnace, ductwork, registers And all necessary hookups.					
Area to 800 S.F., heat only					
Furnace, gas, AGA certified, direct drive, w/gas piping, 44 MBH	1.000	Ea.	600	251.25	851.25
Duct, galvanized steel	312.000	Lb.	262.08	1,669.20	1,931.28
Insulation, blanket type, ductwork	270.000	S.F.	126.90	607.50	734.40
Flexible duct, 6" diameter, insulated	100.000	L.F.	162	310	472
Registers, baseboard, gravity, 12" x 6"	8.000	Ea.	93.60	155.60	249.20
Return, damper, 36" x 18"	1.000	Ea.	142	44.50	186.50
TOTAL		System	1,386.58	3,038.05	4,424.63
COST PER S.F.		S.F.	1.73	3.80	5.53

D3020 124	Heating-Cooling, Gas, Forced Air	COST PER S.F. MAT.	COST PER S.F. INST.	COST PER S.F. TOTAL
1400	For alternate heating systems:			
1500	Gas fired, area to 1000 S.F.	1.57	3.08	4.65
1600	To 1200 S.F.	1.40	2.74	4.14
1700	To 1600 S.F.	1.34	2.55	3.89
1800	To 2000 S.F.	1.27	3.01	4.28
1900	To 3000 S.F.	1.02	2.41	3.43
2000	For combined heating and cooling systems:			
2100	Gas fired, heating and cooling, area to 800 S.F.	3.54	4.56	8.10
2200	To 1000 S.F.	3.49	3.72	7.21
2300	To 1200 S.F.	3	3.27	6.27
2400	To 1600 S.F.	2.61	2.97	5.58
2500	To 2000 S.F.	2.35	3.38	5.73
2600	To 3000 S.F.	1.85	2.68	4.53
2900	Cut & patch to match existing construction, add, minimum	2%	3%	
3000	Maximum	5%	9%	
3100	Dust protection, add, minimum	1%	2%	
3200	Maximum	4%	11%	
3300	Equipment usage curtailment, add, minimum	1%	1%	
3400	Maximum	3%	10%	
3500	Material handling & storage limitation, add, minimum	1%	1%	
3600	Maximum	6%	7%	
3700	Protection of existing work, add, minimum	2%	2%	
3800	Maximum	5%	7%	
3900	Shift work requirements, add, minimum		5%	
4000	Maximum		30%	
		3.66	4.05	

Figure 27.43

15500 | Heat Generation Equipment

15540 | Fuel-Fired Heaters

				DAILY	LABOR-		2002 BARE COSTS				TOTAL	
			CREW	OUTPUT	HOURS	UNIT	MAT.	LABOR	EQUIP.	TOTAL	INCL O&P	
300	1160	375 MBH output	Q-5	1.60	10	Ea.	4,150	325		4,475	5,075	300
	1180	450 MBH output	↓	1.40	11.429	↓	4,350	370		4,720	5,375	
900	0010	**SPACE HEATERS** Cabinet, grilles, fan, controls, burner,										900
	0020	thermostat, no piping. For flue see 15550										
	1000	Gas fired, floor mounted										
	1100	60 MBH output	Q-5	10	1.600	Ea.	545	52		597	680	
	1180	180 MBH output		6	2.667		815	87		902	1,025	
	2000	Suspension mounted, propeller fan, 20 MBH output		8.50	1.882		420	61.50		481.50	555	
	2040	60 MBH output		7	2.286		515	74.50		589.50	680	
	2100	130 MBH output		5	3.200		765	104		869	1,000	
	2240	320 MBH output	↓	2	8		1,575	261		1,836	2,125	
	2500	For powered venter and adapter, add						254			254	279	
	5000	Wall furnace, 17.5 MBH output	Q-5	6	2.667		535	87		622	720	
	5020	24 MBH output		5	3.200		550	104		654	765	
	5040	35 MBH output	↓	4	4	↓	745	130		875	1,025	
	9000	Minimum labor/equipment charge	↓	3.50	4.571	Job		149		149	232	

15550 | Breechings, Chimneys & Stacks

				DAILY	LABOR-		MAT.	LABOR	EQUIP.	TOTAL	INCL O&P	
200	0010	**DRAFT CONTROLS**										200
	1000	Barometric, gas fired system only, 6" size for 5" and 6" pipes	1 Shee	20	.400	Ea.	39.50	14.05		53.55	66	
	1040	8" size, for 7" and 8" pipes	"	18	.444	"	53.50	15.60		69.10	84	
	2000	All fuel, oil, oil/gas, coal										
	2020	10" for 9" and 10" pipes	1 Shee	15	.533	Ea.	91.50	18.70		110.20	131	
	3260	For thermal switch for above, add	"	24	.333		38.50	11.70		50.20	61	
	5000	Vent damper, bi-metal, gas, 3" diameter	Q-9	24	.667		22	21		43	57.50	
	5010	4" diameter		24	.667		22	21		43	57.50	
	5020	5" diameter		23	.696		22	22		44	59	
	5030	6" diameter		22	.727		22	23		45	60.50	
	5040	7" diameter		21	.762		23.50	24		47.50	64.50	
	5050	8" diameter		20	.800		25.50	25.50		51	68.50	
	5101	Electric, automatic, gas, 4" diameter		24	.667		137	21		158	184	
	5110	5" diameter		23	.696		139	22		161	188	
	5121	6" diameter		22	.727		142	23		165	193	
	5130	7" diameter		21	.762		144	24		168	198	
	5140	8" diameter		20	.800		147	25.50		172.50	203	
	5150	9" diameter		20	.800		150	25.50		175.50	206	
	5160	10" diameter		19	.842		155	26.50		181.50	213	
	5170	12" diameter		19	.842		157	26.50		183.50	216	
	5250	Automatic, oil, 4" diameter		24	.667		153	21		174	202	
	5260	5" diameter		23	.696		156	22		178	206	
	5270	6" diameter		22	.727		159	23		182	212	
	5280	7" diameter		21	.762		162	24		186	217	
	5290	8" diameter		20	.800		169	25.50		194.50	227	
	5300	9" diameter		20	.800		171	25.50		196.50	230	
	5310	10" diameter		19	.842		172	26.50		198.50	232	
	5320	12" diameter		19	.842		175	26.50		201.50	235	
	5330	14" diameter		18	.889		234	28		262	300	
	5340	16" diameter		17	.941		271	29.50		300.50	345	
	5350	18" diameter	↓	16	1	↓	305	31.50		336.50	385	
	9000	Minimum labor/equipment charge	1 Shee	4	2	Job		70		70	112	
440	0010	**VENT CHIMNEY** Prefab metal, U.L. listed										440
	0020	Gas, double wall, galvanized steel										
	0080	3" diameter	Q-9	72	.222	V.L.F.	3.48	7		10.48	15.05	
	0100	4" diameter		68	.235		(4.35)	(7.45)		11.80	16.65	
	0120	5" diameter		64	.250		5.05	7.90		12.95	18.20	
	0140	6" diameter	↓	60	.267	↓	6.25	8.45		14.70	20.50	

Figure 27.44

D5010 Electrical Service/Distribution

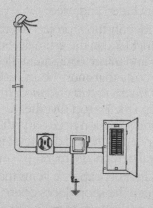

This page illustrates and describes a residential, single phase system including a weather cap, service entrance cable, meter socket, entrance switch, ground rod, ground cable, EMT, and panelboard. Lines with System Components give the unit price and total price on a cost each basis for this system. Prices for an alternate residential, single phase system are also given. Both material quantities and labor costs have been adjusted for the system listed.

Factors: To adjust for job conditions other than normal working situations use Lines D5010 220 2900 thru 4000.

Example: You are to install the system with a minimum equipment usage curtailment. Go to Line D5010 220 3300 and apply these percentages to the appropriate MAT. and INST. costs.

System Components

System Components	QUANTITY	UNIT	MAT.	INST.	TOTAL
100 Amp Service, single phase					
Weathercap	1.000	Ea.	9.15	36.50	45.65
Service entrance cable	.200	C.L.F.	45.40	87	132.40
Meter socket	1.000	Ea.	32	137	169
Entrance disconnect switch	1.000	Ea.	222	230	452
Ground rod, with clamp	1.000	Ea.	19.25	91	110.25
Ground cable	.100	C.L.F.	13.40	27.30	40.70
Panelboard, 12 circuit	1.000	Ea.	182	365	547
TOTAL		Ea.	523.20	973.80	1,497
200 Amp Service , single phase					
Weathercap	1.000	Ea.	20	54.50	74.50
Service entrance cable	.200	C.L.F.	115	125	240
Meter socket	1.000	Ea.	46	230	276
Entrance disconnect switch	1.000	Ea.	480	335	815
Ground rod, with clamp	1.000	Ea.	32.50	99.50	132
Ground cable	.100	C.L.F.	12.40	15.10	27.50
3/4" EMT	10.000	L.F.	6.20	33.60	39.80
Panelboard, 24 circuit	1.000	Ea.	490	570	1,060
TOTAL		Ea.	1,202.10	1,462.70	2,664.80

D5010 220	Residential Service - Single Phase	MAT.	INST.	TOTAL
2800				
2900	Cut & patch to match existing construction, add, minimum	2%	3%	
3000	Maximum	5%	9%	
3100	Dust protection, add, minimum	1%	2%	
3200	Maximum	4%	11%	
3300	Equipment usage curtailment, add, minimum	1%	1%	
3400	Maximum	3%	10%	
3500	Material handling & storage limitation, add, minimum	1%	1%	
3600	Maximum	6%	7%	
3700	Protection of existing work, add, minimum	2%	2%	
3800	Maximum	5%	7%	
3900	Shift work requirements, add, minimum		5%	
4000	Maximum		30%	

Figure 27.45

General Conditions

In the Assemblies estimating format, the General Conditions are typically done after all other portions of the building project have been priced out. The prices shown in the Assemblies section of *Means Repair & Remodeling Cost Data* include the installing contractor's overhead and profit. Chapter 25 of *Means Repair and Remodeling Estimating* provides a complete explanation of how labor rates, overhead, and profit are included and used.

1. **Project Overhead**. It is necessary to identify those project overhead items that have not been previously included in the estimate. Such items include field supervision, tools and minor equipment, field office, sheds, and photos. All have costs that must be included. Division 1 in the Unit Price section of *Means Repair & Remodeling Cost Data*, lists many items that fall into the Project Overhead category. Depending on the estimating precision necessary, the estimator may be as specific or as general as the job dictates. Figure 27.62 may prove beneficial for general percentage markups only.

2. **Office Overhead**. There are certain indirect expense items that are incurred by the requirement to keep the shop doors open and to attract business and bid work. The percentage of main office overhead expense declines with increased annual volume of the contractor. Overhead is not appreciably increased when there is an increase in the volume of work. Typical main office expenses range from 2% to 20%, with the median being about 7% of the total volume (gross billings). Figure 27.63 shows approximate percentages for some of the different items usually included in a general contractor's main office overhead. These percentages may vary with different accounting procedures.

3. **Profit**. The profit assumed in *Means Repair & Remodeling Cost Data* is 10% on material and equipment, and 15% on labor. Since

Electrical Service Calculations			
	Material	**Install.**	**Total**
Service (D5010 220)			
Meter Socket	$ 32.00	$ 137.00	$ 169.00
Disconnect	222.00	230.00	452.00
Panel Board	182.00	365.00	547.00
Subtotal	436.00	732.00	1,168.00
Factors: 5% Materials,			
9% Installation	22.47	61.38	83.85
Total	$ 458.47	$ 793.38	$1,251.85
Entrance, Trough, Feeders	$3,120.00	$4,270.00	$7,390.00
Ground Rod	22.65	47.00	69.65
Ground Wire	10.90	20.50	31.40
Total	$3,153.55	$4,337.50	$7,491.05

Figure 27.46

D50 Electrical

D5020 Lighting and Branch Wiring

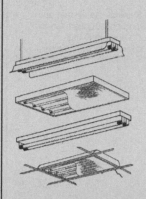

This page illustrates and describes fluorescent lighting systems including a fixture, lamp, outlet box and wiring. Lines within System Components give the unit price and total price on a cost each basis for this system. Prices for alternate fluorescent lighting systems are on Line Items D5020 248 1300 thru 1500. Both material quantities and labor costs have been adjusted for the system listed.

Factors: To adjust for job conditions other than normal working situations use Lines D5020 248 2900 thru 4000.

Example: You are to install the system during evening hours. Go to Line D5020 248 3900 and apply this percentage to the appropriate INST. cost.

System Components		QUANTITY	UNIT	COST EACH		
				MAT.	INST.	TOTAL
Fluorescent lighting, including fixture, lamp, outlet box and wiring.						
Recessed lighting fixture, on suspended system		1.000	Ea.	62	93	155
Outlet box		1.000	Ea.	2.26	24.50	26.76
#12 wire		.660	C.L.F.	3.86	26.40	30.26
Conduit, EMT, 1/2" conduit	Deduct	20.000	L.F.	8.20	51.40	59.60
	TOTAL		Ea.	76.32	195.30	271.62

D5020 248	Lighting, Fluorescent	COST EACH		
		MAT.	INST.	TOTAL
1200	For alternate lighting fixtures:			
1300	Surface mounted, 2' x 4', acrylic prismatic diffuser	115	205	320
1400	Strip fixture, 8' long, two 8' lamps	67.50	193	260.50
1500	Pendant mounted, industrial, 8' long, with reflectors	105	222	327
1600				
1700				
1800		115	205	
1900		- 8.20	- 51.40	
2000		106.80	153.60	
2100				
2200				
2300				
2900	Cut & patch to match existing construction, add, minimum	2%	3%	
3000	Maximum	5%	9%	
3100	Dust protection, add, minimum	1%	2%	
3200	Maximum	4%	11%	
3300	Equipment usage curtailment, add, minimum	1%	1%	
3400	Maximum	3%	10%	
3500	Material handling & storage limitation, add, minimum	1%	1%	
3600	Maximum	6%	7%	
3700	Protection of existing work, add, minimum	2%	2%	
3800	Maximum	5%	7%	
3900	Shift work requirements, add, minimum		5%	
4000	Maximum		30%	

Figure 27.47

D5020 Lighting and Branch Wiring

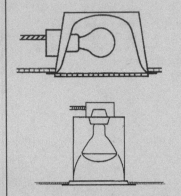

This page illustrates and describes incandescent lighting systems including a fixture, lamp, outlet box, conduit and wiring. Lines within System Components give the unit price and total price on a cost each basis for this system. Prices for an alternate incandescent lighting system are also given. Both material quantities and labor costs have been adjusted for the system listed.

Factors: To adjust for conditions other than normal working situations use Lines D5020 250 2900 thru 4000.

Example: You are to install the system and cut and match existing construction. Go to Line D5020 250 3000 and apply this percentage to the appropriate INST. costs.

System Components	QUANTITY	UNIT	COST EACH		
			MAT.	INST.	TOTAL
Incandescent light fixture, including lamp, outlet box, conduit and wiring.					
Recessed wide reflector with flat glass lens	1.000	Ea.	73.50	65.50	139
Outlet box	1.000	Ea.	2.26	24.50	26.76
Armored cable, 3 wire Deduct	.200	C.L.F.	13.80	39.80	53.60
TOTAL		Ea.	89.56	129.80	219.36
Recessed flood light fixture, including lamp, outlet box, conduit & wire					
Recessed, R-40 flood lamp with refractor	1.000	Ea.	89	54.50	143.50
150 watt R-40 flood lamp	.010	Ea.	8.30	3.35	11.65
Outlet box	1.000	Ea.	2.26	24.50	26.76
Outlet box cover	1.000	Ea.	.77	6.85	7.62
Romex, 12-2 with ground	.200	C.L.F.	4	39.80	43.80
Conduit, 1/2" EMT	20.000	L.F.	8.20	51.40	59.60
TOTAL		Ea.	112.53	180.40	292.93

D5020 250	Lighting, Incandescent	COST EACH		
		MAT.	INST.	TOTAL
2900	Cut & patch to match existing construction, add, minimum	2%	3%	
3000	Maximum	5%	9%	
3100	Dust protection, add, minimum	1%	2%	
3200	Maximum	4%	11%	
3300	Equipment usage curtailment, add, minimum	1%	1%	
3400	Maximum	3%	10%	
3500	Material handling & storage limitation, add, minimum	1%	1%	
3600	Maximum	6%	7%	
3700	Protection of existing work, add, minimum	2%	2%	
3800	Maximum	5%	7%	
3900	Shift work requirements, add, minimum		5%	
4000	Maximum		30%	

	89.56	129.80
	-13.80	-39.80
	75.76	89.00

Figure 27.48

D5020 Lighting and Branch Wiring

D5020 180	Selective Price Sheet	COST EACH		
		MAT.	INST.	TOTAL
0100	Using non-metallic sheathed, cable, air conditioning receptacle	13.05	43.50	56.55
0200	Disposal wiring	10.45	48.50	58.95
0300	Dryer circuit	32	79.50	111.50
0400	Duplex receptacle	13.05	33.50	46.55
0500	Fire alarm or smoke detector	66.50	43.50	110
0600	Furnace circuit & switch	19.60	73	92.60
0700	Ground fault receptacle	42	54.50	96.50
0800	Heater circuit	12.30	54.50	66.80
0900	Lighting wiring	12.95	27.50	40.45
1000	Range circuit	65.50	109	174.50
1100	Switches single pole	12.70	27.50	40.20
1200	3-way	17.05	36.50	53.55
1300	Water heater circuit	19.10	87.50	106.60
1400	Weatherproof receptacle	104	73	177
1500	Using BX cable, air conditioning receptacle	23.50	52.50	76
1600	Disposal wiring	20.50	58.50	79
1700	Dryer circuit	42.50	95	137.50
1800	Duplex receptacle	23.50	40.50	64
1900	Fire alarm or smoke detector	66.50	43.50	110
2000	Furnace circuit & switch	31	87.50	118.50
2100	Ground fault receptacle	52.50	66	118.50
2200	Heater circuit	20	66	86
2300	Lighting wiring	21	33	54
2400	Range circuit	91	132	223
2500	Switches, single pole	23.50	33	56.50
2600	3-way	26.50	43.50	70
2700	Water heater circuit	35	104	139
2800	Weatherproof receptacle	112	87.50	199.50
2900	Using EMT conduit, air conditioning receptacle	26	65.50	91.50
3000	Disposal wiring	24	73	97
3100	Dryer circuit	38.50	118	156.50
3200	Duplex receptacle	26	50.50	76.50
3300	Fire alarm or smoke detector	75.50	65.50	141
3400	Furnace circuit & switch	32.50	109	141.50
3500	Ground fault receptacle	66	81	147
3600	Heater circuit	22.50	81	103.50
3700	Lighting wiring	22	41	63
3800	Range circuit	66	162	228
3900	Switches, single pole	26	41	67
4000	3-way	24.50	54.50	79
4100	Water heater circuit	29	129	158
4200	Weatherproof receptacle	114	109	223
4300	Using aluminum conduit, air conditioning receptacle	29.50	87.50	117
4400	Disposal wiring	27.50	97	124.50
4500	Dryer circuit	49	156	205
4600	Duplex receptacle	28.50	67.50	96
4700	Fire alarm or smoke detector	83	87.50	170.50
4800	Furnace circuit & switch	38.50	146	184.50
4900	Ground fault receptacle	63.50	109	172.50
5000	Heater circuit	28	109	137
5100	Lighting wiring	30.50	54.50	85
5200	Range circuit	80	219	299
5300	Switches, single pole	36.50	54.50	91
5400	3-way	35	73	108
5500	Water heater circuit	39	175	214
5600	Weatherproof receptacle	129	146	275
5700	Using galvanized steel conduit	31	93	124
5800	Disposal wiring	26.50	104	130.50

Figure 27.49

COST ANALYSIS

DESCRIPTION	SOURCE/DIMENSIONS			QUANTITY	UNIT	MATERIAL		LABOR		EQ./TOTAL	
						UNIT COST	TOTAL	UNIT COST	TOTAL	UNIT COST	TOTAL
Division D: Services											
Bathrooms	D2010	959	1700	8	EA	1,982	15,856	2,273	18,184		
Kit. Sink Rough-in	15418	600	4980	8	EA	94	748	377	3,016		
HVAC Systems	D3020	124	2200	7,296	SF	3.66	26,703	4.05	29,549		
Vent Chimney	15550	440	100	96	VLF	4.78	459	11.87	1,140		
Service: Meter socket disconnect & panel board	D5010	220	-	8	EA	436	3,488	732	5,856		
Service: Entrance, feeders & ground	Phone Estimate			1	EA	3,154	3,154	4,338	4,338		
Fixtures: Fluorescent	D5020	248	1300	8	EA	107	856	154	1,232		
Incandescent	D5020	250	-	24	EA	75.76	1,818	89.00	2,136		
Wiring: Disposal	D5020	180	0200	8	EA	10.45	84	48.50	388		
Receptacles	D5020	180	0400	156	EA	13.05	2,036	33.50	5,226		
Smoke detect	D5020	180	0500	10	EA	66.50	665	43.50	435		
Furnace	D5020	180	0600	8	EA	19.60	157	73.00	584		
GFI	D5020	180	0700	8	EA	42.00	336	54.50	436		
Range	D5020	180	1000	8	EA	65.50	524	109.00	872		
Switches	D5020	180	1100	40	EA	12.70	508	27.50	1,100		
Subtotal							57,392		74,492		
G.C. Supervision				10%			5,739		7,449		
Division D: Totals							63,131		81,942		

Figure 27.50

E1090   Other Equipment

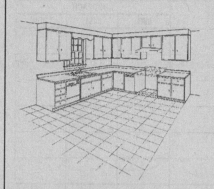

This page illustrates and describes kitchen systems including top and bottom cabinets, custom laminated plastic top, single bowl sink, and appliances. Lines within System Components give the unit price and total price on a cost each basis for this system. Prices for alternate kitchen systems are on Line Items E1090 310 1500 and 1600. Both material quantities and labor costs have been adjusted for the system listed.

Factors: To adjust for job conditions other than normal working situations use Lines E1090 310 2900 thru 4000.

Example: You are to install the system and protect the work area from dust. Go to Line E1090 310 3200 and apply these percentages to the appropriate MAT. and INST. costs.

System Components		QUANTITY	UNIT	COST EACH		
				MAT.	INST.	TOTAL
Kitchen cabinets including wall and base cabinets, custom laminated Plastic top, sink & appliances, no plumbing or electrical rough-in included.						
Prefinished wood cabinets, average quality, wall and base		19 20.000	L.F.	1,880	530	2,410
Custom laminated plastic counter top		19 20.000	L.F.	169	265	434
Stainless steel sink, 22" x 25"		1.000	Ea.	340	144	484
Faucet, top mount		1.000	Ea.	55	45	100
Dishwasher, built-in		1.000	Ea.	279	221	500
Compactor, built-in	Delete	1.000	Ea.	390	79.50	469.50
Range hood, vented, 30" wide		1.000	Ea.	39	168	207
	TOTAL		Ea.	3,152	1,452.50	4,604.50

E1090 310	Kitchens	COST EACH		
		MAT.	INST.	TOTAL
1400	For alternate kitchen systems:			
1500	Prefinished wood cabinets, high quality	6,525	1,675	8,200
1600	Custom cabinets, built in place, high quality	8,475	1,925	10,400
1700				
1800				
1900				
2000				
2100				
2200	NOTE: No plumbing or electric rough-ins are included in the above			
2300	Prices, for plumbing see Division 15, for electric see Division 16.			
2400				
2500				
2600				
2700				
2900	Cut & patch to match existing construction, add, minimum	2%	3%	
3000	Maximum	5%	9%	
3100	Dust protection, add, minimum	1%	2%	
3200	Maximum	4%	11%	
3300	Equipment usage curtailment, add, minimum	1%	1%	
3400	Maximum	3%	10%	
3500	Material handling & storage limitation, add, minimum	1%	1%	
3600	Maximum	6%	7%	
3700	Protection of existing work, add, minimum	2%	2%	
3800	Maximum	5%	7%	
3900	Shift work requirements, add, minimum		5%	
4000	Maximum		30%	

Figure 27.51

E10 Equipment

E1090 Other Equipment

E1090 315	Selective Price Sheet	COST EACH		
		MAT.	INST.	TOTAL
5900	Counter top range 4 burner, maximum	210	73	283
6000	Maximum	485	146	631
6100	Compactor, built-in, minimum	390	79.50	469.50
6200	Maximum	425	133	558
6300	Dishwasher, built-in, minimum	279	221	500
6400	Maximum	310	445	755
6500	Garbage disposer, sink-pipe, minimum	42.50	88.50	131
6600	Maximum	201	88.50	289.50
6700	Range hood, 30″ wide, 2 speed, minimum	39	168	207
6800	Maximum	495	280	775
6900	Refrigerator, no frost, 12 cu. ft.	495	62	557
7000	20 cu. ft.	760	104	864
7100	Plumb. not incl. rough-ins, sinks porc. C.I., single bowl, 21″ x 24″	219	144	363
7200	21″ x 30″	272	144	416
7300	Double bowl, 20″ x 32″	315	168	483
7400				
7500	Stainless steel, single bowl, 19″ x 18″	310	144	454
7600	22″ x 25″	340	144	484
7700				
7800				
7900				
8000				
8100				
8200				
8300				
8400				

Figure 27.52a

E1090 315	Selective Price Sheet	COST EACH		
		MAT.	INST.	TOTAL
0100	Cabinets standard wood, base, one drawer one door, 12" wide	139	32	171
0200	15" wide	191	33	224
0300	18" wide	223	34	257
0400	21" wide	212	35	247
0500	24" wide	246	35.50	281.50
0600				
0700	Two drawers two doors, 27" wide	279	36	315
0800	30" wide	297	37	334
0900	33" wide	299	38	337
1000	36" wide	310	39	349
1100	42" wide	335	40	375
1200	48" wide	355	42	397
1300	Drawer base (4 drawers), 12" wide	340	32	372
1400	15" wide	291	33	324
1500	18" wide	294	34	328
1600	24" wide	320	35.50	355.50
1700	Sink or range base, 30" wide	239	37	276
1800	33" wide	255	38	293
1900	36" wide	268	39	307
2000	42" wide	286	40	326
2100	Corner base, 36" wide	237	44	281
2200	Lazy susan with revolving door	380	48	428
2300	Cabinets standard wood, wall two doors, 12" high, 30" wide	150	32	182
2400	36" wide	174	33	207
2500	15" high, 30" high	160	33	193
2600	36" wide	184	35	219
2700	24" high, 30" wide	194	34	228
2800	36" wide	215	35	250
2900	30" high, 30" wide	221	41	262
3000	36" wide	254	42.50	296.50
3100	42" wide	275	43	318
3200	48" wide	310	43	353
3300	One door, 30" high, 12" wide	132	36	168
3400	15" wide	149	37	186
3500	18" wide	164	38	202
3600	24" wide	182	39	221
3700	Corner, 30" high, 24" wide	152	44	196
3800	36" wide	198	48	246
3900	Broom, 84" high, 24" deep, 18" wide	405	79.50	484.50
4000	Oven, 84" high, 24" deep, 27" wide	595	99.50	694.50
4100	Valance board, 4' long	34	8.05	42.05
4200	6" long	51.50	12.05	63.55
4300	Counter tops, laminated plastic, stock 25" wide w/backsplash, min.	8.45	13.25	21.70
4400	Maximum	16.15	15.90	32.05
4500	Custom, 7/8" thick, no splash	17.30	13.25	30.55
4600	Cove splash	22.50	13.25	35.75
4700	1-1/4" thick, no splash	20.50	14.20	34.70
4800	Square splash	25.50	14.20	39.70
4900	Post formed	9.40	13.25	22.65
5000				
5100	Maple laminated 1-1/2" thick, no splash	34	14.20	48.20
5200	Square splash	38.50	14.20	52.70
5300				
5400				
5500	Appliances, range, free standing, minimum	279	62	341
5600	Maximum	1,325	155	1,480
5700	Built-in, minimum	350	73	423
5800	Maximum	1,175	400	1,575

Figure 27.52b

this is profit margin for the installing contractor, a percentage must be added to cover the profit of the general or prime contractor when subcontractors perform the work. A figure of 10% is used in the appropriate divisions of the sample estimate.

An allowance for the general or prime contractor's project and office overhead, project management, supervision, and markup should be added to the prices in *Means Repair & Remodeling Cost Data*. This markup ranges from approximately 10% to 20%. A figure of 15% is used for the sample project for general conditions in Figure 27.64.

Estimate Summary

It is common practice to include an allowance for contingencies to cover unforeseen construction difficulties or for oversights during the estimating process. Different factors should be used for the various stages of design completion. The following serve as guides to determine contingency factors.

Conceptual Stage: add 15% to 20%

Preliminary Drawings: add 10% to 15%

Working Drawings, 60% Design Complete: add 7% to 10%

Final Working Drawings, 100% Checked Finals: add 2% to 7%

Field Contingencies: add 0% to 3%

As far as the construction contract is concerned, changes in the project can and often will be covered by extras or change orders. The contractor should consider inflationary price trends and possible material shortages during the course of the job. Escalation factors depend on both economic conditions and the anticipated time between the estimate and the actual construction. In the final summary, contingencies are a matter of estimating judgment. Once the estimate is complete, an analysis is required of items such as sales tax on materials and rental equipment, as well as wheel and use taxes for the city, county, and/or state where the project will be constructed. There are some locations that tax construction labor in addition to the above items. It is crucial that the estimator check the local regulations for the area of the project in

Calculations for Kitchen			
	Material	**Install.**	**Total**
Cabinets (adjust for 19 L.F.) (E1090 310)	$1,786.00	$ 503.50	$2,289.50
Counter Top (adjust for 19 L.F.) (E1090 310)	160.55	251.75	412.30
Sink (E1090 310)	340.00	144.00	484.00
Faucet (E1090 310)	55.00	45.00	100.00
Dishwasher (E1090 310)	279.00	221.00	500.00
Range Hood (E1090 310)	39.00	168.00	207.00
Range (E1090 315 5500)	279.00	62.00	341.00
Disposer (E1090 315 6500)	42.50	88.50	131.00
Refrigerator (E1090 310 6900)	495.00	62.00	557.00
Total	$3,476.05	$1,545.75	$5,021.80

Figure 27.53

10530 | Protective Covers

10535 | Awning & Canopies

			CREW	DAILY OUTPUT	LABOR-HOURS	UNIT	2002 BARE COSTS MAT.	LABOR	EQUIP.	TOTAL	TOTAL INCL O&P	
200	5010	Entry or walkway, peak, 12' long, 4' wide	2 Carp	.90	17.778	Ea.	3,500	535		4,035	4,725	200
	5020	6' wide		.60	26.667		5,400	800		6,200	7,275	
	5030	8' wide		.40	40		7,450	1,200		8,650	10,200	
	5060	Radius with dome end, 4' wide		1.10	14.545		2,650	435		3,085	3,650	
	5070	6' wide		.70	22.857		4,275	685		4,960	5,825	
	5080	8' wide	↓	.50	32		6,075	960		7,035	8,275	
	5110	For fire retardant canvas, add					7%					
	5120	For lettering or graphics, add					35%					
	5130	For painted or coated acrylic canvas, deduct					8%					
	5140	For translucent or opaque vinyl canvas, add					10%					
	5150	For 6 or more units, deduct				↓	20%	15%				
	9000	Minimum labor/equipment charge	2 Carp	2	8	Job		240		240	400	

10550 | Postal Specialties

10555 | Mail Delivery Systems

			CREW	DAILY OUTPUT	LABOR-HOURS	UNIT	2002 BARE COSTS MAT.	LABOR	EQUIP.	TOTAL	TOTAL INCL O&P	
600	0010	**MAIL BOXES** Horiz., key lock, 5"H x 6"W x 15"D, alum., rear load	1 Carp	34	.235	Ea.	31	7.05		38.05	45.50	600
	0100	Front loading		34	.235		35	7.05		42.05	50	
	0200	Double, 5"H x 12"W x 15"D, rear loading		26	.308		60	9.25		69.25	81.50	
	0300	Front loading		26	.308		63	9.25		72.25	85	
	0500	Quadruple, 10"H x 12"W x 15"D, rear loading		20	.400		112	12		124	143	
	0600	Front loading		20	.400		107	12		119	138	
	1600	Vault type, horizontal, for apartments, 4" x 5"		34	.235		36.50	7.05		43.55	51.50	
	1700	Alphabetical directories, 120 names		10	.800		111	24		135	162	
	1900	Letter slot, residential		20	.400		56.50	12		68.50	82	
	2000	Post office type		8	1	↓	213	30		243	284	
	9000	Minimum labor/equipment charge	↓	5	1.600	Job		48		48	79.50	
700	0010	**MAIL CHUTES** Aluminum & glass, 14-1/4" wide, 4-5/8" deep	2 Shee	4	4	Floor	645	140		785	935	700
	0100	8-5/8" deep		3.80	4.211	"	720	148		868	1,025	
	0600	Lobby collection boxes, aluminum		5	3.200	Ea.	1,800	112		1,912	2,150	
	0700	Bronze or stainless	↓	4.50	3.556	"	2,300	125		2,425	2,725	

10600 | Partitions

10615 | Demountable Partitions

			CREW	DAILY OUTPUT	LABOR-HOURS	UNIT	2002 BARE COSTS MAT.	LABOR	EQUIP.	TOTAL	TOTAL INCL O&P	
100	0010	**PARTITIONS, MOVABLE OFFICE** Demountable, add for doors										100
	0100	Do not deduct door openings from total L.F.										
	0900	Demountable gypsum system on 2" to 2-1/2"										
	1000	steel studs, 9' high, 3" to 3-3/4" thick										
	1200	Vinyl clad gypsum	2 Carp	48	.333	L.F.	38.50	10		48.50	59	
	1300	Fabric clad gypsum		44	.364		96.50	10.90		107.40	124	
	1500	Steel clad gypsum	↓	40	.400	↓	105	12		117	135	
	1600	1.75 system, aluminum framing, vinyl clad hardboard,										

Figure 27.54

COST ANALYSIS

PROJECT	Apartment Renovation		SHEET NO.		
			ESTIMATE NO.		
ARCHITECT			DATE		2002
TAKE OFF BY: ABC	QUANTITIES BY: ABC	PRICES BY: RSM	EXTENSIONS BY: DEF	CHECKED BY	GHI

DESCRIPTION	SOURCE/DIMENSIONS			QUANTITY	UNIT	MATERIAL		LABOR		EQ./TOTAL	
						UNIT COST	TOTAL	UNIT COST	TOTAL	UNIT COST	TOTAL
Division E: Equipment											
Kitchen	From Table			8	EA	3,476	27,808	1,546	12,368		
Mailboxes	10550	600	0600	2	EA	118	235	20.30	41		
Division E: Totals							28,043		12,409		

Figure 27.55

A20 Basement Construction

A2010 Basement Excavation

This page illustrates and describes foundation excavation systems including a backhoe-loader, operator, equipment rental, fuel, oil, mobilization, hauling material, no backfilling. Lines within System Components give the unit price and total price per cubic yard for this system. Prices for alternate foundation excavation systems are on Line Items A2010 140 1600 thru 2200. Both material quantities and labor costs have been adjusted for the system listed.

Factors: To adjust for job conditions other than normal working situations use Lines A2010 140 2900 thru 4000.

Example: You are to install the system with the use of temporary shoring and bracing. Go to Line A2010 140 3900 and apply these percentages to the appropriate TOTAL costs.

System Components	QUANTITY	UNIT	COST PER C.Y. EQUIP.	COST PER C.Y. LABOR	COST PER C.Y. TOTAL
Foundation excav w/ 3/4 C.Y. backhoe-loader, incl. operator, equip Rental, fuel, oil, and mobilization. Hauling of excavated material is Included. Prices based on one day production of 360 C.Y. in medium soil Without backfilling.					
Equipment operator	8.000	Hr.		395	395
Backhoe-loader, 3/4 C.Y.	1.000	Day	135.30		135.30
Operating expense (fuel, oil)	8.000	Hr.	61.20		61.20
Hauling, 12 C.Y. trucks, 1 mile round trip	360.000	C.Y.	716.40	450	1,166.40
Mobilization	1.000	Ea.	215	90	305
For one full day TOTAL	360.000	C.Y.	931.40	935	1,866.40
COST PER C.Y.		C.Y.	2.59	2.60	5.18
1/2 Day Min.			**465.70**	**467.50**	

A2010 140	Excavation, Foundation	COST PER C.Y. EQUIP.	COST PER C.Y. LABOR	COST PER C.Y. TOTAL
1400	For alternate size excavations:			
1500				
1600	100 C.Y.	4.14	4.37	8.51
1700	200 C.Y.	3.07	2.81	5.88
1800	300 C.Y.	2.71	2.65	5.36
1900	400 C.Y.	2.53	2.57	5.10
2000	500 C.Y.	2.42	2.54	4.96
2100	600 C.Y.	2.35	2.49	4.84
2200	700 C.Y.	2.30	2.45	4.75
2300				
2400				
2500				
2900	Dust protection, add, minimum	1%	2%	
3000	Maximum	4%	11%	
3100	Equipment usage curtailment, add, minimum	1%	1%	
3200	Maximum	3%	10%	
3300	Material handling & storage limitation, add, minimum	1%	1%	
3400	Maximum	6%	7%	
3500	Protection of existing work, add, minimum	2%	2%	
3600	Maximum	5%	7%	
3700	Shift work requirements, add, minimum		5%	
3800	Maximum		30%	
3900	Temporary shoring and bracing, add, minimum	2%	5%	
4000	Maximum	5%	12%	

Figure 27.56

G2030 Pedestrian Paving

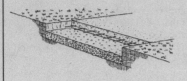

This page illustrates and describes sidewalk systems including concrete, welded wire and broom finish. Lines within System Components give unit price and total price per square foot for this system. Prices for alternate sidewalk systems are on Line Items G2030 220 1700 thru 1900. Both material quantities and labor costs have been adjusted for the system listed.

Factors: To adjust for job conditions other than normal working situations use Lines G2030 220 2900 thru 4000.

Example: You are to install the system and match existing construction. Go to Line G2030 220 2900 and apply these percentages to the appropriate MAT. and INST. costs.

System Components	QUANTITY	UNIT	COST PER S.F. MAT.	COST PER S.F. INST.	COST PER S.F. TOTAL
4" thick concrete sidewalk with welded wire fabric 3000 psi air entrained concrete, broom finish.					
Gravel fill, 4" deep	.012	C.Y.	.19	.06	.25
Compact fill	.012	C.Y.		.02	.02
Hand grade	1.000	S.F.		1.46	1.46
Edge form	.250	L.F.	.11	.63	.74
Welded wire fabric	.011	S.F.	.08	.30	.38
Concrete, 3000 psi air entrained	.012	C.Y.	.91		.91
Place concrete	.012	C.Y.		.22	.22
Broom finish	1.000	S.F.		.58	.58
TOTAL		S.F.	1.29	3.27	4.56

G2030 220	Sidewalks	COST PER S.F. MAT.	COST PER S.F. INST.	COST PER S.F. TOTAL
1600	For alternate sidewalk systems:			
1700	Asphalt (bituminous), 2" thick	.64	1.86	2.50
1800	Brick, on sand, bed, 4.5 brick per S.F.	2.36	7.05	9.41
1900	Flagstone, slate, 1" thick, rectangular	7	7.65	14.65
2000				
2100				
2200		1.29	3.27	
2300		-.11	-.63	
2400		1.18	2.64	
2500				
2600				
2700				
2900	Cut & patch to match existing construction, add, minimum	2%	3%	
3000	Maximum	5%	9%	
3100	Dust protection, add, minimum	1%	2%	
3200	Maximum	4%	11%	
3300	Equipment usage curtailment, add, minimum	1%	1%	
3400	Maximum	3%	10%	
3500	Material handling & storage limitation, add, minimum	1%	1%	
3600	Maximum	6%	7%	
3700	Protection of existing work, add, minimum	2%	2%	
3800	Maximum	5%	7%	
3900	Shift work requirements, add, minimum		5%	
4000	Maximum		30%	

+9%   +17%

Figure 27.57

order to find the necessary tax information. Percentages for contingencies and sales tax are included in the Cost Analysis sheet in Figure 27.65.

A 10% figure for architectural and engineering fees is added to give the owner a better idea of the total project costs. The actual costs would be negotiated by the owner and architect.

The Assemblies Estimate for the sample project could be completed in less than one day, after initial discussions and a site evaluation. When performed with sound judgment and proper estimating practice, the Assemblies Estimate is a valuable budgetary tool.

This book has stressed the importance of the site visit and evaluation. Every existing structure is different and must be treated accordingly. The estimator will be properly prepared to analyze and estimate remodeling and renovation projects only with a thorough understanding of how the existing conditions will affect the work.

G2020 Parking Lots

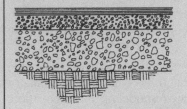

This page illustrates and describes asphalt parking lot systems including asphalt binder, topping, crushed stone base, painted parking stripes and concrete parking blocks. Lines within System Components give the unit price and total price per square yard for this system. Prices for alternate asphalt parking lot systems are on Line Items G2020 230 1500 thru 2500. Both material quantities and labor costs have been adjusted for the system listed.

Factors: To adjust for job conditions other than normal working situations use Lines G2020 230 2900 thru 4000.

Example: You are to install the system and match existing construction. Go to Line G2020 230 2900 and apply these percentages to the appropriate MAT. and INST. costs.

System Components			COST PER S.Y.		
	QUANTITY	UNIT	MAT.	INST.	TOTAL
Parking lot consisting of 2" asphalt binder and 1" topping on 6" Crushed stone base with painted parking stripes and concrete parking blocks					
Fine grade and compact subgrade	1.000	S.Y.		1.37	1.37
6" crushed stone base, stone	.320	Ton	3.34	.92	4.26
2" asphalt binder	1.000	S.Y.	3.03	.87	3.90
1" asphalt topping	1.000	S.Y.	1.68	.58	2.26
Paint parking stripes	.500	L.F.	.07	.06	.13
6" x 10" x 6' precast concrete parking blocks	.020	Ea.	.66	.26	.92
Mobilization of equipment	.005	Ea.		1.53	1.53
TOTAL		S.Y.	8.78	5.59	14.37

G2020 230	Parking Lots, Asphalt	COST PER S.Y.		
		MAT.	INST.	TOTAL
1400	For alternate parking lot systems:			
1500	Above system on 9" crushed stone	10.45	5.65	16.10
1600	12" crushed stone	12.15	5.75	17.90
1700	On bank run gravel, 6" deep	13.95	5.20	19.15
1800	9" deep	18.20	5.30	23.50
1900	12" deep	22.50	5.40	27.90
2000	3" binder plus 1" topping on 6" crushed stone	10.25	5.85	16.10
2100	9" deep crushed stone	11.90	5.95	17.85
2200	12" deep crushed stone	13.60	6.05	19.65
2300	On bank run gravel, 6" deep	15.40	5.45	20.85
2400	9" deep	19.65	5.55	25.20
2500	12" deep	24	5.65	29.65
2600				
2700				
2900	Cut & patch to match existing construction, add, minimum	2%	3%	
3000	Maximum	5%	9%	
3100	Dust protection, add, minimum	1%	2%	
3200	Maximum	4%	11%	
3300	Equipment usage curtailment, add, minimum	1%	1%	
3400	Maximum	3%	10%	
3500	Material handling & storage limitation, add, minimum	1%	1%	
3600	Maximum	6%	7%	
3700	Protection of existing work, add, minimum	2%	2%	
3800	Maximum	5%	7%	
3900	Shift work requirements, add, minimum		5%	
4000	Maximum		30%	
		2.41	2.43	

Figure 27.58

G2050 Landscaping

This page describes landscaping—lawn establishment systems including loam, lime, fertilizer, side and top mulching. Lines within system components give the unit price and total price per square yard for this system. Prices for alternate landscaping—lawn establishment systems are on Line Items G2050 420 1100 and 1200. Both material quantities and labor costs have been adjusted for the system listed.

Factors: To adjust for job conditions other than normal working situations use Lines G2050 420 2900 thru 4000.

Example: You are to install the system and provide dust protection. Go to Line G2050 420 3200 and apply these percentages to the appropriate MAT. and INST. costs.

System Components	QUANTITY	UNIT	COST PER S.Y.		
			MAT.	INST.	TOTAL
Establishing lawns with loam, lime, fertilizer, seed and top mulching **On rough graded areas.**					
Furnish and place loam 4″ deep	.110	C.Y.	2.11	.65	2.76
Fine grade, lime, fertilize and seed	1.000	S.Y.	.28	2.16	2.44
Hay mulch, 1 bale/M.S.F.	1.000	S.Y.	.08	.38	.46
Rolling with hand roller	1.000	S.Y.		.69	.69
TOTAL		S.Y.	2.47	3.88	6.35

G2050 420	Landscaping - Lawn Establishment	COST PER S.Y.		
		MAT.	INST.	TOTAL
1000	For alternate lawn systems:			
1100	Above system with jute mesh in place of hay mulch	3.09	3.97	7.06
1200	Above system with sod in place of seed	4.33	1.40	5.73
1300				
1400				
1500				
1600				
1700				
1800				
1900				
2000				
2100				
2200				
2300				
2400				
2500				
2600				
2700				
2900	Cut & patch to match existing construction, add, minimum	2%	3%	
3000	Maximum	5%	9%	
3100	Dust protection, add, minimum	1%	2%	
3200	Maximum	4%	11%	
3300	Equipment usage curtailment, add, minimum	1%	1%	
3400	Maximum	3%	10%	
3500	Material handling & storage limitation, add, minimum	1%	1%	
3600	Maximum	6%	7%	
3700	Protection of existing work, add, minimum	2%	2%	
3800	Maximum	5%	7%	
3900	Shift work requirements, add, minimum		5%	
4000	Maximum		30%	

Figure 27.59

G1020 Site Demolition and Relocations

G1020 210	Selective Price Sheet	UNIT	TOTAL COST
0100	Cabinets, base	L.F.	7.75
0200	Wall	L.F.	7.75
0300	Carpet, bonded	S.F.	.31
0400	Tackless	S.F.	.07
0500	Ceiling, tile, adhesive bonded	S.F.	.69
0600	On suspension system	S.F.	.82
0700	Sheetrock, on furring	S.F.	.78
0800	On suspension system	S.F.	.86
0900	Plaster, on wire lath	S.F.	1.09
1000	On suspension system	S.F.	1.09
1100	Chimney, brick, 16" x 16"	V.L.F.	21
1200	20" x 20"	V.L.F.	37.50
1300	Concrete, footing, 1' thick, 2' wide	L.F.	11.10
1400	2' thick, 3' wide	L.F.	19
1500	Slab, 6" thick, plain	S.F.	4.73
1600	Mesh reinforced	S.F.	5.20
1700	Wall, interior, 6" thick	S.F.	11.10
1800	12" thick	S.F.	17.75
1900	Door and frame, wood	Ea.	44.50
2000	Hollow metal	Ea.	67
2100	Ducts, small size, 4" x 8"	L.F.	1.55
2200	Large size, 30" x 72"	L.F.	6.20
2300	Fascia, to 6" wide	L.F.	.62
2400	To 10" wide	L.F.	.78
2500	Flooring, brick	S.F.	1.31
2600	Ceramic	S.F.	.99
2700	Linoleum	S.F.	.44
2800	Resilient tile	S.F.	.62
2900	Terrazzo, cast in place	S.F.	2.07
3000			
3100	Wood, strip	S.F.	1.22
3200	Block	S.F.	.99
3300	Subflooring, tongue and groove boards	S.F.	1.22
3400	Plywood	S.F.	.66
3500	Framing, steel girders, 10" x 12"	L.F.	13.90
3600			
3700	Wood, studs, 2" x 4"	L.F.	.31
3800	Rafters, 2" x 8"	L.F.	.74
3900			
4000			
4300	Gutters, attached	L.F.	1.29
4400	Built in	L.F.	3.11
4500	Masonry, veneer, by hand, brick to 4" thick	S.F.	2.70
4600	Marble to 2" thick	S.F.	2.10
4700	Granite to 4" thick	S.F.	2.22
4800	Stone to 8" thick	S.F.	2.16
4900	Walls, brick, 4" thick	S.F.	5
5000	8" thick	S.F.	6.65
5100	12" thick	S.F.	8.10
5200	16" thick	S.F.	9.90
5300	Block, 4" thick	S.F.	1.10
5400	6" thick	S.F.	1.46
5500	8" thick	S.F.	2.19
5600	12" thick	S.F.	2.67
5700	Paneling, plywood	S.F.	.31
5800	Woodboards, tongue and groove	S.F.	.89

Figure 27.60

COST ANALYSIS

PROJECT	**Apartment Renovation**					SHEET NO.		
ARCHITECT						ESTIMATE NO.		
						DATE	**2002**	
TAKE OFF BY: **ABC**	QUANTITIES BY: **ABC**	PRICES BY: **RSM**	EXTENSIONS BY: **DEF**			CHECKED BY	**GHI**	

DESCRIPTION	SOURCE/DIMENSIONS			QUANTITY	UNIT	MATERIAL		LABOR		EQ./TOTAL	
						UNIT COST	TOTAL	UNIT COST	TOTAL	UNIT COST	TOTAL
Division G: Building Site Work											
Excavation (1/2 day)	A2010	140	-	91	CY	-	466	-	468		
Sidewalk	G2030	220	-	356	SF	1.29	459	3.09	1,100		
Asphalt	G2020	230	-	467	SY	2.41	1,125	2.43	1,135		
Landscaping	G2050	420	-	890	SY	2.47	2,198	3.88	3,453		
Demolition	G1020	210	-	1	LS	-	-	-	1,560		
Division G: Totals							4,249		7,716		

Figure 27.61

General Contractor's Overhead		
	% of Direct Costs	
Items of General Contractor's Indirect Costs	**As a Markup of Labor Only**	**As a Markup of both Mat. and Labor**
Field Supervision	6.0%	3.2%
Main Office Expense (see below)	14.7	7.7
Tools and Minor Equipment	1.0	0.5
Field Office, Sheds, Photos, etc.	1.5	0.8
Performance and Payment Bond (0.5 to 1.2% Average)	0.7	0.4

Figure 27.62

General Contractor's Main Office Overhead			
Item	**Typical Range**		**Average**
Managers, clerical and estimator's salaries	40%	to 55%	48%
Profit sharing, pension and bonus plans	2	to 20	12
Insurance	5	to 8	6
Estimating and project management (not including salaries)	5	to 9	7
Legal, accounting and data processing	0.5	to 5	3
Automobile and light truck expenses	2	to 8	5
Depreciation of overhead capital expenditures	2	to 6	4
Maintenance of office equipment	0.1	to 1.5	1
Office rental	3	to 5	4
Utilities including phone and light	1	to 3	2
Miscellaneous	5	to 15	8

Figure 27.63

COST ANALYSIS

PROJECT	**Apartment Renovation**					SHEET NO.	
ARCHITECT						ESTIMATE NO.	
						DATE	**2002**
TAKE OFF BY: **ABC**	QUANTITIES BY: **ABC**	PRICES BY: **RSM**	EXTENSIONS BY: **DEF**	CHECKED BY **GHI**			

DESCRIPTION	SOURCE/DIMENSIONS			QUANTITY	UNIT	MATERIAL		LABOR		EQ./TOTAL	
						UNIT COST	TOTAL	UNIT COST	TOTAL	UNIT COST	TOTAL
Estimate Summary											
Division A: Substructure							2,359		6,228		8,586
Division B: Shell							44,008		36,127		80,134
Division C: Interiors							69,209		75,811		145,020
Division D: Services							63,131		81,942		145,072
Division E: Equipment							28,043		12,409		40,452
Division G: Building Site Work							4,249		7,716		11,965
Subtotal							210,999		220,231		431,230
General Conditions				15%			31,650		33,035		64,685

Figure 27.64

417

COST ANALYSIS

PROJECT	**Apartment Renovation**		SHEET NO.	
ARCHITECT			ESTIMATE NO.	
			DATE	**2002**
TAKE OFF BY: **ABC**	QUANTITIES BY: **ABC**	PRICES BY: **RSM** EXTENSIONS BY: **DEF**	CHECKED BY	**GHI**

DESCRIPTION	SOURCE/DIMENSIONS			QUANTITY	UNIT	MATERIAL		LABOR		EQ./TOTAL	
						UNIT COST	TOTAL	UNIT COST	TOTAL	UNIT COST	TOTAL
Estimate Summary											
Division A: Substructure							2,359		6,228		8,586
Division B: Shell							44,008		36,127		80,134
Division C: Interiors							69,209		75,811		145,020
Division D: Services							63,131		81,942		145,072
Division E: Equipment							28,043		12,409		40,452
Division G: Building Site Work							4,249		7,716		11,965
Subtotal							210,999		220,231		431,230
General Conditions				15%			31,650		33,035		64,685
Subtotal							242,649		253,266		495,915
Contingency				10%			24,265		25,327		49,591
Sales Tax				5%			13,346		13,930		27,275
Total Construction Cost							280,259		292,522		572,781
Architectural and Engineering Fees				10%							57,278
TOTAL BUDGET											630,060

Figure 27.65

Part IV

Disaster Restoration Contracting

Chapter 28

Disaster Restoration Contracting

Origins of Restoration Contracting

During the 1950s, insurers expanded their homeowners' coverage to allow those owners who were victims of fires and other disasters to seek professional assistance in the restoration and cleanup of dwellings and personal property.

With the overwhelming popularity of wall-to-wall carpeting after World War II, carpet cleaning became a lucrative business for many entrepreneurs. Ambitious carpet cleaning companies quickly recognized the potential of providing cleanup services to individuals who had experienced a fire or other specific damages. Painting contractors and remodeling companies also extended their services in what became a new field called Fire Restoration.

These new firms were not just beneficial to property owners. They soon became a valuable tool for insurance companies and played an intricate role in the claims process.

Equipment manufacturers found a new market for a variety of cleaning and restoration products. The production of equipment specifically tailored to the disaster restoration contractor has become a huge industry that is fueled by the need to reduce a loss by restoring property rather than by replacing it.

The Insurance Claims Process

The successful disaster restoration contractor must be informed about the insurance claims process, the distinctive roles of insurance professionals, and the unique aspects of the many different insurance policies. The insurance agent and the claims adjuster are important figures in claims processing. Since there are many different types of insurance policies, a clear knowledge of each policy type, along with its specific damage coverage, can save the disaster restoration contractor days of effort.

Agents and Adjusters

Selling insurance and providing service to clients is the insurance agent's main function. An agent may write business for one insurance company exclusively, or an independent agent may write business for several different insurance carriers. An agent's role during the insurance claims process is an important one. When a disaster occurs, the agent is often the first person contacted. Insurance representatives may have the authority to authorize emergency services, and it is not uncommon for an agent to have up to $5,000 in draft authority that may be issued on behalf of the carrier in emergency situations.

The claims adjuster works exclusively for one carrier or for an independent firm that handles claims for multiple insurance companies. The adjuster's responsibility is twofold: (1) to settle a claim precisely as it is spelled out in the policy, and (2) to work in the best interest of both the insurer and the insured. The adjuster's title and draft authority varies from one carrier to another, and depends on the adjuster's professional experience. A general adjuster has the greatest experience and the most draft authority. A public adjuster works directly for the policyholder and receives a percentage of the total loss (out of any funds paid to the policyholder).

The Claims Process

Following a loss, a policyholder contacts his insurance agent, who then files a claim with the insurance carrier. The agent is responsible for ensuring that emergency measures are taken to mitigate further damage to the property and to protect the insured's immediate well-being. If the insured is without a place to stay or without clothing or food, the agent should assist in addressing these needs.

On a small claim, the agent may have the authority to settle the claim on behalf of the insurance carrier. If the agent is not authorized to settle the claim, he or she is assigned to an adjuster. The adjuster then contacts the insured and inspects the damage. An insurance agent or adjuster often contacts a professional restoration contractor at the onset of a loss to take care of any emergency services and to assist in estimating the damage. The adjuster and the contractor agree on a scope of repairs and negotiate the unit costs. If the policy is in order and the insured is agreeable, the contractor may be given authorization to proceed with repairs before (sometimes weeks before) any final estimate is agreed upon.

Very seldom is a contractor put into a bidding situation with another contractor on an insurance/repair loss. This means that the contractor must have an impeccable reputation among insurance carriers and a strong working relationship with the insurance adjuster.

Insurance Policies

Insurance coverage is available for the dwellings and personal property of homeowners, renters, and owners of commercial premises for damage from fire or other perils. (In the insurance industry, a *peril* is a cause of a loss.) The standard homeowners and commercial policies provide coverage for a basic set of perils, with additional protection available through more comprehensive policies. The amount of coverage provided by a specific insurance policy depends on many factors, including the type of policy, the dollar protection the policy provides, as well as policy limits and exclusions.

The most basic standard homeowners policies are the *Basic HO-1* or the *Older Home HO-8*. These policies cover the following perils:

1. Fire or lightning
2. Loss of property removed from premises
3. Windstorm or hail
4. Explosion
5. Riot or civil upheaval
6. Aircraft
7. Vehicles
8. Smoke
9. Vandalism and malicious mischief
10. Theft
11. Breaking of glass on a part of the building

The *Special HO-3* is the most common of the more comprehensive policies. This policy, along with others such as the *Broad HO-2* and the *Comprehensive HO-5*, includes all of the specific damages listed above and the following additional perils:

12. Falling objects
13. Weight of ice, snow, and sleet
14. Collapse of buildings or any part thereof
15. Sudden and accidental tearing, cracking, burning, or bulging of a steam or hot water heating system or of appliances for heating water.
16. Accidental discharge, leakage, or overflow of water or steam from within a plumbing, heating or air conditioning system or domestic appliance
17. Freezing of plumbing, heating and air conditioning systems and domestic appliances
18. Sudden and accidental injury from artificially generated currents to electrical appliances, devices, fixtures, and wiring (TV and radio tubes not included).

Policies for renters, such as *Renters HO-4* and the *Comprehensive HO-4*, offer coverage for all 18 of the specific damages mentioned above, but only as the perils pertain to personal property (no dwelling coverage is provided). Additionally, the *Special HO-3*, for dwellings only, and the *Comprehensive HO-5*, for dwellings and personal property, offer protection for all perils except flood, earthquake, war, nuclear accident, and any particular damages specified in each policy.

Commercial policies are tailored to specific businesses and industries. In general, these policies are more comprehensive than the standard homeowners' policy, although they cover similar perils.

Distinctive Features of Insurance Policies

To have a thorough understanding of the distinctive features of varied insurance policy coverage and the claims process, the disaster restoration contractor must comprehend the unique aspects of insurance policy underwriting. Some policy contingencies may have an impact on the disaster restoration contractor's work that is not readily apparent.

- **Deductible**—The amount of the claim for which the insured is expected to be responsible is called the *deductible*. A typical homeowners policy might include deductibles that range from $250 to $1,000. Commercial deductibles might be much higher.
- **Coinsurance**—The policyholder (the insured) is commonly required to carry adequate coverage on a property (usually not less than 80% of the total replacement value). In the event that more than one insurance carrier has written coverage on a property, each insurer will be responsible for *only* its portion of the coverage. If the policyholder is underinsured, the property owner may be considered the coinsurer, responsible to pay for a portion of the loss. In some cases, there may be a penalty against a policyholder for not carrying enough property insurance. When a coinsurance penalty is levied against the policyholder, the insurer will not pay the full cost of the claim, and the insured will have to pay a substantial part of the repair bill. The disaster restoration contractor should be aware of this possibility.
- **Actual Cash Value (ACV)**—This figure usually represents the replacement cost less any depreciation. Some ACV policies are only responsible for payment of the depreciated value of a loss. For

example: If damaged carpeting had a life expectancy of 10 years and is already 5 years old, an ACV policy would be responsible to pay out only 50% of the value of the carpet.

- **Recoverable Depreciation**—When a carrier is required to pay out the actual replacement cost, he first pays only the depreciated value (actual cash value) of a claim until the damaged property has been repaired or replaced. At that time, the recoverable depreciated amount that was initially withheld is paid out to the insured.
- **Like Kind and Quality**—This term describes the standard by which replacement costs are estimated. Generally, insurers are expected to replace damaged property with "like kind and quality." This does not necessarily mean that the insurer must provide an identical replacement. The insurer is responsible for putting the property back into its pre-loss condition.
- **Sudden and Accidental**—This term is applied to a covered loss when the damage has occurred suddenly and accidentally.
- **Reserve**—State insurance codes require carriers to set aside adequate liquid funds to cover payouts for claims immediately following occurrence. Upon inspection of the loss, the claims adjuster is responsible for establishing an adequate reserve amount. Disaster restoration contractors are commonly called upon to assist in this estimating process. A reserve is simply an approximate estimate, set by the adjuster, of the amount the carrier will be paying out on a loss to make sure that there is money for that loss.
- **Sebrogation**—This term applies to the right given by a policyholder to the insurance company to act on his or her behalf against another party. If a loss is caused by the negligence of a third party, the insurance company may bring suit on behalf of the insured in an attempt to regain monies paid out on that claim.

The Professional Disaster Restoration Contractor

Disaster restoration contracting is a specialized field that is far different from any other form of contracting, and it requires continued effort to address its ever-changing intricacies. Disaster restoration contractors provide services in two areas of specialty: (1) structural repair, customarily thought of as remodeling, and (2) cleaning, which was traditionally carpet cleaning. Although most disaster restoration contractors specialize in one service or the other, some accept the challenge of offering both structural restoration and cleaning services.

This industry is constantly changing for several reasons, including the following:

- First, manufacturers are regularly coming up with new products and equipment to assist in drying out, cleaning, deodorizing, disinfecting, and repairing damaged property.
- Second, restoration contractors deal with all phases of cleaning and construction, unlike a carpet-cleaning company, a home builder, or a commercial contractor.
- Third, a restoration contractor commonly works on older buildings, often constructed of materials that are no longer available.
- Fourth, every job is different and presents unique challenges.
- Finally, insurance companies and policies are all very different, and the restoration contractor must learn to conform with each on every job.

Structuring the Company

Taking the time to give careful consideration to structuring the company is an important first step for the disaster restoration contractor. Unlike many remodeling niche markets, insurance repair contracting requires a substantial initial investment and maintenance of a considerable overhead. First impressions of the professional and the office are significant, because the insurance industry looks for stability, professionalism, and the ability to perform. Similarly, potential clients—homeowners or owners of commercial space—who have experienced a disaster prefer to be greeted by a friendly voice on the other end of the telephone rather than a message on an answering machine.

Facility needs will vary greatly depending on the size and structure of each individual company. One of the first considerations in structuring the company involves accessibility of the location: It must be easy to reach. Although the possibility of future expansion should always be a consideration, a minimum of 1,000 sq. ft. of office space plus a minimum of 800 sq. ft. for tool storage should be sufficient initially. However, contractors engaged in content cleaning require space for a cleaning station as well as for storage of contents and equipment. They can easily utilize 15,000 sq. ft. Content cleaning is the removal, storage, and physical cleaning of all of an insured's belongings. Contents include furniture, electronics, upholstery, clothing, dishes, books, artwork, and so forth.

An attitude of professionalism must exist throughout the entire restoration company, from first impressions of the office space and the office staff to the sales force and field personnel. Office and management staff should always be professionally attired, and all field personnel should be clean-cut and in uniform. Vehicles need to be clean and should display company door signs. All estimates and other correspondence should be presented in a professional manner. Following are some specific suggestion for the presentation of paperwork:

- All estimates must be computer-generated and presented in a professionally prepared company folder.
- All correspondence should be on company letterhead, or stationery with a company envelope.
- A professional company portfolio might contain "before" and "after" photographs, letters of recommendation, a clear explanation of the warranty, a code of ethics, sample forms, and perhaps some articles that have been written about the company.

The Office Staff

No matter how large or small a company is, a good office staff can make or break a business. The receptionist and the office manager usually hold two of the most important positions in the company (sometimes these two positions are held by the same person). For most customers, the initial telephone conversation with a receptionist provides the first impression of the business. A smoothly run, organized office reflects on all other aspects of a company.

The overall job responsibilities of an owner/manager might include some sales and production management, especially with smaller-volume businesses; but, the owner's primary responsibility must focus on management of the ongoing business operations. The successful disaster restoration contractor must come to terms with the recurring conflict of quality versus quantity. Attempting to juggle too many job responsibilities, by "wearing all of the hats," will stunt the growth of any company.

The financial aspect of any business is an important component and one that the entrepreneur cannot afford to overlook or to shortchange. Today's restoration contractor cannot survive without adequate financing and a substantial line of credit. Unlike most remodeling contractors, insurance repair contractors usually do not receive progress payments and often wait over thirty (30) days for receipt of payment after the completion of work. For this reason, restoration contractors must be able to finance multiple jobs over the space of several months. An accountant who is familiar with the construction industry and who takes an interest in your business can be helpful in setting up and maintaining a financial system that can accommodate the distinct needs of the restoration contractor. Clearly, establishing and maintaining a good relationship with such an accountant will be an asset to your business. Consulting with other contractors, your banker, and networking through trade organizations or your local chamber of commerce are good ways to find a CPA.

Many small businesses make the mistake of undervaluing the year-end financial statement. It is important that the owner/manager has a clear understanding of the difference between the cash and the accrual accounting methods, and the importance of keeping both. This knowledge enables a contractor to report an accurate financial status to the bank, while paying taxes only on income actually received. Forming a good relationship with a bank early on is a must for the disaster restoration contractor.

Business Operations

Changes in technology have become particularly apparent in the 1990s, and any business that is not computerized has struggled because of it. Most business professionals agree that failing to computerize by the turn of the century will be detrimental to the survival of their businesses. This statement is especially true for those involved in the insurance repair industry. Most insurance carriers require estimates that are produced by a recognized estimating software program. These programs allow the contractor to create numerous estimates quickly and accurately. The ability to produce estimates on this basis sets the contractor apart from any competition, who may be preparing estimates that are written out by longhand. Several other business functions, such as accounting, tracking sales, and job costing, mandate the use of computers in today's market.

Today's disaster restoration contractor invests over 15% of gross sales in sales and marketing opportunities. Approximately 7% to 8% is earmarked for sales commission, 3% to 4% for salesman burden and benefits, and from 3% to 5% for marketing and advertising. Disaster restoration contracting requires a substantial marketing and advertising expense. In addition to the typical advertising expenses incurred by most remodelers, the insurance repair contractor must build a good working relationship with numerous insurance professionals. This requires an ongoing marketing campaign.

Assembling a Catastrophe Response Team

Any disaster is an unwelcome tragedy, and it requires immediate attention. Maintaining a state of "readiness" is one of the disaster restoration contractor's most difficult tasks. Scheduling can be a nightmare in disaster reconstruction work, and expectations are often unrealistic. Disasters can occur at any given time and at any given place (often inconvenient) and require immediate response. Restoration contractors are regularly required to pull manpower off of a job-in-progress in order to perform emergency services at a new loss site. Additionally, that new job is expected to get underway the following day.

Insureds' expectations are often unrealistic because all of their attention is focused on getting their lives back to normal as quickly as possible. They are not interested in the effect any other projects may be having on the restoration contractor's workload. During repairs, the insured is often forced to live in a motel or with relatives, which can lead to some irritability and contribute to unrealistic expectations.

In order to move quickly after a disaster, the disaster restoration contractor must be able to assemble a response team from a qualified and reliable staff of multi-skilled craftsman. To maximize his readiness, the contractor must feel confident about the sources of his staff. Contractors have long debated the pros and cons of subcontracting versus maintaining an on-staff workforce. First of all, it is important that the disaster restoration contractor realize that many insurance carriers are reluctant to hire contractors who do not use their own workforce. Without a doubt, a good production manager and a sufficient staff of topnotch carpenters, painters, and cleaning technicians increases the efficiency of the response team. A production manager is responsible for supervising all field employees and scheduling all projects. Even though many contractors have been successful subcontracting the majority of their structural repairs, it can be difficult to supervise a subcontractor's schedule and work habits. It is common for all contractors, however, to subcontract their mechanical services, such as plumbing, electrical, and HVAC.

The specialized field of disaster restoration contracting requires a much higher level of training than most other areas of construction. For example, rebuilding a structure that has been saturated by smoke or water requires special skills. Equally as important as structural soundness is keeping the building free from odor and bacteria. Various organizations specialize in training professionals to deal with moisture, smoke, odor, and bacteria. (ASCR, the Association of Specialists in Cleaning and Restoration, has numerous institutes that deal specifically with all aspects of disaster restoration and provide several certifications.) The disaster restoration contractor who encourages or requires specialty certification of management and field employees will build credibility while ensuring that the work is done with knowledge and foresight.

Maintaining a Safe Environment

Safety is always an issue in contracting. Maintaining a safe environment requires a personal commitment from the owner and management of any organization. An ongoing, strictly enforced, safety policy can save thousands of dollars in workers' compensation premiums and prevent injury to valued employees.

The Psychology of Loss

A disaster, no matter how large or how small, is a tragedy in any individual's life. It is essential that the disaster restoration contractor have an understanding of the emotional effects of disaster trauma on people's lives.

Whether the difficulty is a sewer backup in a basement or a tornado that affects an entire community, the disaster creates human suffering. A catastrophic loss such as a tornado, an earthquake, a flood, or a hurricane can severely elevate the level of trauma. It is common in these situations that the reality is not only loss of a home and all personal belongings, but also the destruction of churches, schools, and places of work. Sometimes a disaster brings with it the greatest tragedy, the loss of life.

Human reaction to loss can involve anxiety, depression, hysteria, anger, physical illness, and post-traumatic stress disorder (PTSD). Typically, there is a tendency among people who are suffering traumatic loss to make emotional and hasty decisions or to fail to take action by beginning to rebuild their lives.

On almost every project, the disaster restoration contractor deals with customers who are undergoing emotional stress. The contractor cannot avoid the need to, in effect, "take them by the hand, walk them through the process," and make them understand that he is part of the solution. The restoration contractor may need to:

- Assist clients as they proceed, step-by-step, through the claims process (this includes contacting their agent if they have not already done so).
- Make sure clients have a place to stay.
- See to it that some of the clients' clothes are immediately cleaned and disinfected, and returned to them.
- Ensure that all of the clients' valuables are protected.
- Work as a mediator between the clients and their insurance adjuster to help expedite the process.
- Explain in detail what clients might expect.

Simply because he is associated with unwanted circumstances, the contractor is often seen as part of the problem. For example, every disaster is unforeseen and unwanted. Even though the restoration contractor is the instrument needed to repair a home following a disaster, he is seldom looked upon as a blessing. This attitude can be difficult to overcome. In contrast, a remodeler who is adding a room for a family has a different relationship with the client. In that case, the client and contractor have carefully thought out the project and are excited about the outcome. It is important for the contractor to realize that he is there because of a serious problem, an unwanted problem, and one that is not quickly or easily resolved.

Disasters occur in any area at any given time, and affect people from all walks of life. The disaster restoration contractor must have the ability to communicate with all kinds of people at any level. Recognizing concerns, situations, and potential problems as they arise can help build a good working relationship between the contractor and the insured.

The disaster restoration contractor benefits from maintaining a clear line of communication with potential clients. Few people understand the cost of restoring a property after a disaster. Because of the nature of a disaster and the need for immediate action, costs are seldom discussed with the insured at the onset of a project. Usually the insurance adjuster and the contractor negotiate them later on. This can often create problems with the insured and cause unwanted tension. However, a decisive explanation of the claims process and restoration process and the costs involved may mitigate this situation.

A Clouded Perspective

More and more often, people look at insurance as an investment. They have the philosophy that since they pay premiums for years and years, the insurance company owes them something in return. When a loss occurs, these people seem to feel that it is an opportunity to cash in on the investment. In a situation like this, the restoration contractor can end up in the middle, between the insurance company and the insured, and benefits from approaching the job with caution.

Cautions for the Restoration Contractor

It is the restoration contractor's responsibility to offer an unbiased professional opinion as to what is pre-existing damage and what might have been caused by an insured peril. This opinion is often unpopular with either the insured or the insurer. It is important for the contractor to stay on neutral ground and not to allow the insured or the insurer to use him in an arrangement against the other.

Some scenarios concerning the identification of damages are quite typical. For example: The restoration contractor will often find that property is described as being in pristine condition prior to a loss. Although many property owners recognize that there was previous damage unrelated to the loss and know that they are simply trying to exploit the situation, some people actually seem to believe that unrelated damage was caused by a disaster. Sometimes these descriptions occur because attention was not previously focused on that structural area. This is often true for cracks in ceilings, walls, and floors, which are often never noticed prior to a fire. However, after a fire, all areas of a building are closely inspected for damage.

There are always those individuals who want "something for nothing" or who convince themselves that they have "something coming to them." These people understand that the contractor and the insurance companies must maintain a good relationship, and they will use that knowledge as leverage to realize their own goals. Dealing with demands such as these puts the contractor in a difficult position, and it should be avoided at all costs.

Emergency Services

Immediately following a disaster, time is of the essence, and a matter of hours can make all the difference. All too often the cost of a claim escalates and valuable items are a total loss due to delays in making the decision to perform emergency services. Property owners are unaware of what decisions to make, and inexperienced adjusters are often reluctant to make decisions.

Loss mitigation is the responsibility of all who are involved in a claim. Insurance policies require the insured to take necessary action to negate further loss or damage. Insurance representatives should assist in the process of securing a property to avoid continued damage.

Services Provided Immediately

Restoration contractors play a significant role in loss reduction. They must be prepared, 24 hours a day and 365 days a year, to provide adequate emergency services, which are needed to prevent further damage and to secure the property. A disaster restoration contractor must be ready to provide the following at a moment's notice:

- Building board-ups and tarp-ups
- Water extraction and dehumidification
- Winterize plumbing
- Secure or demolish dangerous structures
- Debris cleanup
- Content removal and storage
- Temporary power, lighting, and even heat source

As luck would have it, disasters never seem to occur at an opportune moment. Since most emergencies occur after hours and on weekends or holidays, a prescheduled, on-call rotation for available staff and an adequate inventory of materials and supplies will add greatly to the efficiency of the restoration contractor's response.

Securing a Property

The restoration contractor's first step following a disaster is to secure the property from (1) further ravages by the elements, and (2) from intruders. Immediate protection of the property should include the following:

- Cover all openings
- Clean up glass or similar debris around the property
- Determine the necessity of enclosing the property with a temporary fence or barricade

A damaged building must be protected from further destruction as well as from any further liability. Shrinkwrapping is becoming a popular method of securing buildings that are missing part or all of a roof structure. This procedure involves the use of a plastic material (usually 7 or 8 mil) that is placed over the roof and securely fastened to the building. Wood supports and nylon straps support the plastic so it can cover a wide-open area. Often, the plastic is heated with a propane torch until it shrinks to approximately one-third of its original size. The building becomes drum tight and can effectively shed water. The shrinkwrap process can also be used to cover window openings, to enclose rooms, and to build temporary barricades.

Effects of Moisture, Smoke, and Temperature

Every part of a building may be vulnerable to damage by water—whether water was the source of the disaster, whether it came in the form of rain, or whether the water was used to extinguish a fire. Whatever the cause, it is crucial that the water and humidity be removed immediately in order to protect the structure from further damage. Moisture can cause more damage than any other element. Floors soon buckle, wood furniture and molding starts to split and delaminate, carpet backings separate, color dyes bleed, and mold and bacteria set in.

Smoke contamination quickly corrodes metals. All metal fixtures such as faucets, lights, and hardware should be immediately wiped clean of smoke and treated with a lubricant. Computers and electronics are especially vulnerable to smoke contamination. They should be removed from the premises, cleaned with deionized water, dried out and stored in a clean area until the structure has been restored.

Winterization of a structure without heat or utilities will avoid further damage to the building from frozen water pipes and broken toilets. All toilets andp-traps should be filled with antifreeze, and water lines should be turned off and drained. When it is necessary to dry out a building, the disaster restoration contractor may need to arrange for a temporary heat source.

Fundamentals of Estimating for Insurance Repair

Estimating practices for the insurance repair industry are unique. They are both different from and similar to those used in remodeling estimating. Insurance repair contracting is by far the most overhead-intensive of all forms of contracting. For example: Industry demands require that the restoration contractor have an advanced level of knowledge and expertise in all areas of construction and cleaning. Education and training programs must be maintained on an ongoing basis. A substantial workforce is a necessity to the successful restoration contractor. This commitment also increases overhead expenses, because it requires additional accounting personnel and administrative supervisory personnel. A restoration contractor must monitor all construction tools and equipment, and he must also have available an enormous amount of cleaning and dry-out equipment. Vehicles for transporting contents must also be maintained. Adequate storage facilities are a must. Maintaining a good working relationship with several insurance companies requires an ongoing marketing campaign. Managing multiple projects that very seldom issue progress payments requires a substantial line of credit. All of these attributes dictate a tremendous amount of overhead.

Markup and Gross Profit

The line item unit cost estimate, with 10% profit and 10% overhead added to the bottom line, is the traditional format for insurance repair estimating. This practice was adopted over 30 years ago by the insurance industry, and it has never been updated to match current practices. A 10% & 10% markup (20%) represents a 16.6% gross profit, which covers only sales and marketing expenses. This practice leaves no funds in reserve for other overhead expenses and produces a negative profit.

In order to stay in business and to continue serving clients, a restoration contractor requires a gross profit of between 35% and 40%. All overhead expenses and annual budgets are based on a percentage of gross annual sales, rather than on a percentage of the cost of sales. Therefore, the direct cost must be marked up 67%, if a contractor is going to make a 40% gross profit.

It is important to understand, however, that *markup does not automatically equal gross profit*. This concept may be the most misunderstood estimating fundamental in the contracting industry, and it is a contributing factor to the high failure rate of most contractors.

Consider these formulas for calculating gross annual sales and gross profit:

Markup + Direct Costs = Gross Annual Sales

Markup/Gross Annual Sales = Gross Profit

For example, if a contractor does a job that costs $1,000 and marks it up only 40%, the total sale would be $1,400. However, the gross profit would be only 28.5% ($400/$1,400 = .285). In order to achieve a 40% gross profit, a contractor must mark up the cost by 67% ($670/$1670 = .400).

A standard rule of thumb throughout the insurance/restoration contracting industry is: Job cost plus a 67% markup. How then do you explain a lingering adherence to 10% & 10% markup? It is a standard practice in those cases for the line item unit cost to represent the direct cost, plus 39.2%. Then, by adding 10% & 10% (or 20%), a 67% markup is achieved. For example: $1,000 × 1.392 × 1.2 = $1,670 (or a 67% markup).

The disaster restoration contractor must achieve a 67% markup of direct costs, which is the same as making a 40% gross profit, because the contractor's average overhead is 35% of gross sales. This leaves a 5% net profit. To understand these calculations in greater depth, let's look at the overhead expenses, or the annual operating budget, of a typical restoration company with $1 million in gross annual sales.

XYZ Restoration Company—$1 million in Gross Annual Sales Annual Operating Budget

Owner/manager salary (5% of gross sales)	$ 50,000
Sales staff salary or commission (8% of gross sales)	80,000
Production manager salary	30,000
2 Office personnel–secretary & bookkeeper	40,000
Burden & benefits on office personnel (30% of their payroll)	60,000
Office & warehouse payment	24,000
Utilities & telephones (including 3 mobile phones)	20,000
Tools & equipment upkeep	11,000
4 Company vehicles (leased or owned)	20,000
Insurance (building, equipment & vehicles only)	8,000
Office equipment & supplies	7,000
Advertising & marketing (5% of Gross Sales)	50,000
Professional fees	6,000
Educational seminars, conventions, etc.	6,000
Bad debt (3% of sales)	30,000
Interest, penalties, etc.	8,000
Total Overhead	**$450,000**

Let's suppose that this restoration company is conservative, is able to reduce a few salaries, lower its utilities and rent, has only the bare minimum in equipment and vehicles, does little advertising, incurs no bad debt, and never has to borrow money! Let's reduce their overhead by $100,000 to $350,000. If the gross annual sales are $1 million, the overhead is 35% of the Gross Sales. As various contractors will do more or less than $1 million in sales, their overhead will adjust accordingly. However, their percentage of overhead will not deviate from the 35% range. If this company strives to achieve a 40% Gross Profit and their overhead is 35% Gross Annual Sales, their Net Profit will be 5% of Gross Annual Sales.

High markups are common in other industries as well. For example, it is common for many independent adjusting firms to pay their adjusters 50% of the rate charged for their total billable hours (in this case the adjuster would be responsible for their own expenses, etc.). This means that independent adjusting firms commonly charge a 100% markup on their services. The automotive repair industry has been far more organized than the contracting industry. It is common for automotive repair shops to charge a 150% to 200% markup on their services. Automotive mechanics make relatively the same amount of wages as construction workers (and in most cases, much less); however, a mechanic's standard hourly charge would be in excess of $60 per hour. Contractors perform services on a person's primary investment (his home) and make housecalls, while mechanics work on a person's secondary investments (his cars) and their customers have to come to them. It is also standard for restaurants and retail merchandisers to mark their services and products up over 100%.

Markup Considerations for Contractors

For contractors, job cost includes all costs directly pertaining to the completion of a project except overhead expenses. It also includes direct costs, such as employee labor and burden, all materials and subcontractors, permits, fees, debris removal, and so forth.

Contractors must consider burden, which is the additional cost to an employer above the employee's hourly or salaried wage. This cost directly pertains to that employee. It includes: Workers' Compensation insurance and general liability insurance. This insurance is based on a percentage of gross payroll, FICA unemployment, and employee benefits. For example, if a carpenter makes $16 per hour, the actual hourly cost to his employer is $23.55.

Hourly Wage Plus Burden Breakdown—Open Shop

Hourly wage	$16.00
Workers' Comp. insurance (1996 national average of 20.2%)	3.23
General liability (8%)	1.28
FICA (7.65%)	1.22
Medical & Dental	.72
Vacation & Holiday (10 days total)	.86
Retirement/401K	.24
Total wages plus burden	**$23.55**

Using the 67% markup formula to cover a 35% overhead and a 5% net profit, that Carpenter should be charged out at $39.33 per hour (23.55 × 1.67 = $39.33, or $23.55 × 1.392 = $32.78 (plus 10 & 10) × 1.2 = 39.33)

The question often arises as to how much markup should pertain to a subcontractor. Since work that is subcontracted out is still part of a company's gross sales markup for a subcontractor should be treated no different from any other. Not marking it up accordingly would force markup for other non-subcontract items to be increased in order to accommodate the difference.

Most contractors become involved in insurance work under the misconception that the income is guaranteed. In reality this is not always the case. Although many insurance companies issue drafts with the contractor's name on it, some neglect to protect the contractor's interest. Insureds sometimes fail to pay their deductible. Progress payments are seldom issued (unless the project is in excess of $100,000) and when a draft for payment is finally made after the completion of the project, the mortgage company must inspect and then sign off on the draft. This can all be a time consuming and costly process!

Fire and Smoke Damage Considerations

To accurately estimate fire and smoke damage repairs, it is important to understand the characteristics of smoke. Although burned areas of a structure are relatively obvious when exposed, smoke can be difficult to recognize, and even more difficult to neutralize. Understanding the source of the fire, the type of burning product, and the surrounding area is essential in determining the cleanup process and the cost involved.

Characteristics of Smoke

Smoke is made up of unburned by-products of combustion. Three general components of smoke are:

- Solid
- Aerosol
- Gas

The character of smoke depends on (1) the product that burns, (2) the rate and time of combustion, and (3) the nature of the materials. Oxygen-rich fires produce a dry residue, while oxygen-starved fires produce a wet residue. Oxygen-starved fires usually do more smoke damage than oxygen-rich fires do, and they are harder to clean.

Smoke behavior is determined by:

- Temperature of immediate areas
- Temperature of surrounding areas
- Ionization
- Existing pattern of airflow
- General arrangement of space
- HVAC system
- Type of materials burned
- The surface of the structure

Temperature greatly affects the behavior of smoke, which tends to move towards cooler areas, such as windows, and dead air spaces, such as closets or behind drapes. This type of drift is attributed to the *convection principal:* There is less energy in cooler places.

The five categories of smoke residue are:

Substance	Type of Fire	Residue
Wet:	Oxygen-starved, smoldering fire	Heavy, greasy, sticky residue
Dry:	Oxygen-rich, fast-burning	Dry, dusty residue
Plastic:	Example: burning wastebasket	Residue smears easily
Protein:	Example: meat burning	Slight residue, sticky yellow (sometimes invisible), heavy odor
Furnace or Fuel Oil:	Example: Puffback	Heavy, smeary residue

Furnace Puffbacks

A *furnace puffback* is uncontrolled or unbalanced fuel, oxygen, and heat. Following are six causes of furnace puffbacks:

1. Inadequate chimney draft
2. Work ignition spark
3. Incorrect gap on ignition
4. Contaminated fuel

5. Oil vapors collecting in furnace
6. Poor installation

Ionization (Smoke Webs)

Ionization is the polarizing of smoke particles due to a positive or negative charge (opposites attract). *Smoke webs* are ionization of smoke particles, while *cob webs* are ionization of dust particles.

Testing for Smoke Residue

Casual observation can reveal smoke damage by the principle of contrast. Following are three sets of criteria for noticing the presence of smoke residue:

- Soot residue (residue of fuel) is gray and sticky.
- Dust residue is gray and dry.
- Smoke residue is black.

Ideal areas to test for smoke residue are:

- On walls above doors
- Velvet or napped upholstery
- The backs of draperies
- The top linens in a closet
- On carpeting along the walls
- Windows
- Dead airspace
- Areas with contrasting temperatures

Gases

Gases are dangerous during a fire, but there are no known hazards after a fire is out and the building has been ventilated. The most dangerous components to humans are invisible gases. Two of the most harmful gases are carbon monoxide and hydrogen cyanide.

Deodorization

Deodorization of a structure and its contents is one of the restoration contractor's most important tasks following a disaster. Methods of deodorization fall into four general categories, as follows:

- **Removal** by demolition or wiping off
- **Sealing** by encapsulating in a paint product, such as Kilz
- **Chemical change** through use of a disinfectant, ozone, or chemical fog
- **Masking** with a deodorizer

Ozone Facts

Ozone is a reactive form of oxygen that is bluish and irritating, with a pungent odor. Although it is a beneficial component of the upper atmosphere, ozone is a major pollutant of the air at lower levels. The following are other attributes of ozone that the restoration contractor should know:

- The chemical nature of ozone is O_3.
- Ozone oxides any odor pollutant.
- Water extinguishes ozone.
- The TLV (threshold limit value) of ozone is 2 to 3.7 parts per million.
- The odor threshold for ozone is 1 part per million over 8 hours.
- Ozone neutralizes odors with opposite reaction.

Duct Cleaning

Proper air duct cleaning should be performed after most fire and smoke damage. Even if the HVAC system is not operating during the fire, the dead air space and the difference in temperature cause the air ducts to act like a magnet for smoke residue.

Water Damage and Moisture Control

Restoration companies engaging in water extraction and dehumidification must make a substantial investment in dry-out equipment. Moisture damage can be very costly, and accurate assessment will not be revealed until after a fire structure is sufficiently dry and the humidity is stabilized.

It is common for water to find its way into concealed areas such as wall cavities. When this occurs, the wall must be accessed so that wall cavities can be dried out and treated with mildewcide. This process can be achieved by removing the baseboard and drilling a 3/4″ access hole between each stud. The wall cavity can then have dry air forced into it to dry it out, after which it can be fogged with a mildewcide. Insulation may block the passage of dry air and prevent mildewcide from reaching the upper area of a wall cavity. Additionally, if the insulation has been saturated, it may need to be replaced. In this case the walls may need to be at least partially replaced. Failure to adequately dry out and treat wall cavities with a microbial will result in mold and bacteria growth and will contribute to sick building syndrome.

Following a disaster, buildings are particularly vulnerable to damage by moisture. Some form of water extraction and dehumidification is necessary during most disaster response and restoration work. The formation of dew occurs when the temperature falls below the dew point, or the temperature at which moisture in the air begins to condense. A building that has not been sufficiently dried out prior to making repairs will experience cracking wood, peeling paint, mold and mildew growth, bacteria growth, and possibly, sick-building syndrome. Preferred methods of structural drying are the following:

- Extraction
- Air movement
- Dehumidification
- Removal of saturated material
- Warming the air

A structure is sufficiently dry to undertake repairs when normal ambient humidity is present, and when walls, ceilings, and floors are sufficiently dry.

Figure 28.1 shows the water-carrying capacity of air at various temperatures and levels of relative humidity. For example, air at 50°F and relative humidity of 80% holds .20 oz of water. Air at 70°F and RH of 80% holds .40 oz, double the quantity it holds at 50°. Examination of other ranges shows that a temperature increase of 20° doubles the quantity of water vapor present. The maximum water-carrying capacity of air is attained at 100% relative humidity.

The restoration contractor needs to understand the following basic facts about the water-carrying capacity of air:

1. *Water vapor content* in the air is measured with a *hygrometer*.
2. *Absolute humidity* is the actual weight of the humidity (grams of water per cubic meter of air).
3. *Ambient humidity* is the general humidity in a room or area.
4. *Relative humidity* is the ratio of the quantity of water actually present to the amount present in a saturated atmosphere at a given temperature, expressed as a percentage.

5. Relative humidity will decrease if cold air is brought into a warmer building, and it will rise if warm air is brought into a colder building. As the temperature of air rises, so does its capacity to contain water.

Bacteria and mildew require four primary sources to grow: moisture, warmth, lack of air flow, and darkness. If the ambient humidity of a building or space remains high, every effort should be made to find the source of the water vapor. If left unmitigated, it will probably lead to the growth of bacteria and mold.

Cleaning Methodologies

It is impossible to mention disaster restoration without talking about cleaning. The very nature of a disaster involves a substantial cleanup process. The fact that restoration involves cleaning is what makes it such a specialty.

Cleaning may be described as the action to remove foreign residue. Although there are numerous cleaning methodologies, cleaning can be categorized into four basic groups:

1. *Solvent action* is the process of dissolving one substance into another when mixed together. A mechanical action may speed up this process.

2. *Chemical action* takes place between a spotting agent and a stain.

3. *Mechanical action* is the process of removing stains by mechanical means, along with its resulting friction. This method is frequently associated with lubrication or solvent action.

4. *Lubrication* is the process of dislodging foreign particles and soil from a surface by lubricating them to make them smooth and slippery. Cleaning with soap or detergent is a form of lubrication.

Cleaning can also be broken down into three other categories:

- *Dry cleaning* can be either mechanically or manually applied; it involves the use of a dry volatile solvent.

Water-Carrying Capacity of Air at Various Temperatures & Relative Humidity Levels										
F	**10%**	**20%**	**30%**	**40%**	**50%**	**60%**	**70%**	**80%**	**90%**	**100%**
100°	.12	.25	.37	.49	.62	.74	.87	.99	1.05	1.17
90°	.09	.19	.27	.37	.46	.53	.62	.69	.81	.93
80°	.06	.12	.19	.25	.32	.40	.46	.53	.62	.68
70°	.04	.10	.15	.19	.26	.29	.33	.40	.43	.49
60°	.03	.07	.11	.14	.18	.20	.26	.28	.31	.36
50°	.025	.05	.07	.09	.12	.15	.18	.20	.23	.26
40°	.02	.04	.06	.07	.09	.11	.13	.14	.16	.18
30°	.01	.03	.04	.05	.06	.07	.09	.01	.11	.12
20°	.01	.02	.02	.03	.04	.05	.06	.06	.07	.08
10°	.005	.009	.01	.02	.03	.03	.03	.04	.04	.04

Figure 28.1

- *Wet cleaning* can be either mechanically or manually applied and involves the use of water-based cleaning and vacuum extraction.
- *Dry/wet/dry cleaning* is usually mechanically applied and involves the use of odorless mineral spirits, hot water extraction, steam cleaning, hot detergent solution, and a thorough extraction. Only small areas are cleaned at a time to allow rapid drying.

No matter what cleaning methodology is used, it is equally important to kill any bacteria that results from smoke and water. The proper use and application of a disinfectant will serve that purpose.

Chemical and equipment manufacturers offer products for virtually every application, and new ones are introduced to the market every day. The cleaning end of the restoration business requires a substantial investment in equipment, cleaning chemical inventory, and ongoing training. This is one contributing factor to the high cost of overhead.

Catastrophe Response Work

Disaster restoration firms engaging in catastrophe response or working a natural disaster site, often referred to as a *cat loss*, must be prepared for the special challenges of this kind of work. Before taking on disaster restoration work, the contractor must review the present condition of the company.

The answers to these ten initial questions will provide the restoration contractor with an overall understanding of the current status of the company. They will help the contractor decide whether to take on the cat loss job and/or prepare him for the decision-making necessary to meet the demands of the work.

1. What is the current financial condition of my company?
2. What is the current workload of projects for which I am responsible?
3. Do I have adequate staff to double my workload?
4. What resource do I have to increase my staff?
5. Where is the disaster site in proximity to my local office?
6. What are the licensing requirements of that location?
7. What is the availability of lodging?
8. What is the availability of materials and supplies?
9. What insurance contacts do I have at the site?
10. Will taking on a cat loss have a negative effect on my local business?

A Plan
Once it has been determined that you have the resources available and that your current business will not suffer any negative effects, a plan of attack can be formulated. Time is of the essence in response to a disaster, and a decision to act should be made within the first 48 hours following the catastrophe.

Step 1
Contact the state in which the disaster occurred and find out all of the licensing requirements. Many states have stringent requirements that may involve rigorous testing, which can be very time-consuming. Hiring an individual who holds a General Contracting License in that particular state might be an alternative, in this instance.

Step 2
Contact your local insurance representatives and inform them of your intentions. Often they will be able to put you in contact with a representative at the cat loss site. An adjuster with whom you regularly work might even be transferred to that site on a temporary basis. It is extremely important to set up contacts before arriving at the cat loss site.

Step 3

Organize your "cat team," employees who will work at the site, and determine how to hire additional personnel as they are needed. A catastrophe policy for employees should already be in effect.

Step 4

Mobilize your cat team, along with any necessary supplies and equipment. Your team should be prepared to camp out, if necessary.

Step 5

Put one person from your office staff in charge of locating lodging, office and warehouse space, and material suppliers. Use of the Internet can be of great help in locating these necessities. Material supplier accounts should be set up by the time your cat team arrives at the site.

Step 6

Avoid bidding jobs against other contractors. Instead, focus your time on negotiating contracts with serious prospects. A cat loss provides an overwhelming amount of work, so the bid process should be left to less experienced contractors.

Cost Considerations

During a cat loss, normal estimating practices have little meaning. For the same reason that gasoline prices go up on a holiday weekend, when demand is high and supplies are limited, construction costs may also escalate during a catastrophe. For example: Prices will escalate in an area that has experienced a natural disaster or catastrophic loss when there is a limited supply of contractors and the demand for work is high. In addition, contractors incur greater expenses, take greater risks, and seek higher profits.

In addition to experiencing the effects of supply and demand, the restoration contractor who is working a cat loss incurs numerous "hard" costs. These costs will fluctuate greatly depending on the size of the catastrophe, the location, the economic condition of the area, the distance away from a home office, and the availability of manpower and materials. Such costs may include:

- **Per diem expenses**: This is a hard cost that pertains to an employee's additional expense for living out-of-town. It includes lodging, travel, meal expenses, and any additional increase paid as an incentive for living out-of-town temporarily. This added expense may be substantial for employers.

- **Material expenses**: These costs are definitely much more expensive during a catastrophe. Supply and demand greatly affect the retail building material industry, and unless the site is close to a major metropolitan area that has not been plagued by the disaster, materials may be hard to locate.

- **Labor**: Few, if any, skilled workers may be available after a catastrophe. Unless the contractor brings an adequate workforce along, additional labor may have to be hired from surrounding states. This situation may be costly for the employer and has no guarantees. In addition, it is important to keep in mind that prevailing wages fluctuate drastically from one state to another.

- **Subcontractors**: These workers are sometimes too busy during normal times to provide adequate service. During a cat loss, they can be almost unavailable. If they are available, it is at whatever price they want to charge.

- **Communications**: Keeping in touch is crucial on any job, but even more so at a cat loss site. In order to gather information and disseminate decisions in a timely manner and to be able to react quickly to changing site conditions, the restoration contractor may need to set up two-way radios, temporary telephone and computer systems (with e-mail), cellular phone service, and an overnight mail service. It is important to keep in mind that communicating with the home office from a distant site will incur greater expenses than usual.

- **Temporary office facility**: Contractors who operate at a cat loss site for any extended period of time will need a temporary office from which to conduct business. Additionally, a clerical staff may be required, and an adequate storage facility is a must.
- **Necessities**: Following a disaster, lodging as well as basic items that are usually taken for granted (such as electricity, toilets, showers, gasoline, and water) may be in short supply. The restoration contractor may discover that providing these necessities on a temporary basis will require extra resources that may generate added expense.
- **Warranty work**: Potential customers will always be concerned that the work performed is fit for the purpose intended and that it is free from defects. This is especially true when they are dealing with out-of-town contractors. The restoration contractor should make arrangements to address any warranty work upon completion of catastrophe site projects. This is one suggested method: Establish an agreement with a local contractor who will take care of this work.

Other Resources

The Federal Emergency Management Agency (FEMA) can provide valuable information and assistance during catastrophes, losses, and restoration work. Also, the *Appendix* of this book contains a list of state licensing departments across the country. These are agencies through which the contractor may become licensed and from which it is possible to obtain a list of licensed subcontractors.

Specific Disasters　Floods

Floods may be thought of as excess water in areas that are not normally under water, and they typically plague low-lying areas. Floods are caused by severe storms, persistent rains, runoff from melting snow and ice, dam failures, or a body of water that overflows its natural banks.

Brief but sudden flooding, or a flash flood, results from brief, intense rainfall over a localized area. This problem can be compounded by overburdened or inadequate drainage facilities. Even though the rainfall does not last long, flash floods can be dangerous.

A flood catastrophe can be devastating. Floods often cover enormous areas and contaminate everything in their path. There is usually limited funding for the cleanup process.

When handling flood-water losses, the restoration contractor must often face the following questions: What can be salvaged and what should be replaced? A good understanding of the differences between *clean water, gray water,* and *black water* is imperative·in the determination of an appropriate decontamination methodology.

- **Clean water** may result from a broken water supply pipe or similar incident. It may easily be cleaned up by extraction, drying out, and treating with a disinfectant.

- **Gray water** is usually related to sink backup or toilet overflow. It may be taken care of in a similar fashion to clean water.
- **Black water** includes sewage solids. It is classified as *hazardous* or *infectious waste*, and it is by far the most dangerous type of water to handle following flood disaster.

During restoration work following a flood, it is important to track the source and history of water to determine the possibility of comtamination from pathogenic diseases or infectious blood-born and virucidal concerns. Unless the contractor is trained in biotesting, it may be safer to handle all toilet overflows in the same way as black water.

The severe flooding that this country has experienced over the past few years is definitely an example of a black water situation. Contaminants come from many unexpected sources. For example: Flood waters are usually exposed to pesticides on fields, hazardous materials in old buildings (such as lead or asbestos), and diseases and germs of all types that might be found in landfills or cemeteries. Similarly, flood waters from sewage treatment plants, or even a simple sewer backup, contain much more than just raw sewage. Industrial plants that produce sludge from chemicals or acids can legally dump a percentage of it into the sewage collection system. In the same way, hospitals and laboratories can legally depose of blood, excrements, and radiological and pathogenic waste. HIV-contaminated blood and other infectious materials become typical residents of sewage systems.

Owners and managers of deflooding companies should familiarize themselves with the following OSHA regulations: OSHA CFR 910.1030, CAL-OSHA Title #7, and California Senate Bill #198 (pertaining to occupational exposure to blood-born pathogens, and new employee safety regulations).

Some restoration contractors and insurance companies routinely choose to clean and reuse, rather than replace, items such as carpet, garments, sheetrock, and other soft, porous materials that have been contaminated with gray or black water. In so doing, they may very well be putting themselves at great risk. It is essential that the original contaminants of particular materials are known. Otherwise, it seems impossible that a company could guarantee that a carpet or blanket has been cleaned and disinfected adequately. One might ask: Is the potential risk and liability worth it?

Tornados

Although tornados are difficult to predict, scientists have studied and learned about them, especially in the last few years. A tornado is a violently rotating column of air, which extends from a turbulent cloud down to the ground. It is by far the most violent of all storms. A tornado harnesses more destructive power that is directed at a confined area than any other natural force known to man.

Several other types of violent storm clouds are similar to the tornado:

- A funnel cloud is a storm cloud that has not yet reached the ground; it is a tornado in the making.
- A whirlwind is a small, rotating column of air that does little damage as it touches the ground. Whirlwinds are commonly known as *twisters*.
- A microburst is a downdraft that is less than 2.5 miles wide and that causes sudden changes in wind speed and directions. Downdrafts are particularly dangerous to pilots and sailors, and can cause severe property damage. Microbursts were first identified in 1974.

Tornados usually occur in the spring in the central and southern areas of the United States. They generally move from a southwest to a northeast direction at a rate of 35 miles per hour, and usually stay on the ground for approximately 15 minutes. A tornado wind can blow in excess of 250 miles per hour. The path of destruction left in a tornado's wake usually covers between 100 yards to one mile in width, and is approximately 9 miles long. Even though Florida has the greatest frequency of tornados per area, Oklahoma has the strongest, most potentially damaging tornados.

Tornados are rated using the Fujita-Pearson Tornado Intensity Scale (named after its creators). This system provides ratings for each of three areas:

- F = force in wind speed
- PL = path length
- PW = path width

A tornado rating of F-0 would designate the storm as the weakest tornado, one that would cause the least amount of damage. On the other hand, a tornado rated as F-5 would be the most violent and cause the most severe damage.

Hurricanes

A hurricane is a rotating wind system that originates over warm, tropical ocean waters. It is distinguished by a calm, central core, or eye, that is contained within high walls of thick clouds that can reach a height of 7 miles to 9 miles above the ocean. The temperature of these clouds is higher than in the surrounding atmosphere. A hurricane's lifespan may be as long as 30 days or as short as a few hours. It is usually accompanied by rain, thunder, and lightning, and can cover an area up to 600 miles across.

A hurricane may cross oceans and strike locations that are several thousands of miles away from its point of origin. Most of the structural damage from a hurricane occurs along the coastline. Destruction results from the tidal surge at the point of the landfall, very damaging winds, and torrential rains. The normal tropical storm season, starting in June and ending in November, lasts for six months. The height of the hurricane season is in early September when one third of all tropical storms occur.

Several other types of storms are similar to the hurricane. The typhoon, which occurs in the region of the Philippines or the China Sea, is a close relative of the hurricane. Collectively, the hurricane and the typhoon are considered to be the most threatening storms on earth. When the winds are less than 38 miles per hour, a storm is termed a *tropical depression*. If winds range from 39 miles per hour to 73 miles per hour, the storm is termed a *tropical storm*. It is not until winds exceed 74 miles per hour that a storm is called a hurricane.

Earthquakes

Earthquakes may be, by far, the most devastating of all natural disasters and the most costly in rebuilding. Many factors escalate the cost of reconstruction. For example: All utilities, including electricity, gas, and water, may take months to restore after an earthquake. Bridges, roads, and highways are often completely destroyed.

Earthquakes are a trembling of the earth's crust that is caused by sudden movement or displacement of rock. Such movements release energy in the form of seismic waves. This buildup of potential energy may be caused by volcanic eruptions, rapid changes in the water table, other earthquakes, or sometimes man himself (e.g., from a nuclear blast).

The United States has several faults lines within its borders that have experienced earthquakes. For example: California, well known for earthquakes along the San Andreas Fault, has sustained billion of dollars in damage over the last decade. In 1811, the New Madrid Fault, in New Madrid,

Missouri, had an earthquake that sent out tremors covering one million square miles. This earthquake even made the mighty Mississippi River run backwards.

About 60 years ago Dr. Charles Richter devised a system for rating earthquakes that is now referred to as the Richter Scale. The Richter scale is logarithmic, so that each unit jump in magnitude indicates a tenfold increase in seismic wave amplitude and roughly a 30-fold leap in energy. Thus, a magnitude 8 earthquake releases about 30 times as much energy as a magnitude 7, and about 900 times as much as a magnitude 6, and so forth.

According to the Richter scale, the potential for damage in a populated area is:
- Magnitude -3: Lowest quakes ever recorded
- Magnitude 2:   Lowest quake normally felt by humans
- Magnitude 3.5: Quake may cause slight damage
- Magnitude 4:   Quake may cause moderate damage
- Magnitude 5:   Quake may cause considerable damage
- Magnitude 6:   Quake may cause severe damage
- Magnitude 7:   A major earthquake capable of tremendous damage
- Magnitude 8.9: The highest ever recorded

Conclusion

The insurance industry often changes its policies and procedures pertaining to contractors and the handling of claims. The industry's efforts are generally aimed at controlling the cost of claims, but this is very difficult to accomplish. The companies are faced with the question: In an increasingly competitive market, how do insurance carriers control costs and still provide a high level of service? Understanding the contracting business and the role of the contractor are key factors.

Many carriers have instituted "preferred contractor programs," in which an insurance carrier works with a limited number of restoration contractors almost exclusively. In choosing contractors for its preferred contractor program, the insurance company will check out the contractor's quality of workmanship and financial stability. The carrier then recommends contractors who are part of the program to its insureds for restoration services. Some carriers will even guarantee the contractor's work for a period of three years. For the restoration contractor, the drawback to this arrangement is the carrier's desire, in some cases, to become involved in determining the contractor's pricing structure without understanding the extent of the restoration company's overhead costs.

The key to solving the problem of cost control with quality service is to achieve a contractor-insurance carrier relationship in which there is a mutual understanding of needs. The following are two suggestions that would take into consideration the difficulty of controlling overhead without compromising service.

- The insurance carrier should issue regularly scheduled progress payments, in a timely manner.
- The carrier should ensure that the contractor is paid in full for his services.

By adopting these two policies, the insurance carrier could save up to 5% on its claims. When a contractor has to finance a project with his own resources, he must pay interest. This interest is almost never shown as a job cost line item expense, but it substantially increases a contractor's overhead. Additionally, most contractors typically write off approximately 3% of their gross sales to bad debt. This bad debt results from insureds failing to pay their deductibles, or from insurance carriers neglecting to put a contractor's name on a check to secure payment.

Disaster restoration contracting is a challenging business. Contractors who engage in it either love it and excell, or hate it and fail miserably. There really is no in-between. The successful contractor recognizes that this business is unique, becomes a skilled risk manager, and continues to seek education from every available and reliable source. A clear understanding of business and accounting is a necessity, and the contractor must be willing to invest a substantial amount of time, effort, and financial resources in the development and maintenance of the business. Those who can meet these challenges will establish a reputation as reliable providers of an essential service, one that will continue to be in demand no matter how the economy affects the building industry.

Appendix

State Licensing for Disaster Restoration Contractors

Following is a list of state licensing departments across the U.S. These agencies will assist disaster restoration contractors with licensing, and they will provide the contractor with a list of licensed subcontractors.

Alabama
Home Builders Licensure Board
400 South Union St., Suite 195
Montgomery, AL 36130
334/242-2230

Alaska
Department of Commerce and
  Economic Development
Division of Occupational Licensing
P.O. Box 110804
Juneau, AK 99811-0804
907/465-2017

Arizona
Registrar of Contractors
800 West Washington, 6th Floor
Phoenix, AZ 85007
602/542-1525

Arkansas
Contractor's Licensing Board
4100 Richards Rd.
N. Little Rock, AR 72117
501/372-4661

California
Department of Consumer Affairs
Contractor's State License Board
P.O. Box 26000
9821 Business Park Dr.
Sacramento, CA 95827
916/255-3900 or 800/321-2752

Colorado
There is no statewide licensing of
residential construction contractors
or remodelers in Colorado.

Connecticut
Commissioner of Consumer
  Protection Department of
  Occupational Licensing
165 Capital Ave.
Hartford, CT 06106
860/713-6300

Delaware
Department of Finance
Division of Revenue
Carvel State Building
820 North French Street
Wilmington, DE 19801
302/577-3300

District of Columbia
Department of Consumer and
  Regulatory Affairs
John A. Wilson Building
1350 Pennsylvania Ave. NW
Washington, DC 20004
202/727-1000

Florida

Department of Business and
    Professional Regulation
Construction Industry
    Licensing Board
1940 N. Monroe St.
Tallahassee, FL 32339-1027
850/487-1395

Georgia

Most contractors don't need a license
in Georgia. The exception is asbestos
abatement and the mechanical trades
contractors.

Guam

Contractors Licensing Board
Government of Guam
P.O. Box 20339
GMF, Guam 96921
671/646-7262

Hawaii

Department of Commerce and
    Consumer Affairs
Contractor's License Board
P.O. Box 3469
Honolulu, HI 96801
808/586-2700

Idaho

Public Works Contractor's
    License Board
355 N. Orchard St., Suite 107
Boise, ID 83720-0073
208/327-7326

Illinois

There is no statewide licensing of
residential construction contractors
or remodelers in Illinois.

Indiana

There is no statewide licensing of
residential construction contractors
or remodelers in Indiana.

Iowa

Department of Employment Services
Division of Labor
1000 East Grand Avenue
Des Moines, IA 50319-0209
515/218-3606

Kansas

Doesn't license construction
contractors at the state level.

Kentucky

There is no statewide licensing of
residential construction contractors
or remodelers in Kentucky.

Louisiana

State Licensing Board for Contractors
P. O. Box 14419
Baton Rouge, LA 70898-4419
225/765-2301

Maine

General building contractors do not
need a license in Maine.

Maryland

There is no statewide licensing of
residential construction contractors
or remodelers in Maryland.

Massachusetts

State Board of Building Regulations
    and Standards
John W. McCormack State
    Office Building
One Ashburton Place, Room 1301
Boston, MA 02108
617/727-3200

Michigan

Department of Commerce
    Bureau of Occupational and
    Professional Regulation
Residential Builders and Maintenance
    and Alteration Contractors Board
P.O. Box 30245
Lansing, MI 48909
517/241-9254

Minnesota

Minnesota Department of Commerce
Residential Contractor's Licensing
133 East 7th Street
St. Paul, MN 55101
651/296-6319

Mississippi

State Board of Contractors
2001 Airport Rd., #101
Jackson, MS 39208
601/354-6161

Missouri

There is no statewide licensing of
residential construction contractors
or remodelers in Missouri.

Montana

Contractor Registration Unit
Employment Relations Division
Department of Labor & Industry
P.O. Box 8011
Helena, MT 59604-8011
406/444-7734

Nebraska

Workforce Development–Dept. of
    Labor
5404 Cedar St.
Omaha, NE 68106-2365
402/595-3189

Nevada

State Contractors Board
9670 Gateway Dr., Suite 100
Reno, NV 89502
775/688-1141

New Jersey

New home builders:
    Dept. of Community Affairs
    Bureau of Homeowner
    Protection New Home Warranty
    Program
    P.O. Box 805
    Trenton, NJ 08625-0805
    609-530-8800

Home Repair Contractor's License:
    Division of Banking Office of
    Consumer Finance
    P.O. Box 040
    Trenton, NJ 08625
    609/292-5340

New Mexico

Regulation and Licensing Department
Construction Industries Division
725 St. Michaels Drive
P.O. Box 25101
Santa Fe, NM 87504
505/827-7030

New York

There is no statewide licensing of
residential construction contractors
or remodelers in New York.

North Carolina
Licensing Board for
    General Contractors
P.O. Box 17187
Raleigh, NC 27619
919/571-4183

North Dakota
Secretary of State
State Capitol
600 East Boulevard Ave., Dept. 108
Bismarck, ND 58505-0500
701/328-3665

Ohio
There is no statewide licensing of residential construction contractors or remodelers in Ohio.

Oklahoma
Oklahoma doesn't license resident construction contractors except in electrical, mechanical, and plumbing trades.

Oregon
Construction Contractor's Board
700 Summer NE, Suite 300
P.O. Box 14140
Salem, OR 93709-5052
503/378-4621

Pennsylvania
There is no statewide licensing of residential construction contractors or remodelers in Pennsylvania

Rhode Island
Department of Administration
Rhode Island Building Commission
Contractor's Registration Board
One Capitol Hill
Providence, RI 02908-5859
401/222-1268

South Carolina
Contractor's Licensing Board
P.O. Box 11329
Columbia, SC 29211-1329
803/896-4686

South Dakota
There is no statewide licensing of residential construction contractors or remodelers in South Dakota.

Tennessee
Tennessee Board for Licensing
    Contractors
500 James Robertson Pkwy., Ste 110
Nashville, TN 37243
615/741-8307 or 800/544-7693

Texas
There is no statewide licensing of residential construction contractors or remodelers in Texas. Contractors who are performing "residential service contracts" must obtain a license from the Texas Real Estate Commission. A residential service contract includes maintaining, repairing, and replacing structural components, or the electrical, plumbing, heating, cooling, and air conditioning systems of residential properties.

Utah
Department of Commerce
Division of Occupational and
    Professional Licensing
P.O. Box 45805
Salt Lake City, UT 84145
801/530-6628

Vermont
There is no statewide licensing of residential construction contractors or remodelers in Vermont.

Virginia
Department of Professional &
    Occupational Regulation
State Board for Contractors
3600 West Broad St.
Richmond, VA 23230
804/367-8511

Washington
Department of Labor and Industries
Contractor's Section
P.O. Box 44450
Olympia, WA 98504-9689
360/902-5226

West Virginia
West Virginia Contractor
    Licensing Board
319 Bldg. 3., Rm. 319 State Capitol
    Complex
Charleston, WV 25305
304/558-7890

Wisconsin
Department of Industry, Labor, and
    Human Relations Safety &
Buildings
P.O. Box 7969
Madison, WI 53707-7969
608/261-8500

Wyoming
Economic & Community
    Development Department
214 W. 15th St.
Cheyenne, WY 82002
307/777-2800

Index

Index

Notes

Notes

Notes

Notes

Notes

Notes